Praise for *Discovering Mars*

"*Discovering Mars* provides a breathtaking panorama of the human quest to understand our neighboring planet, starting from ancient times through the era of planetary astronomy from Galileo through the 1950s, and through the era of space missions all the way to the 2020s. Authors Sheehan and Bell are the perfect pair to present this journey through time and space, with Sheehan's perspective as a science historian and philosopher, and Bell's perspective as a modern-day explorer leading the camera teams on NASA's rovers."

—Roger C. Wiens, Los Alamos National Laboratory

"An extraordinary chronicle of our centuries-old captivation with Mars, enticingly rich in detail . . . Sheehan and Bell have penned a sweeping, immersive book that takes the reader right to the doorstep of modern exploration."

—Sarah Stewart Johnson, author of *The Sirens of Mars: Searching for Life on Another World*

"Sheehan and Bell expertly distill the history, science, and technology of planetary exploration into a refreshing deep dive into the realm of Mars research. It is a joy to read for both the curious observer and the planetary scientist alike."

—Amy J. Williams, University of Florida, and team member on NASA's Curiosity and Perseverance rover missions

"Bill Sheehan and Jim Bell lead an exhilarating voyage to understand Mars. The journey is as much about human imagination and eccentricity as it is about scientific data and observation, and the team of Mars historian Sheehan and planetary scientist Bell expertly show the way. Never has armchair exploration been so provocative!"

—Kevin Schindler, Lowell Observatory historian

Discovering Mars

"Selfie" taken near the Jezero crater landing site using the NASA Perseverance rover's arm-mounted WATSON camera on April 6, 2021 (Mars 2020 mission Sol 46). The rover had just dropped off the Ingenuity helicopter (seen at left), which had been stowed under the rover's chassis during launch, landing, and early mission operations. The helicopter went on to make history eight sols later, becoming the first aircraft ever to make a powered, controlled flight on another world. NASA/JPL-Caltech/MSSS.

DISCOVERING
MARS

A History of Observation
and Exploration of the Red Planet

WILLIAM SHEEHAN AND JIM BELL

THE UNIVERSITY OF
ARIZONA PRESS

TUCSON

The University of Arizona Press
www.uapress.arizona.edu

ISBN-13: 978-0-8165-3210-0 (hardcover)

Cover design by Leigh McDonald
Cover illustration: Perseverance Guides Itself Towards the Surface, NASA/JPL-Caltech
Designed and typeset by Sara Thaxton in 10.25/15 Minion Pro (text), Montserrat, and Roboto (display)

Library of Congress Cataloging-in-Publication Data
Names: Sheehan, William, 1954– author. | Bell, Jim, 1965– author.
Title: Discovering Mars : a history of observation and exploration of the Red Planet / William Sheehan
    and Jim Bell.
Description: Tucson : University of Arizona Press, 2021. | Includes bibliographical references and index.
Identifiers: LCCN 2021012042 | ISBN 9780816532100 (hardcover)
Subjects: LCSH: Mars (Planet)—Exploration.
Classification: LCC QB641 .S4835 2021 | DDC 523.43—dc23
LC record available at https://lccn.loc.gov/2021012042

Printed in the United States of America
♾ This paper meets the requirements of ANSI/NISO Z39.48-1992 (Permanence of Paper).

To Debb and Jordana, our guiding stars . . .

Contents

Foreword

It's a great pleasure to see this new book on Mars. It fills a need that I've seen growing in the field for over thirty years now, which is for an integral history of humanity's studies of Mars that includes a full description of both the pre-Viking and the post-Viking era. Even though they are parts of the same story, these two eras are very different, and each complicated in itself, so that perhaps there has been no one person who is fully conversant with both. As Sheehan and Bell explain in their introduction, it is the combination of their two angles of approach that allows this book to tell the whole story, and their work of collaboration and integration has given us this special achievement.

Possibly the history of our study of Mars can be compared to looking at a landscape that has been hit by a meteor, by which I mean, the amount of information that came crashing into human knowledge by way of the Mariner and Viking missions to Mars was so huge that it tended to obliterate our awareness of what had existed before. Literally millions of times more data than we had ever had came to us in two great pulses in 1971 and 1976, and much of the work in areology after 1977 was devoted to sorting out what had just come pouring in. There was so much work to do in sorting out what we had been given by these missions that it perhaps did not matter that no further major missions visited Mars for the next two decades; there was already enough to do. But then more missions began to orbit Mars, and to land on it, and there have been enough of these in the years since 1996 that it's very easy to get them confused, and lose sight of just how much more has been added to our knowledge, beyond what Mariner and Viking were able to tell us. By now more exponential leaps in the amount of data have accrued, and that in itself is hard to keep track of and understand.

So the shape of areology over time, in terms of information input, is strange, and needs sorting out. Before Mariner, we first had the long era of naked-eye observations. Certainly Mars has been part of human consciousness for as long as there have been humans, and perhaps even before; its brightness, redness, fluctuating intensity, and wandering course through the stars, with its little hitch, all brought it to the forefront of our attention from the very start, when we were regarding the night sky and wondering what it all meant. Then telescopes revealed the Red Planet as a disc with marks, no doubt a planet like Earth and the other planets; but how much like Earth, or how different, was difficult to tell. The amount we knew grew only a little with improvements in telescopes, but then astronomers brought other instrumentation to bear, so there were notable increases in information, and a number of changes in hypotheses, which make this early era in Mars studies a good demonstration of the scientific method in action, including paradigm shifts and, of course, the tendency to take a little bit of evidence and run a long way with it. This is a great story, and an important one to remember and keep clear in our history of areology, and that's one of the things this book does well.

Then came Mariner and Viking, and these two missions need to have their stories told too, as we also have here. What an extraordinary breakthrough in human exploration! Even now, despite the decades of subsequent missions with their ever-more-refined results, the story of these two landmark missions needs to be recalled and examined for what they can still teach us.

Then in the 45 years since Viking, beginning with the little landers of the late 1990s, we have seen a large number of extremely successful missions to Mars, both in orbit and on the surface. There have been so many of them that it's easy for casual observers to lose track, and also perhaps not everyone has kept up with the latest results. Here too this book serves as a great guide. Even as we stand on the brink of yet more amazing exploration, these great expeditions need to be sorted out and their results explained. It's wonderful to have that.

Important also is the inclusion of the stories of the missions that failed. One thing this book makes clear is that landing on Mars is a very difficult engineering and aeronautical feat. It doesn't always happen, not even close; even in recent years, failures are not uncommon. I was in the Pasadena Convention Center with a huge group gathered by The Planetary Society to celebrate the arrival of the Mars Polar Lander at Planum Australe in 1999; the time of the landing passed, no message came; on that went; finally it was clear the mission had somehow failed. Because this was early on in the post-Viking era, and many young scientists had had no new data to study for their entire careers, I have seldom seen a more dispirited group of planetologists. The rest of us tried to cheer them up, but there was little that could be said. It would be necessary to try again.

And so they did try again, and some successes followed. But the landing problem is severe, and one thing this book reminds us of is the comically elaborate, or even ludicrously simple, methods that the engineers have come up with to finesse this problem. All the easy talk that one hears of landing people on Mars in the near future needs to take on this information, and take the problem seriously; it's not going to be easy.

Meanwhile, we have our superb robot explorers. I want to end by emphasizing how wonderful the work of the Mars scientific and engineering community has been. People tend to obsess about the human as the only thing that matters, as if our physical presence on Mars is what is really important, even though no one expects to be one of those people. Ironically, it's one of the many ways we exist in ideas more than we do in spaces—to think that the idea of being there in person is more important than actually learning the place, by way of our instruments. For me, having spent my own career as a Martian poring over the Viking photos, these new photos from the most recent expeditions are simply stunning—so detailed, fine grained, colorful, informative— it's like looking out a window at the Martian landscape, and given the realities of the Martian atmosphere and surface, looking out a window, even if only the faceplate of a spacesuit, is all we're ever going to be doing (until of course the terraforming is complete, some thousands of years from now). So I say this: we are already on Mars, and it's fantastic. Astronauts actually on the ground there will be very exciting, and yet they won't change the nature of our engagement with the place all that much, for those of us still on Earth. Because of the Martian community, we are really already there. And that's something to celebrate, and tell the story of in detail, so that we really get it. Especially in this era of climate change, comparative planetology is a very important analytic tool for our civilization. And as Mars is the planet most like Earth, by a long shot, studying Mars becomes useful to us, as well as beautiful. In that ongoing study, this book takes its part with distinction and flair. My thanks for it, and congratulations to the authors.

Kim Stanley Robinson
Davis, California
September 2020

Preface

The fascination with Mars for both of us goes back a long way.

One of us (Bill Sheehan) grew up in the early Space Age and first observed Mars as a 10-year-old with a 60-millimeter refractor at the opposition of March 1965—the last of the pre-spacecraft era, when it was still (just) possible to believe there might be canals on Mars or at least some lowly life-forms. The following summer, Mariner 4 made its flyby and seemed to show us a Mars that was dreary and cratered and moonlike, utterly unlike the world of our dreams. The romance was gone, and at least to one disappointed youngster, whose interest had been nurtured by books by Percival Lowell, H. G. Wells, and Edgar Rice Burroughs, it seemed as devastating as if someone had proved that there was no Santa Claus.

That youngster remained fascinated by Mars, and now in high school, but not much better equipped optically, he observed the Great Dust Storm of 1971 as Mariner 9 went into orbit around the Red Planet. A new view of Mars—with shield volcanoes, vast canyons, and dry riverbeds—came into view. Some of the romance that had seemingly been killed by Mariner 4 returned, and so did the interest in Mars that had been early awakened. Perhaps that youngster— uncertain and fumbling toward a career—might have pursued astronomy, and planetary science in particular, as the fulfilment of earlier aspirations, but even had he possessed the ability and determination (which is far from certain), the economy was just then in bad shape owing to such things as the ongoing costs of the Vietnam War and the OPEC oil embargo. The time of "budgetless financing" that NASA had enjoyed during the Apollo era was over—it now seems, probably, forever.

So instead the youngster embarked on a career in medicine, and specifically psychiatry. Strangely, though, even that career led back to Mars, as the soon-to-be medical student became fascinated during a research visit to Lowell Observatory by the question, little considered before then, of how humans perceive detail on a planet like Mars. That led to a book, *Planets & Perception* (University of Arizona Press, 1988), which was an effort at cross-disciplinary studies (blending the history of planetary astronomy and perceptual psychology). In retrospect, it was surely a case of "fools rush in where angels fear to tread." But the University of Arizona Press took a risk with it, and astronomers who reviewed the book liked it (professional historians had more quibbles, as might be expected). The book was mostly about the history of Mars observations, so a few years later the author was approached to write a somewhat more standard work, *The Planet Mars* (University of Arizona Press, 1996), which benefited from a great deal of advance publicity—just as it came out, NASA researchers announced, to great fanfare, the discovery of possible fossils in a Martian meteorite. For a book published by an academic press, the book did very well, and it was accepted as something of a standard history for amateur astronomers that was also of use to professionals (including the second author of this book!).

Over the years, the author of that original book on Mars was approached from time to time about the need for a new edition. It was a tantalizing but impossible prospect. Awareness of the explosion of knowledge and new insights into Mars from the spacecraft era also led to the humbling knowledge that the author would not even know where to begin. The Mars research community had greatly expanded over the years. Planetary scientists—now applying the highly specialized methods and mature techniques of Earth sciences and sophisticated remote-sensing technology—struggled to integrate and comprehend vast quantities of information from long-duration spacecraft, orbiters, landers, and rovers. From the planetary missions perspective, it became impossible to comprehensively describe the entire history of Mars spacecraft exploration results in a single book. The last such attempt by professional Mars scientists was the 1,500-page tome called, simply, *Mars*, published in 1992 by the University of Arizona Press and edited by space scientists Hugh Kieffer, Bruce Jakosky, Conway Snyder, and Mildred Matthews. At that time, the entire history of Mars (and Phobos/Deimos) spacecraft exploration was based mostly on the NASA Mariner 4, 6, and 7 flybys, the Mariner 9 and Viking 1 and 2 orbiter missions, and the Viking 1 and 2 lander missions, with small additional contributions from the Soviet Mars 2 through Mars 6 flyby and orbiter missions. Those missions also informed the spacecraft results that were summarized in the first edition of *The Planet Mars*. Since Viking, however, a stunning 17 additional robotic missions have been partially or fully (and some, wildly) successful in orbiting, landing on, or roving on Mars and—in the process—have completely revolutionized (again) our understanding of the Red Planet's past history and current environment.

It became impossible to think of doing anything with *The Planet Mars* as originally conceived. What was needed was an active professional planetary scientist with a track record not only of interpreting the science but also of expressing the science clearly and compellingly to the wider public. That person was Jim Bell, who not only had trained under one of the legends of solar system astronomy, James Pollack (Carl Sagan's first grad student), but has been a leading scientist on numerous spacecraft missions, has popularized Mars research (as well as other areas of planetary science) through his services to The Planetary Society, and has won accolades for his authorship of many articles and books. So as this new book developed—at first along the lines of becoming a revised *The Planet Mars*, but soon taking on (as such projects tend to do) a life of its own—it became a collaboration between Bill and Jim.

As the new book developed, it became a blend of the history of Mars observations, which is mostly Bill's perspective, and modern and very active current and future-directed spacecraft (including perhaps eventually human) exploration of Mars, which is Jim's expertise. Bill's view is mainly that of a historian who brings as well a psychiatrist's experience and knowledge of human perception and motivation. Jim, as an active planetary scientist who focuses on studying the geology and composition of planetary surfaces, first using telescopes early in his career, and now mostly using robotic spacecraft observations from flybys, orbiters, landers, and rovers, brings primarily a space scientist's perspective and context (amplified by his instrumentation and engineering experience and foundation in observational astronomy).

At times, this duality could make this collaboration seem more like two books instead of one: first, a historical review of what we knew about Mars from the pre-telescopic and telescopic eras, focusing on key individual characters and written in an interpretive style common to historical and social science research writing; and second, a review of what we have learned and currently know about Mars from the Space Age, focusing on large teams conducting "Big Science" (i.e., hundreds to thousands of people working on national or international collaborations) and written in a more objective, fact-based, hypothesis-test style common to modern scientific research writing. Indeed, we recognized that we were taking the risk of essentially working on two separate books when we embarked on this project.

However, we have both read and revised the entire book, and in that process we have tried to weave throughout a common unifying theme that Mars has continually been rediscovered over time, whether through advances in technology, revolutions in the theory or practice of science, or the wakening evolution of our planetary-scale perspective as fellow travelers on Carl Sagan's "pale blue dot." We are aware that the individual voices of the authors continue to be heard across these pages, and we view that as a strength rather than a weakness. We hope that readers find the resulting synthesis of our perspectives and experience to be as enjoyable and informative to read as it was for us to write.

The historical part of the book continues to follow (though with much new research added, as is inevitably the case after 25 years of continuing effort) the lines of *The Planet Mars*, and is concerned largely with the ground-based era of Mars studies and the first few spacecraft missions, including the flyby Mariners, Mariner 9 (which entered Mars orbit coming on 50 years ago already!), and the Vikings. But the second half of the book, devoted to spacecraft exploration, largely breaks new ground. Seen in its totality, the book hopefully bears witness to one of humanity's greatest quests—to explore Mars, the planet beyond Earth that has always been seen as the most Earthlike, and the best place to search for life. That quest, in turn, has been part of an even greater one that we have been fortunate to participate in (or at least witness) in the last two generations or so, which Sagan, again, defined so timelessly: "In all the history of mankind, there will be only one generation that will be the first to explore the Solar System, one generation for which, in childhood, the planets are distant and indistinct disks moving through the night sky, and for which, in old age, the planets are places, diverse new worlds in the course of exploration."[1]

Both authors indeed remember a time when—whether looking through a 60-millimeter department store refractor magnifying 35× and 117× (Bill) or from an 8-inch Newtonian reflector magnifying 400× (Jim), both from the front yards of their houses when they were 10 years old—Mars really was usually little more than a distant and indistinct disc moving through the night sky. In those vastly outmatched small telescopes, it was much as H. G. Wells's narrator in the *War of the Worlds* said of the view through the telescope of Ogilvy, the doomed astronomer who would be one of the first casualties of the invading Martians: "how little it was, so silvery warm—a pin's head of light!" (Odd, but he was not disappointed.) Indeed, it was not much more than that pin's head of light to astronomers peering through their telescopes right up to the beginning of the spacecraft era. That prior time was one in which illusion and reality long seemed equally matched in the human mind and vied mightily for dominance. Illusion, frankly, often had the upper hand.

The Space Age (which is the only era Jim has known, since he came into the world just nine days after the Mariner 4 flyby) saw the at-first gradual and uneven but then increasingly fast-paced and surer-footed progress in which reality has gained the upper hand. We can be quite sure that Mars as we believe we know it today is closer to the real Mars than the Mars of 1965, or 1916, or 1877. The spacecraft part of the book is divided chronologically into chapters focusing on individual missions or individual classes of missions (some of which were contemporaneous, some of which Jim has been directly involved in, and some of which are still ongoing today!). Special attention is given to the "transitional" early flyby missions of Mariners 4, 6, and 7, which put the nails in the coffin of the era of canals, vegetation (still speculated to occur on Mars into the 1960s), and the idea of Mars being a much more Earthlike, more currently habitable world for life as we know it. Special attention is also given to Mariner 9 (the mission that rekindled high-schooler Bill's interest in Mars). It was the first Mars mission of the Space Age to map the planet and reveal the true nature of its geological and climatologic complexity—including the first inklings of the planet's much more Earthlike deep past.

Those inklings spawned the full-fledged biological focus of the Viking missions, especially the landers, which were designed not only to be the first to put a U.S. flag on the surface on Mars, but also to conduct sensitive experiments to search for evidence of living organisms in the soil and in the shallow subsurface. Viking's negative (or at best, ambiguous) biology results arguably set back Mars surface exploration for more than 20 years, although the high-profile failure of the Mars Observer mission was also partly responsible for that gap. (Indeed, we chronicle here both the successes *and* failures of the history of Mars spacecraft exploration—nearly half the missions attempted have not succeeded, despite spectacular engineering prowess and despite the noblest of scientific goals and attempted experiments.)

The chapters on the missions since Viking then summarize their origins, tribulations, and discoveries, but they are necessarily only summaries, since a comprehensive discussion of all the results from and controversies instigated by those missions could easily fill a book for each one (and, indeed, many books, scientific papers, and popular science accounts of those mission results *have* been written, as indicated in the detailed notes and references for each chapter). The armada of orbiters, landers, and rovers sent (or attempted to be sent) to Mars since the early 1990s—by NASA but also by space agencies in Europe, Russia, Japan, India, the United Arab Emirates, and China—is absolutely stunning. Those missions, starting with Mars Global Surveyor but continuing again and again, have revolutionized our understanding of the Red Planet in almost every way: geologically, sedimentologically, meteorologically, climatologically, and even biologically. Was Mars a habitable world? Yes, in many places, early in the planet's history, Mars hosted environments that would have been conducive to life as we know it. Indeed, such environments could *still* exist in the subsurface today. These profound and extraordinary claims—and what they teach us about the history of life, habitability, and sustainability of our own world—are part of the legacy of the past 25 years of robotic exploration of Mars, and we try to capture the origins and justifications for those claims here in this book.

Finally, at the end, how could we resist but to speculate about the future of Mars exploration? Some of the near future of robotic exploration is already well in the works. Three new missions launched to Mars in summer 2020, for example, to take advantage of this latest roughly biannual Earth–Mars launch window, arriving safely in February 2021 just before this book went to press: NASA's Perseverance rover, China's Tianwen-1 orbiter and rover, and the United Arab Emirates' Hope orbiter. Other robotic missions for the next few launch windows are deep into the planning stages, including an exciting but complex scheme to try to return the first samples from Mars (cached by Perseverance) using follow-on missions during the rest of this decade. Beyond that, it is widely acknowledged among the planetary science and human exploration communities (and even among some politicians and sectors of the public) that sending astronaut crews to Mars is inevitable, perhaps by the 2030s or 2040s, and that sustainable research stations or settlements are a realizable long-term goal. But how and when will all that happen? And is there adequate justification for embarking on human exploration versus continuing the safer and much less expensive exploration of Mars using robots? This is an active and exciting topic for discussion and debate, and we frame the arguments here, acknowledging that there is both optimism and pessimism, opportunities and challenges, tall technological hurdles and soaring inspirational potential.

We might say, as used to be said on the death and succession of kings, the planet Mars is dead; long live the planet Mars! We welcome you to *Discovering Mars*. We hope you enjoy the stories told here chronicling the characters, technologies, human (and robotic) failures and successes, and the incredible scientific discoveries that have revealed and continue to reveal the true nature of our most Earthlike of celestial neighbors.

William Sheehan
Flagstaff, Arizona

Jim Bell
Mesa, Arizona

March 13, 2021

Discovering Mars

1

Wanderers and Wonderers

The Red One

Mars looms large in the human imagination. For millennia it has burned its imprint on the human psyche, being among the celestial bodies pondered by humanity's first celestial wonderers. Its influence on the earliest thinkers is assured: a blood red "star," seen wandering among the fixed stars like a wounded animal, would certainly have evoked wonder or fear—especially with the emergence of anthropocentric thinking, when humans first asserted their separateness and dominance over nature, which would have conferred on the wanderer a human role and purpose. As early peoples sought to position themselves in their perceived cosmos, the pulsing light of Mars, mimicking the rise and fall of human emotions, the pulsating beat of the human heart, and the color of blood, must have taken its place in their myths, early traces of which (though only partly intelligible to us who lack the key to their decipherment) we find in the cave art of the Upper Paleolithic, which skillfully depicts in ocher, charcoal, and other natural pigments the great mammals of the Ice Age—antelopes, bison, oxen, horses, woolly mammoths. These were the great beasts that were hunted. Curiously, however, humans are only stick figures rudely sketched in on the margins, less developed than the animals depicted. (It is as if in their own esteem humans filled only a corner of the world dominated by the beasts, of which they stood in awe.) By representing their forms, they hoped, perhaps, in some magic way to acquire power over them.

Mars may well have awakened the curiosity of those Ice Age hunters, but we know nothing about it. The human discovery of Mars is buried in the snowy wastes of time. Perhaps the earliest reference to Mars in human culture is as part of Aboriginal Australians' Dreamtime, a vision from time beyond memory, a mystical part of the culture handed down by legends, songs, and dance for more than 40,000 years. The Aboriginal peoples' view of the cosmos is based in their concept of a distant past when their Spirit Ancestors created the world. To the Aboriginal peoples—some of whom live in the red soil of the desolate outback, a place that looks more like Mars than does almost any other place on Earth—Mars was Waijungari, a newly initiated man who, in ceremony, was covered with red ocher. One day, to escape the wrath of a jealous hunter, Waijungari threw his spear into the sky and, when it stuck, used it to climb into the heavens, where we still see him today.

Though we do not know just when human beings first noticed five bright "stars" moving among the other stars, they must have done so in very early times. The ancient Greeks called them wanderers and gave them their individual names: Hermes, Aphrodite, Ares, Zeus, and Cronos (or Kronos). These were later romanized: Mercury, Venus, Mars, Jupiter, and Saturn.

Each of the planets has a unique personality. Mercury is a shy intruder into our skies, and because of its rapid motion it is well named for the fleet-footed messenger of the gods of Olympus. Venus is the brightest and loveliest of the stars of the night; Jupiter, the most majestic; Saturn, the slowest moving, as befitted the ancient and decrepit god of time.

None of the planets, however, has a more striking personality than Mars.

Mars is one of few conspicuously red objects in the sky, and psychologically, red has more impact than any other color. It is the color, of course, of blood, and as such it is the first color to be added to human vocabularies after black and white. (Among the Inuit, there were only three color words—black, white, and red.) Already the Neanderthals used it for its effects and were making strokes in red ocher on the wall of a cave in Spain 65,000 years ago, tens of thousands of years before members of our species reached the site.[1] Among all the colors, it is unique in its effects on the attention. On the one hand, it is the color of attraction and desire, favored by restaurants and associated with red-light districts (to "paint the town red"), and beloved by artists, some of whom, like Mark Rothko, have used it almost exclusively.[2] It is the commonest color used as an attractant for conspecifics by birds and for insects by flowers. On the other hand, it signals danger, as on stop signs and in the eyes of the poisonous tree frog.[3]

Mars's red color likely seized attention from the first, and from earliest times it has been associated mostly with danger and foreboding. Since fear is said to be the strongest human emotion, it is quite likely that had the planet been some other color, its grip on the human imagination would have been less.

As with the other outer planets, but more conspicuously, Mars is brightest when it stands opposite to the Sun in the sky. It is then said to be at opposition. Oppositions of Mars occur at intervals of approximately two years, two months. It rises when the Sun sets and sets when the Sun rises, reaching its highest point above the horizon at midnight.

We know, as the ancients did not know, that Mars and Earth travel in orbits around the Sun. The separation between the two worlds reaches a minimum at the times of opposition, which is why Mars then appears so bright. Because the orbits of both planets are not perfectly circular, and because that of Mars is especially eccentric, the separation varies depending on where Mars happens to lie in its orbit. At the most favorable oppositions—as in August 2003, when Mars approached closer to Earth than at any time in the last 60,000 years—it can outshine Jupiter, and may even appear whitish rather than reddish, owing to saturation of the color receptors of the eye. It then flares up like a Homeric hero in the throes of his *aristeia*, or a Norse berserker fighting in a furious trance. At more average oppositions it appears wheat colored (which may explain how, in addition to being associated with bloodshed and war, it early acquired a secondary association with agriculture). At its faintest, far from opposition, it appears as a dim blood-red spark no brighter than the seven stars in the Big Dipper.[4] Wildly unstable in its moods, with fiftyfold changes in brightness, it must have seemed to be no inert body but something willful and alive.

In addition to its changes in brightness, it also exhibits dramatic changes in its motion. Around opposition, it reverses direction, as the other outer planets do, and instead of traveling west to east it passes through a giant westward loop before resuming its usual eastward progress. As such, it behaves rather like the Minoan bull jumpers depicted in frescoes in the Great Palace at Knossos, in which youths of both sexes risked serious bodily harm by grabbing the horns of bulls and somersaulting backward. Mars's usual prograde (west to east) movement through the sky is interrupted about a month before opposition, when it begins its somersault. Interestingly, at oppositions that take place when Mars is in the constellation Taurus, the Bull, it can actually make its backflip as if springing off the Bull's horns—thus becoming quite literally a bull jumper in the ancient Minoan fashion.

Mars's color, abrupt changes of direction, and variations in brightness led early sky watchers to impute to it a fiery and impetuous personality. The ancient Egyptians called it "Horus of the Horizon," and later "Har dacher" (Horus the Red). They also gave it the alternative name *sekhed-et-em-khet-khet* (he who moves backward), which proves they were aware of its retrograde motions. The Egyptians of the earliest era seem to have imagined the universe in the form of a box, the bottom of which was narrow, oblong, and slightly concave (and centered, naturally, on Egypt itself). The great river, the Ur-nes, marked the ecliptic, the Sun's path through the heavens, which flowed through the mountains; in the north it was hidden behind them as it ran into a valley, the Daït, where, surrounded in endless night, it became the heavenly Nile— the Milky Way. Along the river floated a boat whose passenger was a disc of fire, the Sun; the same stream carried the bark of the Moon, and the planets. Interestingly, on the ceiling of the tomb of Senenmut, chief architect during the reign of Queen Hatshepsut (15th century BCE), on the West Bank of the Nile, which contains the earliest known star map, all the planets are depicted as figures in boats—except for Mars. This last boat is empty, possibly because Mars's bizarre motion suggested to Egyptians the lack of a pilot.[5]

The Babylonians called it Nergal, for their god of fire, war, and destruction; the Greeks, perhaps borrowing from them, called it Ares after their god of war. In Hellenistic times, it was occasionally referred to, simply, as Pyroeis, "fiery." The Romans, who were notorious for borrowing so much from the Greeks, identified it with their war god, Mars, though it may have originally been identified with Silvanus, their god of vegetation and fertility, later becoming Gradivus, the god of spring honored in the festivals of the Amarvalia, celebrated in Rome each May 29. (The warrior functions may have been added when Gradivus was corrupted into *gradi*, "to march.") The Arabs, Persians, and Turks named it Mirikh, which can signify a torch, iron, or a spear cast to a great distance. In India, it was Angaraka, "the burning coal." In the Far East it was the fire start, a portent for bane, grief, war, and murder: Huo Hsing in China, Kasei in Japan, Hwa-seong in Korea.

Its astrological influence has generally, in all places and in all times, been regarded as baleful and malign. Thus the 19th-century French astronomy writer Camille Flammarion wrote: "Unfortunate Mars! What evil fairy presided at his birth? From antiquity, all curses seem to have fallen upon him. He is the god of war and carnage, the protector of armies, the inspirer of hatred among the peoples, it is he who pours out the blood of humanity in hecatombs of the nations."[6]

Interpreters of Omens

The ancient Egyptians at the time the Great Pyramid was being built (c. 2560–2540 BCE) described a group of circumpolar stars, centered on Meskhethyu, the Bull's Thigh (our Big Dipper). These stars were the "imperishable ones," since they never set below the horizon. The signs of the zodiac and the planets that moved restlessly and endlessly among them, including Mars, were by contrast known as the "unwearying ones." The great astronomical achievement of the Egyptians was their adoption of the first solar calendar in history: it consisted of 12 months of 30 days each, with five additional days at the end of the year. It has been called "the only intelligent calendar which ever existed in human history."[7] However, it was their Near Eastern neighbors between the Tigris and Euphrates, in the area that is now Iraq—the Sumerians, Sumer-Akkadians, and Babylonians—who laid the foundations of astronomical science. Indeed, the discoveries they made have been referred to, by historian Noel M. Swerdlow, as "the most important, the most revolutionary in the entire history of science in antiquity, perhaps in the entire study of the history of science."[8]

In Egypt, the annual inundation of the Nile produced predictably and reliably fertile fields. By contrast, in Mesopotamia conditions for agriculture were always more variable; since the annual rain flow is low, the ground becomes dry and hard and unsuitable for the cultivation of crops for eight months of the year, while the sluggish flow of water in the two rivers deposits large quantities of silt and elevates the bed to the point where the waters overflow the banks or change their course. Mastery of this challenging situation could be achieved only through the creation of an extensive system of artificial canals, a tremendous effort requiring coordination on a hitherto-unattempted scale and leading to vastly larger settlements than the small villages of the early Neolithic.

This great irrigation-based civilization originated in the southern part of the country, the achievement of the still rather mysterious people known as the Sumerians, about 3200 BCE. They created the first cities, produced impressive artworks, built large and differentiated public areas such as that at Uruk around the White Temple of Inanna (the Sumerian Venus), and—most importantly—created the earliest script.[9] This script was needed to keep a large bureaucracy humming, and to keep the royal records in order. The latter included the observations of priest-astronomers of astronomical phenomena, such as the times of the first appearance of the thin crescent Moon, which marked the beginning of each new lunar month. Eventually, in order to get a better view of the horizon, the sky watchers observed from the elevated platforms of seven-level terraced ziggurats. The ziggurat of the ancient Sumerian city of Ur—the biblical "Ur of the Chaldees," the reputed birthplace of Abraham—built in the period 2112–2095 BCE, is the most famous. The priest-astronomers also decided when to add an extra, thirteenth, month to their lunar calendar in order to keep it synchronized with the seasons and religious festivals.

When, later, the Sumerians merged into the Semitic population of Akkad to form the Sumer-Akkadian Empire, in turn succeeded by the Babylonian Empire, the workload of the priest-astronomers seems to have expanded greatly. They began to pay close attention to eclipses, and to phenomena involving the planets: their heliacal risings and settings (their first visibility before or after the Sun); their stationary points, retrograde movements, and changes in their brightness; their conjunctions with the Moon and stars and constellations, and with one another; even their appearances within halos about the Moon, for they were equally interested in meteorological phenomena. They seem to have taken as their purview everything happening in the sky.

This interest was based on a naïve but natural belief leading, in the long run, to science itself. They took for granted that there was a correspondence between what happened in the heavens and what happened on Earth, that, as Shakespeare would express much later in his 15th sonnet, "This huge stage presenteth nought but shows / Whereon the stars in secret influence comment." Confounding correlation with cause and effect, they imagined that if an event of importance, such as the deposition of a king, an uprising, a famine, or a war followed some particular phenomenon they noticed in the sky, then whenever the same sky phenomenon recurred the event would follow also. Thus the priest-astronomers began recording, in cuneiform on clay tablets, a vast quantity of data including, for the first time, records of the planets' irregular motions among the stars, an almost endless variety of their phenomena together with the terrestrial events presumed to follow them. We possess only a small sample of the tablets from the earliest periods, the 70 "Enuma Anu Enlil" tablets that once belonged to the palace library of the Assyrian king Ashurbanipal (668–627 BCE) at Nineveh,[10] and that survived the destruction of the palace and library in 612 BCE, when the Assyrian Empire was overthrown. (Ironically, the fire that gutted the palace partially baked the clay cuneiform tablets and helped them survive until the mid-19th century, when they were rediscovered by the British archaeologist Sir Henry Layard, who had them shipped to the British Museum, where they remain today.) The "Enuma Anu Enlil" tablets contain "omens," signals of the gods to the kings, in which they expressed their pleasure or displeasure on the latter's conduct.

Fortunately, Ashurbanipal was a compulsive collector and had copies made for his library of much earlier tablets. From them we learn that omens related to planetary phenomena were already being recorded as early as the Old Babylonian Empire in the early 18th century BCE. On one of them, the so-called Venus tablet, are recorded omens related to the risings and settings of the planet Venus (known to the Babylonians as Nin-dar-anna, the mistress of the heavens). These notations date far back indeed, and record observations made during the 21-year reign of Ammisaduqa (c. 1646–1626 BCE).

Though presumably beginning with the straightforward idea of connecting an observation of a phenomenon in the heavens with a terrestrial event, the priest-scribes would sooner or later have come to realize that, as Swerdlow says, "the number of permutations is very large, and thus the interpretation of omens was in principle a science of great complexity requiring a high degree of expertise and much specialized knowledge."[11] The complexity meant that the individuals involved in interpreting the commentary of the heavens had to be highly specialized. Note that the Babylonians did not regard the planets as gods; they were *manifestations*—interpreters of the gods—and their wanderings among background stars, reversals of direction, changes of brightness, or combinations with the Moon, the other planets, and the stars provided a cryptic commentary that challenged to the utmost the arts of divination. No doubt the sheer complexity gave the priest-astronomers room to waffle, and this must often have buffered them from the wrath of kings when they got things wrong. Moreover, even an inauspicious omen was not fate. It could be offset by a countervailing auspicious one, or mitigated by rituals and prayers. The text of one of these prayers is the following, a hand-lifting prayer (used to avert bad luck) to Nergal (the Babylonian Mars): "Let them write as follows: 'In the evil of the planet Mars, which exceeded its term [of invisibility] and appeared in the constellation Aries; may its evil not [approach], nor come near, not press upon me, not affect me, my country, the people of me and my army!'"[12]

The one thing we can be sure of is that the validity of the whole scheme of divination by the stars was never doubted, and must have provided a sense of security—including job security for the priest-astronomers—during a time when "seemingly out of nowhere, kings are murdered, armies massacred, nations perish through warfare, famine, and disease, or so would it be were precautions [provided by the omens] not taken."[13] It is a striking example of the way that the brain is a "projective system": it "continually projects its hypotheses onto the external world, putting them to the test of experience. Sometimes, it gives meaning to what has no meaning."[14] In the case of the omens—auspicious, or inauspicious, as the case may be—this must have occurred as often as not. (The brain's projective tendencies were not, as we shall see, limited to the Babylonians, but would form one of the overarching themes in the long history of studies of Mars.)

Human lives, and even the lives of empires, during these long chaotic centuries were nasty, brutish, and short. Even a brief chronicle of the history would require more space than we can devote here.[15] We note only the tenacity—despite all the shocks and changes—of the overall interpretive scheme. Many early observations were, of course, imprecise and did not distinguish between astronomical and meteorological phenomena—thus clouds and halos were on equal footing with eclipses. We also note that omens associated with Mars were usually regarded as inauspicious, as in the following examples:

> If Mars, retrograding, enters Scorpius, do not neglect your guard: the king should not go outdoors on an evil day.[16]

> The planet Mars has gone on into the constellation Capricornus, halted (there), and is shining very brightly. The interpretation is as follows: If [Mars] rides Capricornus, there will be a devastation of Eridu; people will be annihilated. . . . If Mars is bright in the sky there will be an epidemic.[17]

These omens, needless to say, are hardly great literature. To a modern reader they sound rather like the fortunes in fortune cookies. Nor would the observations on which they are based ever have led, as the historian of astronomy Antonie Pannekoek (1873–1960) pointed out, to the discovery of the regularities in the heavens that underlie the development of mathematical astronomy. The priest-astronomers were not looking for regularities; rather, says Pannekoek, the regularities must have gradually "imposed themselves," arousing expectations among the observers, which developed into astronomical prediction.[18]

Though like the later Greeks the Babylonians were concerned with casting horoscopes, there was an important difference in their approaches. According to Swerdlow, "In Greek astronomy, . . . the principal object is to find the location of a body in the heavens at a given time, at any given time, because the time of the horoscope is arbitrary and it is the locations that are significant. . . . In Babylonian astronomy, on the other hand, the principal object is to find the time and the location of a particular phenomenon—because it is the date and the location that are ominous. . . . The time is given in Greek astronomy but must be found in the Babylonian."[19] When the sky was clear, the dates could be determined directly. Not so when the sky was cloudy. For Mars, the priest-astronomers would have known that the planet has approximately 22 heliacal risings in 47 years. However, there is actually a deficit of 8 days from 47 years, and an excess of 1 day over 581 months; an account of the remainder and surplus is crucial, in the Babylonian scheme, to working out the true times of the phenomena to within a day, which is required to interpret the omens correctly. Needless to say, episodes of bad weather could interfere with the observation of a phenomenon such as an eclipse, or the heliacal rising or setting of a planet. When unobservable, the dates had to be determined by calculation. To quote Swerdlow one last time, "All [the] nights of rain and clouds and poor visibility recorded in the Diaries turned out to be good for something after all. When it is clear, observe; when it is cloudy, compute. . . . From bad weather was born good science."[20]

By the last several centuries BCE (the so-called Seleucid period of Babylonian history), the priest-astronomers had discovered highly precise periods in which astronomical phenomena repeated themselves. One important period was the celebrated (but misnamed) 18 year, 11 day, 9 hour "Saros cycle" governing eclipse intervals.[21] Equally important were the periodicities they discovered involving the planets, consisting of equating x intervals of one sort with y intervals of another sort. As summarized by the Alexandrian Greek astronomer Claudius Ptolemy (c. 100–c. 170) in the second century CE, the periodicities were as follows (where *s.p.* stands for synodic period, the interval between successive appearances of the planet in the same position relative to Earth and the Sun):

> Saturn: 57 s.p. = 59 y. + 12/4 d. = 2 rev. + 1°43'
>
> Jupiter: 65 s.p. = 71 y. + 49/10 d. = 6 rev. − 4°50'
>
> Mars: 37 s.p. = 79 y. + 3 13/60 d. = 42 rev. + 3°10'
>
> Venus: 5 s.p. = 8 y. − 2 2/10 d. = 8 rev. − 2°15'
>
> Mercury: 145 s.p. = 46 y. + 1 1/30 d. = 46 rev. + 1°

Among these, the celebrated eight-year "Venus cycle" stands out—the most obvious of these relationships, and one that the Babylonians seem to have discovered as early as the second millennium BCE. Since Venus takes five complete journeys around the zodiac in eight years, whatever Venus is doing tonight it will be doing again eight years from tonight. (In fact, the eight-year period is not quite exact; hence the remainder in the table above. The positions actually slip out of alignment by a little more than 2½ days after each cycle.)

Mars returns to nearly the exact same relative position after 79 years (again with a small remainder), Jupiter after 71 years, Saturn after 59 years. For Mars, this means that oppositions—the alignments of the Sun, Earth, and Mars—

repeat almost exactly each 79 years. Thus, the opposition of June 13, 2001, nearly repeats that of June 10, 1922; the opposition of August 28, 2003, nearly repeats that of August 23, 1924, and so on.[22] (See appendix D for the oppositions of Mars.)

The Greek Miracle

The set of equations above is the distillation of more than a thousand-year-long preoccupation with divination, observation, and calculation, the product of the sustained effort of countless individuals who, without quite realizing what they were about, established the foundations of the first empirical science in human history. The foundations of this astronomy lie in arithmetic. The priest-astronomers used "purely numerical functions—algorithms as it were, described in procedure texts much like computer programs—for intervals of distance and time, without any underlying descriptive models of the motions of the bodies in the heavens."[23] Though we associate this data set mainly with the Babylonians, some of it may have been known to the Greeks long before the classical era;[24] new information about the ancient world continues to turn up, with the potential to profoundly change our views about ancient astronomy (as the Antikythera mechanism, discussed below, has most notably done). It is regrettable that so much has been lost, and that what survives is so often fragmentary and in need of filling in with conjecture and extrapolation. The following is something of a consensus view, and though it may eventually stand in need of significant modifications, at least it is unlikely to prove entirely wrong.

According to this consensus view, after the conquest of Babylonia in 331 BCE by Alexander the Great (356–323 BCE), the Babylonian data set began to fall into Greek hands, and this would lead, within the next century, to an event of singular importance to the history of astronomy, as the spatial and geometric imagination that was uniquely Greek was used to clothe the Babylonian numbers in geometry.

Mention of geometry inevitably brings to mind Euclid, whose *Elements*—which represent the conquest of arithmetic by geometry—was long a foundational text of Western education, and a virtual paradigm of rational argument.

It represented "a codification and arrangement, according to the postulational method, of most of Greek geometry in existence at the end of the fourth century BCE,"[25] but it also exhibited "an imposing coherence and self-consistency that does not appear to exist in any other human creation."[26] It seems to represent (where it applies) a body of truths that cannot be conceived by a rational being as otherwise than true, assertions that are as true today as they were in 300 BCE, that are true for all time. Not surprisingly, the same postulational method was carried over by the Greeks into their astronomy, with two postulates being regarded as axiomatic, also from about 300 BCE:

1. Earth stands at rest at the center of the universe.
2. The only motions allowed in the heavens are simple and uniform circular motions.

These principles are simply asserted, not demonstrated. This is what makes them postulates. The Greek astronomers proceeded to work down from that high ground to the observed phenomena in the heavens, and in so doing attempted something that would never have occurred to the Babylonians with their strictly arithmetical formulations. They attempted by means of geometry to simulate the apparent paths of the Sun, Moon, and planets as projected onto the sky. It is important to emphasize that there was never a presumption that they were revealing the "true" positions of the Sun, Moon, and planets—these could not be known—but rather "only the intersections of lines of sight (between Earth and planet) with the 'fixed-star sphere.'"[27] This, to use their own term for it, was called "saving the phenomena."

The phenomena that were to be "saved" were, of course, the same that had been exhaustively studied for centuries by the Babylonians. Mercury and Venus, the "inferior" planets (so called because they were believed to lie below the Sun relative to Earth), alternately appear as "Morning" and "Evening" Stars; they never venture far from the Sun, and they rise above the horizon and fall back down toward it over the course of weeks, or months, disappear for a time, then reappear. The "superior" planets were those presumed to lie "above" the Sun—Mars, Jupiter, and Saturn—whose retrograde motions around the times when they were lined up opposite to the Sun had simply seemed inscrutable.

At first, the Greek geometers did not attempt anything more than a qualitative, descriptive model of these motions, presumably because they did not yet have very accurate data. A fairly rigorous attempt, for its time, was made by Eudoxus of Cnidus (c. 390–337 BCE), a geometer who was briefly associated with Plato's Academy (whose teachings he seems not to have found very satisfactory), and who was the most able mathematician of the age.[28] Eudoxus's solution was his celebrated model of homocentric spheres, which—though without practical (predictive) value—achieved a degree of success in simulating Mars's retrograde motions. Eudoxus assumed that the planet moved on a sphere attached to another sphere, as if it were a small wheel pinned to the rim of a larger wheel, with the axes tilted toward one another but remaining connected like a compass in a gimbal.[29] Unfortunately, the scheme made the planet retrograde three times instead of the once that was actually observed. Further, it could not explain Mars's drastic changes in brightness. But then why, on such an Earth-centered scheme, should a planet's brightness vary at all?

Plato's most famous pupil, Aristotle (384–322 BCE), seems to have taken the homocentric spheres literally, asserting the actual existence of a heavens full of internested crystalline spheres. When he was 27 or 28 years old, and still studying in Athens under his master, Plato, he observed an occultation of Mars by the Moon, on May 4, 357 BCE, which convinced him that Mars occupies a higher realm in the heavens than the Moon and presumably was ensphered in its own separate crystalline orb. He was followed by Callippus (c. 370–c. 300 BCE), who not only worked with Aristotle in Athens but had personally studied under Eudoxus, and who made further refinements in the scheme. Though the great mathematician Archimedes (c. 287–c. 212 BCE) subsequently produced a working model of the Eudoxan system using glass spheres turned by water power, by then it had become nothing more than a clever toy; the model was no longer taken seriously by Greek astronomers, and new ways of "saving the phenomena" were being asserted.

Disillusioned by the failures of the Eudoxan system, the great astronomer Aristarchus of Samos (c. 310–c. 230 BCE) went so far as to propose a heliocentric (Sun-centered) theory by about 250 BCE, in which Earth was a planet spinning on its axis relative to the fixed stars once in every 24 hours, and moving in a circular orbit around the Sun, completing each circuit in a year. (Aristarchus seems to have made all the planets move in circles.) Since nothing of his work setting forth the heliocentric system has survived, except in echoes of other writers (such as Archimedes), we can only guess how he managed at such an early date to penetrate the two grand illusions presented by the phenomena of the heavens: the first being the apparent diurnal rotation of the heavens around Earth; the second being the apparent retrograde motions of the outer planets, most obstreperous in the case of Mars. This unruliness is an illusion due to the variance of Mars's projected position against the stars as the faster-moving Earth catches up with, passes, and then moves ahead of the slower-moving Mars. His model is certainly impressive, and we naturally wonder why the "ancient Copernicus" found no successors in the next generation, or even in the generation after that. (Not until 150 BCE did Seleucus, a Chaldean of Seleucia on the Tigris, adopt the heliocentric theory, and so far as we know, he was the only one.) Religious prejudice may have played some role; in all times and places it has done that. But there were scientific grounds as well, for Aristarchus seemed to defy common sense by suggesting that the heavens, "as revealed by every passing breeze and flickering fire, was tenuous, unresisting, mobile, and the Earth solid and at rest."[30] Also, his idea implied a universe much larger than most astronomers of the time were willing to allow, and just as importantly, it made a shambles of astrology, which depended on the notion that the planetary motions relative to the stars were commentaries on human destiny in an Earth-centered world. If there were other centers, the motions became relative.

Rather than follow Aristarchus over the cliff, mathematicians rose to put down the heliocentric challenge and to reassert the geocentric system. The greatest was Apollonius of Perga (late third to early second centuries BCE), who is best remembered for his treatise on the conic sections, and who was probably responsible for introducing a number of ingenious and influential geometric devices into astronomy. One was the movable eccentric, in which Earth was placed slightly off-center of the circular path of the planet, and where the center of the circle rotated in turn in a small circle around Earth. Geometrically equivalent, but more convenient, was the celebrated epicycle, whereby a planet's retrograde motion could be explained by having the planet rotate around a smaller circle (the epicycle itself), which in turn was centered on a larger circle (known as the deferent) moving around Earth. The system was flexible and adaptable enough to be able to explain the so-called second-order irregularities in the planets' motions (due mostly, as we now know, to the fact that the actual orbits of the planets are noncircular), which came to be felt as soon as better observational data (at first probably largely Babylonian) became available. Since circular orbits such as had been assumed by Aristarchus were no longer workable, the movable eccentric and epicycle scheme, which effectively offset Earth as a practical center of motions, restored geocentric astronomy to the dominant position it was to hold for 1,700 years.

Resembling a gear mechanism like that used in clockwork, epicycles with deferents for the Sun, Moon, and planets were incarnated, in bronze, in the calculating device known as the Antikythera mechanism, found in an ancient shipwreck discovered by sponge fishermen off the coast of Antikythera (an island between Kythera and Crete) at the beginning of the 20th century.[31] The wreck contains vases in the style of Rhodes, suggesting that the ship may have sank en route from Rhodes, where Hipparchus of Nicaea (c. 190–c. 120 BCE) was active. Unlike earlier astronomers, who had been satisfied with semiqualitative models of the motions of heavenly bodies, Hipparchus seems to have been the first to set out "to construct models with actual predictive value—geometrical models with parameters derived from actual observations."[32] He clearly made use of Babylonian observations, but he also made many observations of his own, though sadly we know nothing about the instruments he used. Like everyone except Aristarchus and Seleucus, he was, of course, a geocentrist.

Everything we know of Hipparchus's work is due to the (much later) summary by Ptolemy (figure 1.1). Ptolemy called his work *Syntaxis*; it was later known as the *Great Syntaxis* to distinguish it from a lesser collection of astronomical writings, becoming, among the Arabs, simply *al-Majesti* (the Greatest). This, translated into Latin, became *Almagest*, the name by which it is generally known.

Ptolemy spent his entire career at Alexandria. The city had been founded by Alexander the Great in the same year he conquered Babylonia, 331 BCE, though ever-restless Alexander himself remained for only a few months before marching on Tyre in Phoenicia. It was owing to his successors—most notably Ptolemy I Soter, one of the three generals who divided his empire among them after the conqueror's death—that the city became the most powerful metropolis in the East. In the astronomer's time, it boasted a population of

FIGURE 1.1 Ptolemy, as imagined by sculptor Jörg Syrlin the Elder (c. 1425–91) for the choir stalls at the cathedral at Ulm. Photo by William Sheehan, 2011.

800,000, second only to Rome itself, but surpassing the capital in industry and commerce. It was celebrated for the Pharos, the great lighthouse that was still standing in Ptolemy's time; its large theater; the Emporium or exchange; temples to Poseidon, Caesar, and Saturn; the Serapeum or Temple of Serapis; and a group of buildings known as the Museum, in which the famous library was housed. Though the library itself was destroyed, in whole or in part, during Julius Caesar's Alexandrian campaign against Pompey (48 or 47 BCE), a daughter library survived that may have been housed in the Serapeum and that was likely home to Ptolemy's observatory. As an observer, Ptolemy was not in Hipparchus's class, but he did record a few of his own observations between March 127 and February 141 CE.

With a mind encyclopedic in scope, Ptolemy worked single-handedly on a "five-foot shelf" of books on every aspect of applied mathematics, including not only astronomy but also harmony, optics, geography, and astrology (which, child of his age, he regarded as a branch of applied mathematics). The *Almagest* was probably the first of the works he completed. His purpose, in his own words, was: "With regard to virtuous conduct in practical actions and character, this science, above all things, could make men see clearly; from the constancy, order, symmetry and calm which are associated with the divine, it makes its followers lovers of this divine beauty."[33]

Ptolemy did not invent but inherited and perfected the system of epicycles and deferents. Judged on its own terms, the complex system was a tremendous success, although it would have its later detractors. Among these, King Alfonso X of Castile, called "the Wise" (1221–84 CE), who sponsored the most famous set of tables based on Ptolemy in use during the Middle Ages, reputedly said of Ptolemy's system, "it seemed a crank machine . . . pity the Creator had not taken advice."[34] Even so, the historian Owen Gingerich has written:

> It is difficult to convey the elegance of Ptolemy's achievement to anyone who has not examined its details. Basically, for the first time in history (so far as we know) an astronomer has shown how to convert specific numerical data into parameters of planetary models, and from the models has constructed a . . . set of tables . . . that employ some admirably clever mathematical simplifications, and from which solar, lunar, and planetary positions and eclipses can be calculated as a function of any given time.[35]

The *Almagest* is not merely a summing up of what others had done; it is in many ways an original work, and nowhere more so than in its planetary theory and, in particular, its theory for Mars. Ptolemy was well aware that Mars's motion around its deferent was strikingly nonuniform: when the planet comes to opposition in Capricornus/Aquarius, it moves twice as fast as it does when on the other side of its orbit, in Cancer/Leo. To account for this, Ptolemy decentered his deferent from Earth. Now he had the speed of Mars's motion right and could account for the dramatic differences in brightness.

But the sizes of the retrograde loops of Mars did not match up with what was observed. They were of different sizes depending on where Mars comes to opposition. Here Ptolemy offered a drastic expedient, and it proved to be a masterstroke: he made Mars move uniformly not around the center of the deferent but around another point called the equant, which he located as equidistant between the center of the deferent and Earth. This solution was called the "bisection of the eccentricity," and it worked brilliantly, even though the construction involved tacit abandonment of the rule of uniform circular motions that had been insisted on by Greek astronomers since the time of Plato. He did not call explicit attention to this. But then it was not his goal to describe the actual paths taken by the planets through space, only to calculate, as accurately as possible, their positions in advance, and this he achieved. His theory allowed the positions of the planets to be calculated, in a direct way, often to an accuracy on the order of $10'$ of arc or so, or well within the limits of accuracy of the measurements possible with the instruments of his time. Indeed, Ptolemy's calculated positions are often better than his actual observations, which are frequently off by as much as a degree!

Among Ptolemy's other books, his *Geography* was the most highly regarded in the Middle Ages, and is remembered today for his mistake of following Strabo (c. 63 BCE–24 CE) instead of Eratosthenes (c. 276–c. 194 BCE) in estimating of the circumference of Earth, which had the effect of making the eastern tip of Asia and the western tip of Europe seem tantalizing close together (and indirectly hastened the European encounter with the New World). Also of interest is the *Tetrabiblos* (Four Books), on the influence of the stars on human life. In the second century CE—and indeed long afterward—astrologers would have been the major consumers of planetary theory. Presumably, however, Ptolemy himself believed firmly in divination by means of the stars.[36] In that regard, he was not so very far from the old Babylonians.

Ptolemy's lifetime overlapped with the reigns of Hadrian and the Antonines—Antoninus Pius and Marcus Aurelius—which the historian of the decline and fall of the Roman Empire, Edward Gibbon, would later describe as the period in the history of the world "during which the condition of the human race was most happy and prosperous."[37] But the very identification of a time when the human race was most happy and prosperous presupposes a downfall into misery and want. Ptolemy himself—living in a time when, as historian Pannekoek elegiacally put it, "the sunset glow of antiquity shed its light radiance upon a worn-out world"[38]—only narrowly escaped, by death, the onset of the squalls that would eventually undermine the power of a world empire. If he sometimes paused from his scholarly endeavors and lowered his eyes from the skies, he would have been surprised by signs of a very different world to come: "Pythagoreanism from Southern Italy, strange gods from Syria, the sun-god Mithras from Parthia, and . . . Yahweh, an uncompromising monotheist from Judea who commanded the most difficult life of piety and regulation. . . . [And] among the Roman Jews who prayed to him some, as yet obscurely distinguished from the rest, who worshiped his incarnate and resurrected son."[39] Ptolemy himself was nominally a pagan. In his personal philosophy, however, he seems to have agreed with the Stoics (including his contemporary Marcus Aurelius) that "to follow Zeus [God] and to obey reason are the same thing."[40] If a Stoic, he seems to have been a cheerful one, and he would doubtless have agreed with the Greek writer Plutarch: "'Does not a good man consider every day a festival?' . . . For the universe is a most holy temple and most worthy of a god; into it man is introduced through birth as a spectator, not of handmade or immovable images, but of those sensible representations of knowable things that the divine mind, says Plato, has revealed, . . . [even the] sun and moon and stars."[41]

There is an evocative wooden sculpture of Ptolemy (by Jörg Syrlin the Elder) in the choir stalls of the cathedral of Ulm: eyes closed, a faint smile on his face, he holds an armillary sphere in one hand while pointing casually upward toward the heavens with the fingers of the other. It is a picture of smugness and complacency, the face of an astronomer so satisfied with the completeness of his theory that he no longer needs to study the actual heavens. That, indeed, is the reputation Ptolemy (unfairly) came to have in later generations, when it became fashionable to criticize his planetary theories, with their complicated gears, and to rail against a heavens "With centric and eccentric scribbled o'er, Cycle and epicycle, orb in orb."[42] A more captivating portrait, however, is one likely penned by Ptolemy himself. In some of the old manuscripts of the *Almagest* appear the words:

> *I know that I am mortal by nature, the creature of a day:*
> *But when I trace at my pleasure the windings to and fro of the heavenly bodies*

I no longer touch the earth with my feet:
I stand in the presence of Zeus himself
And take my fill of ambrosia, the food of the gods.[43]

During the next century after Ptolemy's death, a series of Roman emperors—appointed and then murdered at the pleasure of the army—would watch with increasing alarm the encroaching dangers from the barbarians to the north, and race to build defensive works from the Veneto to Milan. Rome itself would be surrounded with massive walls and fortifications. Instead of marking the furthest limits of the imagination in its indomitable struggle with the unknown, the word "frontier" then acquired the meaning it still has in Europe—the sharp edge of sovereignty, a line to stop at, not an area inviting entrance. It was the outermost wall of fortifications on the verge of fear.

Inevitably, shells of dogma would harden to encase the human mind, and Ptolemy's system would become part of that dogma. Outlasting the Roman Empire in the West (whose last emperor was overthrown in 476 CE) and even that in the East (1453 CE), his system would remain standing—a little cracked and tottering perhaps, but still upright—for over a thousand years. During all that time, Mars, as humans understood it, would trace the prescribed workings of its Ptolemaic machinery, and drop far back into the remote curtain of the sky well beyond the retreating frontier of the human imagination. Instead of a world, it was a mere set of complicated gears, or within those gears, a remote, inaccessible pale red dot.

2

The Warfare with Mars

A Not So Timid Canon

After Ptolemy's death circa 170 CE, Mars for centuries retreated from human grasp. During much of this long period, most advances in astronomy came from the Islamic world, where scholars such as al-Farghani (c. 800–c. 870 CE), al-Battani (c. 850–929 CE), and al-Biruni (973–1048 CE) were active. Meanwhile, in Christian Europe, there was a retrogression: space consisted of the flat Earth with the heavens regarded as an inverted bowl in which the stars were peepholes into the flaming firmament beyond. The Moon, Sun, and planets still moved in epicycles, encompassed by the crystalline spheres of Aristotle and Callippus; even hell still existed as a geographic location within Earth.

With the revival of learning in the West, the commentaries of the Arabic scholars on classical texts (especially Aristotle) and even some of the original texts began to make their way into Europe, where they were translated into Latin. Toledo, a center of learning that was under Islamic Arab rule from the eighth century, was a source of many important ancient manuscripts, including an Arabic version of the *Almagest*, a book long lost to Europe or at best dimly remembered, and translated from Arabic into Latin by Gerard of Cremona (c. 1114–87) in 1175; this translation continued to be consulted for centuries, though it was not printed until 1515. Meanwhile, King Alfonso X of

Castile, the very same person who reputedly called Ptolemy's system a "crank machine,"[1] organized a team of scholars at Toledo to produce new tables based on Ptolemy, the *Alfonsine Tables*. (Tables are a regular series of numbers giving schematically time intervals such as those involving the planetary motions that can be used to generate ephemerides, actual predictions of the positions of the planets for a certain range of dates and times; Alfonso's begin on January 1, 1252, the date of Alfonso's coronation.) Alfonso's tables were based on a pure Ptolemaic theory—that is, with an eccentric, equant, and single epicycle for the superior planets, with parameters almost identical to those originally adopted by Ptolemy.[2]

After the fall of Constantinople in 1453, more ancient manuscripts began to arrive in Europe, as Byzantine scholars, fluent in Greek, migrated across Europe, bearing with them precious Greek texts. Mostly they came to Italy, where these texts were studied and gradually absorbed. Diverse ideas were reintroduced to the human mind, different from those ideas insisted on by the medieval church. Admittedly, old habits of thought do not change overnight, and so people—even scholars—so long used to bending their will to the yoke of authority now began to submit to the authority of texts. But at least the texts were different from Aristotle and the Bible, and, as Bertrand Russell (1872–1970) remarks, that was "a step toward emancipation, since the ancients disagreed with each other, and individual judgment was required to decide which of them to follow."[3]

In due course, translations of the *Almagest* appeared based directly on a Greek manuscript rather than, as hitherto, Arabic intermediaries. From such books the man known in Polish as Mikołaj Kopernik, in German as Niklas Koppernigk, and to the world as Nicolaus Copernicus learned his astronomy, and became familiar both with the intricacies and successes of the Ptolemaic system and with some of its errors and failures.

Copernicus (1473–1543) was born in Toruń (now in northern Poland) on the River Vistula. His father, a prosperous wholesale merchant, died when he was only ten, whereupon he passed into the care of his uncle, Lucas Watzenrode. As bishop of the ecclesiastical state of Warmia, Watzenrode used his powerful position to pull strings, and he obtained for Nicolaus, a sometimes mathematics student at the University of Cracow, a lifetime sinecure as a canon (not a priest) in the cathedral at Frombork, a small town some 40 kilometers east of Gdansk and overlooking the Frisches Haff, a lagoon of the Baltic. Though his future was thus secured, Nicolaus was not in any hurry to settle into this position. He pursued further studies in Italy, which included a stint at the University of Bologna, Europe's oldest university. One of his teachers was Domenico Maria de Novara (1454–1504), who has been called "an enthusiastic Neoplatonist who attacked the Ptolemaic system," an attack presumably lodged on the grounds that "all that pedantic apparatus of epicycles, deferents, equants" was overly complicated and unaesthetic.[4] Under Novara, Copernicus also made his first astronomical observation, an occultation of the star Aldebaran by the Moon, on March 9, 1497.

In 1500, he was in the Rome of the Renaissance pope Alexander VI (1431–1503), lecturing on mathematics. He next studied medicine at the University of Padua and canon law at the University of Ferrara. After finishing a doctorate in canon law, he went in 1506 to Heilsberg (now Lidzbark Warmiński, Poland), where he spent several years as his uncle's personal physician in the Castle of the Warmian Bishops. Only upon his uncle's death in 1512 did he take up his long-deferred residence at Frombork. By then he was almost 40.

As a canon, Copernicus was entrusted with many harassing duties, and his burdens were heavier than sometimes supposed. He was certainly not a man with his head in the clouds. He kept all the accounts of the diocese, saw that the chapter's political interests were protected, collected rents, and settled disputes for the peasants who worked the fields surrounding the cathedral. He was also frequently called on to exercise his medical skills, waged war against the Teutonic Knights, and helped plan the reconstruction of Warmia after the Polish-Teutonic War ended in 1521. The following year he was called to Rome by the pope, Hadrian VI (1459–1523), to make recommendations for the reform of the calendar, and he presented a plan for the reform of the currency to the Diet of Graudenz. (He anticipated the principle now known as Gresham's law: bad currency drives out the good.) He cannot have been the "timid canon" popularized by Arthur Koestler (1905–83).[5] Clearly he was a man not only of towering intellect but of physical and moral courage; we now know, for instance, from examination of his remains, recovered under the cathedral floor in 2005, that he suffered a broken nose on at least one occasion! Though he had quarters in the northwest tower on the cathedral wall, it was hardly an ivory tower: he likely would have used these quarters only in wartime. His main residence was a house outside the cathedral wall, where, by April 1513, he built—or more likely commissioned someone else to build for him—a brick platform for making astronomical observations.[6]

He probably started thinking seriously about planetary theory while still in Italy, as Novara's expressions of dissatisfaction with Ptolemy's system continued to ring in his ears. But like any other man of his time, he dared not make a significant move in astronomy without finding a basis for it in ancient authority. Crucially, he took careful note of several references in ancient texts to heliocentric ideas, which showed that there were precedents, and of which the most notable referred to Aristarchus. For a careful, lawyerly, canonical mind trained to respect authority, this intellectual scaffolding was important: authority must be opposed by other authority. By the time he came to live with his uncle at the Castle of the Warmian Bishops, Copernicus was sufficiently sure of his ground to begin to sketch out the idea of the heliocentric system.[7] He does not tell us directly how he came to formulate his ideas, but we find hints both in his own writings and in the *Narratio Prima* (First Account) written by his sole disciple, Georg Joachim de Porris, better known as Rheticus (1514–74), who tells us (in 1539) that Copernicus was bothered by two things: the difficulties inherent in the Ptolemaic theories of Venus and Mercury, and the large variations of brightness in Mars. The latter, always the bane of geocentrists, made the Red Planet a midwife to heliocentrism, for Rheticus tells us, evidently retracing the master's early thoughts:

In addition to the other difficulties in the correction of its motion, Mars unquestionably shows a parallax sometimes greater than the sun's, and therefore it seems impossible that the Earth should occupy the center of the universe. Although Saturn and Jupiter, as they appear to us at their morning and evening rising, readily yield the same conclusion, it is particularly and especially supported by the variability of Mars when it rises. . . . Whereas at its evening rising [near opposition] Mars seems to equal Jupiter in size, so that it is differentiated only by its fiery splendor, when it rises in the morning just before the sun [near superior conjunction] and is then extinguished in the light of the sun, it can scarcely be distinguished from stars of the second magnitude. Consequently at its evening rising it approaches closest to the earth, while at its morning rising it is furthest away; surely this cannot in any way occur on the theory of an epicycle. Clearly then, in order to restore the motions of Mars and the other planets, a different place must be assigned to the Earth.[8]

That different place was, of course, that of a planet in orbit around the Sun. Instead of Earth, the Sun was the "watchtower." Later, in a moving and uncharacteristically poetic passage of *De revolutionibus*, Copernicus justified this choice: "In the center of all rests the Sun. For who would place the lamp of a very beautiful temple in another or better place than this wherefrom it can illuminate everything at the same time? Not unhappily do some call it the Lamp, the Mind, the Ruler of the Universe: Hermes Trismegistus names him the Visible God, Sophocles' Electra calls him the All-Seeing. So the Sun sits as upon a royal throne, ruling his children, the planets which circle round him."[9]

As soon as he decided to exhume Aristarchus's old idea of the Sun rather than Earth being the "proper center," Copernicus's next step would likely have involved a further analysis of the motions of Mars. He had, moreover—thanks to the heritage of Renaissance art—a distinct advantage over the ancient Greek mathematician. The ancient Greeks appear not to have grasped the theory of perspective, but the artists of the Renaissance, especially Leon Battista Alberti (1404–72), had worked it out in detail. Alberti pointed out that when a viewer sees something through a window pane, a correct image of it can be formed by tracing its outlines, provided that one looks with only one eye and does not move one's head. He also studied how to determine the shape and measurements of the image of a square of known size at a given distance from the beholder. He succeeded, as he writes in his autobiographical *Vita Anonyma*, in creating incredibly convincing illusions, and brought about "things unheard of and that the spectators found unbelievable, and he showed these things [e.g., paintings of tessellated pavements and checkerboards] through a tiny opening that was made in a little closed box. . . . He called these things 'demonstrations,' and they were of such a kind that both artists and laymen questioned whether they saw painted things or natural things themselves."[10] It seems quite conceivable that Copernicus, who must have learned something of artistic technique in Italy and was capable of producing a creditable self-portrait, may have repeated Alberti's experiment, taking an oblong box with an eyehole, stretching strings, and getting out from behind his lines of vision in order to view them from various positions. Thinking of the planetary system, he would have realized at once that when viewed from the Sun, the intertwining strings between Earth and Mars and the other planets magically disentangle themselves, and the apparent backward or retrograde motions—those for which Ptolemy had introduced the largest epicycle—are a mere perspectival illusion produced by Earth's motion around the Sun (figure 2.1). However he came by it, he had reached this point in his thinking at the latest by 1514, when he explained in a draft of his heliocentric theory, *Commentariolus* (Little Commentary, or first sketch), that the retrograde motions of Mars

happen by reason of the motion, not of the planet, but of the earth changing its position in the great circle. For since the earth moves more rapidly than the planet, the line of sight directed toward the firmament regresses, and the earth more than neutralizes the motion of the planet. This regression is most notable when the earth is nearest to the planet, that is, when it comes between the sun and the planet at the evening rising of the planet.[11]

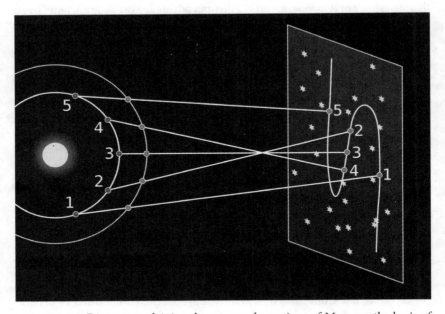

FIGURE 2.1 Diagram explaining the retrograde motions of Mars, on the basis of the heliocentric theory. As the faster-moving Earth catches up and passes the slower-moving Mars, the latter planet's image projected against the background stars traces out the illusion of a snakelike or looped motion. Brian Brondel/GNU Free Documentation License.

This was a brilliant insight, but Copernicus was far from finished. In his own terms, he had hardly begun. He had grasped the first-order relationships but—in contrast to Aristarchus, for whom the observations were still rather primitive and who could thus stop at the first-order relationships—he faced a much more difficult problem. Circular orbits such as those used by Aristarchus were no longer sufficient. He might have made things easier for himself if he had taken up Ptolemy's device of the equant, since, as Ptolemy had shown, it satisfied the motion of Mars so well. However, Copernicus was a purist who found the equant aesthetically displeasing. Instead he insisted on circles, in which "everything would move uniformly about its proper center,"[12] and this meant bringing the epicycles back again.

By about 1530 he had finished the manuscript of his great book *De revolutionibus orbium caelestium* (On the Revolutions of the Heavenly Spheres). It is a difficult book. His greatest successes were with the theory of the Moon—for which he needed only two epicycles, one placed on the other—and the outer planets, including Mars, for which he needed only a single epicycle. But his theory for the inner planets, Mercury and Venus, was at least as complicated as Ptolemy's.[13] He managed to reduce the total number of circles in his system somewhat—according to Johannes Kepler, by ten—but it was not much less cumbersome. He also had no answer to the problem of the parallax of fixed stars, though he did not so much as mention it; wisely, he refused to be distracted from his main purpose, which was to compute the motions of the planets on a heliocentric basis to an accuracy equal to or better than Ptolemy had been able to do. (Indeed, the problem of solar parallax was not fully dealt with until the 18th century.) Though completed by 1530, *De revolutionibus orbium caelestium* did not appear until 1543. In an age of often violent religious strife, the approval of Roman Catholic Church authorities had to be secured. The first copies are said to have reached him shortly before he died on May 21, 1543.

The Observer

Copernicus made a few observations from the platform set up near his house outside the cathedral wall, but he was not, primarily, an observer, and Rheticus once said that he would be as happy as Pythagoras had been on discovering his theorem if he could achieve an accuracy in his observations good to only 10' of arc. Clearly, what was needed now was not more tinkering with theory but better observations. To begin with, this meant better catalogs of stars, since the planetary positions are measured relative to star positions. Star catalogs available at the time inherently contained large margins of error, so the positions of the planets were correspondingly far off. Remedying this situation was to be the great work of the leading astronomer of the generation after Copernicus's death, Tycho Brahe (1546–1601), a Danish nobleman who spared neither pain

nor expense in pushing to the attainable limit the accuracy of naked-eye observations (figure 2.2). Mars was, as ever, a particular challenge. With Tycho—and especially his assistant Johannes Kepler—the planet came to loom ever larger on the stage of human thought.

Shakespeare's *Hamlet* has always merited at least a footnote in astronomical histories. The setting of the play is Elsinore, the site of the 15th-century castle greatly enlarged during the 1580s on the Öresund (Sound) between Denmark and Sweden. It lies on the other end of the Baltic from Copernicus's Frombork and not far from the island of Hven, where Tycho Brahe would build the most splendid observatory of the age. Though Hamlet speaks a few lines that could have been inspired by the peregrinating Dominican scholar-turned-heretic Giordano Bruno (1548–1600), who adopted the heliocentric system and reveled in unorthodoxies such as declaring the universe infinite (e.g., "O God, I could be bounded in a nutshell and count myself a king of infinite space"),

FIGURE 2.2 Tycho's portrait in *Astronomia instauratae mechanica* (Wandesburg, 1598), a book in which he describes his methods and the many instruments he designed and built. The first edition consisted of a run of fewer than 100 copies printed on his own printing press. More unflattering than others, this portrait is very likely the most accurate likeness of him, and it clearly shows his nasal prosthesis. Courtesy of Det Kongelige Bibliotek, Copenhagen.

Tycho is even more definitely alluded to in the play: the names of Hamlet's two school friends, Rosencrantz and Guildenstern, bear the names of Tycho's ancestors as recorded on a portrait Tycho used as the frontispiece of his *Astronomical Letters* of 1596. Presumably Shakespeare saw a copy of Tycho's book and cribbed the names.

Tycho's father, Otto (1518–71), was governor of Helsingborg Castle, across the Sound from Elsinore. Before Tycho's birth, at Knudstrup, Denmark, in 1546, Otto entered into an agreement with his brother, Jørgen (1515–65), according to which, in the event Otto had a son, he would give him up so the latter could adopt him and raise him as his own. (This arrangement was not as unusual at the time as it may seem now.) Though Otto began to regret the pact, Jørgen was not to be denied—he abducted the child and took him to his own castle, at Tostrup. After the initial shock died down, Jørgen was allowed to keep the boy.

Tycho was none the worse off for it. Jørgen was extremely well-to-do, and he doted on his adopted son. Moreover, he could afford to give the boy the best education money could buy. When Tycho was 13, he was sent to the University of Copenhagen to begin the study of law, but soon afterward an event occurred that would forever change the direction of his life. On August 21, 1560, an eclipse of the Sun took place just as astronomers had predicted, and "Tycho thought of it as something divine that men could know the motions of the stars so accurately that they could long before foretell their places and relative positions."[14]

His interest aroused, Tycho purchased a copy of Ptolemy's *Almagest* and worked through it, no mean accomplishment for a teenager. Two years later, he left Copenhagen to continue his legal studies at the University of Leipzig. By then he had a tutor, Anders Sørensen Vedel (1542–1616), hardly older than Tycho himself, who was charged with keeping Tycho's nose to his law books. Vedel did not succeed. Tycho was a headstrong and determined character; he worked at law during the day and stole out at night to study the stars while Vedel was asleep. By then he had acquired a celestial globe "no bigger than a fist," as well as star maps published by Albrecht Dürer (1471–1528), by means of which he learned to recognize all the constellations. But as yet he had no proper instruments, and "could only check the predictions of the ephemerides by lining up a planet and two stars by means of a taut string and estimating the positions of the planet from the positions of the two stars on his little globe."[15]

Then, in August 1563, Tycho's enthusiasm was awakened as never before. He observed the "great" (or triple) conjunction of Jupiter and Saturn. The two planets were so close together as to be almost indistinguishable with the naked eye. Instead of a taut string, he now employed a pair of large compasses, sighting from the vertex along each leg to the two celestial bodies being observed. Even with these rudimentary means, the 16-year-old Tycho was able to establish that the *Alfonsine Tables* were off by a month in their prediction of the date of the conjunction; even the tables meant to replace them, Erasmus Reinhold's *Prutenic Tables*, based on Copernicus's theories and published only in 1551, were off by several days. Tycho then and there resolved to make his life's work the construction of more accurate tables of the planetary motions. Thus, as his biographer J. L. E. Dreyer (1852–1926) writes, "his eyes were opened to the great fact, which seems to us so simple to grasp, but which had escaped the attention of all European astronomers before him, that only through a steadily pursued course of observations would it be possible to obtain a better insight into the motions of the planets, and decide which system of the world was the real one."[16]

Soon after the conjunction, Jørgen died, and Tycho received through inheritance the financial independence he needed to pursue his interests without distraction. From Leipzig, he moved to Wittenberg, and from Wittenberg to Rostock, where he famously lost his nose in a duel; ever resourceful, Tycho promptly had the missing member replaced with a prosthesis, apparently made not of silver and gold as long believed but of mere brass. (It has been suggested that self-consciousness about his now-lopsided appearance explains why he would marry a commoner instead of a woman of noble extraction.) He moved again, to Basel and then to Augsburg, where he at last began to design his own instruments, including a pair of giant compasses with a 30° brass arc and wooden legs one and a half meters long, and—not for his use but for that of a wealthy colleague—a large quadrant, consisting of a 90° arc made of well-seasoned oak to which were attached a brass graduation strip and plumb bob. The quadrant's radius was five and a half meters, and it was so heavy that 40 men were needed to put it in place.

The great quadrant was built in 1570. The next year Tycho left Germany and returned to Denmark. After a brief stay at the family estate at Knudstrup, where he soon became bored to death with the usual nobleman's life of horses, dogs, and luxury, he moved to Herre Vad, where he established a small private observatory on the estate of another well-to-do uncle, Steen Bille. This uncle alone among Tycho's relatives approved of his scientific tastes, dabbling in alchemy himself. It was here, on November 11, 1572, that Tycho, returning from Steen Bille's alchemical laboratory to supper, discovered the "new star" in the constellation Cassiopeia (ever after known as "Tycho's star") that would make him famous and win for him the patronage of the Danish king, Frederik II (1534–88). Frederik offered him the island of Hven (a name Tycho always insisted meant "the island of Venus") in the Sound, between Kattegat and the Baltic Sea, as well as the funds needed to construct a magnificent observatory and to pay for its upkeep. In return, Tycho was tasked only with performing a few nominal duties, such as keeping the cathedral at Roskilde in good repair.

Tycho accepted, and so, above the white cliffs rising from the sea, he built Uraniborg, the "castle of the heavens." The architecture was baroque; the "castle" looked rather like a gingerbread house with an onion dome and cylindrical towers within which Tycho set up a gallery of instruments including sextants and quadrants, every one of which—since they were meant to be used with the naked eye—had open sights. He was a driven and self-centered perfectionist; the same personality traits that later made him an arrogant and much-hated master for his subjects on Hven also made him the most exactingly fastidious observer the world had ever seen. Before Tycho, problems of sighting with instruments had simply been ignored. No complication, however, was too small for Tycho's consideration if it stood in the way of his goal. To eliminate sighting errors he tried various expedients, eventually devising an apparatus with a metal plate having two horizontal slits, one above the other at a distance equal to the diameter of a cylindrical peg placed in the center. If, on looking alternately through the slits, the eye saw equal parts of the star project above and below the peg, then the instrument was exactly pointed on the star. Tycho also improved the accuracy of the divisions of his graduated circles by using transverse rows of points that allowed him to read the smallest subdivisions. He was also the first observer to correct his observations for refraction by Earth's atmosphere, and to make a serious attempt to estimate as far as possible the systematic errors that affected each instrument.

In addition to his astronomical studies, Tycho was devoutly attached to alchemy, and he regularly doled out medicines he concocted to his subjects (and frequently to himself). He was also as astrologically minded as the Babylonians, and he showed the same reverence for the celestial bodies as the old priest-astronomers had done: it is said he never observed except in the splendor of his most luxurious robes. A seemingly "indestructible, blustering social being with an enormous appetite for food and wine," he was "fond of mystery and display, and his observatory at Uraniborg abounded in mechanical devices and imperceptible means of communication with which he liked to mystify his visitors. . . . Attached to the observatory was a dwarf called Jeppe, whom Tycho used to feed with an occasional morsel at table, like a dog. Jeppe was supposed to be clairvoyant and to have made some remarkable prophecies."[17] What dramas must have transpired on Hven in the Baltic, in the heyday of the imperious lord of Uraniborg!

What Tycho called his "virile, precise, and absolutely certain" observations began in 1580. He set out to produce the first major star catalog since the one Ptolemy had published in the *Almagest* (largely based on Hipparchus's work) and measured a thousand stars, many with a tenfold greater positional accuracy than had any of his predecessors. (An abridged version of his catalogue, containing 777 of the most accurately determined positions, was published the year after his death.) He also commenced important studies of the motions of the Sun and Moon. But from the beginning his chief preoccupation was the planets, especially Mars.

Tycho observed Mars carefully at every opposition from 1580 onward, and not only at opposition. A Tychonic innovation—and one that in the end proved crucial—was to keep it under surveillance at other times. Near opposition in 1583, he noticed that Mars was moving retrograde at a rate of nearly a half degree every day. This seemed to prove that the planet could approach much nearer to Earth than could the Sun, as was possible in the Copernican but not in the Ptolemaic system. Nonetheless, despite this evidence, Tycho could not bring himself all the way to accepting Copernicus's heliocentric conclusion. Among other things, the failure to find any stellar parallax weighed on him. Instead it promoted a compromise, known as the Tychonic system: the planets, excepting Earth, travel around the Sun, while the Sun and Moon do so about the stationary Earth.

The 1595 opposition of Mars was the last Tycho would observe at Hven. After suffering his imperious ways for twenty years, his subjects had had enough and wished him gone. He had pursued his exacting course of observations using instruments that were designed with effort and attention to detail, and that came at unbelievable expense, paid for by the heavy taxation of his long-suffering tenants, who rather surprisingly proved to be more interested in bettering their miserable lives on Earth than in discovering the goings-on in the heavens. Nor had Tycho carried out even the light obligations that Frederik had imposed on him. As soon as Frederik died, the Danish court cut off Tycho's funds and refused to allow his wife or children (he had eight in all) to inherit his fortune and his observatory. Suffering from narcissistic rage, Tycho decided to leave. In 1597, he carried off his treasure trove of precious (but haphazardly organized) observations and a few of the most portable of his instruments, and he abandoned not only the observatory of Uraniborg but a second-generation observatory, Stjerneborg ("star castle"), built in 1584, in which the instruments had been located underground so as to be out of the way of wind and weather. Though he wrote a dedication for Stjerneborg in gold lettering on durable porphyry, expressing his hope that his observatory with its precious instruments might not "totter through age nor any other misfortune, nor be transferred elsewhere, nor be violated in any manner," the observatory he had built for all eternity was now reduced to what he could carry with him. Henceforth the work of observations became a rather peregrinating affair.[18]

Eventually he was lured by Rudolf II (1552–1612), the Holy Roman emperor, to Bohemia, where, a year before Tycho's death, he formed a rather unstable but highly productive partnership with a young German mathematician, Johannes Kepler (1571–1630) (figure 2.3). It was Kepler who would complete the reform of astronomy on the foundations of Tycho's observations.

Kepler had been born in Weil der Stadt, a village in Swabia, in southern Germany, and had managed—despite a life of trauma and hardship that had included narrowly escaping death and sustaining serious damage to his eyesight from smallpox, and being pulled out of school to work as a potboy in his quarrelsome father's tavern—to establish himself as a gifted scholar. A devout Lutheran during an age when Lutherans and Catholics in Germany were rapidly descending into the Thirty Years' War, Kepler at first hoped to pursue a religious vocation, but in the end he had to settle for a position as a teacher of mathematics at the Protestant seminary (high school) of Graz. There he had first come to Tycho's attention.

FIGURE 2.3 Detail of a portrait "possibly of Kepler," by Hans von Aachen, the Holy Roman emperor Rudolf II's favorite painter, 1612. Courtesy of Chateau Rychnov nad Kněžnou, Prague.

After several false starts—for both men had serious personality flaws, and could be headstrong and difficult at times—the men began their collaboration in earnest in February 1600. At this time Tycho was living at Benatky Castle (now Benátky nad Jizerou), on the Yser River some 40 kilometers northeast of Prague, and Christian Severin, known as Longomontanus (1562–1647), who had been Tycho's senior assistant at Uraniborg for eight years and had accompanied him on his migration from Denmark to Bohemia, was using Tycho's observations to try to work out motions of Mars from the point of view of the Tychonic system. Kepler, already a confirmed Copernican, realized that Longomontanus had been referring the planetary motions to the *mean* Sun—an imaginary Sun that moves through the sky throughout the year at a constant speed, and that is used in calculating mean solar time—rather than to the *true* Sun. (Both Ptolemy and Copernicus, the latter rather inconsistently, had done this; it was one of the latter's most serious mistakes.) This made the inclinations of the orbits of the planets to that of Earth awkwardly variable. In fact, most of the difficulty Longomontanus was encountering with Mars had to do with its positions not in longitude but in latitude, which depended on the inclination. Kepler reworked the calculations with reference to the true Sun and achieved a first "victory" in his long struggle with Mars: he found that the Martian orbit now had a constant inclination of 1°50′. It was on this occasion that he famously exclaimed that "Copernicus was ignorant of his own riches."[19]

With this boost to his confidence, Kepler rather arrogantly asserted that if only he were allowed to make use of Tycho's observations in his own way he could solve the problem of Mars's orbit in eight days.[20] He was even willing to wager on it. Tycho, though he did not place any wager on it, was at least willing to let him try, and so Kepler instead of Longomontanus was placed in charge of Mars. He was not given an entirely free hand, however; he had to agree to use the Tychonic system rather than the Copernican as the basis of his calculations. Unfortunately, Kepler was hardly getting underway when the two men had a serious falling out. One can understand how it came about: Tycho's household was crowded and disorderly, and Tycho—social, expansive, imperious, self-centered, and domineering as ever—could hardly have presented a stronger contrast with the reclusive and introverted but proud and uncompromising Kepler. Certainly, though Tycho bore much of the blame, the fault was not entirely on his side, and Kepler later admitted that he himself had sometimes behaved like a "mad dog."[21]

For the time being, a separation was needed, and so Kepler moved back to Graz, only to return a year later lured with promises from Tycho that his family would be allowed to join him (Kepler had now married, and had a stepdaughter). He insisted on being provisioned with a certain amount of food and firewood, and to be allowed to set his own work schedule and pursue the problems he personally wanted to study. Remarkably, Tycho agreed to all these conditions. He also agreed to approach Rudolf about granting Kepler a regular appointment at an annual salary of 200 gulden and to make him his associate in calculating new tables of the planetary motions.

Tycho himself had now moved to Prague, where, newly energized, he began setting up his instruments at the emperor's summer palace, the Belvedere. However, disaster struck. Becoming seriously ill during a banquet at which he had had too much to drink, Tycho passed away on October 24, 1601, his work far from done. (Throughout his final delirious night, Kepler later recalled, Tycho kept repeating over and over again, "Let me not seem to have lived in vain.")[22] With his imperious, disturbingly lopsided face and walrus mustache, Tycho could have been well cast as Hamlet senior's ghost in Shakespeare's play, just then premiering at the Globe, with Kepler cast as a Hamlet figure reciting to the ghost of Tycho,

> *. . . Remember thee!*
> *Ay, thou poor ghost, whiles memory holds a seat*
> *In this distracted globe. Remember thee!*
> *Yea, from the table of my memory*
> *I'll wipe away all trivial fond records,*
> *All saws of books, all forms, all pressures past*

That youth and observation copied there,
And thy commandment all alone shall live
Within the book and volume of my brain,
Unmix'd with baser matter. Yes, by heaven![23]

Two days after Tycho's death, Kepler was informed by an advisor to Rudolf that he had been appointed imperial mathematician, with his chief role being to provide astrological advice to the emperor. This meant that at least in principle he had a steady income, though in practice, because Rudolf's finances were always in disarray, worries about money—collecting what was owed him—were to remain a constant to the very end of Kepler's career. More important, from our point of view, was that Tycho's instruments and incomplete work were to be placed in his care, though as usual the path for Kepler was not to be a smooth one, as Tycho's heirs contested the rights to his observations and forced Kepler to take legal action to wrest control of them. Also, Tycho's observations were in disarray, scattered through many pages of his notebooks. Kepler had to spend a great deal of time sorting them out. Even so, he was far from free to use them as he wished. The dead hand of Tycho lay on him: on his deathbed, Tycho had insisted that the reform of astronomy be based on his—Tycho's—principles, not on those of Copernicus or even Kepler's own. (In the end, Kepler would honor this request pro forma by working out the ideas in triplicate.)

Kepler's War with Mars

Now Kepler began what he called the "war" with Mars in earnest. Ever since he had succeeded in showing that the inclination of Mars's orbit became constant by taking the true Sun instead of the mean Sun as the center, Kepler had been analyzing Tycho's Mars observations for the 10 oppositions between 1580 and 1600 in order to plot the planet's orbit in space. At first, he had assumed the orbit would be circular, though he knew—as Ptolemy had known—that its motion in that circle could not be uniform. And at this point he introduced a brilliant stroke: instead of following Copernicus, he followed Ptolemy in using an equant.

His next step was to try to work out the position of the equant for Mars's orbit empirically on the basis of Tycho's observations. The procedure he adopted (which he himself called "wearisome") required the use of four points taken from Tycho's observations of the oppositions, which—after a long series of elaborate and cumbersome calculations—gave a position for the equant. (Anyone wishing to examine a summary of these calculations firsthand is referred to his great book *Astronomia nova . . . Commentaries on the motions of the star Mars*, part 2, chapter 16; the whole set of them in manuscript run to 900 folio pages in Kepler's cramped handwriting.) At the end of this discussion, Kepler asked for the sympathy of the reader, adding, "I have gone through it at least seventy times at the expense of a great deal of time." Such was his frustration that at times he felt, as he recalled Rheticus had felt on one occasion, that a demon was knocking his head against the ceiling while shouting, "These are the motions of Mars!"[24]

Eventually, Kepler's backbreaking calculations allowed him to establish that no circular orbit could ever account for the motions of Mars. Instead, the actual shape had to be an oval of some kind. Along the way he discovered what is now known as Kepler's second law of planetary motion: the line connecting the planet and the Sun sweeps out equal areas in equal times. (It is important to point out that Kepler himself never actually referred to this as a "law.") Thus, Mars travels faster when near the Sun and slower when farther away.

Kepler now used Tycho's observations to painstakingly plot the planet's position at 22 different points along its orbit (thus showing the prescience of Tycho's observing it at other times than around opposition). Finally, on Easter of 1605, Kepler succeeded in achieving his grand result: the orbit of Mars is an ellipse, with the Sun in one focus and the other focus empty. Stated thus, it may not seem like much, and even Kepler wrote, on contemplating the "cloud of husks" that had been discarded along the way, "How small a heap of grain we have gathered from this threshing!"[25] But in this small heap of grain—which we now refer to as Kepler's first law of planetary motion—are contained some of the most prolific seeds in the history of astronomy.

Kepler did not manage to scrape together the funds needed to publish the *Commentaries on the motions of the star Mars*—the book that contained this foundational result and documented his difficult and sometimes haphazard process of winning his way to it—until 1609. As usual, his patron, Rudolf, was chronically short of funds, and did not have the money to fight all his battles on Earth, much less wage planetary warfare of the kind Kepler was fighting. For the rest of his life, Kepler's salary was continually in arrears. He was often obliged to appear at court and to suffer the humiliation of groveling to be paid. There were other hardships. At the end of 1610 his first wife, who seems to have been prone to depression, came down with fever, suffered seizures, and died. His six-year-old son—his favorite—succumbed to smallpox. By 1612, Rudolf was dead, while Prague was on the eve of becoming a battleground in the Thirty Years' War. Kepler fled to Linz, Austria, where, despite the unhappiness of his first marriage, he immediately began searching for another wife. No less than eleven candidates were considered, and Kepler—setting forth the merits and demerits of each with the detachment of a mathematician— chose the poorest, an orphan girl without a dowry. The marriage proved to be much happier than his first had been, and his new wife promptly bore him a succession of seven children. (There are still descendants of one of Kepler's daughters living to this day.)

Through years of ill health, poverty, and worry, he continued to work, and finally, in Linz in May 1618—in a house that still exists—he made yet another significant discovery, his third law of planetary motion, often referred to as the harmonic law since it links the periods of the planets to their distances from the Sun.

One can picture him at this time, laboring over his desk: a small, frail, disheveled figure. At least one admirer found it difficult to comprehend how "such a mass of learning, and knowledge of the most profound secrets, could be locked and concealed in one such small body."[26] He was a mass of infirmities. Nearsighted and suffering from multiple images in one eye, a condition not helped by his habit of incessant night study, he was also prone to bilious attacks from any diet except the blandest, and he habitually kept himself from inanition by chewing on bones and dry bread. Subject to boils and rashes, averse even to baths and washings, he was one whose mind and body existed in a perpetual state of warfare. As a youth, he had thought of himself as resembling a "little house dog," and he always had a dependent and subordinate nature; yet because of his sincere piety, he seems to have felt all the more the leadings of a higher hand that guided his destiny. Nevertheless, his biographer Max Caspar (1880–1956) says, "between his genius and his humanness there remained a latent gap. Although the idea of harmony kept his thoughts busy, he was not harmonic, not adjusted in his nature. He was a restless soul, fluctuating repeatedly between exhilaration and depression."[27] But what other kind of man could ever have hungered so desperately for the harmony of the cosmos?

As a reminder of life's vicissitudes, the following year Kepler had to travel to Württemberg to participate in the final stages of the defense of his elderly mother, Katharina Kepler (1546–1622), against charges of witchcraft. Though the proceedings were drawn out over a number of years, he was ultimately successful in gaining her acquittal. (Witch-hunting was reaching its peak in Germany during this time, with no fewer than 38 put to death in Weil der Stadt between 1615, when charges were first introduced against Katharina, and 1629. No doubt the religious and political turmoil of the period contributed, though it has also been suggested that famines associated with frequent crop failures due to the Little Ice Age played a role.)[28]

In 1626, with Linz coming under siege, Kepler had to flee again, eventually finding refuge at the court of the general-in-chief of the armies of the Holy Roman Empire, Albrecht von Wallenstein (1583–1634), whose horoscope Kepler had cast. At Wallenstein's newly formed Duchy of Sagan, in Silesia, in 1627, Kepler finally published the *Rudolphine Tables*, the tables of planetary motions that had been Tycho's early dream and whose dedication had been promised to the often unreliable but now long dead Rudolf II.

In October 1630, still plagued with money worries, Kepler set out from Sagan to Regensburg in hopes of conferring with the latest Holy Roman emperor, Ferdinand II, about getting paid what was owed him, but the trip proved too much for him. He suffered a chill, and after a short illness, on November 15, 1630, he died. He had already anticipated the event by writing his own epitaph: "Once I measured the heavens, now I measure Earth's shadows."[29] His true epitaph consists of his three immortal laws of planetary motion.

The Motions of Mars

As established by Kepler, Mars's elliptical orbit brings the planet as close as 206.5 million kilometers to the Sun at its closest point (perihelion), and as far as 249.1 million kilometers at its farthest (aphelion). The mean distance is 227.9 million kilometers, and Mars completes each revolution of the Sun in 686.98 days as determined relative to the stars (its sidereal period).

Kepler's laws as established on the basis of Tycho's observations are empirical, but they can be derived and elaborated from the theory of gravitation later developed by Isaac Newton (1642–1727). Because of the gravitational pull of the Sun and planets (especially Jupiter) on the tidal bulge in the equator of Mars, its orbit and spin axis gradually change over time. The position of its perihelion slowly rotates (precesses) in space, and the shape of the ellipse also varies; the current value of the eccentricity is 0.093 (compared with 0.017 for Earth and 0.00 for a perfect circle), but over a period of two million years it ranges between 0.00 and 0.13. The obliquity of Mars—the tilt of its rotation axis relative to the ecliptic plane—is also strongly influenced by the pull of the Sun and planets, varying over hundreds of thousands of years from around 15° to nearly 36° (its present value is about 25°), with significant implications for the planet's past climate.

At opposition, Mars and Earth lie on the same side of their orbits from the Sun, and the two planets come closest together. (Because of the slight 1°51' tilt of Mars's orbit relative to Earth's, discovered by Kepler, the minimum distance between the two planets may actually occur as much as 10 days from the date of opposition; see figure 2.3). Since Earth completes an orbit around the Sun in 365.26 sidereal days, and Mars in 686.98 sidereal days (where a sidereal day is defined as the rotation time measured relative to the stars: 23 hours, 56.1 minutes, for Earth; 24 hours, 37.4 minutes, for Mars). Thus, Earth overtakes and passes Mars on an average of once every 779.74 days (this is known as the *synodic period*, which is also that between successive oppositions; because both Earth and Mars travel at variable speed in their orbits around the Sun, the actual interval between oppositions varies on either side of the average, and may be as little as 764 days and as much as 810 days; see appendix D).

If opposition occurs when Mars is near perihelion, the distance of approach will be only about 56 million kilometers (35 million miles); if it occurs when Mars is near aphelion, it will be rather more than 100 million kilometers (62 million miles). Since the time between oppositions is longer than the Martian year, successive oppositions are displaced at intervals around the orbit of Mars. Hence, there is a series of oppositions going from a favorable perihelic

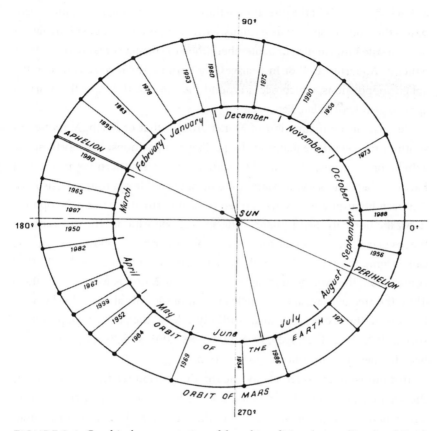

FIGURE 2.4 Graphical representation of the orbits of Mars (outer ellipse) and Earth (inner near-circle), showing the orientation of the Martian perihelion and aphelion relative to Earth's vernal equinox (0° here). Dots and connected lines note the positions of Earth and Mars at example oppositions from 1954 to 1999. The 2003 perihelic opposition is not shown here but was remarkable because it occurred almost exactly where the word "PERIHELION" is on the figure. Adapted from C. Flammarion, *The Flammarion Book of Astronomy* (New York: Simon & Schuster, 1954); see also appendixes D and F.

opposition through less favorable ones to an unfavorable aphelic opposition, then through more favorable ones to another favorable perihelic opposition. The interval between successive perihelic oppositions is 15 to 17 years.

A few other points to keep in mind: The orientation of the Martian orbit in space is, at present, such that the longitude of its perihelion currently lies at 336.06°, which is in the direction of the constellation Aquarius. Since Earth passes this point in space in late August each year, perihelic oppositions always occur in late July, August, or September, when the planet comes to opposition either in Aquarius itself or in nearby Capricornus. The reverse is true of the aphelic oppositions, which occur around the time Earth passes the Martian aphelion (in Leo) in late February.

Because of the inclination of Mars's orbit to that of Earth, Mars travels south of the celestial equator at the perihelic oppositions, and north at the aphelic oppositions; accordingly, locations in the southern hemisphere of Earth are best for studying Mars at the most favorable oppositions. Also, the southern hemisphere of Mars is tilted toward Earth at the perihelic oppositions, the northern hemisphere at the aphelic oppositions. The difference in the angular size of Mars is also significant: at the perihelic oppositions, the disc can become as large as 25.1", at the aphelic ones it may not exceed 13.8". (For comparison, the average angular diameter of the Moon is more than 1800".) Thus, the area of Mars at the aphelic oppositions is only about a third what it is at the perihelic ones, and so, historically at least, the perihelic oppositions, such as 1877, 1892, 1924, 1939, 1956, 1971, 1988, 2003, and 2018, have generally been banner years for Martian exploration.

In contrast to the sidereal day, the rotation period relative to the stars, is the solar day, the period of rotation relative to the Sun. On Earth the mean length of the solar day is 24 hours, with the actual length varying throughout the year by as much as 25 seconds in either direction because of the eccentricity of Earth's orbit; on Mars the mean solar day is 24 hours, 39.6 minutes, which is referred to as a "sol." There are 668.59 sols in a Martian year (see appendix G).

Because the Martian axis is tilted to the plane of its orbit by 25.19° at the present time (vs. 23.44° for Earth), seasonal effects are important on Mars. On Earth, the seasons of spring, summer, autumn, and winter are all of similar length. On Mars, because of the marked eccentricity of its orbit, the seasons differ much more in length (table 2.1). As is evident from the table and the current season of perihelion, the southern hemisphere has short but warmer summers and long but colder winters, compared to the northern hemisphere. This means that the climate in the northern hemisphere is more temperate, and in the southern hemisphere more extreme—a relationship that is reflected

TABLE 2.1. Length of the Seasons on Mars

Martian season	Length in Earth days	Length in Mars days
Northern spring, southern autumn	199 days	194 sols
Northern summer, southern winter	183 days	178 sols
Northern autumn, southern spring	147 days	143 sols
Northern winter, southern summer	158 days	154 sols

in historical telescopic observations, for instance, in the more extreme behavior of the south polar cap compared to the north.

Though various ingenious calendars have been proposed for Mars, which may be useful if and when settlements are established there, in the meantime it is most convenient to refer to the Mars-centric longitude of the Sun (L_s, pronounced "L sub S"), which is 0° at vernal equinox (beginning of northern spring), 90° at summer solstice, 180° at autumnal equinox, and 270° at winter solstice. Because of the eccentricity of Mars's orbit, the planet is at aphelion (moving most slowly) at $L_s = 70.87°$, and at perihelion at $L_s = 25.87°$. The importance of seasonal effects on various observed phenomena (behavior of polar caps, cloud formation, dust storm activity) makes it is useful for Mars observers to know the L_s for a given observation. For instance, major Mars dust storm seasons as recorded in historical telescopic observations typically begin shortly after perihelion, at around $L_s = 260°$ (for more detail, see appendix F).

Though the motions of Mars were worked out accurately by Kepler, nothing had yet been learned about Mars as a world. For that, a telescope was needed (and, eventually, spacecraft). By a strange coincidence, in 1609, the same year Kepler published his *Commentaries on the motions of the star Mars*, the first telescopes were turned toward the sky. A new era of Martian research had begun.

3

The First Telescopic Reconnaissance

Hesper-Venus—were we native to that splendor, or in Mars,
We should see the globe we groan in, fairest of the evening stars.

Could we dream of wars and carnage, craft and madness, lust and spite,
Roaring London, raving Paris, in that point of peaceful light?

Might we not in glancing heavenward on a star so silver-fair,
Yearn, and clasp the hands and murmur, "Would to God that we were there"?

—Alfred, Lord Tennyson, "Locksley Hall Sixty Years After," 1886

Kepler's Forecasts

The ten years dating from 1600 were a time of prodigies. In February and March 1600 a volcano—Huaynaputina, located 70 kilometers east of Arequipa, Peru—erupted in a blast comparable to the celebrated eruptions of Tambora in 1815, Krakatoa in 1883, and Pinatubo in 1991. More than 1,500 people were killed near the volcano, and as far away as Europe the Sun and Moon were dimmed and reddened by ash. Europe, already in the throes of the "Little Ice Age," suffered from the coldest decade in centuries. A new star, known as Kepler's Star, erupted in Ophiuchus in 1604. Rather remarkably, it is so far the last naked-eye visible supernova observed in our galaxy. Shakespeare finished *Hamlet* in 1600 (then in rapid succession *Othello*, *Macbeth*, and *King Lear*).

Prodigies all, yet none greater than the invention of the perspicillum, to use Galileo Galilei's term for the device that Johannes Kepler would call the telescope. Galileo (1564–1642), a professor of mathematics at the University of Padua, made his first wonderful discoveries with such a device in 1609, the same year in which Kepler revealed the secrets of planetary motion.

Of this telescope, Kepler himself would exclaim, "O you much knowing tube, more precious than any scepter. He who holds you in his right hand, is he not appointed king or master over the work of God!"[1]

Kepler envisioned still more illustrious times to come, in which people would teach themselves to fly, and by this means they might reach the Moon or other planets. "Ships and sails proper for the heavenly air should be fashioned," he wrote. "Then there will also be people, who do not shrink from the dreary vastness of space."[2] Of course such sails and ships remained centuries off, but Kepler's wide-ranging imagination conceived of them.

From Perspicilla to Telescopes

Kepler's colleague Christian Severin (Longomontanus), who outlived Tycho Brahe by 47 years and Kepler by 7, would write in 1639: "Astronomy does not so much investigate the heavenly bodies themselves and their causal properties [as] the motions and definite periods. It hands over the peculiarities of the [planets] to physics, which treats them by means of optics."[3]

Indeed, optics, in the form of the telescope, would first bring the planets sufficiently within reach for observers to begin studying them as physical bodies and to show them to be other worlds, even, as was widely assumed at the time, other Earths.

The history of the invention of the telescope is rather long and complicated, and can be traced at least to the development of eyeglasses for reading, thin biconvex glasses mounted in frames and created by Italian craftsmen in the 13th century. Within another two centuries, concave glasses were being worn for myopia. Thus, both concave and convex lenses became available all over Europe by about 1500. It would seem that a telescope—which involves combining a convex and a concave or two convex lenses—could have been invented any time after this. The reason it was not, according to telescope historian Albert van Helden, was owing to the strengths of the lenses. "In order to achieve a significant magnifying effect one must combine a weak convex lens with a strong concave or convex lens,"[4] he writes. Lenses of the right range of strengths were simply not available at the time.

Not until September 1608 was the decisive step taken, in the Netherlands, where members of the provincial government of Zeeland sent to the representative of the States-General in The Hague learned that Hans Lipperhey (or Lippershey, 1570–1619), a spectacle maker in Middleburg, had developed "a certain device by means of which all things at a very great distance can be seen as if they were nearby."[5] Lipperhey applied for a patent; within two weeks, two other claimants to the invention came forward. In the end, though Lipperhey received payment for his device, no patent was granted on the grounds that the invention was already too well-known—examples were already being sold that autumn at a fair by a Dutch peddler in Frankfurt, and the following spring by spectacle makers in Paris.

In Italy, the news pricked the ears of Galileo, who worked out for himself that the device must be based on the principle of refraction. He fit two lenses—one plano-convex to serve as the objective lens, the other plano-concave to serve as the eyepiece—at either end of a lead tube, reputedly made from the sawed-off pipe of a church organ. His first instrument magnified only 3×. By August 1609 he had boosted the magnification to 8× to produce an instrument powerful enough to interest the traders on the Rialto, for whom early knowledge of the approach of wealth-laden "argosies with portly sail" could mean a fortune.[6] By the end of the summer he was busy showing it off to the doge and senators from the highest campaniles in Venice.

At first Galileo seems to have been mainly interested in potential commercial applications of the new instrument rather than its possibilities for pure research. Not until November 30—almost three months after the English mathematician Thomas Harriot (c. 1560–1621) had looked at the Moon with a Dutch-made "perspective tube" magnifying 6×—did Galileo get around to pointing a telescope to the sky. Having carefully sifted through a plethora of lenses to find one optically better than the rest, he improved the definition still further by using an aperture stop that allowed light to pass only from the optically superior central part. The result was an instrument magnifying 20×. He pointed it at the four-day Moon, and immediately was able to make out that the lunar surface was rough and uneven, full of great cavities and mountains. A series of ink-wash drawings recording his observations were a vast improvement over Harriot's crude sketches. By January 1610 he had added to his discoveries the four "Cosmica Sidera" (Cosimo's stars) or "Medicea Sidera" (Medicean stars), the large satellites of Jupiter, named for his patron the grand duke of Tuscany, Cosimo II de' Medici (1590–1621) and then, following Cosimo's own suggestion, for himself and his three brothers. Kepler congratulated him: "I am so far from disbelieving in the discovery . . . that I long for a telescope, to anticipate you, if possible, in discovering two around Mars (as the proportion seems to require)."[7] Galileo described all this in a book, *Sidereus Nuncius* (Starry Messenger), written at white heat and published in March 1610. By the end of that year, he had made the first observations of sunspots, and discovered the phases of Venus, which were consistent with Copernicus's prediction but not Ptolemy's. (They also were compatible with Tycho's system.) It was a remarkable time, during which astronomical discoveries were coming fast and furious—but had nothing to do with Mars.

The reason for this is that Mars is a difficult telescopic object. Even when closest to Earth, it is still 140 times farther than the Moon and appears only about the same size as a medium-sized lunar crater. A telescope magnifying 60× shows it as large as a pea seven millimeters in diameter seen from a distance of one meter, or a little larger than the Moon as seen with the naked eye. Though as a planet orbiting outside Earth it can never show a crescent phase as Venus can, it does show a gibbous phase; indeed, the maximum phase defect is 47°, which occurs when Earth is at its greatest angular distance from the Sun as seen from Mars, and which is about that which the Moon shows three or four days from full. Galileo observed Mars throughout 1610 in the hope of seeing the phase, but he could hardly make out any disc in his small telescope. Despite inconclusive results, he wrote tentatively to Father Benedetto Castelli (1578–1643), one of his former pupils, on December 30, 1610: "I ought not to claim that I can see the phases of Mars; however, unless I am deceiving myself, I believe that I have already seen that it is not perfectly round."[8]

This was the best that Galileo could do with Mars. Nevertheless, his skill, his patience, and his perseverance—and his use of the best telescopes available at the time—had opened the road to future advances. Not without a profound sense of reverence and awe would Camille Flammarion (1842–1925), the French popularizer of astronomy, take up one of Galileo's telescopes (his most powerful, magnifying 32×) in the Academy of Florence and reflect on all that had been accomplished with it:

> After sunset I recaptured the spirit of the Florentine astronomer on one of the beautiful Italian terraces just as the stars were coming out; with feverish impatience I turned this marvelous tube toward the new worlds that he had discovered in the heavens. I recalled that he had shown these sights to those who were incredulous; he still shows them to us today from his grave.[9]

The early telescopes, like those Galileo used, employed only a simple convex lens as an objective and a concave lens as an eyepiece. They had an inconveniently small field of view even with low magnifications, and they suffered badly from spherical and chromatic aberrations.[10] Since these effects are more serious for light passing through the outer parts of a lens, cardboard rings as aperture stops such as Galileo placed in front of his object glasses helped to a certain extent.

Further improvements in telescopic performance waited on a redesign of the instrument. As early as 1611, Kepler, in his book *Dioptrice*, proposed using a convex instead of concave lens for the eyepiece. This was the basic idea of what became known as the astronomical telescope (as opposed to the Dutch or Galilean telescope, which had used a concave lens for the eyepiece). The image was inverted, but this was of little importance in astronomical observations; if one wished, this could be remedied by addition of another lens, though most observers simply tolerated the inverted image—and hence most sketches and photographs made in the pre-spacecraft era are oriented with south at the top. The real advantage of the astronomical telescope was that the field of view was much larger. Kepler, with his poor eyesight, never actually attempted to build and use such a telescope. The first to do so was apparently the Jesuit astronomer Christoph Scheiner (1575–1650), a rival claimant to Galileo for the discovery of sunspots, who did so in 1617. Another early advocate of the astronomical telescope was Francesco Fontana (c. 1585–1656) of Naples. Though a lawyer by profession, he devoted much of his time to the construction of both telescope and microscope lenses. With one of his telescopes he produced two curious drawings of Mars, in 1636 and 1638, showing a black pill or cone in the center of the disc of Mars. As he showed the same in his drawings of Venus, this was certainly an optical effect rather than a sighting of an actual planetary marking. Fontana's drawings are, however, historically significant, and mark the beginning of what Flammarion, in his great book *La Planète Mars*, would call the first period in the telescopic study of the planet, lasting until 1830, in which significant progress was made even though such basic questions as whether Mars has a fixed topography, as does Earth, remained unresolved.

Huygens and Cassini

Better results were obtained within a few years of Fontana's false start. At the perihelic opposition of August 21, 1640, the Jesuit astronomer Niccolò Zucchi (1586–1670), at the Collegio Romano, failed to notice any patches on the disc. However, four years later, at the much less favorable opposition of December 1644, another Jesuit, Daniello Bartoli (1608–85), at Naples, described two patches on the lower part of the disc, which were probably glimpses of actual features of the surface. Further observations of indistinct patches were made in 1651, 1653, and 1655 by the Jesuit astronomers Giovanni Battista Riccioli (1598–1671) and Francesco Grimaldi (1618–83), at the observatory at the College of Saint Lucia in Bologna. Though representing the best work of the time, these results were primitive, so that even today—despite our knowing exactly what part of Mars they were looking at—it is not easy to identify specific features.

   Far superior to any of this was the work of the great Dutch scientist Christiaan Huygens (1629–95), whose father, Constantijn (1596–1687), as secretary of state to the stadtholders, was the highest-ranking civil servant in the Netherlands. At the age of 26, Christiaan, along with his elder brother, Constantijn Jr. (1628–97), began experimenting with new ways of figuring microscope and telescope lenses. Christiaan devised the first compound eyepiece (the Huygenian, still widely used) and by March 1655 was wielding a 5.1-centimeter telescope of 3.2-meter focal length, which he pointed at Saturn through the attic window of his father's house in Het Plein. He discovered Saturn's largest moon, now known as Titan. He also briefly turned his attention to Mars, which passed a perihelic opposition on July 21, 1655, but he did not get around to it until well past opposition. Then he could do no better than Riccioli and

Grimaldi were doing at the time; there was nothing more on the disc than a "somber band."[11] For the next few years he was diverted into other projects, including perfecting the pendulum clock. Not until 1659 did he return to Mars.

On November 28, at 7 p.m., he turned his telescope toward the planet, then near opposition and showing a disc 17.3″ of arc across, and sketched a conspicuous V-shaped marking that we recognize immediately as the Syrtis Major (for a long time rather more colorfully known as the Hourglass Sea; figure 3.1). Here Huygens produced what Percival Lowell (1855–1916) would aptly call, "the first drawing of Mars worthy of the name."[12] It was also, again according to Lowell, "perhaps the most important one of Mars that has ever been made," since Huygens's continuing observations on successive dates allowed him to prove that Mars rotated on its axis, with an Earthlike rotation period of 24

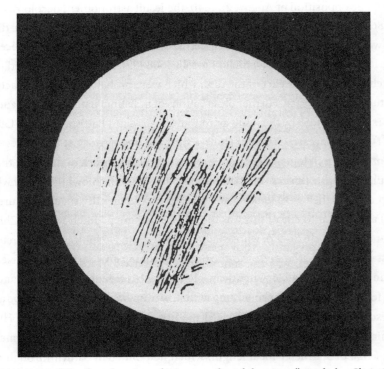

FIGURE 3.1 "The first drawing of Mars worthy of the name," made by Christiaan Huygens at 7 p.m. on the evening of November 28, 1659, which was 305 years to the day before the launch of the Mariner 4 spacecraft. The dark area was long known as the Hourglass Sea but is now known as Syrtis Major. Note that south is at the top. Published in Camille Flammarion, *La Planète Mars*, vol. 1 (1892), figure 9.

hours, while comparison of his drawing to those made by observers centuries later allowed the rotation period to be estimated to within a tenth of a second. In that same year, Huygens published his most famous astronomical discovery, regarding Saturn: the planet was, he concluded after a long and careful study, surrounded by "a thin, flat ring, nowhere touching."

Mars in the Age of Empires

Huygens discovered the first recognizable marking on Mars during the great age of European empires. A century before, Spain had been the greatest power on the globe, owing to the wealth derived from the discovery of the New World. However, much of that wealth was wasted on courtly extravagance, foreign missions, and wars, and under a series of Counter-Reformation papacies Spain had become even more stifling to intellectual inquiry than Italy; an Italian nobleman visiting Spain as late as 1668 noted that "the ignorance is immense and the sciences are held in horror."[13] As Spain and Italy fell behind, the Netherlands, England, and France modernized, with the Netherlands being the first to emerge from the chaos of a turbulent century, not least because, at the beginning of the 17th century, the Free Republic was the only European country to tolerate speculative thought. This toleration was in large part a result of the Netherlands' achievement of both military and commercial supremacy through its navies and its establishment of a commercial presence around the globe. The Netherlands thus took a lead in bringing about the so-called Scientific Revolution of the 17th century.

In 1659, the same year Huygens made out Syrtis Major on Mars, diplomats were concluding a long series of negotiations that ended a long war between Bourbon France and Habsburg Spain. The treaty was the high point of the career of Cardinal Jules Mazarin (1602–61), regent during the minority of France's "visible divinity," Louis XIV (1638–1715). After Mazarin's death, the Sun King began the long era of his personal rule with the world at peace and France militarily and economically the most important country on the Continent. Though he regarded himself as the "decider" and was determined to govern directly, not through a prime minister like Mazarin, Louis had little interest in the details of internal government, and instead surrounded himself with a high council whose most important member was Jean-Baptiste Colbert

(1619–83). Described as "cold, humorless, hardworking, honest, narrow," Colbert became the king's chief minister of finance and Louis's indispensable man—indispensable because Louis, consumed with an unquenchable thirst for *gloire*, never for a moment considered limiting his objectives out of mere concern for money. Colbert's task was to provide it, through the development of "extraordinary" fiscal measures if need be.

On taking charge of the king's finances in 1661, Colbert decided to build up the French navy and rebuild Paris on a heroic scale. Voltaire (1694–1778) claimed that previously in Paris "there was no lighting, police protection, nor cleanliness. Provision had [now] to be made for the continual cleaning of the streets and for lighting them every night with five thousand lamps; the whole town had to be paved; two new gates had to be built and the old ones restored; a permanent guard, both mounted and on foot, was needed for security."[14] Though the king was officially credited with initiating these projects, it was up to Colbert to find ways of paying for them. The Louvre, the château at Saint-Germain, and Versailles (especially Versailles) were greatly built up at this time, and Louis granted his imprimatur to Colbert's plan for the Paris Academy of Sciences, founded in 1666 in large part because of Louis's desire to keep up with England's Charles II (1630–85), whose restoration to the monarchy in 1661 ended the English Civil Wars and whose move to surround himself with expert advice on scientific matters was codified by a royal charter establishing the Royal Society for the Advancement of Learning. Among the first savants recruited for the Paris Academy was Huygens, who came from the Netherlands and took up quarters in the Bibliothèque du Roi.

Huygens was the great man of the time, but he soon had a rival in Giovanni Domenico Cassini (1625–1712). Born in Perinaldo, near Nice, Cassini was, according to Flammarion, "by temperament much more Italian than French."[15] In 1648, he was invited by the Marquis Cornelio Malvasia, a senator and wealthy amateur astronomer, to work at his private observatory near Bologna, in order to produce ephemerides for astrological purposes. There, Cassini completed his scientific education under Riccioli and Grimaldi (neither of whom accepted the Copernican theory) at the College of Saint Lucia, and he was appointed professor of astronomy at the University of Bologna, where he wrote a treatise on a comet in which he set forth his own anti-Copernican views. Meanwhile, he had become acquainted with Giuseppe Campani, a skillful telescope maker in Rome. Beginning in 1664, Cassini turned Campani telescopes of long focal length on the planets, with remarkable results. On Mars, in February and March 1666, he recorded spots, of which the most conspicuous are represented in the form of a dumbbell. From these observations he deduced a rotation period of 24 hours, 40 minutes, within 3 minutes of the modern value. This implied that after a period of 36 or 37 days the entire circumference of the planet would pass in review and the features would return to the same positions at the same hour of the night. (Since Huygens had not yet published his work on Mars, Cassini's determination was completely independent, though curiously, Huygens did not at first accept Cassini's result.) Cassini also sketched bright patches at the poles, which may record the polar caps—though since he also recorded similar patches at the limbs, this is far from certain.

Dark spots were also recorded on Mars during those same months by Robert Hooke (1635–1703), curator of experiments for Charles II's Royal Society as well as professor of astronomy at Gresham College, London, with a telescope of 11-meter focal length. Hooke, by the way, gave an excellent account of how difficult it is to make out detail on Mars, owing to the smallness of the disc and the unsteadiness of the air ("seeing"): "though I often imagin'd, I saw Spots, yet the Inflective veins of the Air (if I may so call those parts, which, being interspers'd up and down in it, have a greater or less Refractive power, than the Air next adjoyning, with which they are mixt) did make it confus'd and glaring, that I could not conclude upon any thing."[16]

In a little-known work written at this time, *De planetetarum maculis* (The Planetary Spots), Cassini made some interesting comments bearing on the nature of the spots observed on Mars.[17] After comparing the planets to Earth, he suggested that when seen from a great distance across space the seas of Earth would appear as dark spots because they absorb sunlight; the continents would appear bright.[18] Thus, Cassini first suggested, by analogy to Earth, that the dark spots on Mars were probably seas.

Mars from the Paris Observatory

Three years after Huygens moved into his apartment in the Bibliothèque du Roi, Cassini came to Paris to take up residence in another of Colbert's projects, the Paris Observatory. The new observatory boasted a palatial building, the plans for which had been designed by Claude Perrault, the architect responsible for designing the much-admired new façade at the Louvre. However, though architecturally elegant, the observatory proved unsuitable for observations using the very long focal-length telescopes in use at the time. (Long telescopes provided the best available solution to the problem of chromatic aberration until the achromatic lens and reflecting telescope were perfected in the next century.) Cassini was resourceful; he simply set up his telescopes in the courtyard outside the building and used them en plein air. During this period, Cassini was by far the most active observer; Huygens observed only occasionally, partly because, in contrast to Cassini, who lived at the observatory, he had to travel a considerable distance in order to make observations, and partly because he began suffering from severe depression at this time, which kept him largely confined to his rooms. The next two decades largely record the progress in planetary astronomy achieved by Cassini (figures 3.2 and 3.3).

At the perihelic opposition of Mars of September 1672, Cassini's major project was an attempt to determine the parallax of Mars (its apparent movement relative to the background stars) as observed from two points widely separated on Earth's surface. Cassini himself remained in Paris, while a col-

FIGURE 3.2 *Colbert Presenting the Members of the Royal Academy of Sciences to Louis XIV in 1667.* The still-unfinished Paris Observatory building is seen through the window. Painting by Henri Testelin after Charles Le Brun, circa 1680, in the Musée National des Château de Versailles et de Trianon. A copy of this painting hangs on the town walls of Perinaldo, where Cassini was born.

FIGURE 3.3 Detail of *Colbert Presenting the Members of the Royal Academy of Sciences to Louis XIV in 1667*. The abbé Jean-Baptiste du Hamel, first secretary of the Academy, appears in blue cassock. Behind him, *right to left*: Pierre de Carcavi, Jean Picard (leaning), Christiaan Huygens (face on), Giovanni Domenico Cassini (three-quarter view), Philippe de La Hire, the abbé Edme Mariotte (in profile), and Jacques Borelly.

league, Jean Richer (1630–96), traveled to the French colony of Cayenne. Despite his general indisposition, Huygens, too, made a few sketches of Mars that year, which are better than any by Cassini or anyone else, and include the first to definitely show the south polar cap.

Huygens's Last Years

Sadly, Huygens and Cassini, though the greatest observers of Mars of the 17th century, seem never to have been on very cordial terms. They were very different types of men. Cassini was ambitious, forceful, and politically astute (for instance, he knew there was something to be said for flattery, naming four satellites of Saturn he discovered the "Louisian stars," after Louis XIV). Huygens, though a much greater scientist, was an introvert, and often depressed. Moreover, as a Protestant, he was vulnerable in a France that was becoming militantly

Catholic. In 1681 his health—which had never been good—broke down, and he returned to the Netherlands hoping that the native air would prove restorative. He planned to return to France eventually, but Colbert, whose absolute protection he had enjoyed, died soon afterward, while in 1685 Louis XIV revoked the Edict of Nantes, which had granted certain liberties to Protestants in France. Under the circumstances Huygens decided it was best to remain in the Netherlands.

He spent his last years in private research at Hofwijck, near The Hague, where he and his brother devised telescopes of the tubeless, or "aerial" type— the object glass was simply mounted on a ball-and-socket joint that could be moved up and down a tall pole. The instrument could be aimed by pulling taut a wire that connected the object glass and eyepiece assembly, which the observer held in his hands while resting his arms on a movable wooden support. Though they were not buffeted by the wind as instruments with long tubes had been, protracted work with aerial telescopes must have sorely tested the observer's patience: when studying an object near the celestial equator, such as Mars at a good opposition, the observer had to move the eyepiece assembly at a uniform rate of several centimeters per minute to compensate for Earth's rotation. This certainly required a steady hand. On very dark nights, the objective lens was difficult to see, so that just pointing the telescope at the object to be studied was a time-consuming task, and utterly impossible without the services of an assistant bearing a lantern. Despite these difficulties, the brothers persisted, and they produced heroic instruments with objective lenses of between 19 and 22 centimeters and focal lengths of between 37 and 63 meters.

The hidden difficulty of these tubeless telescopes was all the stray light that flooded the field of view. (In part, a telescope tube screens out the stray light and renders the field of view darker.) Huygens seems to have placed a large sheet of paper around the objective to act as a screen or baffle and eliminate the worst stray light. Thus equipped, he observed Mars at the perihelic opposition of August 1687. His sketches show no more detail than he had recorded in earlier sketches with more modest instruments. His very last sketch of Mars was made on February 4, 1694, with Mars at an aphelic opposition. Thus ends an era.

Huygens, knowing that he did not have long to live, began writing a truly remarkable book in which he speculated on the physical condition of the planets and the possibility of extraterrestrial life. The inspiration was doubtless a book by the French writer Bernard le Bovier de Fontenelle (1657–1757), *Entretiens sur la pluralité des mondes* (Conversations on the Plurality of Worlds), which enthusiastically embraced the idea that there might be life and inhabitants on other worlds. Fontenelle says a great deal about the Moon, but of Mars he says only: "Mars has nothing curious that I know of; its days are not quite an hour longer than ours, and its years the value of two of ours. It's smaller than the Earth, it sees the Sun a little less large and bright than we see it; in sum, Mars isn't worth the trouble of stopping there."[19]

Though Huygens's book, *Kosmotheoros*, was finished by January 1695, the author's death six months later delayed its publication until 1698. In a valedictory summation of a lifetime of studies, he declared that the planets must have vegetation and animals. Without them, he argued, "we should sink them below the Earth in Beauty and Dignity; a thing that no Reason will permit."[20] Mars, because of its greater distance from the Sun, must be cooler than Earth. And yet its inhabitants could adapt to these conditions. The planet's rotation, he declared, was established from the movements of its spots—and here he finally accepted Cassini's period of 24 hours, 40 minutes. The axis of rotation seemed to be only slightly inclined to the plane of its orbit, so that there would be little difference in the seasons experienced for its inhabitants.

Maraldi's Mars

The aerial telescopes used by Huygens in the last years of the 17th century admitted little further development. Optically they seem to have been better than is often assumed, but they were doubtless unwieldy and difficult to use. It is telling that although Huygens bequeathed them to the Royal Society of London, the Huygenian apparatuses were seldom taken out of storage. According to a note in the *Transactions of the Royal Society* from 1718, "those here that first tried to make use of this Glass, finding for want of Practice, some difficulties in the Management thereof, were the occasion of its being laid aside for some time."[21] Meanwhile, ever grander aerial telescopes were being set up at the Paris Observatory. Cassini used one of these, of which the lens rested on the edge of a deck attached to a wooden tower from the Marly Machine (originally built at Louis XIV's behest as part of a costly and ultimately failed scheme to pump water from the River Seine to the Palace of Versailles). With this telescope, he succeeded in 1684 in discovering two satellites of Saturn, Dione and Tethys. It did not improve his views of Mars.

Near the end of his life Cassini became blind, as Galileo had been, prompting Fontenelle to pay a tribute in which their names were linked together. "These two great men," he wrote, "made so many discoveries in the sky that they resemble Tiresias, who lost his sight for having seen some secrets of the gods."[22] Retiring from astronomical work in 1710, two years before his death, he passed direction of the observatory over to his son Jacques (1677–1756, known as Cassini II), thereby establishing an astronomical dynasty at the Paris Observatory. A nephew, Giacomo Filippo Maraldi (1665–1729), became the leading observer of Mars of the next generation. Using a Campani telescope with a focal length of 10 meters that his uncle had brought from Rome long before, he found at the perihelic opposition of 1704 that the markings seemed rather less defined than normal. (Perhaps the planet was in the throes of one of its intermittent dust storms at the time.)[23] At the perihelic opposition of 1719 Maraldi announced (without explanation) that the south polar cap is eccentric to the pole (sometimes turned toward us, sometimes hidden). As for the dark markings, he failed to convince himself that they were permanent surface features. His drawings of 1719 did not resemble those of 1704, nor those of his uncle from 1666. Could it be, he asked, that the bands on the planet were cloudy and purely atmospheric in nature, as with Jupiter? Surprisingly, it would take another century of scrutiny of the planet, and far better telescopes than any available at this time, before this question was definitively answered.

4

Mappers of Strange Lands and Seas

Reflections

With Giacomo Filippo Maraldi, the study of Mars temporarily came to a dead end. Apart from Charles Messier (1730–1817), the celebrated discoverer of comets, who made a few observations of Mars in 1764 and 1766, the record from Maraldi to the next important observer of Mars, William Herschel (1738–1822), is nearly devoid of results. Camille Flammarion, who in his magisterial *La Planète Mars* devoted 36 pages to the period from 1636 to 1719, managed to cover the entire period from 1719 to 1777 in only two pages.

This period overlaps with the "long night of selenography," when studies of the Moon also went into hibernation. The lack of improvements in the telescope meant that observers of the Moon and planets were unable to make out anything new, and so, inevitably, they lost interest. Observational advances in astronomy have usually followed on technological advances. By 1733, the first achromatic lenses, which combine a concave lens of flint glass with a convex lens of crown glass, were being made; since each type of glass disperses the colors differently, in combination they counterbalance so as to produce a relatively chromatic aberration-free image without resorting to the enormous focal lengths employed in the aerial telescope used at the end of the Huygens-Cassini era. Instead of an aerial telescope 10 meters long, an achromat could be downsized to, say, 1.6 meters, with no loss of performance. The achromatic lens was first patented by the English optician John Dollond (1706–61), and telescopes incorporating such lenses became commercially available. However, they were very expensive. When, in the 1770s, William Herschel—a professional musician who had emigrated from Hanover to England and settled in Bath—became keenly interested in astronomy, he found such telescopes completely out of reach. Instead, he turned his attention to the reflector, where a mirror instead of a lens is used to gather light. There was little advice at the time to guide Herschel's efforts to grind and polish mirrors, but by 1777, after many failures, he succeeded in producing several telescopes; one, with a mirror of 16.5-centimeter aperture and 2.1-meter focal length, was pronounced superior to anything then in use at the Royal Observatory at Greenwich.[1]

Herschel was not only an instrument builder but an excellent observer, with an omnivorous appetite for everything visible in the heavens, including Mars. He studied the planet at the oppositions of 1777, 1779, 1781, and 1783; made many sketches of the dark markings and the polar caps; worked out the rotation period, which he put at 24 hours, 39 minutes, and 21.67 seconds, exactly two minutes too long; and calculated that the axis of rotation was tilted to the plane of the Martian orbit by 28°42'. Although this figure was off by a few degrees, it implied that the Martian seasons must be closely analogous to those of Earth, though nearly twice as long. He not only confirmed Maraldi's discovery that the south polar cap was not exactly centered on the geographic pole (according to his observations in 1783, he estimated that it was off-center by some 8.8° of latitude) but found that the same was true of the north polar cap and offered for the first time a theory as to what the polar caps might be: snow or water ice. In conclusion, he decided that the analogy between Mars and Earth was "by far the greatest in the whole Solar System," and that its inhabitants "probably enjoy a situation in many respects similar to ours."[2]

In March 1781 Herschel discovered the seventh planet of the solar system, now known as Uranus, though Herschel called it "Georgium Sidus" in honor of King George III (1738–1820). This unprecedented achievement made him famous and allowed him to give up his musical career. He was appointed personal astronomer to the king, and he moved from Bath to the neighborhood of Windsor Castle so that he could, whenever summoned, present occasional astronomical entertainments to the royal family and their guests. Though his greatest work—in stellar and nebular astronomy—still lay ahead of him, after the opposition of 1783 he did no more useful work on Mars.

Meanwhile, his achievements inspired the efforts of another, Johann Hieronymus Schroeter (or Schröter, as it is often written in the English-speaking literature, though the astronomer himself never used this form of his name, 1745–1816). He was chief magistrate of the village of Lilienthal, near Bremen. Long interested in astronomy, he became serious about it after Herschel's discovery of Uranus. In a spirit of emulation he acquired from Herschel two mirrors of 12-centimeter and 16.5-centimeter aperture. (The latter cost him 600 reichsthalers, nearly half his annual salary at the time.) Schroeter did not have Herschel's genius, but he equaled the other in enthusiasm, especially for observations of the Moon and planets, including Mars.[3]

Schroeter's 12-centimeter reflector was in use by 1785, when he attempted a first reconnaissance of Mars at its November opposition. He could make out only a "few grey, misty, poorly bounded patches," which he tried to follow from night to night. Though they often appeared similar, he could not convince himself of their identity.[4] At the next opposition in December 1787, he used the 16.5-centimeter telescope for the first time. Oddly, his thoughts about the Martian markings' impermanence now hardened into an idée fixe. "The spots and streaks on the globe of Mars are always changing," he wrote, "even from hour to hour. . . . The same shapes in the same positions develop and pass away again, as one would expect of the variable atmospheric appearances occurring above a solid surface."[5] On the night he wrote this, he sketched Syrtis Major in unmistakably identifiable form!

In addition to acquiring a larger and larger stable of telescopes, including, by the late 1790s, a reflector with a 47-centimeter-diameter mirror—the largest on the Continent at the time—Schroeter published at his own expense a series of lavish monographs on the Moon and each of the planets; even difficult-to-observe Mercury was covered.[6] The most notable omission was Mars. Though he was as zealous and diligent an observer of Mars as he was of any other solar system body, and made 231 drawings of the planet's surface markings, none appeared during his lifetime; perhaps he could not satisfy himself with them. The patches in these drawings appear vague, diffuse, and cloudlike; indeed, they are not infrequently drawn into stripes recalling the dark belts of Jupiter, which, like Maraldi, he seems to have believed they resembled. Even at the excellent perihelic opposition of August 30, 1798, when he was using the great 47-centimeter reflector, his drawings do not represent a decisive step forward from what Christiaan Huygens and Giovanni Domenico Cassini had done, much less Herschel. It is little wonder that with such vaporish views Schroeter remained convinced that on Mars he was seeing nothing than a mere floating shell of clouds.[7]

This only proves, if further proof were needed, that Mars is a very difficult object to observe.

Schroeter's last years were, alas, unhappy. With the outbreak of the Napoleonic Wars, Lilienthal was taken over by the French, and in April 1813, as the French were reeling back from their disastrous winter campaign in Russia, a skirmish took place near Lilienthal between a French detachment under General Dominique-Joseph René Vandamme (1770–1830) and a small band of Russian Cossacks. A French officer was wounded; it was alleged that the local peasantry had fired on the French. In retaliation, Vandamme gave orders to set fire to Lilienthal. Schroeter and his family escaped, but his observatory, though untouched by the fire, was broken into and plundered by the French, who "with a fury the most provoked and irrational destroyed or carried off the most valuable clocks, telescopes, and other astronomical instruments."[8] When soon afterward the French were expelled from Germany, Schroeter was restored to his former position, but he was now a broken man, and his eyesight was beginning to fail. His career as an observer was effectively ended, and though he made a valiant effort to prepare his Mars observations for publication, he had not yet finished doing so when he died in August 1816. These observations were finally published only in 1881; by then they were little more than a historical curiosity, though not without value as some of the observations show changes in the Martian surface features since Schroeter's time. For instance, a number of his drawings show a curved, hooklike marking (see figure 4.1). Herschel had also depicted it in 1783. Located at about 230°W longitude, it was one of the most visible features on Mars during the last two decades of the 18th century, rivaling, and resembling in form, Syrtis Major itself. Much atrophied, it later became the Cyclops "canal"; there is still an oblong dusky marking on the surface at this location. No doubt the prominence of this marking, located about 60° from Syrtis Major and easily confused with the latter, contributed to Schroeter's difficulty in establishing

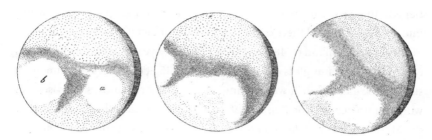

FIGURE 4.1. Three of Schroeter's drawings of Mars, made over an interval of several hours on the night of December 8, 1800. The first drawing shows Syrtis Major; the second shows Syrtis Major rotating off at the terminator, as well as another marking, which the 19th-century astronomer Hendricus G. van de Sande Bakhuyzen (1838–1923) called Spitze B, coming round from the limb; and the third drawing shows Spitze B arriving nearly at the center of the disc and very nearly equaling Syrtis Major itself in prominence. South is at the top. Published in Camille Flammarion, *La Planète Mars*, vol. 1 (1892), figure 54.

the permanency of the markings,[9] but it was hardly the whole explanation. Schroeter's observations of Mars were in the end ambiguous. For definitive results, better telescopes were needed.

The Fraunhofer Era

Those better telescopes were largely to be the contribution of an optician of genius, Joseph von Fraunhofer (1787–1826), who would make the 19th century the "century of the refractor."

Fraunhofer was the son of a master glassmaker. (In fact, the family, on both sides, had been glassmakers for generations.) Orphaned at 11, he was sent to Munich as an apprentice for the court mirror maker and glasscutter, Philipp Anton Weichselberger. Most masters of that time were harsh and unpleasant, and Weichselberger seems to have been no exception. However, in July 1801, his house and workshop suddenly and unexpectedly collapsed. Fraunhofer was pulled alive from the rubble. The prince elector Maximilian Joseph IV of Bavaria (later King Maximilian I, 1756–1825) supervised the rescue operations and took an interest in the unfortunate boy, whose further training in Munich he subsidized. In 1806, Fraunhofer was transferred to

Benediktbeuern, a former Benedictine monastery that had been converted into a glassworks, and there came into fruitful contact with a Swiss technician, Pierre-Louis Guinand (1748–1824), who for some years had been successfully casting flawless flint glass discs of as much as 15 centimeters in diameter by stirring the glass while it was cooling. In Fraunhofer's hands, such discs were worked into achromatic lenses that would produce clear, brilliant images of the Moon and planets, far superior to the faint images produced by the speculum metal mirrors used by Herschel and Schroeter. Already by 1813, the year Schroeter's observatory was wrecked by the French, Fraunhofer had produced a fine achromat of 19-centimeter aperture. When Fraunhofer died, in 1826, at the early age of 39 (either from tuberculosis or from deterioration of his lungs by prolonged exposure to the furnace heat and lead oxide common among glassmakers at the time), he had produced a number of first-rate instruments including, in 1824, his masterpiece: the 24-centimeter refractor at the Dorpat Observatory in Tartu, Estonia, then part of imperial Russia. It was also the first to be equipped with a clock-driven equatorial mounting that could track the stars automatically.

Among the first observers to employ an achromatic refractor for the study of Mars was the French amateur astronomer Honoré Flaugergues (1755–1835), who observed Mars extensively between 1796 and 1809, and also in 1813, from his private observatory at Viviers in the department of Ardèche. The observations from 1813 are of special note: Mars came to the first perihelic opposition of the new century, and Flaugergues made out the more prominent patches on the disc and tried to calculate the times at which the planet's rotation ought to bring the same aspects back into view. However, he found great inconsistencies, and he could not convince himself that he was observing the solid surface of the planet. "The patches seemed to me to be in general confused and badly defined, so that it was difficult to distinguish exactly their outlines and their full extent," he wrote. "I can only say that normally the south part of the disc was the region of Mars which contained the principal patches."[10] Thus, Flaugergues's work failed to advance the study of the Red Planet.

More successful were the efforts of a Berlin attorney and amateur astronomer, George Karl Friedrich Kunowsky (1786–1846). Equipped with an 11-centimeter Fraunhofer refractor, Kunowsky sketched the planet at the winter opposition of 1821–22 and, despite the unfavorable conditions, showed beyond doubt that the markings were fixed patches on the surface.[11] With this result established, it was only a matter of time before the next step—producing a bona fide map—would be taken.

Beer and Mädler Map the Planet

This step, marking the beginning of what Flammarion called "the second period" of telescopic studies of the planet (1830–77), also took place in Berlin. It was the achievement of two amateur astronomers, Wilhelm Beer (1797–1850) and Johann Heinrich von Mädler (1794–1874), who will always be remembered as among the greatest observers of Mars (figure 4.2).

Beer was born into one of the leading Jewish families in Berlin. His father, Judah Herz Beer (1769–1825) was a highly successful entrepreneur whose wealth derived from the sugar refineries he established in Berlin and in Gorizia, Italy; his mother, Amalie (Malka) Wulff (1757–1854), was the daughter of a merchant and banker who was equally well-to-do, having made a fortune delivering supplies to the Prussian army. (Astronomy was still, as it had been for a long time, a plaything of the rich.) The family was, in addition, talented:

FIGURE 4.2 The first mappers of Mars. *Left*: Wilhelm Beer, lithograph by Carl Funke, 1849. *Right*: Johann Heinrich von Mädler, age 40, lithograph after a drawing by Ernst David Schabert. William Sheehan collection.

Wilhelm's older brother was Giacomo Meyerbeer (1791–1864), the most successful operatic composer of his time, while a younger brother was Michael
Beer (1800–1833), a poet and playwright. The magnificent Beer villa, the family residence from 1814, was set in the heart of Berlin, on the Pariser Platz,
opposite the Brandenburg Gate, with the Tiergarten to the west and Unter den
Linden boulevard to the east. Becoming keenly interested in astronomy, Beer
acquired a first-rate instrument, a 95-millimeter Fraunhofer refractor, from
J. W. Pastorff (1767–1833), a landowner and avid solar observer at Buchholz
near Dossen, and set it up on the roof of the villa under an impressive four-
meter-diameter rotatable dome, with a shutter opening on a 20° swath of sky.
The telescope was in place by 1828, and two years later Beer and Mädler—the
latter a rather impecunious schoolteacher at the time, who began supplement-
ing his income by tutoring Beer in the finer points of astronomy—embarked
on one of the most fruitful collaborations in the history of astronomy.[12] They
are best remembered for their great map of the Moon, published in four quadrants between 1834 and 1836.[13] (It has always been recognized, by the way, that
most of the actual observing and mapping was done by Mädler.) At the same
time, they began their epoch-making study of the surface features of Mars,
which included the first efforts to map the planet.

Comparing Mars to a new continent like that accidentally discovered by
Columbus, Flammarion would later write: "Beer and Mädler deserve to be
remembered as the true pioneers in this new conquest. . . . Though theirs was
a relatively modest instrument, it was made by Fraunhofer and was therefore
excellent; moreover, the observers were particularly skillful, meticulous, and
patient."[14]

Beer and Mädler were heavily engaged in the lunar-mapping project when,
with Mars coming to a perihelic opposition on September 19, 1830, they briefly
turned their attention to the planet that was closer to Earth than it had been in
15 years. Mars presented a very different challenge from the Moon, for despite
its nearness and relatively large size, it was only about the size of a small-to-
medium-sized lunar crater. Their first priority was to confirm whether the surface markings were permanent; next was to establish the planet's rotation period.

As most observers of the planet before and since have found on first peering at Mars in a telescope, the markings are weak and ill defined. The small
disc appears, even in a good telescope, "subtle, ethereal, delicate, floating, at

times seemingly without mass or obvious three-dimensionality—even when the seeing is exceptional."[15] At first Beer and Mädler planned to do as they had done with the lunar features, and measure the features with the threads of a micrometer. They soon realized the futility of that approach. As they wrote:

> The use of a micrometer did not seem convenient to us, the thickness of the threads causing more uncertainty in measuring such fine objects than was produced by estimating by the eye alone. The drawings were executed immediately at the telescope. Ordinarily some time elapsed before the indefinite mass of light resolved into an image with recognizable features. We next attempted to estimate the coordinates of the most distinct points, using the white spot at the south pole for the determination of the central meridian, and only then sketched in the remaining detail. . . . Finally, each of us compared the drawing with the telescopic image, so that everything shown was seen by both of us and hopefully can be considered fairly reliable.[16]

Nevertheless, they quickly affirmed Kunowsky's assertion that the markings were permanent. "The hypothesis," they said, "that the spots are similar to our clouds, appears to be entirely disproved."[17]

Right from the beginning, their attention was seized by a small round patch, "suspended as if from an undulating ribbon." Though the same spot had been represented very imperfectly by Herschel in 1783, and on several occasions by Schroeter in 1798, Beer and Mädler were the first to show it clearly. It lay only 8° south of the equator, and its roundness and distinctness made it a convenient reference point for determining the sidereal rotation period, which they set at 24 hours, 37 minutes, and 9.9 seconds. The round patch was designated with the letter *a*. With the exception of the English observer John Phillips (1800–1874), most subsequent students of the planet have concurred in the aptness of their choice, and it has defined the zero meridian—the Martian Greenwich—ever since. (On modern maps, the region is known as Terra Meridiani, the Meridian Land. Its western part, Meridiani Planum or the Meridian Plain, was chosen as the landing site for the Opportunity rover in 2004.)

This being a perihelic opposition, the south polar region was tilted toward Earth, allowing Beer and Mädler to make a careful study of the south polar cap. They followed its rapid shrinking, which continued apace until the Martian season had advanced to a point that would correspond to our mid-July, after which the cap began slowly to increase in size again. These observations lent support to Herschel's idea that it consisted of ice and snow. It was from these observations that Beer and Mädler drew up the first map of Mars ever made, a polar projection overlooking the south pole and showing the markings of the southern hemisphere of the planet.

Further observations with the 95-millimeter refractor were made at the oppositions of 1832 and 1834–35. Then, in 1837, Beer and Mädler were granted permission to use the 24-centimeter refractor of the Royal Observatory of Berlin (a twin to the Dorpat refractor, it would be used a few years later to make the discovery of Neptune). The opposition of 1837 occurred in February, so that not only was Mars's apparent diameter very small, only 14.0" of arc, but the weather in Berlin was almost unprecedentedly bad. Nevertheless, Beer and Mädler calculated a revised rotation period of 24 hours, 37 minutes, 22.7 seconds, agreeing, to the tenth of a second, with the modern value.[18]

Whereas in 1830, it had been the south polar cap that had been tilted toward Earth, in 1837 it was the north, and Beer and Mädler showed that the two caps were remarkably different in their behavior. Though both were centered within a few degrees of the poles, the south polar cap grew to a much greater size; at the same time, its retreat was more rapid and complete. The smallest size Beer and Mädler recorded for the south polar cap was 6°; for the north polar cap, 12° or 14°.

In addition to their studies of the polar caps, Beer and Mädler made many sketches of the surface patches, and they noticed especially rapid changes in the dark area surrounding the north polar cap. In 1837 this dark area was of unequal width and not everywhere equally black. It was, however, darker than the other patches. Two years later, it was faint and narrow. These changes would be satisfactorily explained, they proposed, if the area consisted of marshy soil moistened by melt water from the retreating snow.

In 1840, Beer and Mädler published another polar projection map of Mars, this time centered upon the North Pole (figure 4.3). As with the 1830 map, it represents a giant step forward in the history of Mars studies. One need only compare their drawings made using the small but optically superb Fraunhofer with those by Schroeter using his large reflector to see just how significant had been the advance. The two German observers drew the obvious conclusion

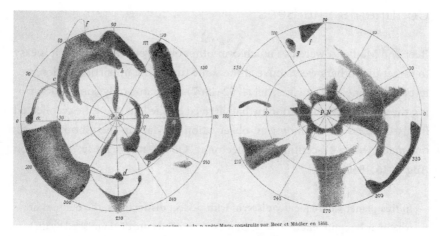

FIGURE 4.3 Polar projection maps of Mars by Wilhelm Beer and Johann Heinrich von Mädler, 1840. The one on the left is very similar to the 1830 map, as it is based mostly on observations from the perihelic opposition of 1830, when the south pole was tilted toward Earth; the one on the right is based on observations at the aphelic oppositions of 1835 and 1837. The fact that these observers used polar projections underscores their particular interest in the Martian polar phenomena. Published in Camille Flammarion, *La Planète Mars*, vol. 1 (1892), figure 68.

as to the nature of the markings they had recorded: "It is not going too far to claim that Mars bears a very strong resemblance to the Earth . . . and it appears as an image of the Earth in the firmament seen from a great distance. The most important differences between Mars and the Earth are the smaller volume of Mars and the greater eccentricity of its orbit. However, the length of the Martian day is practically the same as ours."[19]

That year, the fruitful collaboration of Beer and Mädler ended. Mädler left Berlin to become director of the Dorpat Observatory in Estonia. There, he made a few more sketches of Mars with the 24-centimeter refractor at the opposition of 1841, but he added nothing new. Beer, without the stimulus of his gifted colleague, appears to have given up astronomical work altogether, and spent the last decade of his life until his death in 1850 as a writer and politician. But they had achieved a place of honor in the annals of Martian studies. Flammarion said that "their researches were the most fruitful ever undertaken up to that time, and they really inaugurated our knowledge of Martian geography, or *areography*."[20]

Continents and Seas

Beer and Mädler towered so much over other observers of Mars that it is easy to forget that others were active at the same time. One in particular deserves to be mentioned: John Herschel (1792–1867), William's only son. In describing the state to which astronomy had arrived by the late 1840s, he recollected in his *Outlines of Astronomy* an observation of Mars in the gibbous phase, one made with a 47-centimeter reflector at "Observatory House," Slough, on August 16, 1830:

> In this planet we frequently discern, with perfect distinctness, the outlines of what may be continents and seas. Of these, the former are distinguished by that ruddy colour which characterizes the light of this planet—which always appears red and fiery—and indicates an ochre tinge in the general soil—like what the red sandstone districts on the Earth may possibly offer to the inhabitants of Mars, only more decided. Contrasted with this (by a general law of optics), the seas, as we may well call them, appear greenish. These spots, however, are not always to be seen equally distinct, but, when seen, they offer the appearance of forms considerably definite and highly characteristic, brought successively into view by the rotation of the planet, from the assiduous observation of which it has even been found practicable to construct a rude chart of the surface. . . . The variety in the spots may arise from the planet not being destitute of atmosphere and clouds; and what adds greatly to the probability of this is the appearance of brilliant white spots at its poles—which have been conjectured, with some probability, to be snow; as they disappear when they have been long exposed to the Sun, and are greatest when just emerging from the long night of the polar winter.[21]

The reference to "red sandstone" recalls the rock-bed sequences of sedimentary rock in Great Britain that figured prominently in the Great Devonian Controversy of the early 19th century.[22] In this stratification debate, the older sequence was known as the Old Red Sandstone, and is Devonian, overlying the graywackes from the Silurian on the one hand and underlying the Mountain Limestone of the Carboniferous on the other. The younger sequence, New Red Sandstone, is composed of beds laid down throughout the Permian to the end of the Jurassic. In each case, the sandstone has a reddish color that owes to the presence of iron oxide—to that extent, they are indeed like the soils of Mars.

By throwing in this offhand geological reference, Herschel tacitly suggested that Mars might be susceptible to scrutiny not only in geographic terms but also in geological. In a phrase, astronomy was thus married to geology, joined together as sciences that, as Herschel himself had claimed, ranked first and second in the magnitude and sublimity of their objects. His words proved prophetic, since within a few years the leading interpreter of the Martian landscapes as seen in the telescope would be none other than John Phillips, one of England's leading geologists and a major figure in the great 19th-century debates over stratigraphy, as described below. Indeed, within a little more than a century, the marriage of astronomy and geology (nourished by an ample dowry of mathematics, chemistry, physics, and biology) would lead to the birth of a new field focused on the solar system: planetary science.

The younger Herschel was one of the great polymaths of the 19th century. Though best known as an astronomer, having felt a filial obligation to complete his father's work, he had a particular affinity for optics, and he was well aware that the perception of color depends on context. A grayish patch, seen against an ocher background, appears greenish. This particular color-interaction effect would loom large in the study of Mars since, as already suggested in Herschel's "red sandstone" remark, colors were key to interpreting the nature of the patches on the planet's surface. Though artists had presumably been aware of the color-interaction effect for centuries, the first person to call explicit attention to it was the German poet Johann Wolfgang von Goethe (1749–1832), in his book *Theory of Colours* (1810).[23] A thorough experimental investigation was undertaken a few years later by the French chemist Michel Eugène Chevreul (1786–1889), in his role as director of the dye-works at Gobelins, the royal tapestry works in Paris. Responding to the complaints of weavers that samples of black thread used in the tapestries didn't look right, especially when the black threads were woven next to blue thread, Chevreul found that there was nothing wrong with the colors themselves; they simply looked different when they were seen alongside other colors. Chevreul referred to these perceived interactions as "simultaneous contrast" in his book *De la loi du contraste simultané des couleurs*, published in 1839. (An English translation, *The Principles of Harmony and Contrast of Colours*, appeared in 1854.) Later, Chevreul introduced his celebrated color wheel with 72 colors, in which the complementary colors were set diametrically opposite to one another around the wheel. Red was opposed to green, orange and orange-red to blue, and so on (plates 1 and 2).

When the younger Herschel was writing, the law of simultaneous contrast had already been recognized as a "general law of optics." Whether he discovered it independently or learned of it from Chevreul (or even Goethe), he deserves credit for being the first to apply the concept to the colors seen in telescopic images of Mars.

Against their recognition of the applicability of this general law of optics, astronomers struggled with a strong predisposition, going back at least to Cassini, to regard the image of Mars in the telescope as being like a small Earth, and to identify the patches on its surface with reddish continents and greenish or bluish seas.

The Mountains of Mitchel

The 1845 perihelic opposition, the first since Beer and Mädler had ended their studies of the planet, produced little that was new. There was, however, one notable result, concerning the south polar cap, made by Ormsby MacKnight Mitchel (1810–62), a colorful American popularizer of astronomy and later a general in the Union Army during the Civil War.[24] This was also the first important discovery about Mars to have been made in the United States.

In the years before a national observatory existed in the United States to correspond to the great European institutions of Paris, Greenwich, and Pulkovo (the U.S. Naval Observatory was not founded until 1844), Mitchel, a teacher of mathematics and astronomy at Cincinnati College since 1836, expressed disappointment in the lack of interest in promoting astronomy in the U.S. Congress, which contrasted with the great interest of the general public. A spellbinding lecturer, Mitchel himself was largely responsible for arousing this interest. After the last of a series of lectures given to 2,000 citizens in Cincinnati in winter 1841–42, he remarked:

Up to this time our own country had taken no part in the great movement of astronomical science, which during the present century has been attended with such wonderful discoveries. . . . While the nations of Europe were vying with each other in the career of discovery, the United States, not less deeply interested mentally and commercially than any nation on earth, was lying indifferent, while from the Old World the finger of scorn was pointed at our profound republican ignorance. We had literally done nothing. While Russia with its hordes of barbarians boasted the finest observatory in the world, our own country with all its freedom and intelligence had been recently reported by the Astronomer Royal of England not to possess a single observatory within all its vast extent. It had been even said that the efforts in Congress had forever sealed the fate of astronomical science in this country. The government could never become the patron of astronomy, since no representative would risk in its defence his reputation for sanity. The scientific were too few and too poor to attempt any great enterprise; while the wealthy were too indolent and too indifferent to lend their aid extensively to a matter of which they knew but little.[25]

Instead of waiting on a reluctant Congress or a wealthy donor, Mitchel decided to try a different approach. He devised a public subscription of shares of stock at $25 apiece—more than a month's wages for many laborers at the time—in order raise $7,500 to acquire a first-rate telescope for an observatory to be built in Cincinnati. The plan was a success. Within only a month he had raised the funds and was off to Europe in search of a lens. He found nothing that would meet his needs in the optical shops of London or Paris; but on arriving in Munich, he visited the firm of Merz and Mahler (successors to Fraunhofer), who were undaunted by his requirements and ready to produce the 28-centimeter lens he wanted. It was delivered on schedule in January 1845, but in the meantime Cincinnati College, then Mitchel's sole source of income, had burned to the ground. (It was this catastrophe that forced him to turn to itinerant lecturing during the winters to support his wife, family, and astronomical research.) Despite the college's demise, construction of the observatory went ahead without delay, and that spring the new refractor—at the time the largest telescope in America and second-largest in the world after only the 38-centimeter refractor at Pulkovo in Russia—was put into working order on Mount Adams, just in time for Mars's perihelic opposition on August 18. In search of a favorable object on which he could test the powers of the refractor, he noticed on the evening of August 30 a patch of snow (at 70°S, 320°W), detached from the south polar cap and apparently left behind as the dwindling cap continued its summer retreat (figure 4.4). After the first separation of the patch, the final remnants did not disappear for another twenty or thirty days.[26] By analogy to terrestrial conditions, in which snow remains longest on the tallest peaks, Mitchel supposed the snow to disclose the existence at this point of a ridge of mountains. The feature has been known as the Mountains of Mitchel ever since, although it is not, in fact, an actual mountain range. It always appears at the same time in the late southern hemisphere spring (even to this day). The feature's true nature has been revealed only in the spacecraft era.[27]

A Proliferation of Interest

The next perihelic opposition was due on July 17, 1860. Compared to earlier periods, when only a few enthusiastic observers had paid much attention to Mars, by the 1850s and 1860s there were at least a score. This marked the beginning of more numerous and frenzied activities.

At the opposition of 1856, one of the leading observers was the Englishman Warren De la Rue (1815–89). Attaining to enormous wealth through the family's London printing firm and the invention of an envelope-making machine, he became the "quintessential Victorian wealthy amateur astronomer"[28] and observed Mars with an excellent 33-centimeter reflector on an equatorial mount from his observatory in Islington, North London (later moved to Cranford, Middlesex, which was considerably farther from London's increasingly polluted skies). His drawings, which are qualitatively better than any made previously, show Syrtis Major (still known as the Hourglass Sea at the time) as very narrow, its characteristic form at that period.

Even more important were the Mars observations by the eminent Italian Jesuit astronomer Pietro Angelo Secchi (1818–78). Secchi had been born in Reggio nell'Emilia and attended the local gymnasium (prep school), where his academic promise was duly noted. In order to further his education, he became a Jesuit novice in Rome at the age of 15. In addition to the usual studies of classical literature and philosophy, he received a solid grounding in

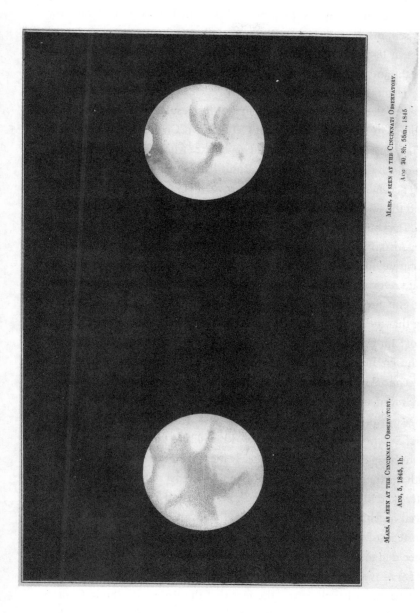

MARS, AS SEEN AT THE CINCINNATI OBSERVATORY.

Aug. 5, 1845, 1h.

MARS, AS SEEN AT THE CINCINNATI OBSERVATORY.

Aug. 30. 8h. 55m., 1845

FIGURE 4.4 Drawings of Mars during the 1845 perihelic opposition, by Ormsby MacKnight Mitchel, using the 28-centimeter Merz and Mahler equatorially mounted refractor of the Cincinnati Observatory. The one on the left shows the Syrtis Major region, that on the right the singular small appendage of the south polar cap now known as the Mountains of Mitchel. Published in *Sidereal Messenger* 1, no. 11 (March 1847): facing 81; courtesy of Trudy E. Bell.

mathematics and physics, which was not so usual at the time. After spending several years teaching physics at Loreto College (which we would refer to now as a high school), he returned to Rome to study theology. He was ordained as a priest in 1847. In the following year, known as the year of revolutions, in which the unrest beginning in Sicily soon spread across Europe, he and the other Jesuits were expelled from Rome. Even Pope Pius IX (1792–1878) was forced to flee. Secchi took refuge first at the Jesuits' Stonyhurst College (high school) in England, and then at Georgetown University in the United States, where he received some practical experience with making astronomical observations. By 1849, the situation in Rome had become safe for the Jesuits, including Secchi, to return. The pope returned after another year and at once appointed the 32-year-old Secchi director of the observatory of the Collegio Romano. The young astronomer found it poorly equipped and crumbling into disrepair. Also, it was located in a tower in the college not stable enough to allow accurate observations. The young director set out to implement some immediate reforms: he acquired a first-rate telescope, a 25-centimeter Merz refractor, equatorially mounted, and scoured the vicinity for a better location. Noticing that there were four massive pillars in the chapel of the college, the Church of Sant'Ignazio, which had originally been planned to support a dome for the church, he surmised that they would likely provide excellent support for a telescope.[29] At Secchi's initiative, a new observatory was duly erected on the church roof, above the massive pillars; it was finished by 1853, and Secchi set at once to work.

At first Secchi rather dutifully devoted himself to conventional activities such as studying double stars and precisely measuring stellar positions. However, with his training in physics, at a time when most of his contemporaries were interested in *where* the planets and stars were and *how* they moved, Secchi proved much more interested in *what* they were. This was the case with Mars. Instead of working out its rotation period to another decimal point, Secchi decided to try to find out what kind of world it was, and even whether it might serve as an abode for life.

Somewhat surprisingly, given the fact that the Roman Catholic Church under Pius IX was turning at this time in increasingly conservative intellectual directions, Secchi was always an enthusiastic advocate of the "plurality of worlds," the once-heretical idea that other planets might be inhabited (an idea that had led the Dominican scholar Giordano Bruno to be burned at the stake in 1600). As early as 1856, Secchi wrote, "It is with a sweet sentiment that man thinks of these worlds without number, where each star is a sun which, as minister of the divine bounty, distributes life and goodness to the other innumerable beings, blessed by the hand of the Omnipotent."[30] As a close friend of the pope, Secchi was completely safe in holding such views,[31] and he spent several years studying the Moon and planets in the hope of advancing his "sweet sentiment." During 1856, he spent six months peering through the Merz refractor, using magnifying powers of 760× and 1000×, in order to complete a detailed drawing of the Moon. "As it was impossible to carry through such a work in a single night," he wrote, "on the first night of good opportunity a general outline was taken, and on the other evenings particular drawings were made, and all these parts, taken in different grades of light and shadow, were afterwards harmonized together and compared against the Moon itself."[32] The resulting drawing, done on a scale of six kilometers to a centimeter, was the most detailed drawing of the Moon made up to that time, and it revealed features, as the English geologist John Phillips wrote, "such as only the larger telescopes can command."[33] That same year he made his first observations of Mars, and finding this preliminary survey encouraging, made a much more careful study of the planet at the opposition of May 15, 1858. It was a better than average opposition, at which Mars attained to a maximum diameter of 18.0" of arc. But it was still only a fifth of the diameter of the lunar crater he had needed six months to sketch. This again underscores just how difficult Mars is as an object of study.

Working together with a Jesuit colleague, Enrico Cappelletti (1831–99), and using a magnifying powers of 300× to 400× on the Merz, Secchi began observing Mars as soon as the seeing was favorable enough. At first he only outlined the main features, which included "a large triangular patch, blue in color." This was, of course, the familiar Hourglass Sea/Syrtis Major. Secchi referred to it as the "Atlantic Canale," and added "it seems to play the role of the Atlantic which, on Earth, separates the Old Continent from the New." This, by the way, was the first appearance of the fateful term *canale*, which was to play such an outsize role in the history of Martian lore. Secchi clearly meant it in the sense of "channel," though another—quite legitimate—translation is "canal," which carries with it the implication of artificial construction.[34]

Secchi found that the best time for observation was two or three hours after sunset, and then only during spells of consistently fine weather. He paid a great deal of attention to the colors of the planet. The "blue-green canal," which often impressed him as having the shape of a scorpion, he wrote, was "followed by a greenish fringe, . . . which extends as far as a yellow patch. At the lowermost boundary of the canal may be seen many very small white strips. These are very remarkable. Are they clouds?"[35] Finding that neither copper engravings nor even chromolithography could give any real idea of these tints, Secchi and Cappelletti employed pastels. In all, they produced 40 color drawings (plate 3). Though Flammarion saw them during a visit to Rome in 1872, they were long regarded as lost. However, they have lately been recovered at the Rome Observatory (the observatory of the Collegio Romano no longer exists). Though they are in damaged condition, one can still discern something of their splendor.

Secchi described the dark areas surrounding the polar caps (which would later become known as the Lowell band) as "ashen colored," but he saw most of the other dark areas as strongly bluish, with an occasional hint of green. This is no doubt a perfectly accurate description of the colors on Mars that the human eye perceives in the telescope. Whether these are genuinely Martian or are produced by the effect of simultaneous contrast described above was uncertain, even to Secchi himself, who wrote: "We cannot decide whether the blue patches are really blue, or whether their color is merely due to contrast. I incline to the view that the color is real, because I have been able to observe small parts . . . separately by using diaphragms; however, when I observed in the daytime I see these areas as almost grey."[36]

Secchi believed that the only possible explanation for the observed variations in the polar caps was the melting of snow or the disappearance of clouds covering these regions. If so, then liquid water and seas must exist on Mars. The reddish regions, like the bluish ones, seemed permanent. "It is probable," he wrote, "that the former are solid, the latter liquid. The tone of the former is not uniform, but markedly *screziato* [speckled, varying in color], as though filled with fine detail, about the nature of which we have no information."[37] This fine speckling would later be interpreted by some observers in terms of linear motifs—the canals of Mars.

Comparing his observations of 1858 with Beer and Mädler's from the 1830s, Secchi thought that changes were likely, though not yet proven. Some, he admitted, might simply be owing to differences in instrumentation and the quality of the atmosphere experienced by different observers. However, at least in some cases he believed the changes to be genuinely Martian, and so— as we now know—they are; many of the specific changes he recorded have now been explained in terms of the windblown movement of dust around the surface. Indeed, Secchi was the first astronomer to associate an increased yellowness of the disc with faintness of the markings, and in retrospect, some of his drawings seem to show obscuration owing to a regional dust storm between June 18 and June 27, 1858.[38]

All in all, Secchi maintained the admirably cautious approach of the scientist, and was careful not to overinterpret his observations. At the conclusion of his studies of 1858, he would not go further than to say: "Whether the dark regions represent water and the reddish areas continents and the white areas clouds—is . . . difficult to answer; one must first decide whether the patches are permanent or variable."[39] So far, the question remained open.

5

Mars Above the Dreaming Spires

End of an Era

Pietro Angelo Secchi's observations of 1858 were excellent, and by far the best up to that time. He had intended them merely as a practice run before the perihelic opposition of July 17, 1860 (when Mars was about as far south of the celestial equator as it can get, and thus low in the sky even from Rome). But Secchi himself did no significant work on Mars that year. Neither did another great Mars observer of oppositions past, Ormsby MacKnight Mitchel. Fifteen years on from the last perihelic opposition, when the great Cincinnati refractor had been the second-largest telescope in the world and Mitchel at the forefront of Martian research, he had fallen on hard times. With his wife in poor health following a series of paralyzing strokes, he had been lured from Cincinnati with the promise of a house and a regular salary to the Dudley Observatory, in Albany, New York. That summer—as the "God of War" loomed once more in the skies, and a divided country prepared to elect a new president, Abraham Lincoln, whose name had not even been on the ballot in the South—Mitchel stood on the veranda of his new home one evening, and pondered the star blazing large and red in the east. His son would write:

> Doubtless memories of 1845, when he was charmed at the admirable performance of the instrument he had been at so much pains to get, when his hair

was not yet flecked with gray, when his children were "wee things toddlin," came up before him sweet but melancholy. At any rate, with a touch of sadness in his voice, he said, "I wonder where we will all be when he comes round again to another opposition in fifteen years."[1]

Ultimately, the only serious observations of Mars in 1860 were made by Emmanuel Lias, an astronomer at the Paris Observatory who was subsequently invited by Dom Pedro II (1825–91), the emperor of Brazil and a strong promoter of science, to Rio de Janeiro, to take charge of the observatory there. Lias emphasized that the reddish color of Mars was due not to its atmosphere, but to the color of its ground; it was, he supposed, due to the vegetation on the planet, which he believed to be reddish-colored rather than greenish like that on Earth.[2] This marks the first appearance of the theory of vegetation on Mars.

New Martians: The Great Opposition of 1862

At the opposition of 1862, Mars was slightly farther from Earth than it had been in 1860, but also farther north of the celestial equator and so better placed for northern hemisphere observers. Secchi now resumed the excellent series of observations that he had begun in 1858. For the first time he drew the feature now known as Solis Lacus, representing it rather curiously in the form of a cyclone. Since he had not noticed it at the previous opposition, he thought it might be a variable—that is, a meteorological feature of the planet—and wrote, "I can note a dark patch different in tone from those to which I am accustomed; I have never seen it before. It seems to be surrounded by a ring, or a spiral cyclone. . . . I believe we are seeing a great squall on Mars." (None of the other observers of Mars in 1862 represented it like this, however.) The south polar cap was tilted toward the Sun during the 1862 opposition, and as he followed its perennial summer retreat, Secchi determined:

> The variations can be explained only by a melting of the snow or a disappearance of the clouds covering the polar regions. . . . These aspects . . . prove that liquid water and seas exist on Mars; this is a natural result of the behavior of the snows. This conclusion is confirmed by the fact that the blue markings which we see in the equatorial regions do not change sensibly in form, whereas the

white fields in the neighborhood of the poles are adjacent to reddish fields which can only be continents. Thus, the existence of seas and continents, and even the alternations of the seasons and the atmospheric variations, have been today conclusively proved.[3]

Unfortunately, this opposition effectively marked Secchi's retreat from the field of Mars studies. He now became absorbed in other work, including the pioneering work in spectroscopy of the Sun and stars for which he will always be remembered. (He did produce some additional color drawings of from as late as 1870–71, which have been recovered in the archives of the Rome Observatory. However, nothing after 1862 was published.)[4]

As Mitchel and then Secchi vanished from the scene, others came to the fore. Frederik Kaiser (1808–72), director of the Leiden Observatory, which since 1860 had been equipped with an 18-centimeter Merz refractor, compared his own drawings of Mars in 1862 with those of earlier observers going back all the way to Christiaan Huygens in 1659 and determined that in the interval Mars had rotated exactly 70,004 times. From this he computed the rotation period to within a thousandth of a second: 24 hours, 37 minutes, 22.735 seconds. Kaiser was an excellent draftsman, and he drew up beautiful maps of Mars, both polar projections like those of Wilhelm Beer and Johann Heinrich von Mädler and—for the first time—a Mercator projection, but he did not attempt to give names to the mapped features.

Joseph Norman Lockyer (1836–1920), at the beginning of a long career in astronomy, also first distinguished himself at this opposition. His father, an early contributor to research regarding the electric telegraph, saw to it that Lockyer received his education at various private schools and on the Continent. "Had he been educated at any of the public schools or universities at the time," says a biographer, "he would have been taught very little, if any, science, and had he taken the career of an astronomer, he would doubtless have been absorbed in the rank and file of those who followed the traditional, and in those days somewhat narrow, grooves."[5] Possessed of an independence of mind, resourcefulness, adaptability, and amateur status during the era of "Grand Amateurs"—those having independence and means to follow their interests wherever they might lead—he was at liberty to select his own fields of research. Since 1857 he had been employed as a clerk in the

British War Office, and he pursued his budding astronomical interest in the evenings after he left his desk. In 1862, while living at Wimbledon, Lockyer set up in the garden in front of his house a 16-centimeter equatorially mounted refractor by Thomas Cooke of York and observed the opposition of Mars. He communicated an account of his observations—evidently his first scientific paper—to the Royal Astronomical Society. His drawings have a thoroughly modern look; indeed, they rather resemble modern photographs (figure 5.1). They certainly rank among the most accurate and attractive images of Mars produced up to that time.

Lockyer's work on Mars represented an early, intense, but ultimately passing phase. With the spectroscope just then becoming available as a powerful instrument of astronomical research, Lockyer was drawn irresistibly into solar physics (making him a keen and sometimes bitter rival of Secchi). He will always be remembered for his spectroscopic discoveries at total eclipses of the Sun, directing no fewer than eight expeditions to observe them and proving himself to be such a dominant personality that one of his friends (probably James Clerk Maxwell, 1831–79) could not resist lampooning him in a few memorable lines of doggerel:

> And Lockyer, and Lockyer,
> Gets cockier and cockier;
> For he thinks hes the owner
> Of the solar corona.[6]

Gradually these studies were laying firm foundations for the study of Martian geography, or *areography*. But Martian geology was also being developed, a testimonial to the widespread interest in geology at the time. The leading figure, and one of the most important researchers of Mars of the era (though comparatively forgotten today) was John Phillips (1800–1874), whose musings on Martian geology provided a distant but legitimate forerunner of the remarkable geological discoveries made by future spacecraft missions.

By the 1860s, the great ages of Earth had been recognized, and the important strata (Cambrian, Silurian, Devonian, etc.) identified in the rock layers. Working against the background of the recently revealed vast canvas of geological time, Charles Darwin (1809–82) published *On the Origin of Species* in

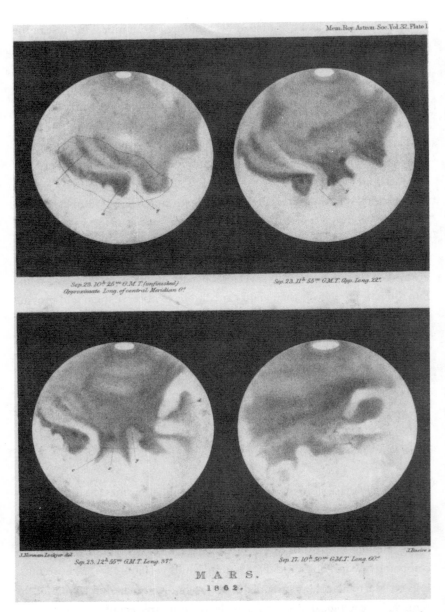

FIGURE 5.1 Joseph Norman Lockyer's drawings of Mars, 1862. South at top. Published in *Memoirs of the Royal Astronomical Society* 32 (1864): plate 1.

1859, in which he demonstrated how new species could arise through natural selection playing up mutations over long periods of time. Seven months after Darwin's book appeared, on June 30, 1860, the British Association for the Advancement of Science (BAAS) held its famous (if overhyped) meeting at the new British Museum (Natural History), in which 700 turned out in a room crowded to suffocation hoping to hear Samuel Wilberforce (1805–73), the silver-tongued bishop of Oxford, "smash Darwin." Darwin himself was not present, but among the small minority arrayed in support of Darwin was Thomas Henry Huxley (1825–95), the former Fullerian professor at the Royal Institution in London. Since no transcript of the meeting was made, it is impossible to know exactly what was said, but Huxley's son pieced together an account from various eyewitnesses, including an correspondent who reminisced about the meeting for *Macmillan's Magazine* in 1898. At the meeting, after a droning and utterly forgettable talk by Dr. John William Draper (1811–82) of New York City (incidentally, the first man to obtain a daguerreotype image of the Moon, in 1840), Wilberforce summoned up all the dignity of his high church robes and distinguished bearing and rose to speak:

It was evident from his mode of handling the subject that [Wilberforce] had been "crammed up to the throat," and knew nothing at first hand. "He ridiculed Darwin badly and Huxley savagely; but," confesses one of his strongest opponents, "all in such dulcet tones." . . .

The Bishop spoke thus "for full half an hour with inimitable spirit, emptiness and unfairness." "In a light, scoffing tone, florid and fluent, he assured us there was nothing in the idea of evolution; rock-pigeons were what rock-pigeons had always been. Then, turning to [Huxley] with a smiling insolence, he begged to know, was it through his grandfather or his grandmother that he claimed his descent from a monkey?" . . .

On this (continues the writer in "Macmillan's Magazine") Mr. Huxley slowly and deliberately arose. A slight tall figure, stern and pale, very quiet and very grave . . . , he stood before us and spoke those tremendous words. . . . He was not ashamed to have a monkey for his ancestor; but he would be ashamed to be connected with a man who used great gifts to obscure the truth. . . . One lady fainted and had to be carried out; I, for one, jumped out of my seat.[7]

It would be interesting to know whether Phillips was present at the meeting. Had he been, he doubtless would have stood with most of those in attendance solidly behind Wilberforce rather than Huxley, since he was a devout Anglican and would never accept Darwin's theory. It would also be interesting to know whether he or anyone else present that night pondered the brilliant burning coal then hovering over Oxford's famous towers and spires, for Mars was then nearing its perihelic opposition of that year. Probably Oxford's ears were ringing too much with thundering words of apes and angels for anyone to pay more than passing attention to the Red Planet. However, two years later, Phillips opened the shutter of a small conical observatory near the museum and pointed toward the planet a 16-centimeter Cooke refractor (an identical instrument to that which Lockyer was wielding at the same time), intending nothing less than a preliminary geological survey of that other world.

By then, Phillips had already had a brilliant career. Born in Marden, Wiltshire, in the southeast of England, he had been orphaned at age seven and taken into the care of his uncle, William "Strata" Smith, a surveyor and self-taught geologist who would have a huge influence on English geology.[8] In his middle teens, Phillips followed his uncle to London, where he received training in surveying. In 1819, shortly after creating the first modern geological map of Britain (a map that was largely ignored by the academic community), Smith was imprisoned for debt and lost his London home. After this disaster he and his young charge spent four years wandering through the north of England, surveying as they went and working on the more detailed geological maps with which Smith came to be regarded as the "father of stratigraphy." Meanwhile, Phillips was establishing his own reputation. He proved so adept at arranging the fossils in the Yorkshire Museum that he was appointed its keeper, and for many years he made York his home. In 1829, he was joined by his sister, Anne, who became his companion and housekeeper for more than thirty years, and he achieved widespread recognition with his *Geology of the Yorkshire Coast*. While still in York, he helped organize the first meeting of BAAS in 1831, and a year later he become assistant secretary, as he was to remain for thirty years. This position expanded his contacts in London. He was elected a fellow of the Royal Society, appointed to the chair of geology at King's College London, and spent several years traveling back and forth between York and London while overseeing the publication of an influential

Guide to Geology and a monograph on the Carboniferous limestone of York-shire. *Palaeozoic Fossils* (1841) followed, in which he introduced the terms Palaeozoic, Mesozoic, and Cenozoic. Gradually he became more and more involved with the British Geological Survey, and—in expectation of becoming director of a new branch of the survey in Ireland—he accepted the position of chair of geology and mineralogy at Trinity College, Dublin. This expected branch did not materialize, but his contacts in Ireland led to one of the turning points of his life.

This was his 1852 visit to Birr Castle, Ireland, where he observed the Moon with the giant reflector of William Parsons, the third Earl of Rosse (1800–1867), whose 1.83-meter mirror would not be surpassed until 1917. Though an interest in astronomy had already begun in his days as keeper of the York-shire Museum, when he sometimes viewed the Moon with a small refractor set up on a stone pillar in his garden, the breathtaking views with the Rosse reflector ignited a real enthusiasm; henceforth he devoted much of his time to astronomy. In 1853 he served as a founding member of the BAAS Moon Committee, helping organize plans for a collaborative effort of 14 volunteer observers to map the Moon on a scale greater than had ever been attempted.[9] That same year he moved to Oxford to assume the position of deputy reader in geology, and as usual was rapidly promoted. He became professor of geology, then was named keeper of the Ashmolean Museum, and finally, in the year of the Wilberforce-Huxley "debate," took charge of the new British Museum (Natural History).

Despite his heavy responsibilities, Phillips remained as keen about astron-omy as ever, and since the facilities for astronomical work were shamefully inadequate in Oxford at the time, he had to provide his own remedies. He applied unsuccessfully for a government grant to obtain a proper telescope. At last, in 1860, he ordered at his own expense the Cooke refractor that was to serve as his workhorse research instrument. It was not delivered for two years, and by then he was grieving the loss of his sister, Anne. The three-story keeper's house behind the museum, where they had lived together for several years, must have been oppressively quiet and full of anguished remembrance. No doubt Phillips's strong Anglican faith helped sustain him through his loss, and so did his friends and committee work; however, his personal and administrative correspondence fell off for a time, as readying the new refractor provided him with much-needed distraction, not to say therapy. By July 1862, it had been set up in its small conical observatory, and Mars was irresistible. The Moon could wait. As an accident of timing, Phillips's first object of serious geological investigation with the new refractor would be not the walled plains of the Moon but the ruddy lands and gray-green seas of this other Earth.

He began by comparing the sketches of previous observers, but he found that they corresponded so little with one another he was not even sure that the features of the planet were permanent. (At the time, the best map of Mars available was still that of Beer and Mädler, now over 20 years old.) At this point he had an inspiration: would not a more satisfactory opinion be formed of the actual arrangement of the lands and seas on Mars if a globe of the planet were constructed? In pursuit of this plan, between September 27 and December 13, 1862, he made fourteen sketches of the planet, which he arranged in order relative to an arbitrary prime meridian. (Oddly, he chose a different one from that used by Beer and Mädler; they had chosen the feature that Richard Anthony Proctor would call Dawes Forked Bay and Giovanni Virginio Schiaparelli would name Sinus Meridiani; Phillips's passed through a point unnamed by Proctor but called Hammonis Cornu by Schiaparelli.) Phillips used these sketches, he says,

> for constructing a globe of Mars. [I] constructed one, and mounted it on a wooden frame. By considering the way in which the globe was presented to the observer on Earth at different periods in the revolution of Mars, I was able to perceive clearly the reason for the very different appearances presented by the drawings of the earlier eminent observers. This was the first example of a globe of Mars on which the main features were laid down.[10]

This globe seems at some point to have first been exhibited for the Royal Society of London. It was apparently quite small, not a globe in the proper sense (i.e., a ball). Instead, it was fashioned around a hexagonal frame of wood, on which Phillips mounted his drawings in the order of the computed longitudes. He subsequently produced two more similar globes. They must have been of rather flimsy fabric, and once they had served their purpose they were either broken up or discarded, perhaps by Phillips himself, or knocked about the rooms of the Ashmolean Society for a few years before being discarded as rubbish. In any case, and sadly, none of them have survived. A number of Mars globes by other astronomers would follow in due course, until, by the end of the 19th century, they had become rather numerous.[11]

Having, by means of these globes, settled the permanency of the Martian features to his satisfaction, Phillips next called attention to the distribution of the surface features. He wrote, "A great part of the northern area appeared bright, and often reddish, as if it were land, while a great part of the southern area was of the grey hue which is considered to indicate water, but relieved by various tracts of a tint more or less approaching to that of the brighter spaces of the northern hemisphere. The principal boundary of light and shade, for the most part very well defined, runs obliquely across the equator of Mars so as to reach latitudes from 20 to 30° N and S of that line."[12]

He commented on the colors of Mars. As we have seen, John Herschel had speculated that the ruddy color might indicate an ocher tinge in the soil, and had noticed (but did not affirm as genuine) the greenish color of the seas.[13] Another leading English observer, James Nasmyth (1808–90), using a large reflector, likewise found the "land" on Mars to be of a decidedly red tint, the "water" green; Lockyer's "overcorrected" achromatic refractor showed no redness. Phillips himself described the "land" as red in some areas, and in others bright and even "silvery"; the "seas" appeared decidedly grayish or greenish.[14]

In addition to his quality observations and the innovation of the globes, Phillips also produced the first British map of Mars (plate 4). It appeared in 1864, the same year as Kaiser's, and was, with the latter, the first Mercator projection of Mars ever published. Based on four of Phillips's best drawings from 1862 and 1864, the original draft is hand painted, and quite beautiful; it was long thought to have been lost, the published version consisting only of very poor black-and-white engravings, but it was recovered by Stella Bicknell, the librarian at the Oxford University Museum of Natural History, during routine cataloging.[15] Phillips's note on the chart shows he was able to satisfy himself that some of the supposed landmasses had definite "edges"; the colors, "Indian red" for the bright areas and "Sea green" for the dark areas, were, he noted, in keeping with the belief that the planet's surface consisted of lands and seas.

Phillips was also apparently the first to perceive that, if indeed the grayish areas were water as generally believed at the time, then whenever the observing geometry was just right brilliant specular reflections of the Sun ought to be visible from Earth using instruments of sufficient power. He wrote:

> Allowing the white [brighter] spaces to be land, which reflects light as the moon in opposition, it seems a natural supposition that the shady spaces should be called sea; and this may be supported by the obvious requirement of water somewhere on Mars, to agree with the alternate gathering and melting of the snow round the poles. Still, every observer remarks no small resemblance of some of these shady tracts with particular parts of the unequally tinted grey surfaces of the moon. A positive proof of ocean on the disk of Mars would be afforded by the star-like image of the sun reflected from the quiet surface, or more diffused light thrown back from the waves; but nothing of this sort has been placed on record, nor is there such a variation in the appearance of these spaces from the centre toward the edges as to give any special reason for thinking them occupied by water.[16]

He determined by actual experimentation that, at least at a vertical incidence of sunlight, the reflective power of water was so much reduced that specular reflections might not be so easily visible as he had at first assumed. That being the case, he surmised, the lack of such observations might not constitute such a compelling argument against the existence of bodies of water on Mars. After mulling these things over, Phillips concluded that the gray areas were, in fact, probably true seas.

So far Phillips had published his observations principally in the specialist *Proceedings of the Royal Society*, but in 1865, he set out to produce a more accessible account. This paper, "The Planet Mars," was, as historian Roger Hutchins has called it, "a comprehensive overview of all that was known of Mars, its physical characteristics, surface geography and atmosphere."[17] It was intended for a somewhat popular audience, appearing in the new *Quarterly Journal of Science*, which had a much wider circulation than the *Proceedings*.[18] It proved to be Phillips's valedictory on the Red Planet, as well as the most interesting paper written on the subject in English up to that time.

In addition to supplying descriptions of the planet's overall appearance in the telescope, and engravings showing his four best drawings and his map, Phillips attempted to systematically compare the physical properties of Mars to those of Earth, thereby anticipating the multidisciplinary and comparative science later referred to as planetology. Thus, he wrote:

> On the whole then, the circumstances already collected regarding the physical aspect of Mars appear to justify the conclusion, that the planet has a larger proportion of land than water, in this respect differing from the earth, and that the land is mostly collected in a broad band, including the intertropical spaces, and the north temperate region, while broad seas encircle the north pole and a large proportion of the south temperate zone. In this last-mentioned large space, the appearances vary much, and in such a manner, as to indicate the overspreading of clouds. . . . We are thus placed by telescopic observation in front of a planet, whose main characters of surface correspond to those of our earth; which has nearly the same density; nearly the same daily and nightly period; and is enveloped like our earth, by an atmosphere partially loaded with scattered clouds.[19]

He considered the way that the radiation of heat from the surface and effects of the Martian atmosphere and ocean would determine the climate of the planet, and suggested that it ought to be comparatively mild, despite the planet's greater distance from the Sun. He did not specifically address the question of life. However, his arguments seemed to be more favorable than opposed to the possibility.

Britain Extends Its Sway

Phillips's influence was more limited than deserved, and after 1868, he published nothing more on astronomy—probably in large part because by then he, like so many Victorians, was grappling with the implications of Darwin's theory, and especially of his 1871 *The Descent of Man*, in which Darwin rather scandalously posited that humans had evolved from subaltern primates. A devout Anglican, Phillips fought Darwin tooth and nail, and given his high rank as a paleontologist, he tried to argue that Darwin had exaggerated the incompleteness of the fossil record. He also maintained that Earth could not possibly be as old as Darwin suggested, basing his opinion on the work of William Thomson, Lord Kelvin (1824–1907), who had calculated the rate of cooling of Earth. Into his 70s, Phillips still seemed to be robust and to have a great deal of work left in him; perhaps he would have taken another tilt at Mars as it moved toward a series of more favorable oppositions, but it was not to be. After falling down the stairs at All Souls College, he never regained consciousness; he died in April 1874.

During these years, British astronomers were by all odds the leaders in Martian research. Better known than either Phillips or Lockyer was the "eagle-eyed" Reverend William Rutter Dawes (1799–1868). Dawes—whose father was a mathematics teacher and onetime astronomer on an expedition to Botany Bay, Australia—studied medicine as a young man but abandoned the consulting room for a career in religion. He became a clergyman with a small independent congregation at Ormskirk, north of Liverpool, but ill health forced him to give it up; henceforth he devoted himself entirely to astronomy. (He was an exceptional observer, noted for the keenness of his sight, though for ordinary purposes he was so nearsighted that he is said to have passed his wife in the street without recognizing her.) He spent several years working at the private observatory of a wealthy wine merchant in London, until his second marriage, to an Ormskirk solicitor's widow, gave him the financial independence to follow his astronomical pursuits without distraction. From 1857 until his death in 1867, he operated a private observatory at Haddenham, Buckinghamshire, equipped with a 21-centimeter Alvan Clark refractor, and observed Mars extensively at the favorable oppositions of 1862 and 1864–65, at which he made 27 drawings (figure 5.2).[20] The astronomy writer Richard Anthony Proctor (1837–88) thought them the best anyone had made so far, and was especially intrigued by the intimation of finer features (perhaps the equivalent of Secchi's *screziato*): "There is an amount of detail in Mr. Dawes' views which renders them superior to any yet taken. I must confess I failed at a first view to see the full value of Mr. Dawes' tracings. Faint marks appeared, which I supposed to be merely intended to represent shadings scarcely seen. A more careful study shewed me that every mark is to be taken as the representative of what Mr. Dawes actually saw."[21]

Proctor would use these drawings, supplemented by other observers' drawings of features not well displayed at that opposition, as the basis of his 1867 chart using a stereographic projection of the planet (figure 5.3). Proctor's is memorable for having been the first chart to introduce a properly worked-

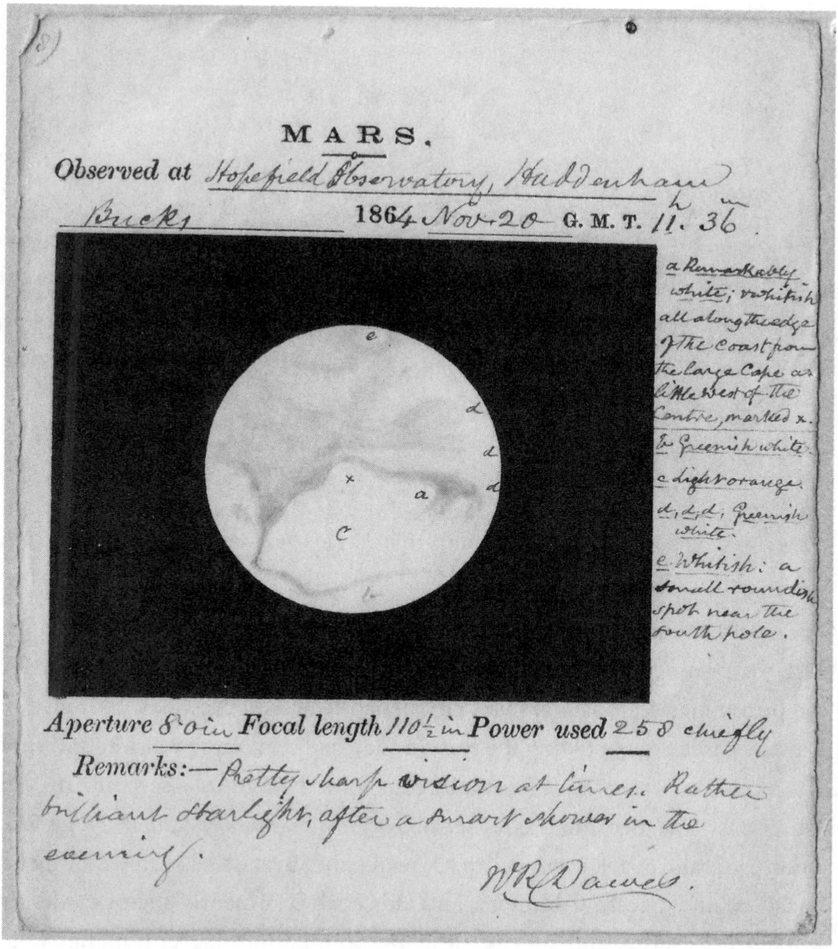

FIGURE 5.2 Mars, November 20, 1864, by the eagle-eyed Reverend William Rutter Dawes, using a 21-centimeter Alvan Clark refractor at his private observatory at Haddenham, Buckinghamshire. This is one of his original drawings and was recorded on an especially prepared observing card. Note the bifurcated feature on the right side of the drawing, which was later referred to as Dawes Forked Bay. South is at top. Courtesy of Richard McKim and the British Astronomical Association.

out system of nomenclature. In line with the widely shared belief at the time that Mars and the other planets were Earthlike and indeed inhabited worlds, Proctor referred to Martian "oceans," "seas," "bays," "straits," "continents," and "lands." In addition, he attempted to immortalize the names of astronomers

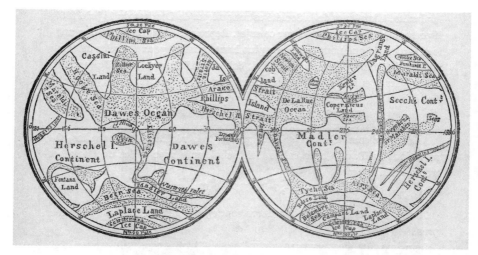

FIGURE 5.3 Richard Anthony Proctor's *Chart of Mars, from 27 drawings by Mr. Dawes*, 1867. Published in Camille Flammarion, *La Planète Mars*, vol. 1 (1892), figure 127.

(many still living) who had paid special attention to the planet by confer-ring their names on the features. The spot chosen by Beer and Mädler for the prime meridian of Mars was christened Dawes Forked Bay, in honor of Dawes's success in resolving the feature they had seen merely as a round spot into a more complex shape resembling the forked tongue of a serpent. The old "Hourglass Sea" became the Kaiser Sea. Other features included De La Rue Ocean, Maraldi Sea, Dawes Ocean, Cassini Land, Beer Sea, Mädler Continent, Lockyer Land, Secchi Continent, and Herschel Continent. Proctor evidently worked with haste, and so—as was soon pointed out—several names were repeats: Dawes appears six times (Dawes Ocean, Dawes Continent, Dawes Sea, Dawes Strait, Dawes Isle, and Dawes Bay), Beer twice (Beer Sea and Beer Bay), Lockyer twice (Lockyer Land and Lockyer Sea), and so on. This might have been overlooked but for the fact that the names of many other important observers of Mars were left out entirely. Also, Proctor had a clear penchant for the names of his own countrymen, and this alone sufficed to doom his system to significant modification within only a few years. When in 1876 Camille Flammarion published a new map, Proctor's basic scheme was retained but with more Continental astronomers' names included; even this, as we shall see, was not the final word.

Though leaving something to be desired, Proctor's chart was a significant development. For the first time a well-defined chart, with a convenient set of names useful in identifying and remembering specific landforms, had been introduced. Later, when the stereographic projection used in his 1867 map was replaced in 1869 and 1872 with versions using the Mercator projection, this too was significant. Historian K. Maria D. Lane notes that the Mercator projection had been "first conceived and used as a navigation map, and had become a standard world-map projection by the 19th century, even though it severely distorts upper latitudes and polar regions. By using this projection, Proctor visually prioritized the Martian equatorial regions and deepened the implication that Mars was essentially mappable, navigable, and controllable in the same way that the imperial powers had come to see the equatorial regions of the Earth."[22] The extension of the European empires' imperial sway, which was then playing out in deadly earnest in Africa, now reached even as far as Mars. The trope was commonplace even at the time. When, a few years later, Giovanni Virginio Schiaparelli (1835–1910), creator of what would become the definitive nomenclatural scheme of the planet, began the exploration of this "New World, this world of Mars . . . which we must conquer little by little," he expressed the hope that it would be "a less difficult and bloody conquest than the exploits of Cortés and Pizarro."[23]

6

The Moons of Mars

On September 5, 1877, Mars came to a perihelic opposition in the constellation Aquarius, approaching to within 56 million kilometers of Earth. Since the perihelic opposition of 1860, many new telescopes had been placed into service, including the 66-centimeter equatorial refractor of the U.S. Naval Observatory, fashioned by Alvan Clark. Asaph Hall (1829–1907), who was in charge of the telescope, set out to use it to make a discovery that Camille Flammarion would call "certainly one of the most curious and most interesting of modern times."[1]

Asaph Hall

Hall, born in Goshen, Connecticut, in 1829, came from an old, distinguished, but no longer prosperous New England family (figure 6.1). His grandfather, the first Asaph Hall, had been captain of the Connecticut militia and was present at the capture of Ticonderoga in the Revolutionary War. He built a large house and accumulated considerable property, so that Asaph II, his only child, born a few months after the elder's death, seemed destined for a life as a gentleman. Instead of allowing him to go off to Yale College as planned, however, his mother refused to be separated from him, and before he came of age he was already making wooden clocks, which he sold in the southern United States. While thus engaged in Georgia, he suddenly died, leaving the family in dire financial straits with an underwater mortgage on a farm. Thirteen-year-old Asaph III (the future astronomer), the eldest of six children, spent the next three years helping his mother, Hannah, try to rescue the farm by operating a commercial cheese factory. Failing, Hall, at 16, was forced to quit school and enter a career as a carpenter. At over six feet in height and possessed of a muscular physique, he was well suited to his new career. He spent three years as an apprentice, then another five as a journeyman carpenter.

After eight years of making doors, blinds, and window sashes, as well as hewing timbers for the frames of houses, Hall, at the unusually mature age of age 24 or 25,

FIGURE 6.1 Asaph Hall, photographed on July 5, 1871, six years before he discovered the satellites of Mars. Courtesy of Brenda Corbin and the U.S. Naval Observatory.

decided to further his formal education. Learning of an arrangement whereby students could defray the cost of tuition and room and board by manual labor, he entered Central College in McGrawville, New York, to begin the study of architecture. Central College had been founded by antislavery Baptists and was noteworthy for accepting "anybody, white or black, man or woman."[2] The college was poor, and there were only about 90 students at the time. "Of this number some were fanatics, many were idealists of exceptionally high character, and some were merely befriended by idealists, their chief virtue being a black skin. A motley group, who cared little for classical education, and everything for political and social reforms."[3] By 1860, weathered by scandal and facing bankruptcy, Central College was forced to close its doors. Nevertheless, it was a noble experiment, and it has been regarded as a forerunner of Cornell University.

Soon after Hall's arrival, he encountered Chloe Angeline Stickney, a petite and determined young woman from an old, large, and rather chaotic family from upstate New York, then completing her senior year. Brought up Baptist and deeply religious but also imbued with radical views, she was a fiery abolitionist and a suffragist who, wanting to liberate women from the oppression of having to wear heavy and constricting skirts, bravely wore the more healthy and comfortable divided garments for the lower body known as the "bloomer" costume.[4] She also taught a course in geometry in which Hall was a pupil, and won his respect. Not long after their meeting, her health broke down: "it was a case of sheer exhaustion, aggravated by a tremendous dose of medicine administered by a well-meaning friend."[5] Her condition remained rather frail for several years, and yet the attachment between her and her carpenter-pupil persisted. It seems, in part, to have been a matter of opposites attracting, and they wed in 1856. Despite their differences, their marriage proved highly successful.

Asaph had decided, meanwhile, to give up the idea of becoming an architect and instead aimed himself at a career as an astronomer. In pursuit of this dream, the couple moved soon after their wedding to the new observatory at Ann Arbor, Michigan, headed by the famous astronomer Franz Brünnow (1821–91), who had just emigrated from Germany, where he had been an assistant at the Berlin Observatory. Though lack of funds forced the Halls to leave after only three months, Asaph nevertheless made a favorable impression. But Angeline's health remained frail; she suffered from sporadic fits of paralysis. Touchingly, Hall sometimes carried her on his back across the fields hunting for wildflowers.

After a year working as schoolteachers in Shalersville, Ohio, during which Asaph decided he preferred carpentry to teaching, the couple resolved that Asaph should resume his studies of astronomy. They were, however, unsure where to go: Brünnow had invited him to return to Ann Arbor, while William Cranch Bond (1789–1859), then director of the Harvard College Observatory and in need of an assistant, encouraged him to come there. Setting out from Shalersville they made it as far as Cleveland before finding the way farther west (to Ann Arbor) barred by a storm on Lake Erie. Fate determined that they would go east, and so to Harvard they went.

While Angeline went back north to join her family for a time, Hall, with only $50 in his pocket saved from work as a carpenter, arrived in Cambridge, Massachusetts, in August 1857. The position he had accepted was not a very lucrative one. He recalled afterward that soon after his arrival, he had a "free talk" with George Phillips Bond (1825–65), the director's son and an astronomer in his own right, "who found out that I had a wife, $25 in cash, and a salary of $3 a week. He told me very frankly that he thought I had better quit astronomy, for he felt sure I would starve. I laughed at this, and told him my wife and I had made up our minds that we were used to sailing close to the wind, and felt sure that we would pull through."[6]

Pull through they did. After six months, Hall's pay was increased to $4 a week. Even better, early in 1858, he got extra work observing culminations of the Moon for Colonel Joseph E. Johnston's (1807–91) army engineers, being paid a dollar for each observation. His wife, like an alarm clock, would wake him out of a sound sleep and send him scurrying off to the observatory. During the month of March alone, he made 23 such observations. At this time, he also began computing almanacs, and he finally won the respect of the younger Bond for his willingness to work long hours for extra pay and to solve problems that Bond himself had been unable to solve. After William Bond's death in 1859, Hall's salary was raised to $400 a year. (It was hardly a lucrative sum, and worth only about $12,000 in today's money.)

By now, Angeline had joined him in Cambridge. Together, says Angelo Hall, their third son and Angeline's biographer, "husband and wife lived on much less than the average college student requires. She mended their old clothes again and again, turning the cloth."[7] The college librarian, "observing [Hall's] shabby overcoat and thin face, exclaimed, 'Young man, don't live on bread and milk!'"[8] In fact, he was not living on bread and milk; he was living on astronomy. A large part of his success was owing to his wife's willingness to sacrifice her own ambitions (as so commonly was the case in those days) and attend to domestic cares while he concentrated on astronomy.

From spring 1859 to the end of their stay in Cambridge three years later, the Halls occupied the little Bond cottage at the top of Observatory Hill and began to start a family. With the outbreak of the Civil War, Hall, like many other men of ability, decided to seek employment in Washington, D.C. The city was then in turmoil. The staff of the U.S. Naval Observatory (located at the time in the miasmic lowlands of Foggy Bottom on the Potomac) had been decimated, with some, including the observatory's first director, Commander Matthew Maury (1806–73), resigning to join the Confederacy. Others were ordered elsewhere by the federal government. Hall accepted a position as assistant astronomer, for which he received a small pay increase from what he had been making at Harvard, though it was rendered less valuable as a result of wartime inflation and the high prices in the city. Less than a month after his arrival, the Second Battle of Bull Run was fought. Hall could hear the roar of cannon and the rattle of musketry from the observatory, and was tasked with going to the battlefield to search for wounded friends. Angeline, who had remained in Cambridge, wrote to him expressing her worries. He reassured her that he was keeping about his business, which included "observing the planet Mars in the morning."[9]

Though from the beginning Hall had wanted to avoid establishing a home in Washington, and continued to regard New England as a more suitable place to live and raise a family, Angeline would not be separated from him. Two of her stepbrothers were wounded in the Second Battle of Bull Run; one of them, not mortally wounded but left in a damp stone church, caught a cold from which he died. A few weeks after this battle came the Battle of Antietam (September 17, 1862), and Hall, finding the climate enervating and exhausting himself with his constant exertions for wounded friends, suffered a breakdown in his health. He spent six weeks confined indoors with jaundice, and did not fully recover for two years. Fortunately, by this time Angeline's own health had improved, and she was able to take charge of the family affairs. Finally, in April 1863, there was an uptick in their fortunes. Upon the resignation of one of the U.S. Naval Observatory astronomers, Angeline wrote a letter without Asaph's knowledge in application for the position, and was successful. Now Hall entered in the service of the U.S. government, and he was henceforth a "professor."

The conditions in the city during the war were very difficult. Angelo Hall later recalled:

> The curse of war was upon the city. Crowded with sick and wounded soldiers, idle officers and immoral women, it was scourged by disease. Forty cases of small-pox were at one time reported within half a mile of the place where the Halls lived. But people had become so reckless as to attend a ball at a small-pox hospital. Most of the native population were Southern sympathizers, and some of the women were very bitter. They hated all Yankees—

people who had lived upon saw-dust, and who came to Washington to take the Government offices away from Southern gentlemen. As Union soldiers were carried, sick and wounded to the hospital, these women would laugh and jeer at them.[10]

They were also, of course, well aware of the fact that history was being made around them. One day Hall saw Ulysses S. Grant (1822–85)—"short, thin, and stoop-shouldered, dressed in his uniform, a slouch hat pulled over his brow"—as the latter was on his way to take command of the Grand Army of the Potomac. On another occasion, while Hall was in the observing tower, he heard a knock at the trapdoor. After leisurely completing his observation, he went to lift the door. He was surprised as "up through the floor the tall President [Abraham Lincoln] raised his head."[11] Lincoln had come unattended through the dark streets of Washington to look at the Moon through the observatory's 24-centimeter refractor, its then-largest telescope and the one on which Hall was assigned as assistant observer at the time.

The Quest for the Satellites

After the war, federal investment in science increased rapidly, and in 1870, the U.S. Congress authorized new construction at the U.S. Naval Observatory, which led three years later to the installation of the great 66-centimeter Clark equatorial refractor, then the largest in the world.[12] In 1875, Hall was placed in charge of it. He later recalled finding "in a drawer in the Eq[uatorial] room a lot of photographs of the planet Mars in 1875. From the handwriting of dates and notes probably [assistant astronomer Edward Singleton] Holden [1846–1914] directed the photographer, but whoever did the pointing of the telescope had . . . satellites under his eye."[13]

By the time he wrote this, Hall also had "satellites under his eye." The possibility of Martian moons had been entertained since the time of Galileo and Kepler. Most famously, Jonathan Swift, in his fictional *Gulliver's Travels*, had made the scientists of Laputa discover, "two lesser stars, or satellites, which revolve about Mars." U.S. Naval Observatory historian Steven J. Dick affirms, "One might have expected that the general background of speculation, and the approaching favorable opposition of Mars [in 1877], coupled with the new [66-centimeter] refractor and the need for an improved mass of Mars, would be all the incentive needed" to turn to the question of moons.[14] However, apparently it was some work Hall did on Saturn that played the decisive role, according to his own account:

> In December, 1876, while observing the satellites of Saturn I noticed a white spot on the ball of the planet, and the observations of this spot gave me the means of determining the time of the rotation of Saturn, or the length of Saturn's day, with considerable accuracy. This was a simple matter, but the resulting time of rotation was nearly a quarter of an hour different from what is generally given in our text books on astronomy: and the discordance, since the error was multiplied by the number of rotations and the ephemeris soon became utterly wrong, set before me in a clearer light than ever before the careless manner in which books are made, showed the necessity of consulting original papers, and made me ready to doubt the assertion one reads so often in the books, "Mars has no moon."[15]

He began to look seriously into the matter in spring 1877 and discovered, from Frederik Kaiser's summary of Martian observations in the *Annals of the Leiden Observatory* for 1872, that there had been very few previous searches for Martian satellites. William Herschel had been on the lookout in 1783, and possibly John Herschel in 1830, though easily the most thorough search had been undertaken by Heinrich Louis d'Arrest (1822–75) in 1862 and 1864, using the 28-centimeter refractor of the Østervold Observatory in Copenhagen. D'Arrest had been guided by rough calculations of the distance from the planet at which a satellite could exist without being wrenched away by the Sun into its own planetary orbit. D'Arrest found this limit to be 70' of arc from the planet at its greatest elongation, but on doing his own calculation, Hall discovered that d'Arrest had made an error; the limit was more like 30', with any Martian satellites likely to exist even closer in to the planet. Perhaps d'Arrest had failed to look close enough to the planet. With this encouragement, and "remembering the power and excellence of our glass," Hall wrote, "there seemed to be a little hope left." And so, in early August 1877, he began his search for satellites.[16]

The Discovery

Naturally, Hall wanted to work alone so as to receive full credit in the event of a discovery. By great good luck, Holden, his ambitious assistant, was invited by amateur astronomer Henry Draper to Dobbs Ferry, New York, "at the very nick of time,"[17] so Hall had the dome to himself. He began by scrutinizing faint stars at some distance from Mars, but each one soon dropped behind the planet, thereby proving itself to be an ordinary field star. He then pressed the search closer and closer inward, "within the glare of light" that surrounds Mars, using special observing techniques to reduce the glare, such as "sliding the eyepiece so as to keep the planet just outside the field of view, and then turning the eyepiece in order to pass completely around the planet." On the night of August 10, the first on which Hall began to examine the inner space near Mars, the seeing on the banks of the Potomac was horrible, and he gave up early. The next night he began by making routine measurements of the satellites of Saturn, followed by examining Mars. He recorded in the observing log of the Great Equatorial: "Seeing good for Mars. The edge of the white spot has two notches near the center of its outline. (A faint star near Mars)." (He would later add a note: "This proves to be satellite 1 . . . Deimos.")[18] There was scarcely time to secure the star's position before the Potomac fog began rolling in, and he shuttered the dome at half past two. The next few nights were overcast. On August 15, Hall slept at the observatory, waiting for the sky to clear; it did, at 11 o'clock, and Hall observed Saturn's ring and its satellites Titan and Iapetus before turning to Mars. Unfortunately, the atmosphere remained "in a very bad condition," and Hall left the dome tired and discouraged, returning to his residence in Georgetown. Angeline was well aware of his search, and she remained "full of enthusiasm. . . . Each night she sent her husband to the observatory supplied with a nourishing lunch, and each night she awaited developments with eager interest." According to her son and biographer, she more or less willed her husband to discover the satellites of Mars.[19]

The next night, August 16, Hall again recorded the "star near Mars," and he succeeded in making first a rough measurement and later a more precise series of chronometer measurements establishing that it was no mere field star—it was moving with the planet. So far he had only confided his findings to his family. However, he now showed the "star near Mars" to an assistant,

George Anderson (who was actually a laborer rather than an astronomer at the observatory). He strongly suspected it to be a satellite, but he asked Anderson to keep quiet about it until the matter was settled beyond doubt. Though Anderson kept quiet, Hall did not. In his own words, he "spilled the beans" to Simon Newcomb (1835–1909), the well-known mathematical astronomer who was soon to be appointed superintendent of the U.S. Nautical Almanac Office. On the basis of Hall's measures, Newcomb estimated the presumed satellite's orbital period and announced that it would be too close to Mars to be visible the following night, August 17. Hall's observing logbook for that night records a "Mars Star" that was a *"fixed star* and not the object observed last night." He expected the satellite (the outer one, Deimos) to emerge from behind the planet about four o'clock; it did. While he had been expecting its reappearance, he was in for a surprise—the inner satellite (Phobos). He closed the pages for August 17 with the remark: "Both the above objects faint but distinctly seen both by G. Anderson and [me]" (figure 6.2).

On the night of August 18, Hall and Anderson were joined by a veritable multitude, including Newcomb, David Peck Todd (1855–1939), and Wil-

FIGURE 6.2 Mars and its moons. Charge-coupled device image by Martin Stangl and Rolf Winkler, July 23, 2018, with a portable C14 at *f*/11 and a ZWO ASI 178m monochrome camera. At the time, the Great Dust Storm of 2018 was in its full fury, but because a near-infrared filter was used, the surface markings are faintly visible. South at top. Courtesy of Martin Stangl and Rolf Winkler.

liam Harkness (1839–1903). On August 18 seeing was as usual very bad, but Hall "saw the satellite immediately." Todd was also successful, adding: "'Halo' around the planet very bright, and the satellite was visible in this 'halo.'"[20] Only then did Hall announce the discovery.

Then the intrigue began. Two days after Hall's announcement, Newcomb alleged in an article in the *New York Tribune* that Hall had not completely grasped what he had found until Newcomb had worked out the period of revolution from preliminary observations. Meanwhile, Holden was also getting into the act, claiming on August 28 that using Henry Draper's 71-centimeter reflector at Dobbs Ferry, he had discovered a third satellite, followed after his return to Washington by a fourth. Hall was skeptical and wrote to Arthur Searle of Harvard College Observatory, "I think it will turn out that the Draper-Holden moon and the recent Holden moon do not exist."[21] Hall attempted to confirm these alleged discoveries with the great Washington refractor without success, and later computations showed that Holden's moon did not even obey Kepler's laws of motion. "Its existence was therefore a mathematical impossibility," Hall wrote to Edward C. Pickering of Harvard College Observatory, adding: "If I were to go through this experience again other people would verify their own moons."[22] Rumors of Holden's moons continued to circulate in the astronomical community for years, and Holden became known as the man "who had set all Washington astronomers laughing by detecting a . . . satellite of Mars with an impossible period and distance, and remaining deceived by it for months!"[23]

Holden, however, at least in Hall's view, had behaved more admirably than Newcomb. Though many years later Newcomb—by then the most honored American scientist of his day, and a founder and first president of the American Astronomical Society—attempted to apologize to Hall for what he now claimed was "entirely a misapprehension on my part,"[24] Hall never forgave him. As late as 1904 he was still writing bitterly to Boston astronomer Seth Carlo Chandler Jr.: "Newcomb was greatly excited over my discovery. Holden was away, and he and Draper made a blunder, and afterwards Holden behaved very well. Newcomb felt disappointed and sore, and something is to be allowed for human nature under such circumstances. He was always greedy for money and glory."[25] This assessment sounds harsh, but Hall was not the only one to regard Newcomb as possessed of a rather despicable streak. He had a reputation for denigrating the work of others and even ruining their careers, and it was not merely Hall's personal view that Newcomb was a man driven by greed. An exemplar of the Gilded Age, Newcomb considered himself to be an expert economist, and he wrote books advocating an unbridled capitalism that some termed (not to his face, of course) the "gospel of greed." Perhaps it should come as no surprise that this rather unpleasant man helped inspire the Moriarty character in Arthur Conan Doyle's Sherlock Holmes stories.[26]

Unfortunately, even this is not quite the final humiliating chapter of the story of attempts to rob Hall of some of the luster of his discovery. Todd, waiting until after Hall's death, would repeatedly assert, "Mine was the first eye that ever saw [the inner satellite] recognizing it as a satellite."[27] This claim, too, can be easily dismissed: the testimony of the observing logbook for that night is unequivocal and proves there is no truth whatever to his claim. Credit for the discovery of both satellites belongs to Hall and Hall alone.

Hurtling Moons

In the end, there were but two satellites, with Hall acclaimed as discoverer of both. (There are now, in addition, a plethora of artificial satellites—put in orbit by us—around the planet.) They were given the names Phobos ("Flight") and Deimos ("Fear"), for the horses of Mars mentioned in the 15th book of the *Iliad*, as suggested by Henry Madan (1838–1901), science master of Eton College.[28]

They were certainly strange little worlds, and more like those discovered by the Laputans than like any of the other satellites of the solar system known at the time. They are now believed to be either captured asteroids or chunks of Mars's surface ejected by a powerful impact early in Mars's history. They are certainly tiny, and they are located very close to Mars—thus accounting, at least in part, for d'Arrest's failure to find them.[29]

Phobos lies at a distance of 9,400 kilometers from the center of Mars, or only 6,000 kilometers from the surface of the planet. As seen from Phobos, Mars would be an astounding sight, with a disc subtending an angle of 43°—covering half the sky from horizon to zenith. It completes each revolution around Mars in 7 hours, 39 minutes, and 13.84 seconds, so that it completes three full revolutions in the time that Mars takes to rotate once. Because of its rapid motion, it rises in the west and sets in the east, and it remains above the horizon (for a viewer at the equator) for only 5½ hours at a time. It was aptly described, by Edgar Rice Burroughs, as a "hurtling moon of Barsoom" (Barsoom being Burroughs's fictional name for the planet).

Because Phobos's orbital inclination to the Martian equator is only 1.08°, and it lies so close to Mars, observers on the surface above 70° north and south latitude would never see it at all; it would never clear the horizon. For observers below these high latitudes, it would be eclipsed by Mars's shadow every time it passes around the planet (except during brief periods around the times of summer and winter solstice). Because of its rapid motion, there are 1,330 Phobos eclipses every Martian year. Deimos lies 23,500 kilometers from the center of Mars, and its orbit, too, is nearly equatorial (the inclination is 1.8°). The period of revolution is about 30 hours, and it undergoes eclipse once every revolution. However, it remains above the Martian horizon 60 hours at a time. It is never visible in the polar regions above 82° north or south, since it never clears the horizon from those latitudes.

In addition to eclipses in which the satellites pass into the shadow of Mars, they also pass in front of the Sun and then cast shadows on the Martian surface (these passes are called "transits" rather than "solar eclipses" because both are too small to ever completely cover the disc of the Sun). Because of their proximity to the planet and the near-equatorial planes of their orbits, together with the high axial tilt of Mars, successive eclipse series are centered on Mars's solstices, which average about 343 Earth days apart. Phobos's event series average 31 weeks in length; those of Deimos, 11 weeks. In between these event series are periods without events.[30]

The most important immediate result of the satellites' discovery was the calculation of a much more accurate value for the mass of Mars, 0.1076 times that of Earth (very close to the modern value, 0.1074). This prompted the French mathematical astronomer Urbain Jean Joseph Le Verrier (1811–77), who was responsible for the previous best estimate of Mars's mass based on its gravitational pull on the other planets, to hail Hall's discovery as "one of the most important discoveries of modern astronomy."[31]

It was certainly the hallmark of Hall's career.[32] He labored honorably for another 15 years at the quiet, systematic, routine but worthy kind of work that attracts little attention from the wider public, until, in 1891, he turned 62, the mandatory age of retirement. However, such was the regard in which he was held that with Mars about to come to another favorable opposition, the then superintendent of the U.S. Naval Observatory, Frederick V. McNair (1839–1900), granted him permission to use the 66-centimeter refractor to observe Mars and its satellites another year. He then moved back to Goshen, Connecticut, and remained there—except for the five years of 1896–1901, when he taught celestial mechanics at Harvard—until his death in 1907.

It is pleasant to note that Angeline, who died in 1892, has a share in Asaph's immortality. As a young woman she had worried that her ambition

> *To write [my] name*
> *In everlasting characters upon*
> *The gate of Fame's fair dome*

would interfere with her nobler wish to have it "writ in Heaven in the Lamb's book of Life."[33] The largest crater on Phobos has been named, in her honor, Stickney. Hall, too, has a crater, as do several characters from *Gulliver's Travels* (Gulliver himself, as well as Clustril, Drunlo, Flimnap, Grildrig, Limtoc, and Reldresal). And even Todd has a crater named for him, based on his false claim of being the first to see Phobos, though there is none for either Holden or Newcomb.

7

A Tale of Two Observers

Further Discoveries in 1877

The Martian satellites were the first sensational discoveries of 1877, during what would prove to be a singularly productive season in the history of Martian telescopic studies. In addition to the region of circum-Martian space, the planet's surface was subject to closer scrutiny than ever before (figure 7.1). Despite the planet's declination 12° south of the equator, most of these studies were carried out in the northern hemisphere, where the vast majority of observatories were stationed.

The practice of keeping Mars under constant surveillance for months before and after opposition was still a thing of the future. Most observers began to study the planet only weeks before the September 5 opposition date, doubtless in part because it then became conveniently accessible in the evening hours rather than requiring rising at an unseemly hour to catch it in the morning sky. The exception was the French artist-astronomer Étienne Léopold Trouvelot (1827–95), who has been one of a mere handful of exceptional artists to try their hand at drawing Mars.[1] A native of Guyancourt, in northeastern France, during a period of political instability associated with Louis-Napoléon Bonaparte's (1808–73) successful bid to regain power, Trouvelot emigrated to the United States in his early 20s in order to pursue a career

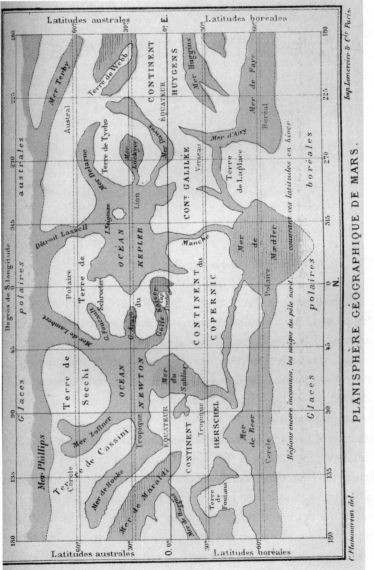

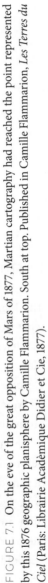

C. Flammarion del.

Imp.Lemercier & Cᵉ Paris.

PLANISPHÈRE GÉOGRAPHIQUE DE MARS.

FIGURE 7.1 On the eve of the great opposition of Mars of 1877, Martian cartography had reached the point represented by this 1876 geographic planisphere by Camille Flammarion. South at top. Published in Camille Flammarion, *Les Terres du Ciel* (Paris: Librairie Académique Didier et Cie, 1877).

as a scientific illustrator. He and his wife, Uranie, settled in the Boston suburb of Medford, Massachusetts, not far from the Harvard College Observatory. Perhaps inspired by the vivid auroral displays of the sunspot maximum of 1870, he acquired a 16-centimeter refractor with which he began to make a long series of exquisite solar and planetary drawings. He was never more active than at the opposition of Mars of 1877, and he began to observe the planet in April, long before anyone else; the next observer to record any observations, Camille Flammarion, did not do so until July 30. Even though the planet was still five months from opposition, and showed as a small gibbous disc only 8.8″ of arc across, there was enough detail to commence drawing. Having the planet to himself, Trouvelot, alone in all the world, witnessed and documented what would now be classified as a planet-encircling dust storm, the first ever recorded in the annals of the planet (figure 7.2).[2]

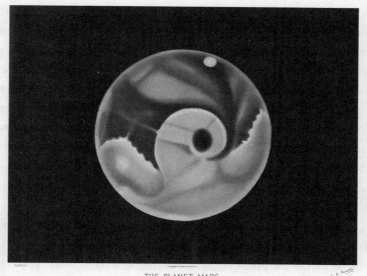

THE PLANET MARS.

FIGURE 7.2 Mars, on September 3, 1877, based on a sketch by Étienne Léopold Trouvelot, with a 16-centimeter refractor at his private observatory at Medford, Massachusetts. Dust from a planet-encircling dust storm that Trouvelot had documented in his observing logbook between May and August had by now largely cleared. Courtesy of David Wilson and the Museum of Jurassic Technology, Culver City, California.

Green

The dust storm largely obliterated familiar details from the first part of May until the first part of August. By then, almost everything had returned to normal (though Sinus Meridiani remained pale). As Trouvelot continued to diligently sketch the planet on every available night, he had competition from another exceptional artist, Nathaniel Green (1823–99). A professional artist and art teacher who had trained at the Royal Academy of Arts (then located in the east wing of the recently completed National Gallery at Trafalgar Square), Green by 1847 had established his home and studio at 39 Circus Road, St. John's Wood, London, then an area notable for its quiet repose and green spaces.[3] His chief interest was always landscape painting, in both oils and watercolors. After several years of painting terrestrial landscapes he discovered there were equally stunning landscapes to be depicted in the heavens. His first observations of Mars were made with a 23-centimeter reflector at the opposition of 1873, and were worked up into a rough map of the planet.

For the opposition of 1877, Green set off for the Portuguese island of Madeira, in the Atlantic some 900 kilometers from Morocco. It was a place that had played a prominent role in the history of the European discovery of the terrestrial globe. From here in 1497, Vasco da Gama (c. 1460s–1524) had set out on his famous voyage around Africa to reach India, and now from here in 1877, Green set out across the ocean of interplanetary space to reach Mars. His telescope was a 33-centimeter Newtonian reflector, employing what was then a rather newfangled silver-on-glass mirror;[4] the assembly included an open tube and alt-azimuth mount, which could be readily dismantled for transport up the island's steep roads.

By August 19, Green, assisted by Charles Blandy, who owned the largest stock of Madeira wines on the island, had established his observing post in the hills to the east of the island's capital, Funchal, and set to work. This was the first of 47 nights he would spend on the island, watching and drawing Mars whenever the seeing permitted; 26 of those nights were favorable for drawing, but only two were "superb." Of 41 exquisite pastels of the planet, 12 were lithographed and published in color by the Royal Astronomical Society, with Green himself doing the drawing on stone for the lithography (figure 7.3). In addition to recording the surface markings, Green noted brightenings at the

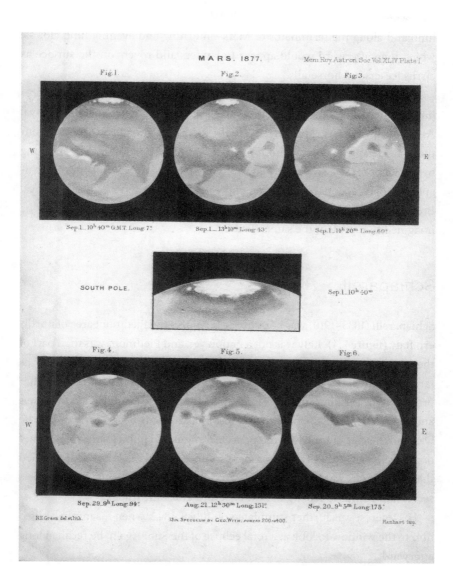

MARS. 1877.

Mem Roy Astron Soc Vol.XLIV Plate I

Fig.1.

Fig.2.

Fig.3.

W

E

Sep.1. 10ʰ 40ᵐ G.M.T. Long 7?

Sep.1. 13ʰ 10ᵐ Long 43?

Sep.1. 14ʰ 20ᵐ Long 60?

SOUTH POLE.

Sep.1. 10ʰ 40ᵐ

Fig.4.

Fig.5.

Fig.6.

W

E

Sep. 29. 9ʰ Long 94?

Aug. 21. 12ʰ 30ᵐ Long 131?

Sep. 20. 9ʰ 5ᵐ Long 175?

R.E.Green del et lith.

13ɪɴ. Sᴘᴇᴄᴜʟᴜᴍ ʙʏ Gᴇᴏ.Wɪᴛʜ. ᴘᴏᴡᴇʀs 200 ᴛᴏ 400.

Hanhart imp.

FIGURE 7.3 Mars from Madeira, as observed by Nathaniel Green. Though Green made 41 exquisite pastel drawings of the planet, which are preserved in the library of the Royal Astronomical Society at Burlington House, only 12, including these 6, were lithographed and published. The drawing in the center shows the Mountains of Mitchel. Green described the dark markings, widely believed at the time to be water, as various shades of greenish-gray, though professional artist that he was he realized that this tint was likely to be at least in part due to contrast with the orange. Published in *Memoirs of the Royal Astronomical Society* 44 (1879): plate 1. Courtesy of Richard McKim.

limb and along the terminator of Mars—morning and evening limb clouds, which long afterward would appear to landers and rovers on the surface as cirrus ice-crystal horsetails.

Green returned to England by the end of September, and he gave a preliminary account of his observations six weeks later at a meeting of the Royal Astronomical Society at Burlington House, Piccadilly. Though he planned eventually to complete a new map of the planet, he was a slow worker, and long before it appeared a rival map had been drawn up by Giovanni Virginio Schiaparelli, a professional astronomer of great international renown at the Royal Brera Observatory in Milan. Set side by side, the two maps could hardly be imagined to represent the same planet.

Schiaparelli

Schiaparelli (1835–1910) was born in Savigliano, in the Piedmont area of northern Italy (figure 7.4). Italy was not a nation yet, and Piedmont was then part of the Kingdom of Sardinia, under King Carlo Alberto of Savoy (1798–1849); as far back as the 17th century, Schiaparelli's family had operated kilns in which it produced for the kings of Savoy and others tiles, bricks, and terracotta vessels and figures. Young Schiaparelli's interest in astronomy began early. When he was four, his father took him out on a clear night and pointed out some meteors. "I asked what they were," Schiaparelli later recalled, "and my father replied that such things were only known to God Almighty. There thus arose in me a secret and confused feeling of immense and awesome things."[5]

His interest was further stimulated on July 8, 1842, when his mother called him to the window to look at a total eclipse of the Sun. Again, he recalled long afterward,

> In my school book I had read about the fact that sometimes the moon hides the sun and brings complete darkness in the middle of the day. Now I was indeed seeing the moon as a black disk covering the sun, surrounded by a beautiful halo. After following the various phases of the phenomenon, I wanted to keep memory of the event by drawing a color picture. And my amazement grew even more when I was told that men existed who could predict such phenomena to

the day, hour, and even the minute. It was at
that moment that the desire arose in me of
becoming one of those men daring to share
in the discovery of the laws governing the
universe.[6]

He received formal encouragement in the
achievement of his dream from a priest
of the church of Santa Maria della Pieve,
Paolo Dovo, who showed him the phases
of Venus, the moons of Jupiter, and the
rings of Saturn through a small telescope
set up in the campanile. He was already a
good scholar, and he built two sundials on
the south façade of the church (these still
exist). At age 15, and interested in mathematics, he was sent to the prestigious Royal
University of Turin. His four years were difficult. He afterward recalled, "At that time

FIGURE 7.4 Schiaparelli's best known
portrait, taken in 1895. Photograph by
G. B. Ganzini, Milan. Courtesy of William Sheehan.

there was no other way to study mathematics, at the University of Turin,
than to follow a career in engineering, so I resigned myself to becoming an
engineer."[7] Despite his rather poor family's willingness to draw down all their
meager resources to make it possible for him to remain at the university,
Schiaparelli suffered from conditions of severe deprivation. In a diary entry
he wrote just after Christmas 1855, he complained, "Now it's the cold, now
it's hunger, now the need to find something to provide solace for the mind—
which, however, I cannot do, because I haven't the means. . . . My relatives
will have reason to complain that they have spent a not inconsiderable part
of their fortune supporting me, to no purpose, since here I am unhappy and
unable to do anything to help myself. . . . In leading such a poor life, study
becomes impossible."[8]

Eventually he rallied, graduating with a degree in architecture and hydraulic engineering. By now, however, he was determined to pursue a career in
astronomy. His initial attempts were rebuffed, so he treaded water for a while
as a teacher of elementary mathematics—a position for which he seems

to have been ill suited, and which he resigned as soon as he could. Finally, Schiaparelli's mathematics teacher at the University of Turin, Quintino Sella (1827–84), succeeded in getting a fellowship for him from the Ministry of Public Education to continue his studies in Berlin, then a city of 450,000 inhabitants and the leading cultural center of Europe, where he learned practical astronomy under Johann Franz Encke (1791–1825) at the Berlin Observatory. He kept thorough diaries in Berlin, from which it appears that, besides studying science and mathematics, he picked up several languages; in addition to English, French, and German, he learned Russian, Greek, and Arabic. His main efforts at this time were devoted to calculating the orbits of comets and measuring and computing the orbits of double stars.[9] A few years later he continued his training at Pulkovo Observatory in Russia. He returned to Italy in 1859, just as the struggle for independence from Austria was entering a critical phase.[10]

In June 1860, a month after the sailing of the Thousand under Giuseppe Garibaldi (1807–82)—the turning point of the war—Schiaparelli accepted a position at the Brera Observatory in Milan. He was *secondo astronomo* to the director, Francesco Carlini (1783–1862), and when the latter died two years later, Schiaparelli at 27 became director of what was now the Royal Brera Observatory. The observatory had been founded a century before, in the old Brera Palace, originally a Jesuit college, by the Croatian priest-astronomer Roger Boscovich (1711–87). When Schiaparelli arrived, the instruments were woefully antiquated. Though Vittorio Emmanuele II (1820–78), who became the first king of a united Italy in March 1861, was utterly without intellectual interests—Lord Clarendon (1800–1870), the British secretary of state for foreign affairs, referred to him (privately) as an "imbecile"—he had some capable ministers in his cabinet. Sella, Schiaparelli's former mathematics teacher, was appointed minister of finance, and Schiaparelli was granted funds for a first-rate telescope, a 22-centimeter Merz refractor (figure 7.5). Unfortunately, delivery was long delayed, in large part because of the king's disorderly finances, and it was only installed on the roof of Brera in 1874. Schiaparelli, however, was hardly idle in the meantime, and among other achievements showed that the August meteors (the Perseids) followed the same orbit as the bright Swift-Tuttle Comet, which had passed near Earth in 1862.[11] Thus he had answered the very question he had long before put to his father.

FIGURE 7.5 A recent picture of the 22-centimeter refractor used by Schiaparelli, in its refurbished state in its dome on the roof of the Palazzo Brera. The observatory shares its history and its main building with the Accademia di Brera, a state-run tertiary academy of fine arts in Milan, and the Pinacoteca di Brera, Milan's main public museum for art. Courtesy of William Sheehan.

On the arrival of the new telescope, Schiaparelli devoted much effort to work he had learned at Berlin and Pulkovo: the measurement of close double stars. He was clearly a workaholic, and even after marrying Maria Comotti in 1865,[12] and becoming father to two sons and three daughters, he never relaxed his pace. "In my robust years, from 25 to 60," he said, "I usually worked ten hours a day. When I planned to observe I did not have dinner, but slept a while before going up to the dome as I felt it necessary to have a fresh mind and clear eyes, in order to make good observations."[13]

Though Schiaparelli is best known today for his studies of Mars, they were, in fact, something of an afterthought. Before August 1877, he had made only one sketch of Mars, in 1862, with one of the small, antiquated instruments at the observatory at the time. On August 13, 1877, after completing and entering his routine double-star measures for the night, he recorded at 3:30 a.m.: "Very handsome Zodiacal light rising in lenticular form directly toward the Pleiades. I see the gegenschein 15 to 18 degrees west of Mars."[14]

Ten days later, there was an eclipse of the Moon. During the early phases of the eclipse, Schiaparelli noted, "a horrible storm prevented us from observing double stars, and even after it ceased, the cold air, wind, and terrible images prevented us from working." The sky cleared. As the eclipse advanced, the Moon, turning blood red, stood high in the sky, with Mars nearby glowing like a red coal. Shortly before midnight, Schiaparelli swung the telescope from the Moon to Mars, "only to see," he recalled, "whether the Merz refractor which had given such good performances on double stars, possessed the necessary optical qualities to permit the study of the surfaces of the planets."[15] His first view was taken in poor air, and like most novice Martian observers, he found the detail confusing. "I must confess," he said, "that on comparing the aspects of the planet in view with recently published maps, this first attempt did not seem very encouraging." He continued:

> I had the misfortune of making my first observations on those parts of the surface of Mars that had ever been the most difficult and doubtful: the region designated in this memoir with the name of Mare Erythraeum, also that which, according to the diurnal rotation of the planet, immediately follows [it] onto the disk. At first, I didn't know how to orient myself at all. Only later, and then with difficulty, did I begin to recognize the forms on the planet which are shown in my drawings. But when I began to examine closely the very handsome sketches made by Professor Kaiser and Mr. Lockyer at the opposition of 1862, I found that the configurations they showed were almost identical to those in 1877, and in essential respects in agreement with my own. I was thus able to convince myself that . . . I saw the planet as others had seen it, that the apparent differences were due to the various ways observers have of representing things, and that on the whole much remained to be done on the topography of the planet, even with my limited means. I therefore resolved, on September 12, 1877, despite the

fact the opposition had already passed on September 5, to make observations whenever possible.[16]

So began Schiaparelli's great campaign on Mars. Painstakingly trained in precision astronomy at Berlin and Pulkovo, he resolved to do as no other student of the planet had ever done: rely exclusively on quantitative, geometric methods. His first step was to rigorously determine the direction of the axis of rotation and the position of the south polar cap. Instead of merely using dead reckoning with the eye to estimate the locations of the surface features, he employed a micrometer to measure the longitudes and latitudes of 62 fundamental points on the surface. These points formed a precise grid within which he could capture the rest of the details by sketching them in directly at the eyepiece. Though he started work after opposition, he found that the atmospheric conditions, indifferent during most of September, actually improved during the month of October, and despite the planet's increasing distance, his views of surface detail got better and better.

Careful astronomer that he was, he set forth at the outset the limits of his proposed investigation. He wrote, "I intended to proceed in the description of Mars not depicting its portrait as it appears to the eye, but following instead a geometrical scheme . . . in order to produce a map exactly as in our geographical maps, not to draw pictures imitating the appearance of the planet, but just to facilitate the description."[17]

His usual magnification in this work was 322×. By the end of September, he had laid down the fundamental points. Now he began the quest for a better nomenclature for the surface features of the planet (the earlier schemes proposed by Richard Anthony Proctor and Camille Flammarion no longer sufficed with all the new details that were being revealed). For convenience, he decided to adopt categories inspired by terrestrial features:

> One may wish to restrict oneself to what is seen without making any interpretations at all; but this is difficult in practice, since interpretations necessarily impose themselves upon the descriptions themselves. I have found, indeed, that in order to record the things seen at the telescope, I had to attach names to each of the variously shaded areas, lines, and points observed on the planet. In general the configurations present such a striking analogy to those of the terrestrial

map that it is doubtful whether any other class of names would have been preferable. And do not brevity and clarity compel us to make use of words such as island, isthmus, strait, channel, peninsula, cape, etc.? Each of which provides a description and notation of what could otherwise be expressed only by means of a lengthy paraphrase, and one, at that, which would need to be repeated each time one spoke about the corresponding. . . . We speak in a similar way of the maria of the Moon, knowing full well that they do not consist of liquid masses. If understood in this way, the names I have adopted can do no harm, nor will they interfere with the rigorous discussion of the facts.[18]

Specific names for the seas, lands, and more thus designated were taken from the *Odyssey*, from the legends of the Argonauts, from the Old Testament. After those began to run out, he drew on Herodotus's *Histories*. He chose better than he knew. Anyone who has had experience with nomenclatural issues in astronomy knows how problematic they can be, and how great an achievement it is to produce a workable—and acceptable—scheme. In proposing what Schiaparelli called a "chimera of euphonic names, whose sounds awaken in the mind so many beautiful memories," he had unwittingly set about wrapping the Martian surface in ancient dreams and nostalgic yearnings that have proved ever since to be siren calls stirring the spirit to adventure.[19] (Additional information about the nomenclature of Mars is found in appendix E.)

On the same map on which appeared this romantic and evocative nomenclature, Schiaparelli also introduced a set of new Martian markings. Hints of them had appeared on the drawings of Pietro Angelo Secchi and William Rutter Dawes, but they were not very well defined. To Schiaparelli's eye, they appeared as sinuous streaky features, originating at the margins of the "seas" and winding across the "continents." Recalling the term Secchi had first introduced for a few of the features, Schiaparelli called them *canali* (Italian for grooves, streams, channels, canals, and so forth). Indeed, the two broadest and most prominent of Secchi's *canali* from 1858 were still obvious, though somewhat attenuated, in 1877; Schiaparelli called them Nilosyrtis and Hydaspes. Such features were very obvious. Of them, Schiaparelli would later affirm, "It is [as] impossible to doubt their existence as that of the Rhine on the surface of the Earth."[20]

For that matter, as we now know, Mars is a windswept planet, with variable streaks and regional tonal boundaries in many places exactly where Schiaparelli and others mapped *canali*. The heated controversy that developed over these strange features during the following three or four decades was not, however, so much about these obvious features but about finer, much more regular, filaments—veritable spiderwebs covering the disc, later alleged by some to be nothing less than the works of a highly intelligent race of Martians. There was at least some basis for these filaments in the filigree of fine dust, and their forms and locations varied with the Martian winds.

Schiaparelli himself dramatically described the process by which he came to recognize the *canali*:

> In most cases [a] presence . . . is first detected in a very vague and indeterminate manner, as a light shading which extends over the surface. This state of affairs is hard to describe exactly, because we are concerned with the limit between visibility and invisibility. Sometimes it seems that the shadings are mere reinforcements of the reddish color which dominates the continents—reinforcements which are at first of low intensity. . . . At other times, the appearance may be more that of a grey, shaded band. . . . It was in one or the other of these indeterminate forms that, in 1877, I began to recognize the existence of the [*canali* called] Phison (October 4), Ambrosia (September 22), Cyclops (September 15), Enostos (October 20) and many more.[21]

On October 2, on a disc that with the usual power of 322× appeared four times as large as the Moon with the naked eye, Schiaparelli experienced "superlative" seeing and detected a tiny "inland lake" to which he conferred the name Juventae Fons, the Fountain of Youth. Its apparent diameter would have been less than 1 second of arc at the time. This gives some idea of the resolution Schiaparelli was getting. (We now know that, pointlike though Juventae Fons appeared in Schiaparelli's telescope, it is actually an enormous box canyon, 250 kilometers by 100 kilometers on Mars.)

The Case of Schiaparelli versus Green

Schiaparelli continued sketching Mars until March 1878, by which time the disc had dwindled to only 6 seconds of arc. He was a rapid and tireless worker, and within two months, in May 1878, he had published his map along with his great memoir on the opposition. Green did not finish his map until December 1878, and published it only in October 1879 (figures 7.6 and 7.7). Schiaparelli's map was therefore seen and its details assimilated by astronomers long before Green's even appeared.

These two maps furnish a remarkable case of what has been called the "personal equation": the individual bias of one observer compared to another in measurements and observations. Green was a portrait artist with a brain acutely sensitive to nuanced forms and color. He depicted a delicately tinted, almost dreamy landscape.[22] Schiaparelli was a mathematically trained draftsman whose instruments were not the crayon or pastel but the rule, the compass, and the draftsman's pencil. Building up his image from points determined with micrometric precision, he formed a graticule onto which spots and lines were entered so as to convey in exact terms all relative positions, angles, and distances.

Pondering the two maps several years later, and not without a hint of bafflement, the Reverend Thomas William Webb (1807–85), vicar of Hardwick and author of the well-known *Celestial Objects for Common Telescopes*, wrote:

At first sight there is more apparent difference in their results than might have been expected. It is not surprising that in the case of minute details each should have caught something peculiarly his own; but there is a general want of resemblance not easily explained, till, on careful comparison, we find that much may be due to the different mode of viewing the same objects, to the different training of the observers, and to the different principles on which the delineation was undertaken. Green, an accomplished master of form and colour, has given a portraiture, the resemblance of which as a whole, commends itself to every eye familiar with the original. The Italian professor, on the other hand, inconvenienced by colour-blindness, but of micrometric vision, commenced by actual measurement of sixty-two fundamental points, and carrying on his work with most commendable pertinacity, has plotted a sharply-outlined chart, which, whatever may be its fidelity, no one would at first imagine to be intended as a representation of Mars. His style is as unpleasantly conventional as that of Green indicates the pencil of the artist; the one has produced a picture, the other a plan. The discordance arising from such opposite modes of treatment would naturally be less than apparent; still, a good deal remains that is not easy to harmonise.[23]

The gauntlet had been thrown down, and two schools of Mars observers—Schiaparelli on the one side, Green on the other—would vie with each other like the realists and nominalists of old. At stake was nothing less than the scientific possession of another world.

The Great Debate

The next Martian opposition took place on November 12, 1879. Mars, approaching only within 73 million kilometers, was not quite as close as in 1877 but was more favorably placed for northern hemisphere observers. Among nonastronomers captivated by the planet's unusual prominence in the night sky was the American poet Walt Whitman, who in Saint Louis, Missouri, watched it rise over the Mississippi, and wrote in his diary: "All along, these nights, nothing can exceed the calm, fierce, golden, glistening domination of Mars over all the stars in the sky."[24]

Schiaparelli, finding the air at Milan unusually calm and transparent, enjoyed excellent images. Green was also active, though observing not at Madeira but at St. John's Wood, London, where he found that the quality of the atmosphere left something to be desired, complaining to his Italian rival that "the definition afforded . . . has barely sufficed to identify the details of the Madeira observations."[25]

At this opposition, Schiaparelli introduced a number of innovative observational techniques. He illuminated the telescope field to suppress the effects of contrast between the bright planet and the surrounding sky, and by keeping his eye at the eyepiece no longer than necessary to obtain the best views, he managed to avoid eye fatigue and thus could work effectively for several hours when the atmospheric conditions were very good. He also used a yellow filter in front of the eyepiece to improve the contrast of the shadings against the yellowish disc.[26] By such means, he added significant results, including micrometric determinations of 114 points on the surface, which provided the basis of a new map. One of his measured points was a small whitish patch

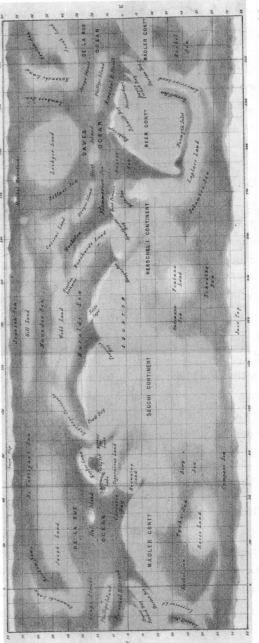
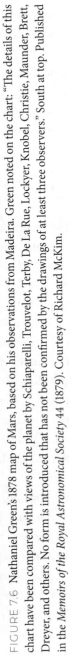

FIGURE 7.6 Nathaniel Green's 1878 map of Mars, based on his observations from Madeira. Green noted on the chart: "The details of this chart have been compared with views of the planet by Schiaparelli, Trouvelot, Terby, De La Rue, Lockyer, Knobel, Christie, Maunder, Brett, Dreyer, and others. No form is introduced that has not been confirmed by the drawings of at least three observers." South at top. Published in the *Memoirs of the Royal Astronomical Society* 44 (1879). Courtesy of Richard McKim.

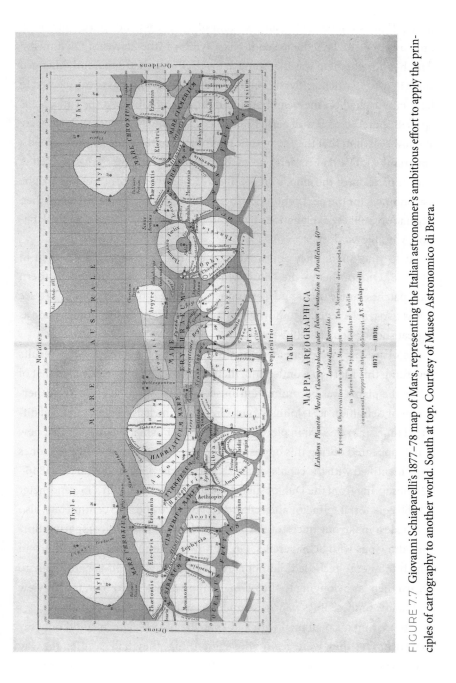

FIGURE 7.7 Giovanni Schiaparelli's 1877–78 map of Mars, representing the Italian astronomer's ambitious effort to apply the principles of cartography to another world. South at top. Courtesy of Museo Astronomico di Brera.

in the Tharsis region, only half a second of arc across and noted only once, on November 10, 1879; he called it Nix Olympica (the Snows of Olympus).[27] Another development was that some of the dark areas had apparently changed since 1877: in particular, Syrtis Major (formerly the Hourglass Sea) seemed to have encroached on the neighboring bright area of Libya. At the time, Schiaparelli shared the view of earlier observers that the dark areas of Mars were shallow seas, and that they sometimes inundated parts of the adjoining lands; apparently, Libya had undergone such an inundation. The changing Etch A Sketch patterns of the albedo (reflective) features would continue to intrigue astronomers for the next century, as they considered first inundation and then vegetation as likely explanations before finally hitting on the correct explanation: windblown dust.

As for the *canali*, they appeared as finer and more regular lines than in 1877–78. One of them—the Nilus, which tracked between the far northern dark patches Lunae Lacus and Ceraunius—appeared strangely double. "To see it was two tracks regular, uniform in appearance, and exactly parallel came as a great shock," Schiaparelli wrote.[28]

As observers elsewhere attempted to absorb Schiaparelli's latest results regarding this Martian wonderland, their reaction generally resembled that of Alice: the planet was becoming "curiouser and curiouser." Despite the generally poor definition at St. John's Wood, Green acknowledged that "faint and diffused tones may be seen in places where Professor Schiaparelli states that new canals appeared during this opposition."[29] Were these faint and diffused tones to be identified with the canals? An Irish astronomer, Charles E. Burton (1846–82), observing from the Dublin suburb of Loughlinstown with a 15-centimeter Grubb equatorial refractor and reflectors of both 20 and 30.5 centimeters, also claimed to make out traces of the Schiaparellian canals. On comparing Burton's drawings with Schiaparelli's, Green however found that the positions in which they were recorded did not agree, and concluded, "It is hardly safe to regard them as belonging to the permanent markings." He reiterated his view that "some of these lines may be boundaries of faint tones of shade."[30]

Comparing his own map with Schiaparelli's of 1877, Green could only say that, if in fact they were seeing the same thing, they had "a different way of expressing it."[31] Webb imputed this difference partly to Schiaparelli's color blindness, which made him—as Schiaparelli himself confessed to his closest astronomical friend, François Joseph Charles Terby (1846–1911), a professor of physics and chemistry at the Collège communal in Louvain, Belgium—"only slightly sensitive to the nuances of color."[32] The same weakness for nuances of color arguably made him more sensitive to subtle contrasts, perhaps helping explain why, as Green noted disapprovingly, Schiaparelli had "turned soft and indefinite pieces of shadings into clear, sharp lines."[33]

Schiaparelli was well aware of such criticisms. In early 1879, he confided to Otto von Struve (1819–1905) at the Pulkovo Observatory: "I am prepared to await calmly and confidently the corroborations of other observers. . . . I don't understand why Mr. Green should worry so much. His drawings made at Madeira are excellent, and accord with mine as far as one could expect of a reflector. . . . All that he saw, I saw also: he should be pleased, as I am, with the wonderful agreement, instead of talking about optical illusions!"[34] A year later he told Terby: "Rather than oppose himself to me, [Green] would have done better to suspend his judgment. . . . Happily, it is not possible that the lens of Milan is the only one in the world capable of showing so many of these details."[35]

The debate would rage on, inconclusively, for many years. About one thing Schiaparelli would prove correct: other instruments would in due course show many of the disputed details. But even before the first set of findings had been confirmed, a new and even more startling phase of the canal controversy was getting underway, with Schiaparelli again serving as agent provocateur. At the opposition of December 1881—an unfavorable one, with the planet's apparent diameter never larger than 15.5″—Schiaparelli, as Flammarion put it, "continued his series of astonishing observations, and this time passed from marvel to marvel."[36] His observations began in October. While the months of October and November were not favorable, during the period after the opposition, from December 26, 1881, to February 13, 1882, the seeing at Milan, always best in the winter months, improved, and Schiaparelli enjoyed 50 good nights, some of which were "superb." What Schiaparelli called geminations—apparent doublings of the canali—had gone from a single case in 1879 to an apparently widespread phenomenon, affecting many features and giving to the planet a strange, even bizarre, appearance. In his memoir on his Mars observations of 1881–82, Schiaparelli wrote:

On the planet, crossing the continents, there are a large number of dark lines to which I have given the name *canali* [channels] though we do not know what they are. . . . Their positions appear invariable and permanent, at least insofar as I can judge. . . . However, their aspect and their conspicuousness are not always the same. . . . Sometimes the channels show up as shaded vague lines, while at other times they are as clear and precise as pen-strokes. . . . Each channel terminates, at either end, in a sea or in another channel; there is not a single example of a channel stopping short of a land-mass.

This is not all. At certain seasons these channels are doubled, or, more accurately, duplicate themselves.

This phenomenon seems to come at a definite time, and to occur almost simultaneously over the whole extent of the continents of the planet. No indication of it was seen in 1877, during the weeks before and after the southern solstice on Mars. A single, isolated case was noted in 1879. . . . I was able to see the same phenomenon [involving this channel] on January 11, 1882, a month after the spring equinox. . . . At this same date . . . another doubling had already occurred. . . . After January 19 I had surprise after surprise; in succession the Orontes, Euphrates, Phison, Ganges, and the majority of the other channels showed up as clear and incontestable doubles. . . .

These are the observed facts. . . . It is premature to make conjectures about the nature of these channels. With regard to their existence, I need hardly say that I have taken all possible precautions to avoid all chance of illusion; I am absolutely certain of what was observed.[37]

Though Schiaparelli always insisted on the reality of the phenomenon, contending that such possible causes as eye fatigue or squinting had been ruled out, the geminations do look rather suspiciously like examples of what ophthalmologists call monocular diplopia or double vision, a condition that can be caused in one or both eyes by retinal damage, cataracts, strokes, or other diseases. It is at least suspicious that in winter 1882—just as the geminations began to appear in such perfusion and Schiaparelli's drawings went from merely strange to bizarre—Schiaparelli was deeply involved in a long series of daylight observations of Mercury, the innermost planet to the Sun.[38] He would continue to observe Mercury intensively for the next seven years, before coming to the realization, as he told Terby, that these daytime observations were dangerous, harming his eyesight and even affecting his Mars observations.[39]

What Larger Telescopes Showed

Meanwhile, Schiaparelli was eagerly awaiting the delivery of a new and more powerful telescope. Aware that the 22-centimeter refractor at his disposal was increasingly falling behind the instruments at observatories elsewhere, he took steps as early as 1878 to obtain a larger telescope. He put the case directly to the Chamber of Deputies in Rome. One of the deputies argued ironically that the province of Milan, which had been exempted from taxation on the grounds that two-thirds of its agricultural population was so poor the government had to supplement its diet with maize, could not afford such an extravagance. Nevertheless, the chamber voted on the question. The secret ballot was overwhelmingly in favor, according to the decree, "because there was an astronomer who was worth it." The telescope, a 49-centimeter Merz-Repsold refractor, was installed only in 1886, and at the opposition of that year Schiaparelli first trained it on Mars. His results were not dramatically better than those he had achieved with the smaller Merz. However, at the next opposition, the telescope proved its value. Thus, he wrote to Terby on June 7, 1888:

> I believe that I saw the planet well enough on May 9, 25, and 27, and I began to be almost satisfied, having confirmed at least three or four geminations. But I had a happy surprise on June 2 and 4; and only then did I have any idea of the power of a 49-centimeter aperture for Mars! I then saw that the unforgettable days of 1879–80 and 1882 had come back for the first time, and that I could again see those prodigious images presented in the telescope field as an engraving on steel; again there was all the magic of the details, and my only regret was that the size of the disc was only 12" of arc. Not only could I confirm the gemination of the Nepenthes (*quantum mutatus ab illa!*) and the reappearance of the Triton of 1877, but I could again see Lacus Moeris, reduced to a very small point, but sometimes perfectly visible and scarcely separated from the Syrtis Major.[40]

However, the views through the larger telescope did not entirely clarify the picture. At times, everything seemed more confusing than it had been before. In the same letter to Terby, Schiaparelli described his great difficulty in making out the exact nature of the markings in the region of Boreo-Syrtis (as he called the continuation of the then-prominent northward extension of the pigtail tip of Syrtis Major). He wrote: "What strange confusion! What can all this mean? Evidently the planet has some fixed geographical details, similar to those of the Earth. . . . There comes a certain moment, and all this disappears, to be replaced by grotesque polygonations and geminations which, evidently, seem to attach themselves to represent apparently the previous state, but it is a gross mask, and I say almost ridiculous."

These tantalizing glimpses of a possibly quite different Martian reality beneath the mask of illusion were not to be sustained, for unfortunately, the opposition of 1888 was the last one at which Schiaparelli enjoyed truly satisfactory views of Mars. Mars dropped far south of the equator for the opposition of 1890 (its declination was −23°11′) and even farther south (−24°) for the perihelic one of 1892, and so it had to be viewed through the murk and turbulence of an atmosphere increasingly compromised by the growth and industrialization of the city of Milan in all directions around the observatory. Schiaparelli sadly noted the change. "Smoke from the burning of coal makes the atmosphere increasingly opaque," he said, "and the excessive use of electric lights takes away much of the darkness of the night."[41] At the same time he was aware that his eyesight was deteriorating. In 1890, 1892, and 1894, he confided to Terby, "I have had hardly any success [with Mars] at all . . . To tell you the truth, I am afraid it may be the result of an unfortunate decrease in my sensitivity to weak illuminations the cause of which I assign to the observations of Mercury near the Sun. . . . I have now abandoned this kind of dangerous observations completely."[42] His right eye had always been defective in some respects, but his left eye had been perfect; now the latter's field of vision was becoming darker, and the images deformed. For all intents and purposes, Schiaparelli's career as a Mars observer ended in 1890. Thereafter the surveillance of the planet he had maintained for so many years had to be given over to fresher talents.

In addition to his memoirs of the oppositions of 1877–78 and 1879–80, Schiaparelli published memoirs (and maps) of the oppositions of 1884, 1886, 1888, and 1890.[43] These monographs are rather Baconian in method; they are rich in data, but they are rather dry and contain few interpretations. Only after further observing had become futile did he allow himself to speculate freely on what the phenomena he had so carefully documented might mean. Thus, in February 1893, in a four-part essay titled "Il Pianeta Marte," published in the journal *Natura ed Arte*, he wrote regarding the *canali*: "Their singular aspect, and their being drawn with absolute geometrical precision, as if they were the work of rule or compass, has led some to see in them the work of intelligent beings, inhabitants of the planet. I am very careful not to combat this supposition, which includes nothing impossible."[44] To say that it was not impossible was not the same as to say that it was the most probable, and he hastened to add—taking a page from his friend Sella, who, apart from his role as finance minister, had long been a keen student of crystallography—there are many perfectly regular and geometric structures in nature that require no such means of explanation:

> The perfect spheroids of the heavenly bodies and the ring of Saturn were not constructed in a turning lathe, and not with compasses has Iris described within the clouds her beautiful and regular arch. And what shall we say of the infinite variety of those exquisite and regular polyhedrons in which the world of crystals is so rich! In the organic world, also, is not that geometry most wonderful which presides over the distribution of the foliage upon certain plants, which orders the nearly symmetrical, starlike figures of the followers of the field, as well as of the animals of the sea, and which produces in the shell such an exquisite conical spiral, that excels the most beautiful masterpieces of gothic architecture? In all these objects the geometrical form is the simple and necessary consequence of the principles and laws which govern the physical and physiological world.[45]

Pending, he said, the intervention of what Galileo once referred to as "the courtesy of Nature" (*la courtesia della natura*), when an unexpected ray of light illuminates a truth that had previously seemed inaccessible to the investigator, he saw no recourse at the moment other than to "hope and study."[46]

This was hardly his last word on Mars.[47] He continued to offer up, sometimes facetiously, various ideas about the planet until the year before his death. A particularly noteworthy effort, published in November 1895, was "La vita sul pianeta Marte" (Life on the Planet Mars). By then Percival Lowell's theory that the *canali* were actual artificial canals, built by intelligent beings in order to irrigate their drought-stricken planet, was attracting a great deal of attention (see chapter 9). Schiaparelli quipped that "Mars must certainly be a plumbers' paradise!" He included some other speculations, and even utilized his long-ago training as a hydraulic engineer to work out a system of locks and dikes for controlling the flow of water in the canals of Mars. He left no doubt that the whole exercise was intended mischievously, for he ended with the comment, "As for me, I am descended from the hippogriff."[48]

His observational work on double stars ended in 1899, and the following year he reached the mandatory retirement age of 65. But he was not the kind of man to rest on his laurels. "Doing nothing has always been a torment to me," he later wrote. "Until my sixty-fifth year my main amusement was taking long walks. Later these diminished, and now when I'm in town I don't go out anymore. The noise of the city hurts me and the streets of Milan have become impractical for me for this reason."[49]

Certainly, the grueling 10-hour workdays had to be given up, and he now spent three months every summer in a large Napoleonic-era villa at Sorino di Monticello (now Monticello Brianza) that his wife, who had died in 1893, had inherited from her father. It still exists, and enjoys a magnificent prospect, looking out to the south and east on the Lombardy Plain and the hills of Montevecchia, toward the north and west to the undulating Lambro Valley, the pre-alpine summits of Valsassina, and the eastern branch of Lake Como.[50] There, the greatest Martian of the age suffered a stroke on June 22, 1910, from which he died on July 4, 1910. Like Mark Twain, he had entered the world with Halley's Comet, and had left on its following return.

8

Mars in the Gilded Age

Prelude to the Mars Furor

Though Giovanni Virginio Schiaparelli's attempts to study Mars were defeated in later years by deteriorating eyesight, industrial development in Milan, and a series of far southerly oppositions, his systematic observations of changes on the planet had dominated an era. Meanwhile, his reports of the "canals" had so greatly increased interest in the planet that the number of serious observers would grow from a mere handful in 1877 to at least several score in 1890. This increase was not a little abetted by organizations such as the Société Astronomique de France (SAF), founded by Camille Flammarion in 1887, and the British Astronomical Association (BAA), established by Edward W. Maunder (1851–1928) in 1890.

Apart from Schiaparelli's own observations, a turning point in the history of the canal controversy was in much-vaunted apparent confirmations by the French astronomers Henri Perrotin (1845–1904) and his assistant Louis Thol-lon (1829–87), using the great 76-centimeter refractor of the Nice Observatory in 1886. Henceforth Schiaparelli's maps—both their nomenclature and canals—would acquire almost unquestioned authority, and many imitators. Also, Perrotin and Thollon verified Schiaparelli's earlier report of changes in Libya (the brightish region tucked into the notch between the southeastern side of Syrtis Major and Mare Tyrrhenum). Schiaparelli had earlier witnessed a partial inundation of the area by the encroaching sea; the French astronomers found subsequently that the entire area was apparently submerged. Schiaparelli exclaimed, "The planet is not a desert of arid rocks. It lives; the development of its life is revealed by a whole system of very complicated transformations, of which some cover areas extreme enough to be visible to the inhabitants of the Earth."[1]

Historian K. Maria D. Lane points out in *Geographies of Mars* that most astronomers of this era apparently concluded that Schiaparelli had indeed discovered something truly new and important on the surface of Mars, while his maps essentially touched off a race to discover who could make out the most canals.[2] Indeed, even Nathaniel Green himself, who took on a leading role in the BAA and served as its president in 1897, two years before his death, apparently acquiesced to the reality of the canals, though he still insisted that the term "channel" better described them.[3]

Another perihelic opposition—the next after the Grand Opposition of 1877—took place in 1892. Again the planet would approach within the 56-million kilometer mark, though as opposition occurred on August 4 rather than September 5, Mars was inconveniently farther south of the equator (24° versus 12°). This, naturally, put northern hemisphere observatories at a great disadvantage. Typical of their results were those obtained at Lick Observatory, on Mount Hamilton in California, whose 91-centimeter refractor, installed in 1888, was the largest in the world. The director, Edward Singleton Holden (1846–1914), and staff astronomers Edward Emerson Barnard (1857–1923) and John Martin Schaeberle (1853–1924), enjoyed limited success despite the power of their instrument because of Mars's low elevation. Holden wrote: "Most of the drawings . . . have been made with a magnifying power of 350 diameters; a few with the power of 260. . . . How unfavorable the circumstances have been can be estimated when it is remembered that powers of 1000 and even more have been employed on Jupiter and Saturn with good result."[4]

Naturally, better results were expected from the handful of observers based in the southern hemisphere. By far the most anticipated were made by William Henry Pickering (1858–1939) of the Harvard College Observatory (and the younger brother of the director, Edward C. Pickering, 1846–1919), who was then in charge of the Harvard College Observatory's Boyden Station at

Arequipa, Peru (elevation 2,469 meters). Though Pickering had been sent to the southern hemisphere to push forward with the observatory's main line of work, photographing the southern stars, he and his assistant, Andrew Ellicott Douglass (1867–1962), a recent graduate of Trinity College in Connecticut, had other ideas. They spent most of their time making visual studies of the Moon and planets with a 30-centimeter Clark refractor.

It is hard to blame them. In August 1892 Mars was a stunning sight indeed from Arequipa, culminating almost at the zenith, and in the superb mountain seeing Pickering made out extraordinary amounts of detail, as he described in a series of telegrams to the *New York Herald*, for which he was a paid correspondent. One telegram reads: "Mars has two mountain ranges near the south pole. Melted snow has collected between them before flowing northward. In the equatorial mountain range, to the north of the gray regions, snow fell on the two summits on August 5 and melted on August 7."[5] According to another: "Prof. Pickering . . . says that he has discovered forty small lakes in Mars." Holden, from Lick, where the results had been so sharply disappointing, asked almost plaintively, "How does he know the dark markings are lakes? Why does he not simply call them dark spots?"[6] Edward Pickering wrote to his brother, acknowledging that the cables "have given you a colossal newspaper reputation," but adding an admonishment: "In my own case I should have restricted myself more distinctly to the facts."[7] Edward's concerns that William's sensationalist reports would bring discredit to the Harvard College Observatory— together with the latter's utter neglect of what was seen as his highest priority, photography of the southern stars, and his utter inability to stay decently within his budget—led to William's recall from Peru at the beginning of 1893.[8]

Lowell

On his return, William stopped in Chile in April 1893 to observe a solar eclipse with amateur meteorologist Abbott Lawrence Rotch (1861–1912). He was no sooner back at Harvard College Observatory than he immediately began devising a new Mars expedition. This time, he planned to remain closer to home, taking advantage of what he perceived to be the favorable climate of the Arizona Territory (where the U.S. Weather Bureau had indicated desert conditions that might rival those in Peru) to set up a Mars observing station there. He had just begun making inquiries for funding and support of the project in January 1894, when he was approached by fellow Bostonian and budding Mars enthusiast Percival Lowell (1855–1916). At age 39, Lowell was undertaking a remarkable case of personal reinvention (figure 8.1).

A fascinating figure from a fascinating family, Lowell has been more thoroughly biographed than any other astronomer of his time, so it is necessary here to give only a little flavor of the man.[9] The founding of an observatory—like any good deed (or any crime, for that matter)—requires both means and motive. Lowell had both. His means came from the fact that he was the elder scion of the senior and most affluent branch of the leading capitalist family in New England at the time (its fortune largely based in textiles). His motive was his eagerness, as Lick Observatory astronomer William Wallace Campbell (1862–1938) would put it, to "take the most popular side of the most popular scientific question about . . . the discovery of intelligent life on Mars."[10]

Percival's father, Augustus (1800–1900), an aristocrat to the core, had been first treasurer of the Merrimack Manufacturing Company, Boott Cotton Mills, and more. He was a "monarch of mercantilism" who considered himself to be "the last of the Union Whigs"; in other words, "of that species easily swayed by big-business and social connections with the great planters of the Old South . . . an advocate of hard money and laissez-faire capitalism and [favoring] the time-tested Federalist principles associated with strong, centralized government."[11] He was a "Cotton Whig" rather than a "Conscience Whig," and his self-interest made him quite happy with the Compromise of 1850, which had for the time being slowed the abolitionist movement and facilitated the continuation of the booming business of weaving into saleable cloth the cotton picked by slaves in the South. During his reign the family prosperity reached its peak. Percival's mother, Katherine Bigelow Lawrence (1832–96)—whose father, Abbott Lawrence (1792–1855), was the business partner of Augustus's father and had served as ambassador to the Court of St. James's—was a woman of talent who, however, languished, as many women of society did at the time, as an invalid. Though Percival clashed with his strong-willed father, he was devoted to his mother (and she to him; in her letters, she always began with "My Darling Percy"). One of his sisters, Katherine (who became Mrs. Alfred Roosevelt) recalled that Percival was "peculiarly fond of his mother . . .

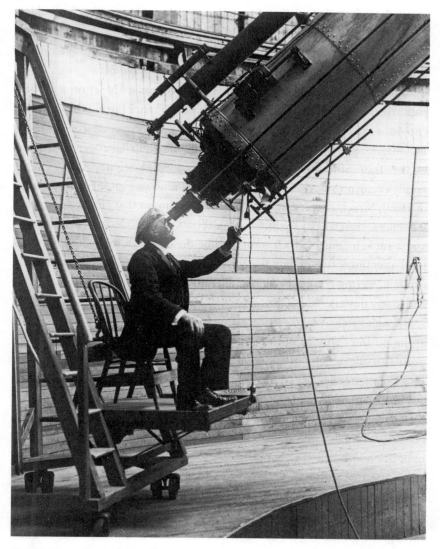

FIGURE 8.1 Percival Lowell in an iconic image showing him peering not at Mars but at Venus during the daytime on October 17, 1914. The photograph was taken by Philip Fox of Northwestern Observatory during a visit to Flagstaff. Courtesy of Lauren Amundson and Lowell Observatory Archives.

his tender solicitude for her was unfailing. . . . We used to say that he could do anything with her, and she spoiled him."[12] His other siblings included Abbott Lawrence, a future president of Harvard; Amy, a cigar-smoking avant-garde poet; and Elizabeth, who married highly successful financiers.[13]

Percival, possessed of considerable precocity, was early set on a brilliant path. He learned French at a boarding school in France during the Civil War (Augustus had taken Percival's mother to France to recuperate from a malady first diagnosed as "nervous exhaustion," and thought to have been occasioned by the strain of too many childbirths spaced too closely together). On the family's return to Boston, Augustus acquired an estate on the western slope of Heath Street, Brookline, on which stood a house that "in its magisterial whiteness, looked like a wedding cake, set back of a wooded park, with stables and a sunken garden which soon became the object of Augustus's horticultural attentions. Before long the acreage was covered . . . with beautiful and exotic flowering shrubs, brought from all parts of the world, and many rare and lovely flowers."[14]

Though Percival would share an interest in horticulture, and as an exercise at the Boston Latin School would give an impressive performance by composing an ode in Latin hexameters on the wreck of a toy boat (perhaps in the Frog Pond), he soon found himself in the grasp of an even keener interest. At age 15, he received a gift from his mother (which made it all the more special), a 6-centimeter refractor mounted on a pillar-and-claw stand that he set up on the flat roof of the wedding-cake mansion and used to observe Mars at its opposition in 1870.

Needless to say, he went to all the best schools: the Boston Latin School, Noble's Classical School (now Noble and Greenough), and finally Harvard (class of 1876). His Harvard career was singularly spectacular. Adhering to then-president Charles W. Eliot's (1834–1926) ideal of a harmonious and balanced education by taking an equal number of courses in sciences and humanities, he excelled in all of them, turning in superior performances in physics, math, English composition, and history. Presaging future preoccupations, at his graduation in 1876, he gave a short talk on the nebular hypothesis, based largely on Richard A. Proctor's *Saturn and His System* (1865). His mathematics professor, Benjamin Peirce (1809–80), called him the most gifted mathematician of all those to have come under his observation, while, according to

Lowell family biographer Ferris Greenslet, his cousin James Russell Lowell, professor of languages at Harvard, was even more laudatory: "Not always enthusiastic about the promise of young kinsmen, . . . the Professor referred to him in conversation as the most brilliant young man in Boston."[15]

After Percival's graduation, the expectation was that he would take up an investment-centered life in Boston. He resisted, instead embarking on a series of long moratoria on commitments and final decisions that included an eight-month grand tour of Europe as far as Syria, and he narrowly avoiding getting himself enlisted in a war between Turkey and Serbia. Toward its conclusion, he wrote, on January 16, 1877, to Harvard chum Barrett Wendell (1855–1921), "I hope that Chastity Hall is, as in old times, the scene of many a good time and that you like my old qualities. . . . [As for myself,] I have . . . become decidedly misanthropic and, with the exception of a few friends, should not feel many pangs at migrating to another planet—or ceasing to exist—were either plan practicable."[16]

There is no record that Percival paid any particular attention to the great opposition of Mars in 1877; he was probably more concerned with the financial crash of that year. For the next several years, he disappeared into the affairs of business, serving in the family's State Street office and taking roles as treasurer (executive head) of the Merrimack Manufacturing Company and Boott Cotton Mills, trustee of bleachery and electric companies, and so forth. Despite his apparent conformity to the expected role, beneath the surface all was not well. At 27, he did something shocking: he broke off an engagement to Rose Lee (1860–1953), daughter of investment banker George Cabot Lee (1830–1910) (and sister to the wife of up-and-coming New York assemblyman Theodore Roosevelt, 1858–1919), and resigned from the family business. The family correspondence contains scarcely a mention of this earth-shattering sequence of events. For a long time, it was completely (and effectively) covered up. One of the rare instances in which it is mentioned is in a letter from Percival to his sister Elizabeth (1862–1935), then in Europe with their parents: "I am so blue myself that I would I could speak with some of you. It grieves me to have brought so much sadness to all of you. But I have had it too. . . . Everyone seems to have been happy while I alone was wretched."[17] Though Augustus in particular was livid, reacting to what he perceived as his son's rebellion by exerting an emasculating surveillance over Percival's financial affairs until well past his 30th birthday, he still bestowed on him (as he did on all his children) a sum of $100,000—the equivalent of millions today—so that Percival was not without means to pursue whatever course he wished. (And Percival was a shrewd-enough businessman to enlarge it with mostly conservative investments, so that by 1900, the year of his father's death, his fortune had grown to $500,000, more than $15 million today.)

From the standpoint of Brahmin society, Percival had committed virtual social suicide, something of the attitude of elite Boston society being expressed in a letter that Harvard president Eliot wrote a decade later to Edward C. Pickering: "Mr. Percival Lowell is undoubtedly an intensely egoistic and unreasonable person. . . . Fortunately he is generally regarded in Boston among his contemporaries as a man without good judgment. So strong was this feeling a few years ago that it was really impossible for him to live in Boston with any comfort."[18] Percival seems to have sensed this himself, but rather than remain in Boston moping, he took his page from Shakespeare's Coriolanus: ". . . I turn my back; / There is a world elsewhere."[19]

The world elsewhere was not yet quite as exotic as Mars, but it was exotic enough. Lowell was lured to the Far East, a part of the world little known in the West, by a series of lectures at the Lowell Institute (founded by another relative). Given by zoologist Edward Sylvester Morse (1838–1925), the lectures were devoted to the culture and traditional ways of life of feudal Japan, which were fast vanishing with the modernization of the Mejii Restoration. Lowell spent much of the next ten years in the Far East and wrote four books that were generally highly regarded when they first appeared (though over time, their Western chauvinism has lessened their reputation): *Chosön: The Land of the Morning Calm* (1885), *The Soul of the Far East* (1888), *Noto* (1891), and *Occult Japan* (1894). At last, as the novelty of the Far East began to wear off, Lowell searched for fresh stimulation. He would exchange the Far East for the Far Out, and find it in Mars.

Though generally his sudden career change took friends and family by surprise, there had been at least a few premonitory rumblings. He had talked with William H. Pickering about the planet in Cambridge even before the latter set out for Arequipa, and had mooted the possibility of a visit. While waiting to sail out from San Francisco to Yokohama Bay on his last journey to the Far East in December 1892, he had gone to the trouble of looking up noted planetary observer Edward Emerson Barnard of the Lick Observatory, and had left his forwarding address in Tokyo.[20] He had also taken with him a 15-centimeter Clark refractor, which he set up in the garden of his 18-room rented house and used to observe Saturn. Nevertheless, as late as October 1893, he was apparently still undecided about his next move, and was tentatively discussing with his old Harvard chum Ralph Curtis (1854–1922) the possibility of spending the following Easter in Seville.

What happened between then and the end of January—and the event that sealed his fate—seems to have been a Christmas gift from his cousin Mary Traill Spence Lowell Putnam (1810–98), sister of the poet James Russell Lowell. It was none other than volume 1 of Camille Flammarion's *La Planète Mars*, the French astronomer's monumental compilation of all Mars observations made up to that time. Flammarion was an enthusiast, and nothing captivated him more than the idea that Mars and the other planets might be living worlds. As evidence, he included lengthy summaries of Schiaparelli's Mars memoirs up through that of 1888. He also included many of Schiaparelli's drawings and maps—so that Lowell was, like so many others, biased at the outset in favor of a Schiaparellian view of the planet. This was to be decisive, for it seems that a drawing or painting (of a planet or anything else) owes as much to other drawings or paintings as it does to direct observation. According to the art historian E. H. Gombrich (1909–2001), "The familiar will always remain the likely starting point for the rendering of the unfamiliar; and existing representation will always exert its spell over the artist even while he strives to record the truth."[21]

Lowell seems to have read Flammarion at white heat. His imagination was fired, and probably through Rotch, a cousin, he arranged to get together with William H. Pickering at Harvard College Observatory before the end of January 1894. With astounding speed the two men formulated a plan in which they would pool their resources. Pickering would provide the professional expertise, and Lowell the financing, for an expedition (at first expected to be temporary) to the Arizona Territory for the express purpose of observing Mars at its October 20, 1894, opposition, when it would be some five degrees north of the celestial equator and more favorably placed for observation than it would be for another 15 years. For telescopes, Pickering used his contacts to borrow a 46-centimeter Brashear refractor from Pittsburgh optician John

A. Brashear (1830–1920) and a 30-centimeter Clark from Harvard. Pickering, who had considerable talent for mechanical engineering, also designed a pre-fab dome that could be shipped west by rail in time for the opposition.

As a next practical step, Pickering's Arequipa assistant, Andrew Ellicott Douglass, was sent to the Arizona Territory with the same 6-centimeter refractor Lowell had used in Tokyo, ostensibly to test the seeing conditions there. What followed in six weeks of March and early April, as solar astronomer and Douglass scholar John A. Eddy (1931–2009) has noted, "was a comic opera of a one-man, whirlwind site survey—made by rail and horse-drawn wagon with a small telescope and a large stack of Western Union Telegraph blanks, the latter to keep his anxious employer advised, in real time, of nightly measures of the skies."[22]

Douglass tested 10 sites in all, beginning in the southern part of the state, in Tombstone and Tucson, he worked his way north to Tempe, Phoenix, and Prescott. Finally, in early April, he reached Flagstaff, a site that had not even been on his original list but was added at the end of March. The place was then a rather rough-and-tumble town of several hundred whose main industries were lumber, livestock (mostly sheep), and a railroad station (located on the main Santa Fe Railroad line to California). Douglass found the seeing conditions good in the few observations he made in an "opening in the woods" on the mesa west of town that was afterward to be known as Mars Hill. The elevation, 2,134 meters, appealed to Lowell, who, chomping at the bit to get started, with the opposition only six months away, telegraphed his decision to Douglass on April 16, 1894: Flagstaff.

Things continued to move with lightning speed. Just over a month later, on May 22, Lowell, still in Boston, gave a talk to the Boston Scientific Society about the planned expedition. Regarding the thin lines first observed by Schiaparelli, he suggested "the most self-evident explanation . . . is probably the true one; namely, that in them we are looking upon the result of the work of some sort of intelligent beings." He also explained his broader purpose in establishing an observatory:

> This can be put popularly as an investigation into the condition of life on other worlds, including last but not least, their habitability by beings like [or] unlike man. . . . If the nebular hypothesis is correct . . . then to develop life more or

less distinctly resembling our own must be the destiny of every member of the solar family which is not prevented by purely physical conditions, size and so forth, doing so.[23]

The program was set.

Mars at Last!

On May 28, 1894, Lowell arrived by train in Flagstaff, and immediately wrote to his mother: "Here on the day. Telescope ready for use tonight for its Arizonian virgin view. . . . Today has been cloudy but now shows signs of a beautiful night and so, not to bed, but to post and then to gaze."[24] At first the astronomers took quarters in the Bank Hotel, just across from the rail station. However, on that first night an eager Lowell and Pickering roughed it by camping in the still-uncanvassed dome to be ready for "early-rising Mars"—still a very long way from Earth, as opposition would not take place until October 20. Clouds moved in, and it began to rain, but on May 31, the sky proved more cooperative, and the 30-centimeter Clark was used for a first view of the planet. The next night, Lowell entered a sketch with the 46-centimeter Brashear into the observing log with the note: "Southern Sea at end first and Hourglass Sea . . . about equally intense. . . . Terminator shaded, limb sharp and mist-covered forked-bay vanishes like river in desert."[25]

In terms of practical experience in planetary astronomy, Lowell was very much a beginner, though on the morning of June 7, 1894, only a week after commencing work, he seemed to enjoy an unexpectedly intimate Martian revelation, later set down in some of his most evocative prose:

On that morning, at about a quarter of six . . . as I was watching the planet, I saw suddenly two points like stars flash out in the midst of the polar cap. . . . The seeing at the time was very good. It is at once evident what the other-world apparitions were,—not the fabled signal-lights of Martian folk, but the glint of ice-slopes flashing for a moment earthward as the rotation of the planet had turned the slope to the proper angle. . . . It had taken them nine minutes to make the journey; nine minutes before they reached Earth they had ceased to be on Mars, and, after their travel of one hundred millions of miles, found to note them but one watcher, alone on a hill-top with the dawn.[26]

Most likely all he saw was a brighter patch of the polar cap set quivering in the fluctuating seeing. But though Lowell's inference regarding what he had seen is doubtful, the passage conveys something of the excitement of a great adventure.

With this possible exception, Lowell did not enjoy immediate success. His logbook shows that the canals he "glimpsed" were rather broad streaky features, like Nilosyrtis, which extended from the tip of Syrtis Major like a pig-tail, or the Lethes. On June 9 he described the latter as "like an old friend now." It seemed to be "very broad and glimpsed double," though he hastened to add: "These sudden revelation peeps may or may not be the truth." On June 19, one entry reads: "With the best will in the world I can certainly see no canals." Another, 15 minutes later, reports, "the canals came out distinctly and four good drawings were made of them." Clearly, the canals (as Lowell called them without apology or reservation) were, apart from those of the thick and stumpy variety, proving more elusive than expected. But repetition of fleeting impressions produced confidence, and in any case, he could always eke out the slender data with his imagination. In a never-published poem, begun perhaps even before he arrived in Flagstaff and set eye to telescope at all, he wrote:

> One voyage there is I fain would take
> While yet a man in mortal make;
> Voyage beyond the compassed bound
> Of our own Earth's returning round . . .
> My far-off goal seems strangely near,
> Luring imagination on,
> Beckoning body to be gone
> To ruddy-earthed, blue-oceaned Mars.[27]

The Theory

This passage makes clear that Lowell at first still embraced the Schiaparellian notion that the blue- or greenish-gray areas were bodies of water, actual seas.

He still subscribed to that notion when, at the beginning of July, he returned to Boston. It was there, at Sevenels, and not in Flagstaff, where he conceived his theory of intelligent life on Mars. A critical underpinning was the discovery by his colleagues, who remained glued to the telescope during his absence, that the dark areas could not be seas after all. Analyzing light from the dark areas on Mars with an Arago polariscope, Pickering found that it was not polarized as it ought to have been if reflected from a watery surface, while Douglass had discovered canals crisscrossing the dark areas. *Presto!* The canals must be artificial waterways built by a civilization of intelligent Martians to irrigate their evolutionarily advanced and progressively dried-up rind of a planet. The historian William Graves Hoyt notes that though Lowell "had already arrived at some positive and quite sensational conclusions about life on Mars in particular and extraterrestrial life in general before he ever looked through a telescope from Flagstaff... on his own testimony at least, he did not formalize his thinking into what he considered to be a full-blown scientific theory until late in July of 1894."[28]

"Full-blown" might be a bit of an exaggeration, for Lowell's theory was largely an exercise in speculation, and straightforward enough for a child to grasp. Even Lowell could later sketch it in a short paragraph to Douglass, as follows:

> Roughly speaking the evidence seems to be that Mars has (1) some but not much atmosphere; (2) is an aged world with no water to speak of except what makes the polar caps; (3) is provided with an elaborate system of line markings which are best explained by artificial construction. . . . (4) shows what seems to be artificially produced oases as the termini of the canals—what we see and call canals being merely strips of vegetation watered by the canals, the canals themselves being too narrow to be seen.[29]

In a nutshell, this was the position Lowell was to spend the remaining 22 years of his life tirelessly bolstering and popularizing to a fascinated and generally sympathetic public, while defending it against a growing onslaught of attacks from the professional scientific community.

Apart from briefly returning to Flagstaff at the end of August, Lowell spent most of the summer in Boston. He did not return for a serious stint of observing activity until early October, as opposition date approached (and Mars became accessible in the earlier and more convenient evening hours). His continuing observations consolidated his earlier views, and also revealed changes in the intensity and color of the dark areas that seemed to be synchronized with the melting down of the polar cap.

He noted on resuming his observations in mid-October that the blue-green areas were fading. The first affected had been far-southerly dark regions adjacent to the melting polar cap (which was surrounded by a prominent dark band Lowell attributed to release of moisture from the cap, giving rise to temporary lakes and marshes; it is still called the Lowell band). Not only were the dark areas affected but so were the canals, which appeared to be strengthened in intensity as a "wave of color" (later as a "wave of darkening") swept down the disc (figure 8.2).[30] Lowell would devote much effort to the elucidation of this phenomenon, qualitatively in 1894 and quantitatively at succeeding oppositions when its rate would be estimated as approximately that of a leisurely walk, three kilometers per hour. By the end of October, the fading had reached Mare Cimmerium and Mare Sirenum; by November, Mare Erythraeum. All this seemed to fit with Lowell's theory, assuming that the fading of the dark areas was due to the autumnal yellowing (dying) of Martian vegetation. As Lowell later wrote, "We see that if the blue-green tint were due to vegetation, that blue-green color should have been most pronounced at the earlier date, since then the vegetation would have been most luxuriant, and should then have changed to ochre as the crops got into their sere and yellow leaf. And these tints were the very ones observed, and observed in this very order."[31] In addition, the darkening of the canals also fit with Lowell's theory, suggesting to Lowell that "water must almost certainly be the *deus ex machina* in the matter, and, secondly, that the phenomenon must be due not to its direct transference but to its indirect effects in causing vegetation to sprout."[32] In other words, the canals behaved just as would be expected if they really were irrigation channels, used by the Martians on their drought-stricken planet to pump water from the melting polar caps to the deserts equatorward. Even the geminations could be fitted to the theory, for whenever the flow of water exceeded the capacity of one canal, the Martians could be presumed to open the locks to permit diversion of flow into a second, parallel canal.

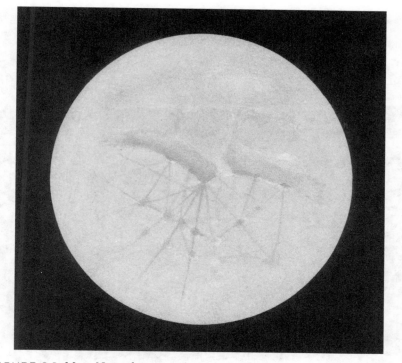

FIGURE 8.2 Mars, November 1894, showing the network of canals and oases that Lowell would later describe as a "mesh of lines and dots like a lady's veil." This is the original watercolor of the figure Lowell used as the frontispiece for his book *Mars* (Boston: Houghton Mifflin, 1895). Courtesy of Lauren Amundson and Lowell Observatory Archives.

Needless to say, this was only one possible explanation for a phenomenon so far studied only during a single Martian apparition (as the old saying goes, one swallow does not a summer make, and neither does one wave of color a theory prove). Even as Lowell and his colleagues were testing the limits of perception from Flagstaff, a much more experienced and skillful observer, Edward Emerson Barnard, was observing from Lick Observatory on Mount Hamilton, near San Jose, California (figure 8.3). As early as 1880, when Barnard was a 23-year-old amateur astronomer in Nashville, Tennessee, he had produced the manuscript of a book on Mars and its recently discovered moons in which he emphasized that there are always multiple possible explanations for any phenomenon that is indistinctly seen. In this never-published text, he wrote:

FIGURE 8.3 Edward Emerson Barnard, posing for this "selfie" at the eyepiece end of the 91-centimeter Clark refractor at Lick Observatory. At the time he posed for this photograph in September 1892, he had just discovered the fifth satellite of Jupiter. In contrast to Lowell, he was no dandy, as can be deduced from the stain on his left pant leg. Soon after this photograph was taken, Lowell met Barnard in San Francisco, before shipping out to Yokohama Bay on his last Far East trip. Lick Observatory Records. Courtesy of Special Collections, University Library, University of California, Santa Cruz.

It is well to fetter the wings of our fancy and restrain its flights. It is quite possible we may have formed entirely erroneous ideas of what we actually see. The greenish gray patches [on Mars] may not be seas at all, nor the ruddy continents, solid land. Neither may the obscuring patches be clouds of vapor. Man is too quick at forming conclusions. Let him but indistinctly see a thing, or even be undecided as to whether he does actually see it and he will then and there set himself to theorizing, and build immense castles of conjecture on a foundation, of whose existence he is by no means certain.[33]

Now, 14 years later, he commanded the world's largest refractor, the 91-centimeter Clark refractor, and trained it on Mars every Friday night (when he was assigned to use the telescope), often doing so at the very same moments that Lowell and the other Flagstaff observers were doing so a thousand kilometers to the southeast.

Barnard had begun scrutinizing the still-tiny disc as far back as May 21, 1894, when he recorded a dusky patch in the midst of the south polar cap before it began its rapid melting. By the end of July and throughout August he had produced an excellent series of drawings of the progressive changes in the south polar cap, and he was beginning to enjoy breathtakingly detailed views of the planet's surface. However, only in early September—during the dry late summer conditions on Mount Hamilton, when the grass turns yellow-brown and the seeing is often exquisite—did he experience what might be described as a true revelation of that other world. He was using magnifications of more than 1000× on Mars (compared to only 310× by the Flagstaff observers), and he kept the planet under scrutiny from sunset until dawn. Instead of drawing the planet at the customary scale of four or five centimeters to the diameter of the planet, as Lowell, Pickering, and Douglass did, he found this scale to be too small for the quantity of details he was seeing, and he was now using discs of 13 centimeters to the diameter of the planet.

On September 2–3, 1894, with the Solis Lacus region in view, he entered into his logbook several remarkable sketches as well as the following note: "There is a vast amount of detail. . . . I however have failed to see any of Schiaparelli's canals as straight narrow lines. In the regions of some of the canals near Lacus Solis there are details—some of a streaky nature but they are broad, diffused and irregular and under the best conditions could never be taken for the so-called canals."[34] A week later, with Syrtis Major coming onto the disc and the planet on the meridian shortly after sunrise, he wrote to Simon Newcomb at the Nautical Almanac Office in Washington, D.C.:

> I have been watching and drawing the surface of Mars. It is wonderfully full of detail. There is certainly no question about there being mountains and large greatly elevated plateaus. To save my soul I can't believe in the canals as Schiaparelli draws them. I see details where some of his canals are, but they are not straight lines at all. When best seen these details are very irregular and broken up—that is, some of the regions of his canals; I verily believe—for all the verification—that the canals as depicted by Schiaparelli are a fallacy and that they will be so proved before many oppositions are past.[35]

Rather more detail was visible in the "seas" than in the "continental regions," which Barnard would compare to the mountainous country around Mount Hamilton. "I can imagine," he wrote, "that, as viewed from a very great elevation, this region, broken by canyon and slope and ridge, would look just like the surface of these Martian 'seas.' During these observations the impression seemed to force itself upon me that I was actually looking down from a great altitude upon just such a surface as that in which our observatory was placed."[36]

In October, Barnard witnessed something that Lowell, away in Boston, had missed. A large regional dust storm was rising to obscure this part of the planet. It began about midmonth in either Libya or Northern Ausonia, covered Mare Cimmerium, spread through the southern desert of Ausonia to Phaethontis and part of Hellas, and at last extended thinly over Mare Sirenum, Mare Australe, and Aonius Sinus.[37] These, of course, were regions most affected by the fading of the dark areas noted by Lowell. However, as Barnard's observations make clear, these changes were owing not to the dying of vegetation but to the obliteration of a whole hemisphere by the ubiquitous dust (figure 8.4). (Such dust storms are now, of course, well known, but Lowell—though eventually conceding the existence of yellow dust clouds now and then—would never acknowledge their occurrence on such a grand scale.) Unfortunately, Barnard, who was a conservative when it came to the interpretation of his observations and in any case was busy with other work including the wide-angle photography of the Milky Way, did not publish his

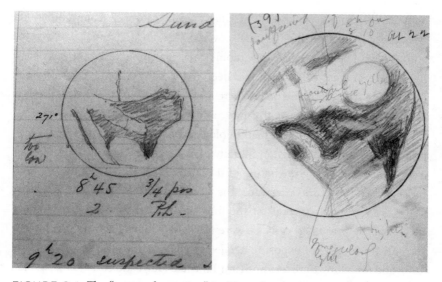

FIGURE 8.4 The "personal equation" in Mars observations: nearly simultaneous sketches of Mars made by two famous observers. The anomalous darkening extending from the westerly end of Mare Cimmerium marks the boundary of a large regional dust storm, and is shown as a streak by Barnard and as a "canal" by Lowell. *Left*: Lowell used the 46-centimeter Brashear refractor at Flagstaff on October 21, 1894, to make this sketch. Courtesy of Lauren Amundson and Lowell Observatory Archives. *Right*: At almost the exact same moment, Barnard used the 91-centimeter Clark refractor at Lick Observatory to make his sketch. Lick Observatory Records. Courtesy of Special Collections, University Library, University of California, Santa Cruz.

Mars observations at the time, allowing Lowell, unchallenged, to sweep all before him as he continued to popularize the idea of intelligent life on Mars.

A Point about Drawings and Maps

Lowell's campaign had been more a blitzkrieg than a protracted siege. His own observations had only covered the month of June, a little bit of late July and early August, and October and November. This seems like a rather short apprenticeship to one of the most difficult observing challenges then known to astronomy. But he was largely able to take shortcuts in building up his perception of the planet by relying on Schiaparelli's earlier authority, and especially

Schiaparelli's maps, which as Mars historian K. Maria D. Lane points out, "succeeded in legitimizing the canal-covered Martian landscape. . . . Given the authority and nature of the cartographic data-recording format, it was nearly impossible to erase canals that had been mapped by a credible astronomer. Just as was true for many of the terrestrial expeditions of the day, prestige inhered in putting things *on* the map, not taking them off."[38]

Certain aspects of Lowell's methodology exposed the limitations of the eye-brain-hand system. Though eventually planetary photography would prove useful in documenting the nature of the surface markings, in the 1890s, and indeed for a long time afterward, the slow emulsions of the photographic plates were not able to register the finest planetary details. Fine details—including the presumed canals—were hopelessly blurred in the two to four seconds (or longer) of exposure needed to capture any kind of image at all. Then and for a long time to come (until the late 1980s and early 1990s, in fact, with the advent of digital charge-coupled devices, or CCDs), the eye remained superior to the plate for planetary work. (For the study of stars and nebulae, it was another matter, and photography quickly asserted its predominance.) Because the planet always needed to be viewed through the streaming and eddying ocean of air, details like the canals of Mars typically appeared even to visual observers in little more than a flash of sudden revelation, lasting a fraction of a second.[39] On nights of steadier seeing, of course, the distortions might be less, and the "revelation peeps" (as Lowell called them) or "tachistoscope flashes" (Sheehan's term) might be somewhat longer.[40] On horrid nights, the image might jump around so badly and be so blurred as to allow no useful work to be done whatever. On those nights, the only recourse was to shutter the dome and try again later.

The difficulties of observation under such conditions would hardly be evident, however, to anyone examining published drawings and maps of Mars. One might imagine that the astronomers who produced them were rendering the object before them as if it were a still life. In fact, these drawings and maps were composite records, based on fleeting impressions. The detail seen and the forms under which it appeared were severely limited by the finite speed of the human eye-brain-hand system.

Though not widely appreciated by the public, the point was understood by a few writers. Thus, University of Wisconsin astronomer George C. Comstock (1855–1934), in his 1901 *Text-Book of Astronomy*, included drawings by Lowell and his map of Mars showing the "faint dark lines which are generally called canals, . . . running in narrow dusky streaks across the face of the planet according to a pattern almost as geometrical as that of a spider's web."[41] Although Comstock himself did not doubt even the geminations of the canals, he reminded the reader that composite maps "must not be taken for a picture of the planet's appearance in the telescope. No man ever saw Mars look like this, but the map is useful as a plain representation of things dimly seen."[42] Similarly, Edward Sylvester Morse, the man whose lectures on the Far East had proved so inspiring to Lowell and who spent May 1905 with him in Flagstaff scanning Mars, recounted in his 1906 book *Mars and Its Mystery* his experience in coming to see the canals: "For years I had been familiar with different representations of Mars in which the surface features had been strongly depicted in black and white; in other words, photo-reliefs, or engravings incorporated with the printed page. I had unwittingly come to believe that these features were equally distinct when one observed Mars through the telescope."[43] Nothing could be further from the truth.

The Public's Attention Aroused

The opposition was hardly over and Lowell once more back in Boston than he began to make a great deal of noise about what he had accomplished, launching a public relations effort the likes of which the astronomical world had never seen before. A brief, largely descriptive article had appeared in the August 1894 issue of *Astronomy and Astro-Physics*, a journal co-edited by William Wallace Payne (1837–1928) and George Ellery Hale (1868–1938). It was followed by a six-part series in Payne's *Popular Astronomy* running between September 1894 and April 1895. By January 1895, the public interest was already running high, as Lowell gave four lectures in Boston's 1,000-seat capacity Huntington Hall (on the MIT campus), at which "standing room was nil, and demands for admission were so numerous and insistent that repetitions were arranged for the evenings. At these repeated lectures the streets near by were filled with motors and carriages as if it were grand opera night."[44] These lectures served as the basis for a four-part series of articles running from May to August in

Atlantic Monthly, in turn providing much of the material that Lowell mined for his best-selling book *Mars*, largely written in the three-story brownstone rowhouse at 11 West Cedar on Beacon Hill he purchased after his mother's death in April 1895 and intended for his life en garçon. A young editor and publisher from New York who passed by each day on the way to his own modest establishment on the next block, "used to observe him every weekday at five-thirty. His handsome head was to be seen vis-à-vis the *Boston Evening Transcript* beneath a life-sized plaster Venus similar to those that infest the Athenaeum. Visibility was perfect, for the shade was always raised to the very top of the window as if to admit no impediment to a message from Mars."[45]

Mars appeared in December 1895. It was a literary tour de force, whatever one might think of its standing as a work of science. Its basic ideas were set forth with lawyerly logic:

> To review, now, the chain of reasoning by which we have been led to regard it probable that upon the surface of Mars we see the effects of local intelligence. We find, in the first place, that the broad physical conditions of the planet are not antagonistic to some form of life; secondly, that there is an apparent dearth of water upon the planet's surface, and therefore, if beings of sufficient intelligence inhabited it, they would have to resort to irrigation to support life; thirdly, that there turns out to be a network of markings covering the disk precisely counterparting what a system of irrigation would look like; and, lastly, that there is a set of spots placed where we should expect to find the lands thus artificially fertilized, and behaving as such constructed oases should. All this, of course, may be a set of coincidences, signifying nothing; but the probability points the other way. As to details of explanation, any we may adopt will undoubtedly be found, on closer acquaintance, to vary from the actual Martian state of things; for any Martian life must differ markedly from our own.[46]

About one thing Lowell was certainly mistaken. He supposed that his readers would recoil from the "possibility of peers," and that "like the savage who fears nothing so much as a strange man, like Crusoe who grows pale at the sight of footprints not his own, the civilized thinker [would] instinctively [turn] from the thought of mind other than the one he himself knows."[47] In fact, however, not only did Lowell find a vast audience of readers who were electrified by his idea of intelligent life on Mars, his book immediately began to gather around itself a voluminous body of what might today be described as "fan fiction."

The first entry came from Herbert George Wells (1866–1946), a 30-year-old schoolteacher-turned-writer who in May 1895 had moved with his mistress, Amy Catherine "Jane" Robbins (1872–1927), from London to "Lynton," Maybury Road, Woking, south of London. (He was still married at the time, but he divorced his wife and married Jane in October 1895.) In place of Lowell's intelligent, benign, and peace-loving folk, struggling for existence on their doomed world, Wells gave them a menacing twist that made them more interesting and far more humanly believable. Glancing across space from their dying world they set their faces like flint toward ours:

> No one would have believed in the last years of the nineteenth century that this world was being watched keenly and closely by intelligences greater than man's and yet as mortal as his own: that as men busied themselves about their various concerns they were scrutinized and studied, perhaps almost as narrowly as a man with a microscope might scrutinize the transient creatures that swarm and multiply in a drop of water. With infinite complacency men went to and fro over this globe about their little affairs, serene in their assurance of their empire over matter. It is possible that the infusoria under the microscope do the same. No one gave a thought to the older worlds of space as sources of human danger. . . . Yet across the gulf of space, minds that are to our minds as ours are to those of the beasts that perish, intellects vast and cool and unsympathetic, regarded this earth with envious eyes, and slowly and surely drew their plans against us.[48]

Equipped with a giant cannon, Wells's Martians shoot a series of 10 missiles across the intervening void, of which the first lands in the area on Horsell Common (right behind the house where Wells was living at the time; indeed, a visitor to the area, equipped with late 1890s versions of the Ordnance Survey maps, can still follow exactly where the Martians went, and even which houses they destroyed.) Set against a complacent view of the greatness of empire in the twilight of Victoria's long reign, Wells's book reminded its readers of the damage that our own species has "wrought, not only upon animals, such as the bison and the dodo, but upon inferior races, such as the Tasmanians who in spite of their human likeness, were entirely swept out of existence in a war of extermination waged by European immigrants, in the space of fifty years. Are we such apostles of mercy as to complain if the Martians warred in the same spirit?"[49]

Serialized in 1897 and published between covers in 1898, *The War of the Worlds* was, according to cultural historian Robert Markley, the first of a voluminous body of works of science fiction, "concerned with depicting the social, political, and economic consequences on Lowell's dying planet."[50] Among Wells's no doubt thousands of readers, one particularly receptive individual was a sickly 16-year-old, Robert Hutchings Goddard (1882–1945), of Worcester, Massachusetts, who, while recuperating from a kidney ailment, came across a serialized version in the *Boston Post*. The story, Goddard later recalled, "gripped my imagination tremendously."[51] Goddard's own Martian epiphany occurred a year later, on October 19, 1899. While climbing a cherry tree in order to clip off dead limbs he began to wonder "how wonderful it would be to make some device which had even the possibility of ascending to Mars." He would recall being "a different boy when I descended the tree from when I ascended." Years later, after successfully launching the world's first liquid-fuel rocket from his Aunt Effie's farm in 1926, Goddard wrote to Wells to explain that the "spell [of the book] was complete about a year afterward, and I decided that what might conservatively be called 'high altitude research' was the most fascinating problem in existence."[52] Goddard's modest vehicle was only a beginning. Only 50 years later, two NASA spacecraft, Vikings 1 and 2, would achieve in fact the landings about which Goddard could only dream. The line from Lowell through Wells and Goddard to Mars runs true.

9

The Rise and Fall of the Canals

Only the superficial never changes its expression; the appearance of the solid varies with the standpoint of the observer. In dreamland alone does everything seem plain, and there all is unsubstantial.

—Percival Lowell, *The Soul of the Far East*, 1888

The Tacubaya Expedition

As soon as he had seen *Mars* off to the press, Percival Lowell set sail for Europe to meet some of the leading European students of Mars, including Camille Flammarion and Giovanni Virginio Schiaparelli. He also scouted additional potential observing sites in Algeria. With Charles Trépied (1845–1907), director of the Algiers Observatory; Ralph Curtis; and Wrexie Louise Leonard (1867–1937), a young Boston woman whom he had just hired as his secretary and who would remain on his staff until his death, he visited Boghari and Biskra as well as sites on the northern fringe of the Sahara Desert. Conditions in the Sahara proved unfavorable.

His was driven by the fact that after the successful 1894 observing campaign, he began to doubt Flagstaff as a site for an observatory. Though conditions had been excellent in the spring and early summer, they turned rainy in the July to September "monsoon season," and in the winter they were almost useless, "a fraud," as he told Andrew Ellicott Douglass, who remained in Lowell's employ even though William Henry Pickering had returned to Harvard.[1] In search of something better, Douglass was dispatched to Mexico to test winter observing sites between Chihuahua and Mexico City, as a possible destination for the next Mars opposition, on December 1, 1896.

Meanwhile, as the 46-centimeter borrowed refractor had been returned to Pittsburgh, Lowell ordered as a replacement a 61-centimeter refractor from the Cambridgeport, Massachusetts, firm of Alvan Clark and Sons, which was delivered to Flagstaff in August 1896. A consideration that entered into the design of the telescope was to have significant unforeseen consequences. Wanting, presumably, to save money, Lowell specified that the focal ratio of the new telescope be short enough that it could fit inside the same dome that had housed the Brashear (*f*/17.4). The Clark firm thus produced a refractor with a focal ratio of only 16.0, meaning that the Clark, though optically superb, suffered from rather severe chromatic aberration, also known as "color fringing" or "purple fringing."[2] Stopping down the aperture by means of an iris diaphragm located in front of the lens is one option to counter this problem, and this is what Lowell and his colleagues routinely did (usually to 30 or 46 centimeters). Viewing the image through a color filter was also found helpful. Following Schiaparelli's practice, Lowell at least sometimes used a yellow- or neutral-tinted glass in front of the eyepiece.

The use of an iris diaphragm can also be helpful in matching the aperture to the seeing conditions. One of the claimed innovations of Lowell's first few years, though not entirely original with him, had been his decision to site the observatory where he hoped it might enjoy the best possible atmospheric conditions for planetary observation. Since the observer is in fact situated at the bottom of an ocean of air, telescopic images of planets frequently appear as if viewed through streaming water, with large and small ripples moving swiftly across. As described earlier, this turbulence is referred to as the "seeing," and it varies with location, season, and local climate.

The superiority of Flagstaff's seeing for planetary observation, claimed by Lowell, was generally conceded by observers elsewhere, and remained a standard defense against anyone who doubted the reality of the markings seen there. Thus Reginald L. Waterfield (1900–1986) of the British Astronomical Association (BAA) asserted in 1923, "the atmospheric conditions at Flagstaff are probably superior to those of any other observatory."[3] But even at Flagstaff the "seeing" is far from perfect much of the time, as Lowell himself had begun to realize at the end of the 1894 Mars campaign (though his efforts to find a better site, such as in the Sahara, ended in disappointment). If the air is very turbulent, the image is blurred and the eye cannot resolve features as small as should be theoretically possible, as Douglass himself had shown in a series of experiments.[4] He employed the technique of centering the telescope on a bright star, then removing the eyepiece to observe parallel bands of light and shadow crossing the illuminated exit pupil. By watching these light-and-shadow bands he found that atmospheric turbulence was more likely due to temperature variations than to rapid "gusts," and that even "the slightest change of temperature in the dome or in the telescope [was] harmful." As for atmospheric streams passing overhead, he found that multiple currents at different altitudes, frequencies, and lateral speeds combined to produce "vibrating" confusion in the image. Subsequent astronomers have largely confirmed these early results. As summarized by two modern planetary observers,

> Turbulence may be caused by winds and convection high in the atmosphere or by unequal temperatures in the immediate vicinity of the observer or even within the telescope tube. Depending on the cause, the turbulence may be rapid and small scale, distorting the image differentially so that the whole image becomes fuzzy; or it may be so slow and large scale that the image moves as a whole and the eye is able to follow it. In the latter case, the seeing conditions might be considered excellent despite the turbulence, since the eye will be able to see fine details despite the slow excursions of the image. The usual case is between these extremes.[5]

This is what is meant by matching the aperture to the seeing. Under most conditions, a small telescope will give good definition. Because perfect conditions allowing a large aperture to be used to advantage are rare, in general an iris diaphragm is used to stop down the aperture of large telescopes to decrease the blurring (as well as chromatic aberration). Lowell himself recognized that the finest detail often appeared on a disc that moved bodily, and that with the lens stopped down to apertures of 30 to 46 centimeters the detail was usually much sharper than with the full aperture (61 centimeters), which reflects the fact that even at Flagstaff, the seeing is often "mediocre," though the seeing can be very good at times.[6]

Of course, this is also true at most other observatories. Eugène Michel Antoniadi (1870–1944), during twenty years of work at the Meudon Observatory near Paris, found that only on one in 50 nights was the seeing "perfect."[7]

All this also means that one is consigned to spend a lot of one's observing time making the best of the mediocre nights, on which detail might appear sharply defined for a second or two and then vanish. But determination of what precisely is revealed under such circumstances is extremely challenging for the eye-brain-hand system. One sees something in a sudden flash, and tries to guess at what it might be. Psychology becomes very important here. It used to be supposed that the way the brain builds its pictures of the world (or of *a* world, such as Mars) was from the bottom up, through the amassing of low-level cues supplied by line detectors, edge maps, and so on in the eye and brain. More recently, however, the concept of the "predictive" or "Bayesian" brain has become fashionable, according to which perception involves a bidirectional cascade of cortical processing. In this view, the brain is a statistical organ that generates hypotheses or "fantasies" that it then attempts to test (and if necessary correct) against sensory evidence. "In other words," write two modern neuroscience researchers, "the brain is trying to infer the hidden causes and states of the world generating sensory information, using predictions based upon a generative model that includes prior beliefs."[8]

If a percept or inference turns out to be different from the true causes generating the stimuli, the inference is said to be illusory or false. However, it is not always easy for an observer to accept this, for once a definite expectation is established, "it is inevitable that one will see something of what one expects. This reinforces and refines one's expectations in a continuing process until finally one may end up seeing an exact and detailed—but ultimately fictitious—picture."[9] In the case of the canals, some observers—not only Lowell but Schiaparelli as well—never did succeed in liberating themselves from the fictitious picture (figure 9.1).

Troubles with Venus

When the new Clark refractor arrived in Flagstaff, Mars was not in a favorable position for observation, so Lowell, Douglass, Leonard, and several other assistants bided their time putting it through its paces on Mercury and Venus, well placed for daylight observation. Lowell confirmed two (now known to be incorrect) Schiaparellian results: the synchronous rotation of both Mercury

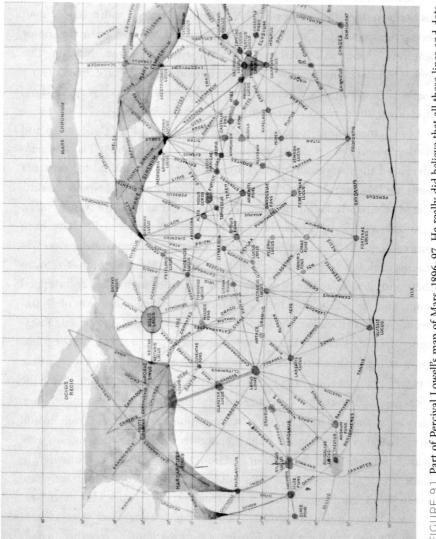

FIGURE 9.1 Part of Percival Lowell's map of Mars, 1896–97. He really did believe that all these lines and dots existed on the Martian surface. Courtesy of Lauren Amundson and Lowell Observatory Archives.

(in which the planet kept one side perpetually toward the Sun and other away) and Venus. Schiaparelli had been quite confident in the case of Mercury, more tentative in that of Venus. Lowell's Mercury observations created little stir, but his Venus observations immediately landed him in a quagmire. Rather than seeing the vague and elusive markings described by most astronomers for centuries, he found the markings to be "surprisingly distinct," radiating like spokes from a common center. They were so evident, he claimed, that they put the synchronous rotation beyond doubt: Venus rotated once every 225 days, he claimed, the period of its revolution around the Sun.

Lowell, meanwhile, commissioned the brothers Godfrey (1861–1948) and Stanley Sykes (1865–1956), operators of a bicycle repair shop in Flagstaff who advertised themselves as "makers and menders of anything," to construct a unique 13-meter diameter wooden cylindrical dome that would be suitable for dismantling and shipment elsewhere. (Thus Lowell's plan to make a snub-nosed Clark that would fit in the earlier dome proved to have been unnecessary.) At the end of September 1896, as the late monsoon season was making itself felt in Flagstaff, Lowell decided on the Mexico option. The Sykes dome and Clark telescope were disassembled, loaded on flatcars, and shipped south, and reassembled by local workmen under the supervision of Douglass at Tacubaya, outside Mexico City and not far from the National Observatory. Despite Lowell's characteristic impatience and a stream of telegrams emphasizing the need for haste, the installation of the dome took longer than expected mainly because of difficulty adjusting the double track on which the Sykes dome revolved. It was not in readiness until the end of December, when Mars was already nearly a month past opposition. Lowell arrived from Boston three days after Christmas with the precious lens in hand. Observations were made from then until the following March, 1897, but there was little new to report about Mars—only that the results of the 1894–95 opposition were said to have been "confirmed." Most of the Tacubaya observations involved Mercury and Venus. Again the Venusian hub-and-spoke system was recorded, and Lowell claimed that these were permanent features of the surface that, instead of being concealed by an all-but-impenetrable shroud of clouds as generally believed, were visible through a diaphanous veil (figure 9.2).

On his return from Mexico, Lowell published relatively little about Mars, but he blitzed the astronomical world with publications about Mercury and Venus.

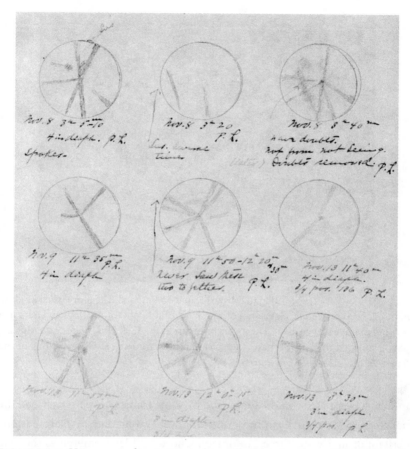

FIGURE 9.2 Venus, according to Percival Lowell. These drawings, showing the hub-and-spoke system first recorded at Flagstaff and Tacubaya in 1896–97, were made in November 1899, during the period in which Lowell was still recovering from his nervous breakdown. He used the 18-centimeter refractor at the Amherst College Observatory, diaphragmed to 7.75- or 10-centimeter aperture. Courtesy of Lauren Amundson and Lowell Observatory Archives.

Though his Mercury observations and map (bizarre though it was) elicited hardly a gasp, his Venus publications provoked a torrent of criticism that apparently took Lowell completely by surprise, as they described markings utterly unlike anything seen before (figure 9.3). Commenting on Lowell's map of Venus at a meeting of the Royal Astronomical Society, Captain William Noble (1828–1904) could not help but wonder whether "Mr. Lowell has been looking at Mars

FIGURE 9.3 Venus, according to Edward Emerson Barnard. The more usual kinds of markings seen on Venus, consisting of vague nebulous patches, are depicted here in a drawing made with the 30.5-centimeter refractor at the Lick Observatory on May 29, 1889. Published in E. E. Barnard, "Physical and Micrometrical Observations of the Planet Venus," *Astrophysical Journal* 5 (1897): plate 19.

until he has got Mars on the brain, and by some transference . . . ascribed the markings to Venus."[10] Even more scathing were comments by Flammarion's assistant at Juvisy-sur-Orge, Antoniadi, who would later emerge as the bête noire of Lowell's Mars theories: "It is to be hoped . . . [the] canals of Venus, though negatively advancing our scanty Aphroditographical knowledge, will advance optical science in a positive manner, and enable us, in a near future, to have a clearer grasp of the canaliform illusion which so violently agitated of late the public mind."[11] So caustic were Antoniadi's criticisms that all communication between the two men was broken off for many years.

The Invalid Years

Already while in Mexico, Lowell had begun to experience nervousness and fatigue. After a brief respite in Boston in April, he set out again for Flagstaff, but he made it only as far as Chicago before he suffered a complete nervous collapse and had to turn back. No doubt strain and nervous exhaustion played a role, but the Venus criticisms must also have been a factor. These criticisms, as Lowell's biographer David Strauss has noted, "cast doubts about the validity of his observations and his future prospects in astronomy" and precipitated "one of those instabilities of mood and nerve that tended to complicate Lowell's life at important junctures."[12] Whatever the cause of his breakdown, Lowell retreated from the field, at first hibernating in his brownstone house on Beacon Hill, and then, when he failed to recover, returning to Sevenels to live with his father. Examined by his father's physician, he was put under strict orders of "absolute rest," with neither visitors nor work being allowed. Lowell tried this for a month, found it intolerable, and in the company of his doctor set out on the time-honored regimen for nervous breakdowns for those who could afford it: travel, exercise, and sunshine. Recovery was slow, and the next four years were marked by periods of anxiety, melancholy, sleeplessness, and apathy.

During his absence, his brother-in-law William Lowell Putnam II (1861–1923) took charge of the observatory's business affairs, while Douglass did his best to keep the astronomical work going.[13] Though publicly championing Lowell's work—even that on Venus—Douglass had privately come to harbor serious doubts. Suspecting that many of the details seen on the planets might be illusory, he began a series of experiments in which he viewed artificial planet discs set at a distance through the telescope, and this only deepened his doubts. The only assistant astronomer in residence on Mars Hill for the 1901 opposition of Mars, he was still engaged in these experiments when Lowell was sufficiently improved to return to take full charge of the observatory's work. Lowell himself tried his hand at the artificial planet experiments, and promptly drew in place of a shaded area a double canal!

Douglass had by now gotten used to a degree of independence and was resistant to submitting to Lowell's autocratic style. He complained to W. H. Pickering about Lowell's "strong personality, consisting chiefly of immensely strong convictions," and questioned his methods of research.[14] On the same day, he also voiced a similar complaint to Putnam, asking that it be considered confidential. But the writing was on the wall. Douglass had feared a "break-up" with Lowell since summer 1896, he told William Wallace Campbell of Lick

Observatory; his fears were now realized. Inevitably, Lowell regarded Douglass as insubordinate and dismissed him from the staff in early July 1901.[15]

Thereafter, Lowell resolved to soldier on alone in his Mars Hill redoubt. He confided to his sister Elizabeth, "I am so much at home here . . . and yet no one I know knows it."[16] Eventually he relented and hired as an assistant Vesto Melvin Slipher (1875–1969), recently graduated from Indiana University, to operate a spectrograph Lowell had ordered from John A. Brashear in the hope of rebutting the criticisms of his Venus work (by showing its rotation to be synchronous). Lowell followed with two other hires from Indiana, Carl Otto Lampland (1873–1951) and Vesto's brother Earl C. Slipher (1883–1964). These men, during long careers at the observatory, would prove unflinchingly loyal to their employer, never challenging his ideas about Mars while buttressing his theories and sketching methods with data from more modern instruments, the spectrograph and camera. Even at Lowell Observatory, a more astrophysically informed direction in Martian research began to emerge in the first decade of the 20th century.

Views from Elsewhere

Mars, of course, was shining in other places than Flagstaff and Mount Hamilton, and scores of observers swung their telescopes toward Mars during each opposition's favorable window for observations. At the 1896–97 opposition, which had largely been a wash for Lowell and his staff, the leading observers included the Catalan astronomer Josep Comas i Solà (1868–1937), who used the 15-centimeter Grubb refractor at the Barcelona Observatory; Flammarion and Antoniadi, who used the 23-centimeter Bardou refractor at the Juvisy Observatory; Henri Perrotin (1845–1904), who was on invitation from the director of the Meudon Observatory and used the 83-centimeter Henry Brothers refractor; Vincenzo Cerulli (1859–1927), who used the 41-centimeter Cooke refractor at his private Collurania Observatory, near Teramo, Italy; and Percy Braybrooke Molesworth (1867–1908), a captain in the Corps of Royal Engineers, who used a 24-centimeter reflector at the near-equatorial site of Trincomalee, Ceylon (now Sri Lanka).[17] Allowing for the inevitable "personal equation" that enters into all human observations, the results in general showed reasonable with Schiaparelli's school of representation generally prevailing over Nathaniel Green's. However, the interpretation of the features varied markedly. On the one hand were those who believed that canals—that is, actual linear surface markings—existed more or less in the form in which they appeared. On the other hand were those who, though sometimes glimpsing linear markings, were inclined to regard them as illusions.

Antoniadi, employed as Flammarion's assistant at Juvisy-sur-Orge and also since 1896 serving as director of the BAA Mars Section, summarized the observations made in 1896–97 in his first BAA *Memoir*. Though Antoniadi's own map compiled on the basis of his own and others' observations shows canals in the grand Schiaparellian style, his critical judgment is already evident. He says, "The [Mars Section] Director's experience . . . is that the canals are very difficult objects, visible only by rare glimpses, and had it not been for Prof. Schiaparelli's wonderful discoveries, and the fore-knowledge that 'the canals are there,' he would have missed three quarters at least of those seen now."[18]

One BAA member, the Reverend P. H. Kempthorne (d. 1920), opined that a considerable number of the canals were merely the edges of diffuse shadings. Molesworth, the most prolific recorder of canals in 1896–97 and in 1899–1900, would find, upon graduating to a larger telescope (a 32-centimeter reflector) in 1903, "myriads of small details far beyond anyone's ability to adequately depict in a drawing." He added: "Those who have never had the privilege of studying Mars with a considerable aperture, high powers, and really good seeing can have no conception of the amount of detail visible. . . . What has struck me most during this apparition is that almost all the canals appeared as streaks, and not as lines. . . . I cannot regard the delicate spider's-web appearance on Lowell's drawings as being in the least a true rendering of the actual appearance of the canals."[19]

Similarly, the Reverend Theodore Evelyn Reece Phillips (1868–1942), best remembered today as a Jupiter observer but for a time an equally keen student of Mars, using a 24-centimeter reflector in 1896–97, wrote, "My experience of Martian observation this winter has led me to believe that Mars is not nearly so difficult an object as is commonly supposed, and that many of the canals are easy."[20] By 1898–99, he was less confident. "Faint markings have been glimpsed now and then" he wrote, "and it would be easy, by the 'scientific use of the imagination,' to conjure them into lines and streaks harmonizing with the charts." He added, "I am careful to represent on my drawings only what I feel I can see with certainty and hold with tolerable steadiness."[21] The very experienced George D. Hirst (1846–1915) of New South Wales, who had observed Mars with a variety of telescopes going back to 1877, wrote in 1905, "I see no canals. In fact, I have never seen anything on this occasion, or in years past [even] when using larger apertures, that could by any stretch of the imagination support the wonderful observations of these marvels made by other observers, unless the few streaks depicted in my drawings may be taken as indications of them. If so, the canal champions are welcome to them."[22]

The Illusion Theory

The whole question turned on whether the lines that appeared at times (often in glimpses) corresponded to the actual features of the Martian surface. As far back as 1894, E. Walter Maunder, a cofounder of the BAA and a photographic and spectroscopic assistant who specialized in solar observations at the Royal Observatory, Greenwich, had noted in an interesting article titled "The Tenuity of the Sun's Surroundings" that even "the smallest portion of the sun's surface visible to us as a separate entity, even as a mathematical point, is yet really a wide extended area."[23] However, he added:

> Now this fact has an important bearing on some of our theories. We easily fall into the mistake of supposing that the most delicate details which we can see really form the ultimate structure of the solar surface; but it is not possible that they can do so. The finest granule, the smallest pore, as we see it, is only the integration of a vast aggregation of details far too delicate for us to detect; and a minute speck of brighter or duller material may, and probably does, contain within itself a wide range of brilliancy, not to speak of varieties of temperature, of pressure, of motion, and of chemical composition.[24]

But the same, of course, was true mutatis mutandis of Mars, as Maunder explicitly stated in a later article. "We have no right to assume," he said, "and yet we do habitually assume, that our telescopes reveal to us the ultimate structure of the planet."[25] He added in 1903, "We need not for one moment make the assumption that the various observers of the canals have not actually seen what they have represented, or that they have badly represented what they have seen. The question is one simply of the limitations of our sight. Under certain conditions the impression as of lines is given to us as by objects not linear."[26]

A similar view was taken by Cerulli, who came from one of the wealthiest and most prominent families of the Abruzzo region of Italy and could afford (much like Lowell) to set up his own observatory dedicated to the study of Mars. His 41-centimeter Cooke refractor at Teramo was the second largest in Italy after only the 49-centimeter refractor used by Schiaparelli at Brera. At the opposition of 1896–97, Cerulli confirmed the existence of what appeared to be linear markings but suspected that if they were more fully resolved they would present more complex forms (figure 9.4). Thus, on January 4, 1897, during "some moments of perfect definition [in which] Mars appeared perfectly free from undulation," he watched with astonishment as the canal Lethes "lost its form of a line and altered itself into a complex and indecipherable system of minute patches."[27] Henceforth he doubted the reality of the whole network.[28] He believed that, as in the case of the Lethes, the actual surface of the planet was probably marked by small spots, which, roughly aligned, the eye joined into lines. He was skeptical in the canal appearances so common in the small-ish telescopes available at the time. He was confident that in larger telescopes a more accurate picture of the planet would emerge, coming asymptotically closer to what Maunder had called "the ultimate structure of the planet."

Pursuing the same idea, Maunder himself in 1903 collaborated with Joseph Edward Evans (1855–1938), a teacher at the Greenwich Royal Hospital School, in an experiment in which students naïve to the matter had been challenged to draw what they saw on artificial discs on which no canals were present. Many recorded nonexistent canals! Lowell, mocking what he called "small boy theory," insisted that the observations of trained astronomers at the telescope were to be trusted over the impressions of students in a reform school.[29] He had a point, though he ignored the fact that a number of highly trained astron-omers had come to similar conclusions based on direct observations through the telescope. By 1903, their number included Gaston Millochau, one of the few of this era, apart from Edward Emerson Barnard, to examine Mars with a really large telescope, the 83-centimeter Henry Brothers refractor at Meudon Observatory, near Paris.[30] Millochau observed Mars at the oppositions of 1899,

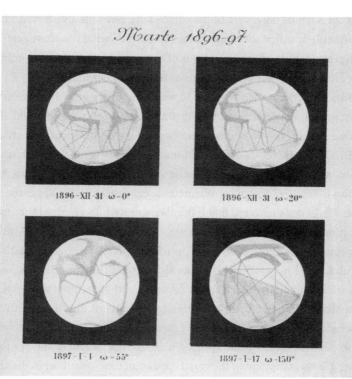

FIGURE 9.4 Drawings of Mars in 1896–97 by Vincenzo Cerulli, using a 41-centimeter Cooke refractor at his private observatory of Collurania, at Teramo, Italy. Published in *Marte nel 1896–97* (Teramo: Pubblicazioni dell'Osservatorio privato di Colurania, 1898), no. 1.

1901, and 1903, and was able to make out, in the positions of some of the canals, "small dark patches, with ragged borders, separated by non-tinted spaces." These tended, he found, to be joined up by the eye into lines "whenever the vision is not concentrated on one point."[31]

Seeing by Glimpses

Against such testimony, Lowell could call on that of his old friend Edward Sylvester Morse, who came to Flagstaff in May 1905 and took regular turns on the 61-centimeter Clark when Mars was near opposition, in order to "see for himself." At first he saw hardly anything, but with practice he acquired the skill of seeing the canals much as Lowell saw them. His conclusion: "the delicate lines, known as canals, do exist."[32] Acute eyesight, training in the special arts of planetary observation, a good glass, excellent air, and the continuation of observations through the right season were all necessary adjuncts to their perception. Even so, however, the delicate lines were far from easy and—significantly—showed themselves only in "glimpses" lasting a fraction of a second. Lowell, for that matter, often emphasized the fugitive nature of the canals' apparitions. For instance, in a memo headed "Hints on 'Seeing'" sent to David Peck Todd, he wrote: "Fine detail on the planets only comes by glimpses. Experience enables one to tell the stamp of the true from the imaginary and experience further shows that of the few doubtful cases almost all stand for a reality not the reverse."[33]

The Canals—Photographed or Not?

The fact that the canals appeared only in "glimpses" was, in fact, one of the greatest challenges to Lowell's efforts to prove their objective reality. The eye, with what seemed to be nearly instantaneous perception, was fleet enough to capture momentary impressions, but in transferring these impressions from eye to brain to sketch pad, an element of unavoidable subjectivity was introduced. Photography promised greater objectivity. In Lowell's words, "a photograph can be scanned by everybody, and the observation repeated until one is convinced."[34] For stars and nebulae, photography had already proved indispensable, through the ability to build up over many hours a cumulative record of the light. However, the photographic plates available at the time were simply too slow to register any but the coarsest planetary detail. As far back as 1879 Mars had had its portrait taken by Benjamin Apthorp Gould (1824–96) at the Córdoba Observatory in Argentina; it showed as a mere featureless blob. W. H. Pickering at Arequipa in 1890 and 1892 had marginally improved the result by recording the polar caps and limb clouds. The detail even in the best of these images was, however, far less than could be grasped by a visual observer.

By 1903, Lowell's assistant Carl Otto Lampland had designed a specialized planetary camera that enlarged the image by means of a secondary lens tube.

With the use of a sliding plate holder, the operator could obtain a series of 20 or more images in rapid succession on a single plate, simply by turning a ratchet and squeezing a shutter bulb. This increased the likelihood that at least some exposures were made during moments of steady seeing. A yellow or other-tinted filter was used to enhance the detail. Some of these filters had been made by James Wallace (1868–1945), an assistant at Yerkes Observatory (who quipped to George Ellery Hale, the director of Yerkes, "Lowell wants me to make a screen for his telescope to photograph the Canals in Mars—for 'a consideration' I could put them in the screen").[35]

Not to denigrate Lampland's achievement, for the quality of his images was far superior to anything achieved before, but the detail was still severely limited by coarse-grained and insensitive emulsions, with exposures of two to four seconds needed to register Mars at all. Thus they were, as Lowell said, "incommunicable of canals."[36] (They did, however, provide excellent records of the configurations of the larger, darker areas—a noteworthy achievement in itself.)

Lampland kept refining his technique, until, by Mars's opposition of May 1905, he had obtained images revealing not only the larger dark areas but, as Lowell exulted, "the grosser of those lines that had so piqued human curiosity, the canals of Mars. . . . Thus did the canals at last speak for their own reality themselves."[37] Unfortunately, since the original images were on a scale of only 50 millimeters to the disc of the planet, what showed on them was subject to dispute. Barnard, who late that year passed through Flagstaff on his return from Mount Wilson, where he had been photographing the Milky Way, carefully examined Lampland's negatives and also chatted with Douglass, then temporarily serving as a probate judge in Prescott. Noting defects in the focus of the photographs, Barnard told Hale, "I am perfectly sure you would have agreed with me that they did not show the canals as claimed."[38]

The photographic campaign continued in 1907, when Mars was 62 million kilometers from Earth (closer than it had been since 1892) but was also 28° south of the equator, well placed only at southern observatories. Lowell discussed an expedition to observe and photograph Mars with his old crony David Peck Todd, professor of astronomy at Amherst College, who had just acquired a 46-centimeter Clark refractor for the college observatory and obligingly offered to disassemble it, crate it up, and ship it to South America over land via Panama (the canal would not open until 1914). Lowell agreed to pay all expenses for what became known as the "Lowell Expedition to the Andes." Such was the celebrity of the undertaking that this address alone was sufficient to assure mail delivery to the astronomers en route. In addition to Todd and his wife, Mabel, the team consisted of 24-year-old Earl C. Slipher, armed with a duplicate of Lampland's planetary camera and carefully coached by Lampland in the intricacies of its use. On the eve of departure, Todd gave an interview in the *Boston Transcript* in which he declared: "We are going to South America in hope of getting the facts, and to obtain information on the question of the Martian canals. . . . Is Mars inhabited? This is a question my wife has often asked me, but do you know I've never been able to answer it? Maybe when I come back."[39]

The seven-ton telescope was shipped from New York via Panama to the Chilean port city of Iquique; from thence taken by train 65 kilometers farther inland to the nitrate-mining center of Alianza, located 1,200 meters above sea level; and there set up in open-air splendor under burnished, rainless skies. Todd evoked the otherworldly bleakness of the scene:

The region is an utter desert; the moon itself could not reveal greater barren-
ness—not a tree or a flower or a blade of grass for miles, not even moss or
lichens. . . . The clustered dwellings of the workers in the nitrate fields . . . gave
an almost populous effect to the barren landscape. . . . Around the whole settlement stretched the solemn, brown, impressive pampa, undulating to the great
mountain border, the Andes, its peaks here and there snow-capped, lofty, and
magnificent. Here . . . [is] one of the best astronomical stations ever occupied.[40]

Rain at the site was unknown, and distant thunder heard several times a day proved to be owing not to thunderstorms but to earthquakes rumbling in the mountains. Despite occasional fog and haze, most of the time the air was sub-limely still and clear. Under these conditions, with the planet high overhead, the image in the eyepiece was unforgettable. According to Todd:

Every observer, whether professional or [amateur], was amazed at the wealth of
detailed markings that the great reddish disk exhibited. Its clear-cut lines and
areas were positively startling in their certainty: the splendor of the first visual

glimpse in steady air can never be forgotten. . . . Nearly everybody who went to the eyepiece saw canals; and once I fancied I heard even the bats, as they winged their flight down the pampa, crying, "Canali, canali, canali!"[41]

Great hopes were held out for the expedition's photographs. Historian K. Maria D. Lane has pointed out that as of 1907, "photography supplanted cartography . . . as the proper standard of proof for Mars representations. The buildup of expectations regarding the Lowell photographs focused on their purely objective quality and their ability to resolve long-standing disputes among astronomers over the existence of the canals."[42] Though Todd was the official leader of the expedition, most of the real work was done by Slipher. He kept busy night and day taking and developing images of the planet, as well as making his own series of sketches from visual inspection of the planet. In the end, 13,000 images of the planet were obtained. As some of the choicest prints began to trickle back to Flagstaff, Lowell found them "beyond expectation fine. . . . The canal stock . . . has already risen in consequence."[43] They were certainly remarkable by standards of the day, and Lowell expected they would finally deliver a crushing blow to the skeptics. Unfortunately, they fell short just as—and for the same reasons—those in 1905 had done. After a fierce bidding war, some of them were published in the *Century Magazine*, where they proved to be just as maddeningly subject to bias and to differences in interpretation as were visual drawings of the planet. Those who believed in canals saw canals; those who did not saw randomly aligned grains in the photographic emulsion.

Eager to win over Schiaparelli once and for all, Lowell sent pages and pages of photographic reprints, which the Italian astronomer—despite being nearly blind—dutifully examined with a specially designed microscope of low magnification fitted with a micrometric reticle for drawing and measuring. With the aid of this instrument, Schiaparelli was disconcerted to find that some of the lines shown on the images were mere cracks in the emulsion. Some of the details, however, didn't seem to be owing to such faults, in which case, he told Lowell, "We are saved."[44] At the end of his life, Schiaparelli was apparently strongly leaning toward Lowell's side—but he never quite made it all the way.

A Counterreaction Forms

The Lowell Expedition to the Andes generated a level of public interest in Mars that would not be surpassed at least until the spacecraft era. At the end of the year, the staid *Wall Street Journal* pronounced that the outstanding event of 1907 had been not the financial panic but "the proof afforded by astronomical observations . . . that conscious, intelligent life exists upon the planet Mars."[45] Meanwhile, Cerulli and Simon Newcomb had put forward new versions of the optical illusion theory.[46] The following year, 1908—and even more so 1909—would in fact prove the decisive turning point regarding the canals. According to David Strauss, "Lowell's apparent success in photographing the canals . . . and attracting the support of prestigious intellectuals like . . . Edward S. Morse galvanized the American astronomical establishment to launch a counterattack to refute Lowell's claims for the canals."[47] Members of the establishment included William Wallace Campbell of Lick Observatory, Hale (now at Mount Wilson), and Barnard, who decided that the time had finally come to step on Lowell's leash and mount a concerted effort to thwart what they were convinced had developed into "a serious threat to the public's faith in science."[48]

The counterattack was directed not at the canals themselves but rather at Lowell's assertions that the conditions of Mars were such as would theoretically support intelligent life. Thus, at the end of 1906—the year that saw the publication of his Martian magnum opus, *Mars and Its Canals*—Lowell had written, "up to the present time the chief obstacle to crediting Mars with the possibility of life has lain in accounting for sufficient heat on the surface of the planet."[49] From the planet's albedo (the ratio of light reflected from the surface relative to that incident on it), he attempted to work out approximately the amount of solar radiation received, reflected, and retained. There were large uncertainties, but his result was that the planet's maximum temperature was 22°C (72°F), with a mean of 9°C (48°F).[50] This, it turned out, was about the same as the mean temperature of the South of England. Such a warm climate seemed intuitively implausible to Alfred Russel Wallace (1823–1913), who had co-discovered natural selection with Charles Darwin, and who, in a review of *Mars and Its Canals*, showed how bitterly cold Mars would actually have to be given the thinness of its atmosphere, which Lowell had not attempted to calculate.[51] Lowell, in an effort to refute this criticism, then published a calculation of the thickness of the Martian atmosphere, again utilizing Mars's albedo.[52] His result, 87 millibars, was off by an order of magnitude, mainly because he failed to take into account the amount of dust in suspension in Mars's atmosphere, leading to an overestimation of its thickness. But it would prove surprisingly durable over the next half century of studies of the planet.[53]

And so the back and forth continued. In January 1908, V. M. Slipher claimed the spectrographic detection of water vapor in the atmosphere of Mars, based on a side-by-side comparison of the spectra of Mars and the Moon when the air over Flagstaff was unusually dry. One of the most prominent water vapor bands, called the *a* band, appeared strengthened in Mars's spectrum compared to the Moon's. However, Campbell doubted the result after personally examining the original spectrograms, and he went so far as to mount an extraordinary expedition to the top of 4,419-meter Mount Whitney, California's highest peak, to obtain further spectrograms of Mars during the opposition of 1909. Assisted by Charles Greeley Abbott (1872–1973) of the Smithsonian Institution, Campbell showed that there was in fact no difference in the *a* band in the Moon and Mars spectra, thereby handing Lowell a decisive defeat and suggesting a very low upper limit for water vapor on Mars.[54] Campbell's result would stand the test of time.

The 1909 Opposition: The Canals Routed

In his review of *Mars and Its Canals*, Wallace expressed admiration for "the extreme perseverance in long continued and successful observation" that Schiaparelli, Lowell, and others had demonstrated in their studies of Mars. He accepted "unreservedly the substantial accuracy" of the whole series of their observations, but he added as a caveat:

> It must however always be remembered that the growth of knowledge of the detailed markings has been very gradual, and that much of it has only been seen under very rare and exceptional conditions. It is therefore quite possible that, if at some future time a further considerable advance in instrumental power should be made, or a still more favorable locality be found, the new discoveries

might so modify present appearances as to render a satisfactory explanation of them more easy than it is at present.[55]

Schiaparelli had put forth a similar view to Cerulli, comparing the different stages of the history of Mars observations to the view of a printed page as seen from various distances.[56] In a first stage, A, the vision was confused, and the page appeared as nothing more than a gray square; at a next stage, B, this view was replaced with one of geometric lines; at a third stage, C, breaks and irregularities were suspected; finally, at stage D, individual letters could be described. He described this pattern as a stairway of perception, and he anticipated the arrival to stage D along that difficult stairway.

Until 1909, very few observers—Barnard, most notably—had stood on stage D of the stairway. But at that year's perihelic opposition Mars not only approached within 58.6 million kilometers but stood, at opposition, only 4° south of the celestial equator, so that it was reasonably well placed for northern hemisphere observers. A number of observers with large telescopes would contribute to the aforementioned Lowell counterattack, including Barnard and Hale. However, the chief protagonist was Eugène Michel Antoniadi, whose observations with the 83-centimeter Henry Brothers refractor (the "Grand Lunette") of the Meudon Observatory near Paris would deliver a blow from which the canal system would never recover.

The early life and education of Antoniadi—who was born of a Greek family in Constantinople (now Istanbul) and christened Eugenios Mihail Andoniadis, a name later naturalized to the French—have largely eluded the efforts of his biographers.[57] His family appears to have been part of a repressed Greek community of the time, living in a part of Constantinople (Tatavala, today known as Kurtuluş) that was referred to as "Little Athens" because of its large Greek population. Antoniadi's family, like most Greeks, despised the ruling sultan, Abdul Hamid Khan II (1842–1918), known as the "Red Sultan" or "Abdul the Damned" because of the massacres of Greeks and other minorities during his rule, and his extensive use of secret police.[58] There were serious tensions between Greeks and Turks during these closing decades of the once-great Ottoman Empire. Antoniadi must have felt them, but he did not suffer personally from any kind of persecution, probably because of the family's connections with merchants, bankers, and even the notorious arms dealer Basil Zaharoff

(1849–1936), known as the "merchant of death" and "mystery man of Europe," who was one of the wealthiest men in the world and also Antoniadi's cousin. The Antoniadis were even related to at least two European royal families. Obviously, the family possessed means, and Antoniadi received a solid education, presumably from tutors. He did not go to university, but at some point he seems to have received first-rate training as an artist and architect, which was to stand him in good stead for the rest of his life. By his late teens, his interest in astronomy had appeared. He acquired a 75-millimeter refractor, set it up both in Constantinople and on the island of Prinkipo in the Sea of Marmara, and made a number of drawings of sunspots and planets, some of which he sent to Flammarion. The latter had founded the Société Astronomique de France (SAF) in 1887 and edited its bulletin, to which Antoniadi began to contribute. He also joined the BAA immediately on its founding in 1890.

At this time Antoniadi became obsessed with Hagia Sophia, one of the world's monumental buildings, notable for its massive dome. It had been a Greek Orthodox basilica for a thousand years, later a mosque, for a while a museum, and now a mosque again. Antoniadi published an article on it in the journal *Knowledge*, and in 1907 a three-volume *Atlas of the Mosque of St. Sophia*, written in Greek and illustrated with numerous sketches and watercolors made with exclusive permission of the Red Sultan himself over a period of 4½ months in 1904 (figure 9.5). It is an exquisite work and unfortunately extremely rare; the authors have never seen a copy. Its relevance to Antoniadi's career as an astronomer is in showing the development of his unique artistic style, which later found such magnificent expression in his drawings of the Red Planet.

Antoniadi was still in Constantinople in 1893, when he received an invitation to join Flammarion as an observing assistant at the latter's private observatory at Juvisy-sur-Orge, a railroad junction 18 kilometers southeast of Paris. Antoniadi eagerly accepted. One presumes he made the long journey on the Orient Express.

When Antoniadi arrived at Juvisy-sur-Orge, Flammarion had just completed his encyclopedic résumé of Mars observations, *La Planète Mars*, on which we have relied heavily in our own research into Martian history and which is still an indispensable work for anyone interested in the field.[59] At least at first, Antoniadi must have found Flammarion to be extremely inspiring to work for. In addition to observing with Flammarion's 23-centimeter Bar-

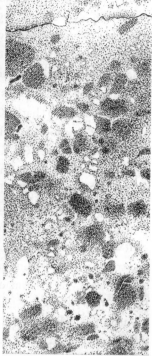

FIGURE 9.5 Examples of some of Antoniadi's work from his *Atlas of the Mosque of St. Sophia*, showing the development of the style he was to use so effectively in rendering detail on Mars. *Left*: Antoniadi's graphic reconstruction of the Byzantine narthex of the cathedral. *Right*: His stipple rendering of the appearance of the surface of a marble revetment. Published in Eugène Michel Antoniadi, *Ekphrasis tēs Hagias Sophias*, 3 vols. (Istanbul: P. D. Sakellarios, 1907–9). Courtesy of Randall A. Rosenfeld, Royal Astronomical Society of Canada.

dou refractor, Antoniadi had access to Flammarion's extensive library, whose extraordinary holdings continue to turn up surprises to the present day (a copy of the *Nuremberg Chronicle*, worth millions, was found during a recent inventory; it was immediately deposited for safekeeping in the Bibliothèque nationale de France!). In addition to his services for Flammarion, Antoniadi in 1896 became director of the BAA Mars Section, in which capacity he published his memoir of the section's work done in 1896–97. Over time, alas, his relationship with Flammarion began to show signs of strains. They were both highly methodical, rather fastidious men, with strong personalities, so perhaps this was inevitable. In large part the problem seems to have had to do with the contract Antoniadi had signed, which allowed the older man to take credit for what was actually Antoniadi's work. Antoniadi came to resent the arrangement, and he began to suffer in his health, probably mostly owing to nerves.[60] For a time he considered moving to England. At one point he confided that he "could not stand M. Flammarion any longer."[61] There were other troubles as well, but fortunately he was rescued by marriage in 1902 to one Katherine Sevastopolos (1879–1952), a member of a wealthy Greek family living in Paris who shared Constantinople roots and connections. Following his marriage, Antoniadi no longer needed to work for a living. He resigned his position at Juvisy-sur-Orge, finished writing the BAA *Memoir* for 1901, and— while occasionally observing with a 22-centimeter Calver reflector he kept at his flat at 74 Rue Jouffroy, Paris—largely devoted himself to working on the three volumes on St. Sophia and studying chess, which he had played in his Constantinople days. (He would eventually achieve near grand master status.)

It was characteristic of Antoniadi's intense personality to blow hot and cold, and so, after leaving Flammarion, he seems to have almost abandoned astronomy for eight years. (He made only two drawings of Mars in 1905, three in 1907.) Even in July 1909, with Mars heading toward its very favorable late September perihelic opposition and caught in the throes of a planet-encircling dust storm, he was nowhere near a telescope; instead he had gone to Athens for a chess tournament. On his return to Paris, he found a letter, dated August 4, from René Jarry-Desloges (1868–1951), another man of independent means who devoted a significant part of his wealth to studies of the Moon and planets and who, by 1909, had established two observing stations in southeastern France equipped with powerful instruments: one at Mont Revard, with a 29-centimeter Merz refractor, the other at Le Massegros, with a 37-centimeter Schaer refractor. Jarry-Desloges had written to Antoniadi about the recent discovery of a yellowish veil on Mars. Eager to see for himself, Antoniadi fetched the Calver reflector from storage and set it up at Juvisy-sur-Orge. On August 11 he found the markings on Mars to be unusually pale, as if seen through a veil of "pale lemon" haze or possibly very thin cirrus.[62] The pale-yellow appearance continued up to August 23, after which the markings regained their normal strength.

Meanwhile, another letter arrived, of greater consequence, from Henri Deslandres (1853–1948), the director of the Meudon Observatory, inviting Antoniadi to come to Meudon for Mars's opposition, to use the Grand Lunette.

Obviously, despite the hiatus in Antoniadi's career, his talent had not been forgotten, and so at age 39, the same as that at which Lowell had taken up the study of the planet, Antoniadi found himself summoned to a stage worthy of his great abilities.

The building housing the Grand Lunette is built on the foundations of an ancient castle that was almost completely destroyed by the Prussians during the Siege of Paris in 1871, and it rises on the edge of a high terrace that slips unimpeded to the Meudon Park below. The sheer drop on that side, overlooking the city of Paris, causes the seeing to be generally very good for objects to the east of the meridian, even those that are low and rising; to the west, where there is no such drop, air currents from the ground cause the seeing to deteriorate rapidly as objects approach the meridian.[63] With Mars rising above the city of Paris, Antoniadi's first evening on the Grand Lunette was on September 20, 1909. As Richard McKim points out, it would have been rather tricky to use the instrument in 1909.[64] There was no rising elevator floor like those installed for the great refractors at Lick or Yerkes; instead, a small, unsteady-looking platform was hoisted up and down the dome wall by a system of giant rails and pulleys. Riding the platform off the floor, Antoniadi would have seen only a long vertical strip of starlight through the dome's narrow slit as the eyepiece end of the Grand Lunette approached. Darkness would have swallowed everything else around him (figure 9.6).

FIGURE 9.6 A rare image of Antoniadi, the great "Astronome volontaire à l'Observatoire de Meudon," in action at the 83-centimeter Henry Brothers refractor, about 1927. At the time Antoniadi was carrying out a long daylight study of Mercury's surface markings. He is taking notes on the viewing platform, while his colleague, Fernard Baldet, peers into the eyepiece. Prior to the modernization of the 1950s, there was no rising floor at Meudon. Courtesy of Audouin Dollfus and the Paris Observatory (Meudon section).

On ascending to his lofty perch, and preparing to put eye to the eyepiece, Antoniadi found that the options were rather limited: someone had borrowed most of the eyepieces, and 320× was the highest magnification on hand. (He would immediately regret he had not more.) Though as with any large refractor, chromatic aberration could not be entirely avoided—revealing itself in a bluish-violet ring that fringed the image when the eye end of the focusing tube was pushed in, and a purple-red fringe when it was drawn out—Antoniadi claimed he did not find it greatly troubling in the case of Mars. Moreover, he recounted, "notwithstanding the great length of the refractor [16 meters], its focusing varied but little during the observations through the passage in front of the object-glass of masses of air of differing densities. Indeed, the right hand of the observer was more free for drawing with the [83-centimeter] refractor than was the case with the [22-centimeter] Calver on the lawn at Juvisy."[65] Clearly, the Henry brothers had taken great care in constructing the telescope; its performance was nearly as perfect as could be expected for an instrument of such great size.

The sky transparency over Paris that night was not very good; there was a fog over the city. But this proved, in fact, highly advantageous, for in Antoniadi's experience it was almost always during light mist that the images were steadiest at Meudon and the full aperture could be used to advantage. By contrast, on nights when the sky was transparent, with brilliant, strongly twinkling stars, he usually found the images highly agitated, and the Grand Lunette showed Martian surface patches no better than a 30-centimeter or smaller instrument. September 20 thus conformed to the rule, and indeed, Antoniadi, during his entire career, would never again experience better seeing conditions. Moreover, the conditions remained nearly perfect for a period of several hours. Experiencing something akin to an astronomical miracle, Antoniadi could hardly believe his eyes. "The first glance cast on the planet," he wrote, "was a revelation. . . . [I] did not believe that our present means could ever yield us such images of Mars." He continued,

The planet appeared covered with a vast and incredible amount of detail held steadily, all natural and logical, irregular and chequered, from which geometry was conspicuous by its complete absence. Syrtis Major was approaching the central meridian, and seemed expanded into a huge cornucopia, twice severed by dusky bridges. . . . A gigantic triple bay (Deltoton Sinus) extended N. of Hammonis Cornu. Lacus Mœris . . . [enclosed] a "peninsula" and an "island," constantly visible. Libya and Hesperia appeared shaded, Mare Tyrrhenum like a leopard skin! . . . A maze of complex markings covered the S. part of Syrtis Major, and, although these were held quite steadily, no trace whatever of "canals" in the dark regions could be detected.[66]

Antoniadi's practice was to concentrate on a small region on the shimmering disc and await the most favorable instants to catch the fine details and engrave them in his memory. He did the same in turn with each adjacent region until, having stored all the shapes in their relative positions, he would withdraw from the eyepiece and sit in front of a table with a lit control panel on the observation deck. He then proceeded to set down on paper the impressions before the memory of them began to fade.

"The greatest thing a human soul ever does in this world is to see something, and to tell what is saw in a plain way," wrote John Ruskin (1819–1900) in *Modern Painters*. "Hundreds of people can talk for one who can think, but thousands think for one who can see. To see clearly is poetry, prophecy, and religion, all in one."[67] Antoniadi had learned to see, and to represent exactly and effortlessly the forms before him in order to build an image from component shapes. He was thus uniquely qualified for the task before him, and he succeeded so greatly that his drawings of Mars at Meudon have become legendary.[68]

For all that, Antoniadi was a touchy personality to deal with. Even in chess, according to McKim, he often published sharply personal and idiosyncratic opinions about the game, and he regularly crossed swords with other writers in the chess-playing literature.[69] In what might be described as the chess match over Mars, following the magic revelations of that first night, he emerged as an aggressive champion of the anticanal school. Lowell, who could be "a rather agile party to deal with," had finally met his match. For Antoniadi, too, was a "rather agile party."[70] On the day after his Mars revelations, he dispatched notices to a newspaper in Athens and to the Council of the BAA, confidently asserting that "narrow black canals . . . were not visible at all, although details much smaller . . . were quite plain in that giant telescope. I conclude that those geometrical spider's webs . . . do not exist, as vanishing in the great instrument, which shows us this neighbor world much more resembling the Earth than was hitherto believed to be the case" (plate 5).[71]

Later, he could not resist repeating the account of the magical night over and over. To William Henry Wesley (1841–1922), the assistant secretary of the Royal Astronomical Society, he wrote, "I have seen Mars more detailed than ever": "Mars appeared in the giant telescope very much like the Moon, or even like the aspect of the Earth's surface such as I saw it in 1900 from a balloon at a great height (3,660 meters)."[72] He recounted that "at first glance" through the Grand Lunette he "thought he was dreaming and scanning Mars from his outer satellite."[73] Notwithstanding the "prodigious and bewildering amount of sharp or diffused natural, irregular, detail, all held steadily," he could not see a straight line anywhere on the disc. Instead, he declared, "it was at once obvious that the geometrical network of single and double canals discovered by Schiaparelli was a gross illusion."[74] He added, "we have never seen a *single* genuine canal on Mars, nor should we see any from Phobos, the nearest satellite to the planet, on a disk of 42°."[75]

Though Antoniadi never again experienced seeing like that of September 20—and in fact usually found the seeing at Meudon to be poor, with images that he referred to as "boiling"—there were occasional exceptions. The night of October 6, 1909, was "glorious," and Antoniadi made out an unevenness of tone in Mare Sirenum while also resolving the faint and then pear-shaped Solis Lacus into at least six irregular patches. The night of October 11 was also good, allowing confirmation of much of what was seen on October 6. Of nine observing nights between September 20 and October 23, three were entirely satisfactory; the others yielded unsteady, agitated images, though even under these conditions he found that during the less agitated moments, "the large aperture outperformed smaller instruments." Good views were "preceded by a period of slight rippling of the disk, very detrimental to the detection of fine detail. The undulations would then cease *suddenly*, when the perfectly calm image of Mars revealed a host of bewildering irregularities."[76] He found nothing to support the usual Flagstaff practice of diaphragming the lens, even when the seeing was indifferent; while on the best nights, when the image was quite steady, "Mars appeared covered with such a bewildering amount of detail that its perfect representation was evidently beyond the power of man."[77] Incidentally, it appears that his unusually retentive eidetic memory allowed him to grasp and retain detail that would have left most observers in utter confusion, far exceeding anything accessible in even the best photographs of the time (figure 9.7).

At a time when the majority of French, German, and British Mars observers were arguing like medieval scholars about the number of angels able to dance on a pinhead—debating which canals had doubled and which ones crossed which oases—Antoniadi moved the discussion to a different level. He did not take credit for being the first to see the reality of the Martian surface details; instead he gave that credit to Barnard, who in 1894 had delivered "a rebuke from which the spider's webs have never recovered."[78] After hearing of Antoniadi's success, Barnard wrote: "I am particularly glad that you have had the same view of the planet, and have drawn it with the great Meudon refractor. I note in your drawings the canals have become broader, more diffuse and more irregular than most people show them. This is in accord with my own observations... made with large instruments."[79] Similarly, Hale, using the 1.52-meter reflector at Mount Wilson Observatory (diaphragmed to 1.12 meters) and a magnifying power of 800×, concurred. "I was able," he wrote, "to see a vast amount of intricate detail—much more than has been shown on any

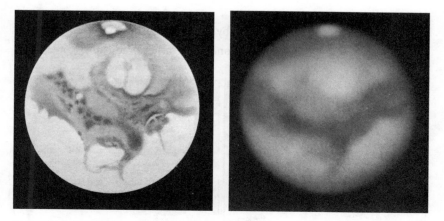

FIGURE 9.7 *Left*: Antoniadi's drawing with the 83-centimeter refractor at Meudon Observatory, September 20, 1909. Published in his "Fourth Interim Report," *JBAA* 20 (November 1909). *Right*: The best photograph of Mars taken up to that time, by E. E. Barnard with the 1.02-meter refractor of Yerkes Observatory, September 29, 1909. Courtesy Richard McKim and G. P. Kuiper and M. R. Calvert, *Astrophysical Journal* 105 (1947): 215.

drawings with which I am acquainted. In spite of the very fine seeing on certain occasions . . . no trace of narrow straight lines, or geometrical structure, was observed. A few of the larger 'canals' of Schiaparelli were seen, but these were neither narrow nor straight . . . I am thus inclined to [think] . . . that the so-called 'canals' of Schiaparelli are made up of small irregular dark regions."[80]

Indeed, this was Antoniadi's view as well. Although he had absolutely no faith whatsoever in the Lowellian spiderwebs, he did believe that a basis in reality existed for most of the Schiaparellian canals (especially as shown in Schiaparelli's 1877–78 map, in which they are broad and winding rather than as later geometrically straight, and often double, lines; figure 9.8). Antoniadi later wrote:

> In the positions of each of [Schiaparelli's canals], single or double, on the surface of the planet, there is present an irregular trail, a jagged edge of halftone, an isolated lake, in a word, something complex. These details are extremely varied. . . . Due to the exceptional acuity of his sight, Schiaparelli surpassed all the observers who worked with instruments of equal size; but his modest refractors did not permit him a glimpse of these complex details in any other form than that of fleeting lines. To do so he would have had to overcome the immutable barrier of diffraction.[81]

Despite Antoniadi's endorsement by Barnard, Hale, and others, who concurred that the canals were illusory, one man, rather predictably, took rather

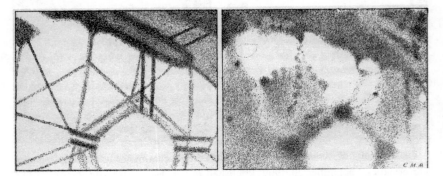

FIGURE 9.8 The "canals" as seen by Schiaparelli (*left*), and the same area of the planet as recorded by Antoniadi. Elysium is the somewhat polygonal form near the bottom of each image. Published in E. M. Antoniadi, *La Planète Mars* (Paris: Librairie Scientique Hermann et Cie, 1930), figure 7.

violent exception to that conclusion: Percival Lowell. Just before the 1909 opposition, Antoniadi had renewed a correspondence with Lowell that had been dormant ever since Antoniadi's sharp attacks on Lowell's Venus work. He offered as an olive branch his remarkable skill in drawing. "I draw well, and if I can be of any use to you in that line, I hope you shall not hesitate to apply to me in need."[82] Lowell responded by sending Antoniadi copies of Lowell Observatory publications and lecturing him on the need to diaphragm the aperture of the Grand Lunette.[83] By the time this letter of counsel arrived, Antoniadi had already experienced his night of revelation—using the full aperture of the telescope.

Deploying the formidable artistic skill he had mentioned to Lowell, Antoniadi sent four exquisitely executed pencil copies of his Mars drawings, made on September 20 ("splendid definition"), September 25 ("bad definition"), October 6 ("glorious definition"), and October 11 ("moderate definition"). The drawings spoke for themselves, but Antoniadi added a few comments. "Of course, the Flagstaff skies must give you an ideal definition," he conceded; "here in N. France, we seldom have really good images." However, he insisted that his work with the Meudon refractor had surpassed all his expectations:

> I hasten . . . to send you 4 of my best drawings, obtained with the full aperture (83 cm) of that splendid refractor. These drawings are by far the best I have ever made. . . . With the exception of [several] "linear canals" . . . all seen by flashes of ⅓ of a second, all the other markings I show were held steadily, and the tremendous difficulty was not to see the detail, but accurately to represent it. Here, my experience in drawing proved of immense assistance, as, after my excitement, at the bewildering amount of detail visible, was over, I sat down and drew correctly, both with regard to form and intensity, all the markings visible. On Sept. 20 . . . however, one third of the minute features I could not draw; the task being above my means.[84]

In addition to the four drawings just mentioned, Antoniadi also sent a second drawing made on September 20, this one showing how the fine details were smudged in "tremulous definition." This drawing showed fine Lowellian spider threads. Lowell's response to Antoniadi shows why Campbell had once worried that it was impossible to argue with him "along honest scientific lines."[85] Lowell told Antoniadi, "The one you marked tremulous definition strikes me as best.

It is capital." The others, made under much better conditions, Lowell thought were not so good; somehow he managed to convince himself that details such as canals had been blurred by Antoniadi's use of the full aperture.[86] Privately, Lowell told science journalist Waldemar Kaempffert (1877–1956) that Antoniadi was "a man without knowledge of how to observe."[87]

Antoniadi's last letter to Lowell is dated November 15, 1909. There was absolutely no possibility, he wrote, that the incredible detail he had seen with the Grand Lunette had been due to "blurring." On September 20, 1909, Mare Tyrrhenum had appeared mottled "like a leopard skin," and instead of being laced with canals, the Martian "deserts" had shown "a maze of knotted, irregular, chequered streaks and spots."[88] Indeed, in the desert known as Amazonis, where in bad seeing "hideous lines" were wont to appear in brief glimpses, he had on October 6 and again on November 9 been vouchsafed nothing less than "an elementary view of the true structure of the Martian deserts." The image had been slightly tremulous on each date. However, suddenly definition had become perfect, and a wonderful sight had presented itself for a dozen seconds on both occasions, the "soil of the planet becoming covered with a vast number of dark knots and checquered fields, diversified with the faintest imaginable dusky areas, and marbled with irregular, undulating filaments, the representation of which was evidently beyond the powers of any artist."[89] Even held thus steadily—and despite his extraordinary powers as a draftsman—the details were too intricate to draw. Antoniadi could only produce an impressionistic sketch (figure 9.9). It was not published until 1988, when the detail was found to strikingly resemble the pattern of windblown streaks around the cratered terrain shown in Mariner 9 and Viking spacecraft images.[90] Such was Antoniadi's skill as an observer, the retentiveness of his eidetic memory, which he had honed during his years under the vast vault of Hagia Sophia and now deployed to the elucidation of the detail on another world.

Anticlimax and a Death

After the great 1909 opposition, Mars began to retreat into a series of less favorable oppositions, and inevitably public interest faded. At the opposition of November 1911, Antoniadi again observed Mars from Meudon, but thereafter

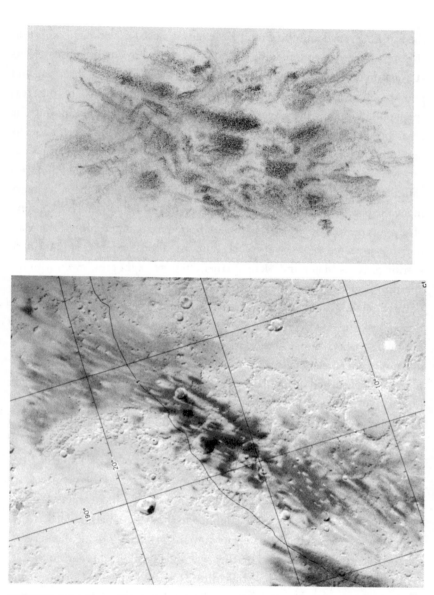

FIGURE 9.9 *Above*: The impressionistic sketch included in the letter from Antoniadi to Percival Lowell, November 15, 1909, as described in the text. The area shown is a swatch in Amazonis. Courtesy of Lauren Amundson and Lowell Observatory Archives. *Below*: Detail from airbrush Mariner 9 topographic relief map MC-15, of the Elysium quadrangle, orientated to make a suggestive comparison with Antoniadi's sketch. Published in R. M. Batson, P. M. Bridges, and J. L. Inge, *Atlas of Mars* (Washington, D.C.: NASA, 1979).

the war intervened and the Grand Lunette was closed down owing to extensive repairs needed to the dome. (He did not make any further observations
at Meudon until 1924.) In 1911, Barnard was at Mount Wilson, observing and
photographing Mars with the 1.52-meter reflector. Though most of his time was
spent photographing the planet, he also made visual observations, in which he
favorably compared the image in the great reflector to that in the refractors he
had hitherto used. In his observing logbook he noted that images in refractors were inevitably degraded by chromatic aberration and seemed to have a
"muddy or dirty look"; the color-free image in the reflector, on the other hand,
made the planet appear "as if cut out of paper and pasted on [the] background
of sky. It is perfectly hard and sharp with no softening of the edges." The Martian disc appeared to him "very feeble salmon—almost free of color"; the dark
markings Syrtis Major and Mare Tyrrhenum were "light grey." Even with the
aperture stopped down to only 30-centimeter because of indifferent seeing,
Barnard saw Mars better than he had ever seen it before. Syrtis Major in particular broke up "into a great number of wispy masses. . . . No trace of anything
resembling a canal either in the dark or the bright regions could be seen."[91]

Lowell, nonetheless, gave no ground. Despite what the professional astronomers said, he still had the public on his side. Though interest in Mars had fallen
off after 1907, even in 1910, during a visit to England, Lowell found that the British public "is simply mad over Mars," and that "hundreds of people were turned
away from the lecture at the Royal Institution." He added that the *Daily Mail*
had been receiving hundreds of letters on the subject of Mars, while he himself had been mobbed by reporters and photographers.[92] It is hardly necessary
to add that his observing logbooks for the increasingly unfavorable series of
oppositions of 1911, 1914, and 1916 continue to show the canals. Indeed, if anything they appear more obstinately narrow and artificial-looking than ever.[93]

Lowell continued to defend his theory of intelligent life on Mars to the bitter end—including on an exhausting lecture tour of the Pacific Northwest and
the West Coast of the United States in September and October 1916, during
which he railed, with increasing stridency, against the "troglodytes," the astronomical "conservatives" who failed to acknowledge the Flagstaff discoveries.

One wonders whether, had he only taken up the challenge of looking through
Antoniadi's or Barnard's telescopes—as some suggested at the time—he would
have changed his mind.[94] But somehow this sort of collaborative work didn't

seem to be part of his repertoire. He had very little of what the poet John Keats (his sister Amy's favorite) called "Negative Capability"—"that is when [a] man is capable of being in uncertainties, [m]ysteries, doubts, without any irritable reaching after fact & reason."[95] Rather than having an open-minded cognitive style and tolerance of ambiguity, he possessed a rigid, obsessional personality structure with great need for closure. He always wanted things sharply defined as black and white, as appears even in his planetary drawings (including those of Venus), where everything is depicted as hard, sharp, clearly bounded, well defined. Perhaps this need for definiteness and certainty had something to do with the Puritanism of New England culture.[96] Whatever its basis, it was part and parcel of his "personal equation." For a personality like Lowell, Mars, which to most observers appears blurry and unclear and well defined only in glimpses, appeared "hard and clear, never blurred or indefinite."[97] It was inevitable that for such a personality continued research only reinforced the same conclusions reached during his flurried first half year of observations:

> I have said enough to show how our knowledge of Mars steadily progresses. Each opposition as it comes round adds something to what we knew before. It adds without subtracting. For since the theory of intelligent life on the planet was first enunciated 21 years ago, every new fact discovered has been found to be accordant with it. Not a single thing has been detected which it does not explain. This is really a remarkable record for a theory. It has, of course, met the fate of any new idea, which has both the fortune and the misfortune to be ahead of the times and has risen above it. New facts have but buttressed the old, while every year adds to the number of those who have seen the evidence for themselves.[98]

At the end of his lecture tour, Lowell's train pulled into Flagstaff on the evening of October 19, 1916. To all appearances still in good health and only 61 years old, he was as usual high strung, but rather than allowing himself the rest he needed he threw himself at once into a new program of heavy observational work, teaming up with E. C. Slipher in taking nightly measures of the fifth satellite of Jupiter, Amalthea (discovered by Barnard in 1892), with the Clark refractor. He was chasing his latest notion—that the interior of Jupiter might consist (as he believed Saturn's to consist) of a series of differentially rotating shells, "like an onion in partitive motion." His last observing entry recorded the results of the night of November 11. The following morning, a Sunday, shortly after becoming angry with one of the drivers or a member of the household staff, he suffered a massive stroke. By 10 o'clock that evening he was gone. Constance Lowell (1863–1954), whom Lowell had married in 1908, recorded a chalk inscription on the wall of the room in which he expired: "Percival Lowell's earthly existence terminated in this chamber upon the green couch."[99]

Lowell was gone, but like Caesar's ghost, his influence would be mighty yet. William Graves Hoyt has said, "No one before him or since has presented a case for intelligent life on Mars so logically, so lucidly, and thus so compellingly."[100] And so the Flagstaff astronomer would continue to cast a long shadow over Mars studies right up until the spacecraft era. Even now, at a time in which we have spacecraft in orbit around the planet, and landers and rovers on the surface, it is impossible to entirely avoid seeing Mars through Lowell's tessellated eyes.

10

The Martian Sublime

A Matter of Brain-Directed Vision

Eugène Michel Antoniadi's observations at the 1909 opposition suggested to many astronomers that canals, at least in the form of the grand Lowellian irrigation system, were most likely a case of the emperor's new clothes. Historian Robert Markley has pointed out, "enthusiasm [for them] slid into agnosticism, and with interruptions, declined into skepticism."[1] The slide was hastened by the death of their greatest champion, and there remained only a few defenders, notably astronomers on Percival Lowell's own staff. But the dream of the canals lingered in science fiction, and would figure in the paintings of space artists like Howard Russell Butler (1856–1934) and Chesley Bonestell (1888–1986), and the novels of Edgar Rice Burroughs (1875–1950), Ray Bradbury (1920–2012), and many more (figure 10.1). Only in *The Sands of Mars* of 1950 did Arthur C. Clarke (1917–2008) break with Lowell's legacy and make Mars a canal-less world.

The final death knell came from Mariner 4 during its July 1965 flyby, which showed craters like those on the Moon but no canals. After that, there was nothing left to do but write their requiem.

But if the canals do not exist, how did so many astronomers, especially in the 1890s and 1900s, believe they did? We can say they were only illusions, but even illusions need explanations.

FIGURE 10.1 A Lowell-inspired vision: *Mars as Seen from Deimos*. Oil on canvas by Howard Russell Butler. Princeton University, gift of H. Russell Butler Jr. Courtesy of Princeton University Art Museum.

In an obituary of Lowell, Princeton University astronomer Henry Norris Russell (1877–1957), a longtime friend and supporter of Lowell Observatory, suggested that ability to see the canals—or inability to see them—was an example of the "personal equation." This had first been recognized in the early 19th century among observers trying to determine the moment a star

crossed the wire of a transit instrument. It was found that observers who
attended to the clock before switching attention to the wire differed in their
estimation from those attending to the wire before switching to the clock. The
former underestimated the time; the latter overestimated. Something similar,
said Russell, seemed to occur with Mars. Observers who attended first to the
desert areas differed from those who attended first to the dark areas, with the
former more apt to see the canals.

> Under ordinary circumstances, Mars is a heartbreaking object for the observer.
> The larger and less interesting details upon his ruddy disc are indeed visible
> telescopically on any good night, but the finer markings are very delicate, and
> flash into view only by glimpses in the too rare moments when the ceaseless
> turmoil of the atmosphere through which we must look dies down and permits
> us to see the planet with relatively little blurring of its finer lineaments. . . . There
> is, however, a remarkable diversity in the descriptions of their appearance given
> by those who have studied them at first hand. . . .
>
> Between the entrance of light into an observer's eye and his record of his
> observation, whatever this may be, intervenes a process of extreme complexity
> going on in the recesses of the brain, which we can follow only with difficulty,
> and mainly by its effects. In so simple a case as that when a man merely presses
> a telegraph key to record the time when he sees a thing happen one may be
> two or three tenths of a second ahead of another—not once, or by chance, but
> again and again—by almost the same amount. One man's nervous system works
> slowly, while his neighbor, perhaps, anticipates the event. . . .
>
> In the far more complex case of observations of planetary details, concern-
> ing which all observers agree that they can be seen only in the best moments
> for [at most] a few seconds at a time, similar principles are doubtless to be
> applied.[2]

Russell himself did not work out the details, though he certainly grasped the
gist of the matter. In fact, the neurophysiology involved in the visual observation of a planet (or anything for that matter) is dauntingly complex; whole
books could be written about the topic.[3]

A key aspect of the canals' visitations was their tendency to be seen only in
"glimpses" and "flashes." Many observers, including Antoniadi, were emphatic

about this.[4] Nor would Lowell himself have disagreed. In *Mars and Its Canals* (1906), he wrote:

> When a fairly acute-eyed observer sets himself to scan the telescopic disk of the planet in steady air, he will, after noting the dazzling contour of the white polar cap and the sharp outlines of the blue-green seas, of a sudden be made aware of a vision as of thread stretched somewhere from the blue-green across the orange areas of the disk. Gone as quickly as it came, he will instinctively doubt his own eyesight, and credit to illusion what can so unaccountably disappear. Gaze as hard as he will, no power of his can recall it, when, with the same startling abruptness, the thing stands before his eyes again. . . .
>
> By persistent watch . . . for the best instants of definition, backed by knowledge of what he is to see, he will find its comings more frequent, more certain and more detailed. At last some particularly propitious moment will disclose its relation to well-known points and its position be assured. First one such thread and then another will make its presence evident; and then he will note that each always appears in place. Repetition in situ will convince him that these strange visitants are as real as the main markings, and as permanent as they.[5]

Though vision appears nearly instantaneous when compared to the slow photographic plates available in those days, in fact the eye-brain combination has—as we would express it in the era of charge-coupled device (CCD) imaging—a fairly slow capture rate of 5 to 15 images per second.[6] (The exact rate, dependent on the brightness of the image and variations between different observers, defines the threshold for "flicker fusion," the term used by perceptual psychologists to refer to the way the eye-brain connection fills in portions of a picture to smooth out and average noise so as to create the perception of a sharp, vivid picture.) The eye-brain combination thus resembles a digital device with rather limited capacity, so whenever movements occur in increments with duration shorter than, say, an 8th or a 10th of a second, it is simply too quick for the eye-brain combination to isolate individual images.[7]

Though early attempts to understand the psychology of planetary observation were made by Andrew Ellicott Douglass and others, not enough was known to get very far along that road. Of course, we now know a great deal about perceptual processes. Vision begins with the retina's photoreceptors, some of which are sensitive only to dark-light differences (rods), others to wavelength differences (cones). The cones are packed in a small (five-millimeter) pit in the macula lutea of the retina, also called the fovea. Here resolution of detail is greatest, but foveal vision subtends only 1.5° of the visual field, so that in order to take in all of the visual environment the fovea is constantly reorienting itself by means of jerky movements known as saccades. Saccades, occurring at intervals of about one-third of a second, redirect the fovea to whatever is most interesting. During the saccades—the intervals when the eye is roving—inputs are suppressed. Thus, the apparently seamless visual world we experience is perceived mostly during fixations of the eyes, short intervals (approximately 200–300 milliseconds) between saccades when the eyes are stationary.

In the case of Mars observations, a brief interlude of good seeing as a larger air cell of uniform refractive index passes in front of the telescope produces a sudden flash of detail—a "revelation peep" or "tachistoscope flash" (figure 10.2). Becoming conscious of this detail, the observer unconsciously initiates a saccadic movement to try to direct the fovea onto it. Once the detail is registered on the retina, the information flows into the brain, where data about shape, color, and orientation is processed; the brain attempts to "build up the scene," storing the information in short-term visual memory (which is very poor for exposures of less than 100–200 milliseconds). Thus, instead of stitching information together from every fixation to build up a detailed picture, the brain only extracts conceptual (schematic) information from visual input. Here the observer's "priors"—expectations and beliefs—enter in, and the brain produces a schematic version or "scene gist." This is rather like a rough draft of the scene, and it is just complete and detailed enough to allow visual comprehension. This process, too, is extremely rapid, requiring as little as 13 milliseconds (usually longer).[8] The important point is that the "scene gist" is not a facsimile, and its creation is very sensitive to attentional effects. An observer trying to focus on a detail such as a dark shading isolated in the desert will, with each flash of good seeing, subconsciously move the eyes to follow the direction of the more conspicuous lines in the field of view, so that, as the Italian neuroscientist Giovanni Berlucchi explained at a 2010 conference dedicated to Giovanni Virginio Schiaparelli, "The subject tends to see only those parts of the scene that are in spatial register with the direction of attention. In unattended parts of the visual field, observers report many

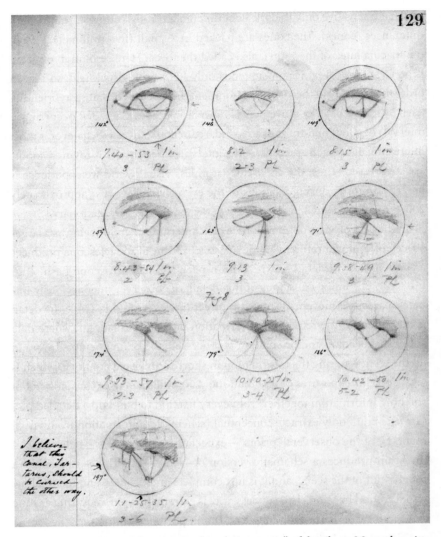

FIGURE 10.2 Moment-by-moment "revelation peeps" of details on Mars, changing with the seeing conditions. These sketches by Percival Lowell made on November 3, 1894, illustrate what has been referred to as the "tachistoscope effect" and show how a highly complex map of Mars was built up by the accumulation of detail seen only by glimpses. Courtesy of Lauren Amundson and Lowell Observatory Archives.

illusory conjunctions."[9] These "illusory conjunctions" were the components of the Lowellian spiderwebs. (Note: we are not claiming here that the handful of actual broad streaky features corresponding to fields covered in windblown dust were illusory; a significant number, most notably the Agathodaemon canal, which corresponds to the Valles Marineris and can be seen in very small telescopes, are quite real; we are referring here rather to the ubiquitous spider threads.)

Putting it more simply: *You see with your brain, not with your eyes*. As long ago as 1892, the Irish writer Ellen Clerke (1840–1906) called Schiaparelli "a gazer gifted with the supreme power of brain-directed vision."[10] This was, of course, a great insight, to which must be added another one: the eye-brain-hand system did not evolve for the purpose of making planetary observations but for survival, as on the African savanna where our hominid ancestors had to very quickly decide whether a rustle in the grass was a lion or the wind blowing. This situation has been described as "processing under pressure."[11] If we assume the rustle is a lion and it turns out to be merely wind, we have made a Type I error in cognition (a false positive). We startle, move away, and realize our mistake. But if we assume that the rustle is due to the wind when it is really a lion, we have made a Type II error (a false negative), and the consequences are fatal. Thus, we are all biased toward making Type I errors, which involves seeing meaning in meaningless noise, what psychologist Michael Shermer has described as "patternicity."[12]

The arrangement of actual surface features on Mars is closer to meaningless noise than to the kinds of geometric patterns recorded by Schiaparelli and Lowell. And yet, under the difficult conditions of observation in flashes ("processing under pressure"), many observers were guilty of making Type I errors regarding the nature of the Martian surface features. As with assuming the rustle due to wind to be a lion, in the case of possible irrigation channels on Mars—evidence of inhabitants of the planet—it was preferable to make a Type I error (a false positive) and grasp something of startling and even potentially momentous importance than to make a Type II error (a false negative) by assuming one was seeing nothing more than bare rock and sand. The Martian canals, where marginal perceptual data was pushed to support conclusions greater than it could possibly bear, stand among the best exemplars of Type I errors in the history of science.[13]

Antoniadi wrote in 1909:

We are thus led to consider the dangers of glimpsing. A glimpsed object is not as certain as an object held steadily, and, however self-evident or trite such a remark may be, yet it is a very important one to make here. A great many real objects are doubtless glimpsed with inadequate means; but a great many subjective appearances are also glimpsed, and, unfortunately, not recognised as such; and, were due regard to have been paid to the treacherous character of glimpsing, the existence of many celestial marvels would never have been foisted on the scientific world.[14]

Mars Without Canals

With or without canals, Mars continued to be envisaged in Lowellian terms as a dying, desert world. The timing of his writings about Mars was perfect, for, as George Basalla points out, he promulgated his theory just as the public was becoming increasingly preoccupied with water issues in the American West.[15] The years in which the Martian canals were being discussed saw the passage of the Desert Land Act of 1877, which mandated irrigation as part of a homestead claim in the West, and the National Irrigation Congress in 1891. A few years later the Newlands Act of 1902 was passed, authorizing the formation of the U.S. Bureau of Reclamation, which "remade the face of the West."[16] Putting Arizona in the forefront of its program, the bureau chose as its first project the damming of the Salt River below its confluence with Tonto Creek, in order to store water for later release into the Phoenix area. The dam project got underway in 1903 and was completed in 1911. Bringing water to central Arizona was the crucial step in the move to statehood in 1912. Against such contexts, Lowell's attempt to see in the Martian scenery the effects of water reclamation on a planetwide scale would have seemed entirely plausible. He was, as historian Michael Amundsen has put it, "seeing Arizona, imagining Mars."[17]

Robert Markley has observed:

Even Lowell's critics accepted the broad outlines of [his] evolutionary theory and appropriated [his] metaphor of a dying world to describe their views of

Mars. . . . The quest to understand the vast time scales of planetary development has been at the forefront of comparative planetology since Lowell's time, and his canal theory owed much of its popularity to his skill in relating an impersonal geological history to the compelling picture he evoked of the struggle for existence on a desert world. In this respect, the canals allowed Lowell's readers to meditate on relationships between the human experience of scarcity and the larger cosmic forces that, in the late nineteenth century, seemed to indicate that Earth was "going the way of Mars."[18]

In setting up his observatory in the Arizona Territory, Lowell practically predestined himself to seeing analogies not only to Earth as a whole, as previous astronomers had done, but to the particularly rich and suggestive part of it that became his adopted home. His visions of the far-off planet were incomplete and fragmentary but could be eked out by experience in the arid geological wonderland around Flagstaff. Indeed, he and his friends made regular forays to Sunset Crater; to Wupatki Ruins and the Grand Falls of the Little Colorado River; to Meteor Crater and the Painted Desert and the Petrified Forest.[19] These explorations added local color to flesh out his visions of distant Mars.

They also added color in the literal sense. No scene impressed itself as more Martian than that of the ruddy washes of color of the distant Painted Desert as seen from atop the San Francisco Peaks. In *Mars and Its Canals*, Lowell described the ocher regions of the Red Planet as:

Nothing but ground, or, in other words, deserts. Their color first points them out for such. The pale salmon hue, which best reproduces in drawings the general tint of their surface, is that which our own deserts wear. The Sahara has this look; still more it finds its counterpart in the far aspect of the Painted Desert of Arizona. To one standing on the summit of the San Francisco Peaks and gazing off from the isolated height upon this other isolation of aridity, the resemblance of its lambent saffron to the telescopic tints of the Martian globe is strikingly impressive. . . . Even in its mottlings the one expanse recalls the other. To the Painted Desert its predominating tint is given by the new red sandstone of the Trias, the stratum here exposed; and this shows in all its pristine nakedness because of the lack of water to clothe it with any but the sparsest growth.

Limestones that crop out beside it are lighter yellow. . . . Seen from afar they have rather the tint of sand; and the one effect, like the other, is Martian.[20]

The pitiless yet still lovely deserts Lowell saw near Flagstaff and imagined on Mars were rapidly seized on by science fiction writers beginning with Burroughs, who in 1912 published his first Mars story, "Under the Moons of Mars" (serialized in six parts in *All-Story Magazine*, then issued in book form as *The Princess of Mars*), which introduced his hero John Carter of Virginia on a planet the Martians themselves called Barsoom. Before taking up writing, Burroughs had as a young man enlisted and been assigned to the Seventh Cavalry at Fort Grant, Arizona Territory. There, amid the scorching Sun-baked lands of the Arizona desert, and in the presence of an artificial lake with waterways, Burroughs took part in hard labor consisting of road work, ditch digging, and "boulevard building." While he was discharged from the army because of heart trouble and left Arizona after only a year, when he came to write his Martian fantasies he could hardly do otherwise than conjure up recollected Arizona desert scenery:

Few western wonders are more inspiring than the beauties of an Arizona moonlit landscape; the silvered mountains in the distance, the strange lights and shadows upon the hog back and arroyo, and the grotesque details of the stiff, yet beautiful cacti form a picture at once enchanting and inspiring; as though one were catching for the first time a glimpse of some dead and forgotten world, so different is it from the aspect of any other spot upon our earth.[21]

Many later science fiction writers, artists, and indeed scientists right up to the spacecraft era would envisage Mars in essentially Lowellian terms, imagining the "really terrible reality" of its pitiless deserts and the inevitable last gasp of the "unspeakable death-grip" to which the nebular hypothesis had consigned it from the first (and would one day in the future consign our own).[22] They also saw it as an otherworldly Arizona. As late as 1955, when movie producer George Pal (1908–80) charged space artist Chesley Bonestell with painting a panorama of the surface of Mars as seen from a landed spacecraft for the film *Conquest of Space*, the movie's director, Byron Haskin (1899–1984), vetoed the work on the grounds it looked too much like Arizona! Bonestell later recalled how Robert S. Richardson (1902–81), a Mount Wilson Observatory astronomer and consultant to the film, rose to his defense. "Well, you know," Richardson dryly noted, "Mars looks a lot like Arizona."[23]

The Post-Lowellian Era

With or without the canals, Mars remained good newspaper copy and continued to excite the general public as the likeliest place beyond Earth to harbor life—perhaps not intelligent life but at least some form of plant life. Admittedly, during the 1920s and 1930s professional astronomers increasingly had more pressing business to occupy them than arguing about life on Mars. The endless arguments over the canals, in the end inconclusive, produced a sense of fatigue. Writing in 1938—the year of the infamous Orson Welles invasion of Mars radio broadcast—Reginald L. Waterfield of the British Astronomical Association (BAA) admitted that the story of the canals was "a long and sad one, fraught with back-bitings and slanders," and that "many would have preferred that the whole theory of them had never been invented."[24] The large telescopes at Mount Wilson, together with the increasing sophistication of spectrographs, inveigled most professional astronomers to the study of problems of the larger cosmos, the stars and the distant nebulae, to the relative neglect of solar system studies, which right up almost to the dawn of the Space Age was relegated largely to amateurs.[25]

There were a few prominent voices during this period still concerned with Mars. One was William Henry Pickering. Believing that "any theory, even a false one, is better than none at all,"[26] he put forward theory after theory to explain the canals, ranging from strips of vegetation lining the banks of actual canals too small to be seen, to strips of vegetation growing along the borders of volcanic cracks similar to those he observed in Hawaiʻi, to bushes or trees planted or left standing in cleared fields, to simple marshes. Increasingly alienated from the professional astronomical community and forced into retirement from Harvard after his brother Edward's death in 1919, William spent his last decades on a plantation at Mandeville, Jamaica, where he observed Mars through the 1924 opposition with a 28-centimeter refractor belonging to Harvard.[27] After this telescope was repossessed by Harvard, Pickering replaced it with a 32-centimeter Newtonian reflector, with which he continued to observe the Moon, Mars, and the other planets at least sporadically. Still respected by amateurs who read his Reports on Mars in the pages of *Popular Astronomy*—44 of them appeared between 1914 and 1930—he was widely regarded among his peers as a crank. Walter H. Haas (1917–2015), a recent high school graduate from Ohio who, in 1935, three years before Pickering's death, had spent several months observing and studying in Jamaica, remarked of the experience, "One could sense a certain disappointment that his achievements had gone unappreciated. . . . By then, the professor's astronomy was perhaps behind the times by a decade or two."[28]

Another figure from the past who briefly returned to the limelight was David Peck Todd. Veteran of the Lowell Expedition to the Andes of 1907, Todd promoted a plan in 1909 to ascend by balloon high into the atmosphere in an attempt to communicate with Mars using Hertzian waves. Such ideas led to his being forced into early retirement from Amherst in 1917. Nevertheless, he continued to be preoccupied with the possibility of communicating with the Martians. Institutionalized for insanity in 1922, he was temporarily released during the close opposition of 1924, and he succeeded in gaining the cooperation of the U.S. military to organize periods of radio silence on August 23, when Mars approached closer to Earth than it had in 120 years. Alas, no wireless messages from Mars were received, and Todd spent the rest of the last years of his long life in hospitals and nursing homes working on what he called "Vital Engineering," a program for achieving eternal life, while also worrying about the prospect that the Sun was about to split in two and bring about the end of the world. He died in 1939.

Pickering and Todd were faded glories. Now the study of the planet passed to younger observers. Among the most talented was Georges Fournier (1881–1954) of France, who in 1907 began to work for René Jarry-Desloges at several of the latter's observing stations in France and at Sétif, Algeria. Fournier, along with his younger brother Valentin (1884–1978), who was also a keen observer of the planets until he was called to the war, depicted the planet with thin wispy lines between regions of different tone rather as Lowell had done. In addition, they made many valuable observations of Martian clouds. The most renowned observer of this period was, of course, Antoniadi, who seems to have forsaken astronomy almost completely for chess between 1917 and

1924. Only in 1924 was his work at Meudon resumed, which would lead to the publication of his classic work, *La Planète Mars*, in 1930.[29] Here, he promoted a view of Mars that can be regarded as, in many ways, a canal-less version of Lowell's. Though, in contrast to Lowell's world of artificial constructions, the features on Antoniadi's Mars were natural, he agreed with Lowell in believing the planet to be in a state of "advanced decrepitude." The reddish tracts were deserts, on an enormous scale, and covered an area "almost ten times larger than that of our Sahara together with the deserts of Libya and Nubia."[30] Some of Antoniadi's greatest contributions were to the understanding of Martian meteorology. He described the two main types of clouds (yellow and white) and was able to give good estimates of their heights. The white clouds were higher in the atmosphere and presumed to consist of ice crystals like the cirrus clouds of Earth; the yellow clouds were either windblown sand or dust. Antoniadi noted that at times the yellow clouds could grow in size and cover a vast extent of the surface so as to virtually obliterate the usual markings. This was the case in December 1924, when Antoniadi could make out only a single dark spot on the entire disc. Its position is now known to be that of one of the great Martian shield volcanoes in this part of the planet. Indeed, his drawing is very reminiscent of photos by Mariner 9 during the 1971 global dust storm.[31]

Antoniadi accepted, of course, that the white spots at the poles were snow, but he doubted there could be much water flowing on Mars. He affirmed, on the basis of his observations of what Lowell had called the "wave of color" but was later called the "brown wave" or "wave of darkening," the vegetation hypothesis. Regarding higher forms of life, he made the following interesting remarks:

> The observations of clouds at considerable heights show that though the atmosphere of Mars is very thin, it is still able to hold ice needles or water droplets in suspension—or even particles of dust. Therefore the atmosphere does not seem to be too rarefied for living things. . . . In this connection we must bear in mind that the people of Tibet have no difficulty in breathing in an atmosphere twice as rarefied as ours. If, then, we consider also life's marvelous power of adaptation, one of the aims of the Creator of the Universe, we can see that the presence of animals or even human beings on Mars is far from improbable. . . . However, it

seems that advanced life must have been confined to the past, when there was more water on Mars than there is now; today we can expect nothing more than vegetation around the vast red wildernesses of the planet.[32]

Cameras, Spectrographs, and Thermocouples

Though Percival Lowell's will had made generous provision for the continuation of the observatory's work after his death, his body was hardly cold when his wife, Constance, set out to circumvent it. In fact, she intended to break it completely, so that she would become the sole heir and beneficiary. During the years of their marriage (1908–16), she had from all indications been a devoted, agreeable, and compatible companion. However, immediately after Percival's death, she showed an unpleasantly argumentative, contentious, and litigious aspect to her personality. According to William Lowell Putnam III (1924–2014), Percival's great-nephew, the consequence of this would be that "seen in hindsight . . . [the] marriage was a disaster of the first magnitude for the pursuit of astronomy."[33]

Mrs. Lowell proceeded to exert a stranglehold on her husband's estate that was resolved only after a costly legal struggle dragging on for more than 10 years, during which the observatory was forced to keep its operating budget to a bare minimum while seeing the resources of the estate progressively sapped through court costs and what Vesto Melvin Slipher, appointed director after the founder's death, called "very excessive" attorneys' fees. Instead of being so well endowed that it could, as Antoniadi once quipped to Gabrielle Flammarion (1877–1962), "talk for a long time about canals on Mars,"[34] the observatory was broke by the time the lawsuit was finally settled at the end of 1925. During these financially difficult years, the senior staff—V. M. Slipher, Carl Otto Lampland, and Earl C. Slipher (younger brother of V. M.)—kept their noses to the grindstone and made a point of doing such work as would speak for itself while, as V. M. put it, "keep[ing] away from telegraphic announcements."[35] Though privately the Lowellian orthodoxy about Mars still reigned, none of the three men was particularly keen to be drawn back into damaging public controversies over the canals.

Despite William Wallace Campbell's rebuttal from the summit of Mount Whitney, V. M. Slipher still maintained that his spectrograph results from 1908, when the air over Flagstaff had been unusually dry, had stood the test of time. However, he was unable to confirm the results in 1924. He also failed to detect evidence of chlorophyll in the spectrum of the planet—a negative result that, needless to say, *has* stood the test of time.[36] Despite these setbacks, he remained "a cautious but convinced Lowellian."[37] When in 1926 a student asked him for a disclaimer of life on Mars to use in a class debate, he replied: "Unfortunately for your side, recent investigations tend more and more to confirm Lowell's conclusions . . . by adding further evidence of atmosphere and water and temperature such as would sustain organic life. While the *canali* markings are best interpreted by assuming Mars possesses intelligent beings, yet the great distance between us and Mars renders the telescope incapable of showing directly objective evidence of living beings."[38]

Lampland was also loyal to his former employer's memory, and he teamed up with the pioneering infrared physicist William W. Coblentz (1873–1962), chief of the Radiometry Section of the U.S. National Bureau of Standards, to use a new instrument for investigating conditions on the planet, the thermo-couple. (The same line of investigations was pursued independently, at the same time, by Edison Pettit [1889–1962] and Seth B. Nicholson [1891–1963], with a 2.54-meter reflector at Mount Wilson Observatory.)

The story of infrared astronomy begins in 1800, with William Herschel. While investigating why wedged glasses placed in the focus of his telescope in order to observe the Sun kept cracking from the heat, he discovered that beyond the red end of the visible spectrum lay a region of invisible "heat rays," much later called the infrared (from *infra*, "below," and red).[39] The idea of using some kind of sensor to detect infrared radiation from the Moon and hence determine its temperature became the preoccupation of Lawrence Parsons, the fourth Earl of Rosse (1840–1908), whose astronomical research built on that begun by his father, William, the third Earl of Rosse (1800–1867), at Birr Castle, Ireland.[40] Though much of this radiation is absorbed by water vapor and carbon dioxide, a window between 8 and 13 microns is sufficiently transparent to allow it through, and in the 1870s Rosse used a thermophile—a device that could convert thermal energy into an electric current—with the

91-centimeter reflector at Birr Castle to measure, roughly, the temperature of the Moon, a remarkable achievement at the time.[41]

Since the planets are so much farther away than the Moon—Mars is always at least 140 times the distance—and are also (at least in the case of the outer planets) much cooler, the detection of thermal emission then requires much more sensitive instruments, which had become available by the 1920s, when Coblentz and Lampland conducted their investigations. Using a 1.2-meter reflector that Lowell had set up on Mars Hill in 1909, Coblentz and Lampland focused light and heat emanating from Mars onto a tiny thermocouple junction (consisting of two dissimilar metal alloys, such as bismuth-antimony or bismuth-tin, in contact) to produce a small electric current, which was amplified and measured in an ironclad galvanometer located at the base of the dome.[42] Lampland's job was to position the telescopic image of Mars or another planet on the thermocouple, while Coblentz, in a little underground room adjoining the reflector dome, recorded the galvanometer readings. By interposing various filter screens that cut off radiation transmission beyond certain wavelengths, they could parse the spectrum into its visible and infrared (heat) components. In this way the approximate temperature of a swatch of the planet's surface could be estimated. Coblentz and Lampland published measures made at the oppositions of 1922, 1924, and 1926—the first ever made of thermal radiation from the surface of Mars.[43]

The measured planetary radiation was actually recorded as percentages of the total radiation. However, these figures were (illegitimately, in Lampland's view) converted into actual thermometric degrees by the young Princeton astronomer Donald Menzel (1901–76), suggesting a degree of accuracy greater than was experimentally warranted. Reported in this form, the two collaborators' results were found to be "in harmony with other observations of Mars," presumably meaning Lowell's 1906 calculations of the temperature based on Mars's albedo (which had yielded a temperature of 22°C [72°F]). Coblentz and Lampland found the temperate zones of Mars to be "not unlike those of a cool bright day on Earth, with ranges from 8° to 18°C or 45° to 65°F."[44] In addition, the dark areas were warmer than the desert areas, which led Coblentz (in an article not co-authored with Lampland) to assume "the existence of plant life in the form of tussocks, whether grass or moss."[45]

Similar results were obtained in 1924 by Pettit and Nicholson. They found that the temperature rose to as high as 20°C for the dark areas and ranged from −10°C to +5°C for the desert areas at the equator. Unlike Coblentz, however, they did not comment on the relevance of their data to the existence of Martian vegetation, whether grass, moss, or other.[46]

Though these values might be interpreted to mean that at least at times and in places, liquid water might exist on the planet so that the inhabitants might enjoy shirtsleeve weather, it should be pointed out that these were *ground* temperatures. In his 1907 critique of Lowell's theories, Alfred Russel Wallace had gone to great lengths to insist that because of the extreme thinness of the Martian atmosphere, air temperatures (which are what are usually referred to by meteorologists) would be much lower. Lampland knew this, and in an unpublished manuscript he wrote, "means of ground temperature near the surface and the air temperature a foot or more above the ground on the Painted Desert may give a difference of 46°F (25°C), and for the middle of summer would no doubt greatly exceed this value."[47] On Mars, this meant that even if ground temperatures were as high as Lowell had calculated, one's feet could be warm even as one's head was freezing. Also, though daytime ground temperatures might be warm on Mars, the thin air provided little ability to retain heat after sunset, so that even in midsummer at the equator nighttime temperatures were bound to be extremely bitter. (All these effects have been subsequently verified by landers and rovers on the surface.) Even so, there seemed little reason to doubt that very tough lower forms of plant life, similar to terrestrial lichens, could survive and even thrive in such harsh conditions.

Meanwhile, during the 1920s, Earl C. Slipher continued the photography of Mars he had begun on the Lowell Expedition to the Andes in 1907. He would image Mars at every opposition through that of 1963, a year before his death, creating a unique photographic record that eventually included some 200,000 photographs of Mars, and that added a great deal to the understanding of seasonal effects on phenomena such as polar caps, clouds, and dust storms.[48] Slipher, and independently the great French astronomer Bernard Lyot (1897–1952) at the Pic du Midi Observatory, pioneered the technique of stacking multiple images, which is now a mainstay of modern digital astro-imagers. By selecting the eight or ten best images for each plate, and stacking them under an enlarger to produce a single averaged image, the photographer reduces graininess of the images, eliminates small defects, and enhances sharp but faint features. Slipher also used the Kodak Dye Transfer process to produce some of the first color images of the planet from tricolor black-and-white negatives taken through blue, yellow-green, and red transmission filters.

The Era of Lingering Romance

All the staff astronomers at Lowell hired by the founder—the Slipher brothers and Lampland—were conservative, and did not stray very far from Lowellian orthodoxy. There was little original initiative in the observatory's Mars program, though this was hardly unique to the situation at Flagstaff. Worldwide, Mars research was rather plodding, at least in part because of the Great Depression, followed by World War II. Many of the most important researchers were in France, where there had always been a great deal of skepticism about Lowell's views—not only, of course, from Antoniadi but also from other well-known figures such as the Abbé Théophile Moreux (1867–1954), who directed the Bourges Observatory at the seminary St. Célestin at Bourges. Among the younger generation a leader was Gérard de Vaucouleurs (1918–95), who, at the oppositions of 1939 and 1941, observed Mars with a 20-centimeter refractor at Julien Péridier's (1882–1967) observatory at Le Houga. De Vaucouleurs would later estimate that the average growth rate in the number of published research papers on Mars was about 6 percent per year during the half century before the spacecraft era, with bulges following the perihelic oppositions of 1909, 1924, and 1956, when the close approaches of the planet rekindled public and scientific interest.[49] (The 1939 bulge was dampened by the outbreak of World War II.) Despite exponential growth in the number of papers published, de Vaucouleurs lamented that many of these papers merely rehashed older research, and claimed that reliable knowledge of the planet had not increased proportionately.

Indeed, though skepticism about the canals was quite firmly established, views about conditions on the planet remained strikingly Lowellian until the early 1960s. For instance, estimates of the thickness of the Martian atmosphere hovered right around Lowell's estimate of 87 millibars, published in

1908. As late as 1954 de Vaucouleurs still argued for a value of 80–90 milli-bars.[50] Attempts to determine the composition of the Martian atmosphere were also stymied. The principal constituent of Earth's atmosphere, nitrogen, is notoriously difficult to detect with the spectroscope. On the other hand, some of the lines of oxygen (A and B lines) and the *a* band of water vapor are in the red part of the spectrum, so though faint, almost invisible, they are in principle detectable—if present on Mars in sufficient quantities—though, as always, the signal for Martian oxygen and water vapor had to be disentangled from the abundant constituents in Earth's atmosphere. Yet another problem was that plates such as those used by Slipher were very sensitive to blue but almost totally insensitive to red. Not until the 1920s were red-sensitive plates suitable for astronomical purposes developed. This, of course, opened up new prospects. While the old method of searching for oxygen and water vapor on Mars was simply to compare the spectrum of Mars with that of the airless Moon, in the 1930s Walter Sydney Adams (1876–1956) and Theodore Dunham Jr. (1897–1984) of the Mount Wilson Observatory used the new red-sensitive plates to apply for the first time a method (first suggested in fact by Lowell) that depended on the Doppler effect: if a body is approaching Earth, the lines in its spectrum will be shifted slightly toward the violet end of the spectrum, if receding, toward the red. As Lowell had noted, Mars's orbital motion relative to Earth would produce displacements in the lines of the spectrum that would be largest when Mars was near quadrature—that is, at right angles relative to Earth and the Sun—allowing, at least in principle, Martian lines to be differentiated from terrestrial ones. The practical application of the method was for a long time limited by the small size of the effect, only 0.03 nanometers, but the 2.54-meter Mount Wilson reflector, the largest in the world at the time, had sufficient resolution for Adams and Dunham to achieve the requisite level of sensitivity. Nevertheless, they failed to detect any lines in Mars's spectrum indicative of the presence of either oxygen or water vapor.[51] Thus, they could only affirm a negative: if any of these gases were present, their abundances had to be below certain very low limits.

Not until after World War II did astronomers gain access to spectrometers sensitive to the near-infrared part of the spectrum. This region, between about 0.75 and 5 microns, is of the greatest interest to astronomers, since it contains the emission and absorption bands of many important chemical compounds found in the atmospheres of planets and stars. Chemists in the early 1900s had developed sensors sensitive to these wavelengths, which revealed a great deal about molecular structure, but the kind of sensors used by chemists in the laboratory were useless at the telescope for astronomical purposes, since beyond 1.5 microns infrared radiation is largely absorbed by water vapor and carbon dioxide in Earth's atmosphere. In this region the infrared radiation of planets and stars is extremely feeble, and in order to explore it, more sensitive detectors and large, specialized telescopes on high mountaintops, above much of the infrared-absorbing water vapor in Earth's atmosphere, are needed. Until the 1940s, none of this was available.

The leading figure in developing infrared astronomy for the study of planets was the Dutch-born astronomer Gerard Peter Kuiper (1905–73), a staff astronomer at Yerkes Observatory of the University of Chicago and later director of the Lunar and Planetary Laboratory at the University of Arizona. His original interest as a student in Leiden had been the origin of the solar system, but during the first part of his professional career, he specialized in double stars. While pondering how double stars formed he was drawn to the grand question of solar system formation. By the early 1940s, when most astronomers were primarily interested in studying stars and galaxies, Kuiper was one of the only professional astronomers to devote a significant amount of his time to the solar system. During two months' leave from war service in winter 1943–44, he used the 2.08-meter reflector at McDonald Observatory, located at an elevation of 2,070 meters on Mount Locke in the Davis Mountains of West Texas, to record the 6.19-micron band of methane in the spectrum of Saturn's moon Titan—the first detection of an atmosphere of a satellite. (Technically, this band was not in the near infrared but in the mid-infrared, which lies between 5 microns and 25–40 microns.) At the end of the war, being fluent in Dutch, German, French, and English, and having contacts with scientists and research facilities throughout Western Europe, he was charged with reading and analyzing captured or intercepted enemy reports and debriefing French and Belgian scientists who were liberated by the Allied armies. In 1945, he joined the Alsos mission, whose purpose was to search behind enemy lines for personnel, records, material, and sites related to wartime German scientific research. Though it had been feared that German scientists were close to developing an atomic bomb, in fact they had made little progress in that direction. They did, however, make a great deal of progress in rockets, producing the much-feared V-2, and near the end of the war were planning to place scientific instruments aboard such a rocket to study the upper atmosphere. In addition, they had been developing lead-sulfide cells for use in infrared detectors, and Kuiper either got the recipe for them or obtained some of the actual detectors while in Germany. On his return to Yerkes, he learned from records still classified at the time but quickly declassified after the war that American efforts to build similar detectors had been underway in the Northwestern University laboratory of physicist Robert J. Cashman (1906–88). Kuiper, keenly aware that no other American astronomer at the time had his insider knowledge, realized that the lead-sulfide cells had made possible an entirely new branch of research: infrared astronomy.

On October 7, 1947, Kuiper used a lead-sulfide cell on the 2.08-meter reflector at McDonald Observatory to obtain simultaneous spectra of the Moon and Mars in the 1–3 micron range. (This region had not been accessible to earlier investigators such as V. M. Slipher or even Pettit and Nicholson.) He used the old method of comparing Mars's spectrum to that of the Moon and detected two small dips in Mars's spectrum at wavelengths of 1.57 and 1.60 microns, as well as indications of others at 1.96, 2.01, and 2.96 microns, not present in the lunar spectrum. These absorption wavelengths are characteristic of carbon dioxide. Astronomers had hitherto expected, by analogy with Earth, that carbon dioxide would be a minor constituent of Mars's atmosphere, so it came as something of a surprise that carbon dioxide—not water vapor or oxygen—was the first gas detected on Mars. Kuiper concluded there was probably about twice as much carbon dioxide on Mars as on Earth.[52] However, he and other astronomers continued to believe that other constituents were present in roughly terrestrial proportions. Indeed, well into the early 1950s, the best guess regarding the Martian atmosphere's composition—given by de Vaucouleurs—was nitrogen 98.5 percent, argon 1.2 percent, carbon dioxide 0.25 percent, and oxygen less than 0.1 percent.[53] All these estimates would prove to be very wide of the mark.

Incidentally, despite the early false positive by V. M. Slipher in 1908 (and others even earlier), the first completely undisputed detection of water vapor on Mars would not occur until 1963, as will be discussed below. Oxygen, which is present in very limited quantities in the atmosphere of Mars and also requires observations above the majority of Earth's atmosphere and highly sensitive instruments able to detect specific wavelengths, is even harder to detect than water vapor, and was not actually found in the atmosphere of Mars until the spacecraft missions of the 1970s.

Thin as the Martian atmosphere is—even a pressure of 80–90 millibars (much higher than the real value, as we now know) is equal to that in Earth's atmosphere at 17,000 meters, an altitude not reached by humans until the stratospheric balloon flights of the early 1930s—it evidently sufficed to support clouds. Lowell had supposed Mars had no mountains and hardly any clouds. He wrote, "the Martian sky is perfectly clear; like that of a dry and desert land."[54] But Antoniadi had shown and E. C. Slipher's photographs confirmed that there are both whitish clouds (sometimes referred to as "bluish" clouds, because they appear most brilliant when the planet is observed in a blue filter) and yellow clouds, which are really dust clouds. The term "yellow cloud" was first introduced by the Irish astronomer Charles E. Burton (1846–82) as far back as the 1870s, though it has become somewhat anachronistic since almost everything on Mars looks bright in yellow light; nowadays such clouds are generally just referred to as dust clouds.[55] As shown by Antoniadi and others in 1909 and 1924, these sometimes develop into enormous storms, covering millions of square kilometers and sometimes growing into monsters that encircle the entire planet.

Martian Atmospheric Phenomena, as Seen from Earth: Clouds and Dust Storms

In the early years dust clouds were not expected, and many—including the large regional storm of May–August 1877, witnessed by Étienne Léopold Trouvelot, and that of October–November 1894, seen by E. E. Barnard but missed by Lowell—were widely overlooked. Discrete clouds, which slightly decrease definition of local boundaries when the planet is viewed with a yellow filter, are not uncommon, forming rapidly whenever Martian winds raise finely divided surface materials to altitudes of three to five kilometers above the surface. They typically last only for a few days. On the other hand, they are an example of a chaotic phenomenon, and under the right conditions they may develop into larger systems, blanketing a region, encircling the planet, or even, at their most extreme, engulfing the entire globe.

The nomenclature for dust storms was introduced by Leonard J. Martin (1930–97) of Lowell Observatory, as follows:

1. A "local" storm is one whose major axis does not exceed 2,000 kilometers
2. A "regional" storm exceeds 2,000 kilometers in at least one dimension, but does not encircle the planet
3. An "encircling" storm is one that spans the entire circumference of Mars

Martin also distinguished planet-encircling dust storms from what he called *global* dust storms. In the latter case, the entire planet is covered, including even the polar caps. These are rare. For a long time, the only truly global dust storm, in this technical sense, was that of 1971, though for reasons that are not entirely clear, they have been occurring with greater frequency in recent years; thus the great dust storms of 2001, 2007, and 2018 were all global (table 10.1).[56]

Apart from the planet-encircling storm recorded by Trouvelot and no one else in 1877, no planet-encircling or global storm seems to have been observed during the height of the Martian canal furor. (The 1894 storm observed by Barnard was a large regional event.) The planet-encircling storm of 1909 began in early June in Hellas. By August of that year, a whole hemisphere of the planet was, as Antoniadi described, "citron yellow," and the usual dark areas such as Mare Tyrrhenum, Syrtis Major, and Sinus Sabaeus had disappeared from view. There was a large regional storm in 1922 (photographed by E. C. Slipher), and then a planet-encircling event in 1924, which Antoniadi observed from Meudon. Antoniadi seems to have been the first to point out that these large storms, which are often followed by striking changes in the forms of the albedo features, always belong to the southern hemisphere, around the time of southern summer solstice, when Mars is near perihelion. This was hardly unexpected, since at perihelion the solar irradiance is 40 percent greater than at aphelion (appendix C), and it produces stronger winds that have a greater capacity to pick up the lighter sand and dust from the Martian deserts. Antoniadi was also right in guessing that the frequency of large dust storms is highest near perihelion, though it is now known that the telescopically observed Martian dust storm season actually lasts from the southern

TABLE 10.1 Great Dust Storms, 1877–2018

Year	Terrestrial date limits	Time for encirclement (days)	Duration (days)	L_s at start (degrees)	Initiation site	Classification
1877	May 8 to Aug. 6	?	91	184	NW Hellas / E. Noachis?	encircling
1909	before June 3 to Sept. 20	13?	119	204	Hellas	encircling
1924–25	Dec. 9 to mid-Feb.	<21	68	311	NW Hellas	encircling
1956	Aug. 19 to Oct. 31	20	73	249	NE Noachis / Iapigia	encircling
1971–72	Sept. 22 to late Feb.	16–17[a]	161	260	Iapigia / NW Hellas	global
1973–74	Oct. 13 to Jan. 11	9	91	300	NW Solis Lacus	encircling
1975	Jul. 14 to Oct. 21	29	100	270	S. Thaumasia	encircling
1977	Feb. 15 to mid-April	19	60	204	Thaumasia Fossae (SW Thaumasia)	encircling
1977	May 27 to late Oct.	<24	158	268	Valles Marineris	encircling
1982–83	Oct. 14 to late Jan.	?	110	208	Southern midlatitudes	encircling
2001	June 26 to Dec. 1	15	159	185	N and NW Hellas	global
2007	June 23 to Oct. 7	15	107	263	NE Noachis	global
2018	May 30 to late Sept.	22	124	184	SE of Niliacus Lacus / Mare Acidalium	global

[a]The time for encirclement was only 12 days according to polarimetric data.

hemisphere spring equinox until the southern hemisphere autumnal equinox. Moreover, spacecraft observations have shown that local and regional dust storms occur throughout the Martian year, although somewhat unexpectedly (and still unexplained), a number of planet-encircling storms have occurred early in the dust storm season, when solar irradiance is much less than at solstice. (This was the case in 1877.)[57]

A holdover from the Lowell era, when the dark areas were believed to be dry sea bottoms, was the assumption that they were areas depressed below the level of the ocher deserts. In 1950, the Estonian-born astronomer Ernst Julius Öpik (1893–1985) suggested that if the dark areas were low lying, they would be quickly covered by yellowish dust unless they had some means of regenerating themselves. At the time, this was seen as a strong argument in favor of the vegetation hypothesis. (Of course, if the dark expanses were actually elevated areas, or simply areas of higher prevailing winds, there was no reason to believe this, as the dust might just as well be removed by wind scouring, as has proved to be the case.)

A great—indeed, paradigm-shifting—planet-encircling dust storm occurred in 1956. As the planet began to come under scrutiny in May, several months before the perihelic opposition, Kuiper found the markings to be of "remarkably low contrast," as though dust were already widely dispersed in the atmosphere. With hindsight, it appears that there had been an earlier regional dust storm, probably originating in Hellas. This, however, was only a prelude to what was to come.

On August 19, Alan Pennell Lenham (1930–66) and Kuiper were observing with the 2.08-meter reflector at McDonald Observatory in Texas. Lenham was first to notice a bright yellow cloud over Noachis, in a region that Antoniadi had depicted as dark in 1928. The following night, two Japanese astronomers, Shotaro Miyamoto (1912–92) at the Kwasan Observatory in Kyoto and Shiro Ebisawa in Tokyo, saw the cloud expanding in an east-to-west direction; initially Miyamoto found it whitish, a few nights later he saw it as yellowish or even red. From this origin, the dust rapidly spread to the west and more slowly to the east. Observers noted an "anomalous darkening" of the ground to the east as the cloud continued to develop. A secondary outbreak in northern Phaethontis/Mare Sirenum occurred on August 30, which enabled the combined dust to develop into a planet-encircling storm. By mid-September, the entire disc had become virtually blank. By October the dust slowly began to clear, until by December the contrast of the features had returned to normal (and were in fact stronger than they had been in May).

Though the storm remained confined mainly to the southern hemisphere and thus never became "global" in a strict sense, BAA Mars Section director Richard McKim says of it, "It was one of the 'great' storms, and the first to be systematically observed visually, photographically, and polarimetrically."[58] Indeed, so significant was this dust storm in the history of Mars research that the year 1956 has been used to define Mars Year 1 in Mars's calendar (see appendix F).

A number of those who later became leading students of the Red Planet were initiated into the pleasures (and frustrations) of Mars observing by that dust storm. Among these were planetary scientist William K. Hartmann, who was using a home-built telescope from his backyard that summer. Looking back, he says, "When I could see hardly any of the markings, I thought I must not be a very good observer, but discovered later that there was a significant dust storm."[59] .

Another distinction achieved by the 1956 dust storm was that it was the first to be recorded on color film. A number of excellent Kodachrome photographs were taken by the noted double-star observer William Stephen Finsen (1905–79) with the 67-centimeter visual refractor at the now-defunct Union Observatory in Johannesburg. Spectacular as they are, they are nevertheless surpassed by those obtained by Caltech physicist Robert B. Leighton (1919–97) with the 1.52-meter reflector (stopped to 0.53 meters) at Mount Wilson (plate 6). Realizing that "solving the technical problem of holding the image steady was what was needed to make progress in photographing the planet,"[60] Leighton devised a photodetector-guided movable lens to compensate for gross image motion during exposures of as much as 20 seconds, and he recorded astonishing and unprecedented photographic detail.

Since 1956, Earth-based observations (and even more spacecraft observations) have added a great deal to our understanding of how dust storms form and develop, and how the redistribution of even tiny amounts of dust is sufficient to dramatically alter albedo features. Local dust clouds can appear during virtually any season and in both hemispheres, though naturally they have been most commonly observed during the southern spring and summer, when Mars is closest to Earth and solar insolation is greatest. Local clouds (the classic "yellow clouds") seem to develop as fairly straightforward consequences of the interplay between topographic and seasonal winds and other features (like polar cap sublimation), which explains why storms tend to initiate in certain sites, such as the great Hellas basin, where the local topography changes abruptly, or the broad zones off the seasonal polar cap edge in both hemispheres. According to the classical explanation, local clouds develop into larger storms as dust traps sunlight, leading to further heating of the atmosphere. Then instability of the atmosphere near the surface generates local winds, raising more dust and trapping more heat, until, by means of a positive feedback loop, the local dust cloud expands into a regional or larger storm. In recent years, the leading sites of dust storm initiation have been Chryse-Acidalia, Isidis–Syrtis Major, and Cerberus in the northern hemisphere, and Hellas, Noachis-Hellespontus, Argyre, and the Solis, Sinai, and Syria Plani regions in the southern hemisphere. However, until 1963, Libya-Isidis was the most frequent initiation site, after which it became telescopically inactive until 2003. In addition to changes in specific initiation sites over time, it is now clear that the distribution of dust in recent decades has involved more fallout in the north, as attested by the gradual fading of Cerberus and Trivium Charontis since 1988 or so. Indeed, the global dust storm of 2018 began in the northern hemisphere—a first.[61]

Though regional storms spread continuously from a location, planet-encircling storms, like that of 1956, are different. They appear to spread through the quickening and merging of activity at a number of cores. Moreover, they do not seem to be very predictable.[62] A satisfactory account would not really become possible until the spacecraft era and the maintenance of high-resolution surveillance over several Martian years (see chapters 13–19).

Another important Martian meteorological phenomenon are the whitish or bluish clouds, which consist of crystals of water ice and carbon dioxide. Their formation is determined by the temperature and water condensation level. In addition, limb hazes are commonly seen in spring and summer at the planet's morning limb. Still other whitish clouds form predominantly in the late afternoon, disperse during the night, and reform the following afternoon. Frequently appearing over the same region for days at a time, they are sometimes referred to as "recurrent" clouds. Often they consist of nothing more than a light haze, but at other times they become so bright as to rival the polar caps.

Clouds form readily above the polar caps during the autumn and winter. These clouds are associated with the Martian polar vortex that dominates the circulation in high latitudes, and are often extensive enough to develop into vast "hoods" covering millions of square kilometers. E. C. Slipher recalled on the basis of his long photographic study beginning in 1907, "Observation shows that there appears about the poles of Mars near the time of autumnal equinox a great hood of dull white which usually varies much from day to day. These white patches are generally more prevalent on the morning limb of the planet. [They] spread with advancing winter and finally cover the whole area."[63]

As expected from its much longer and colder winter, the polar hood in the north is far greater than its southern counterpart in extent and duration. North polar hood clouds are present between $L_s = 150°$ (early autumn) and $L_s = 30°$ (early spring). The hood extends from the pole and usually reaches as far as latitude 60°N, sometimes as far as 50°N or even 40°N when it obscures much of Mare Acidalium.

Another interesting cloud feature is the Syrtis Blue Cloud, a belt of equatorial clouds that recurs every Martian year around the time of northern hemisphere summer and persists throughout the early summer—it has thus been referred to as the "aphelion cloud belt" by modern Mars atmospheric scientists. The cloud circulates around the Libya basin and across Syrtis Major, turning the color of this prominent albedo feature intensely blue. Though most of the blues and bluish greens seen by past observers of the planet (especially those using refractors) are certainly effects of simultaneous contrast, this is one case where the blues are really blue. (Note, however, that Secchi's "Blue Scorpion" of 1858 came too late in the season to have captured the feature. The blues he saw were thus illusory.)

In addition to Syrtis Major, several other areas are also commonly affected by recurrent clouds. One is the area now known as Arcadia Planitia. Another is Elysium Planitia, an area where orographic clouds of water-ice crystals form as moist air rises up the windward slopes of the Elysium shield volcanoes. The orographic clouds that form over Elysium are particularly bright during the late northern hemisphere spring, and at times have been mistaken for the north polar cap.[64] The most celebrated of the orographic clouds are the "W" or "domino" clouds that form over the towering Tharsis shield volcanoes (Olympus Mons, Ascraeus Mons, Pavonis Mons, and Arsia Mons). Their behavior resembles that of orographic clouds on Earth, such as those often seen over the San Francisco Peaks near Flagstaff. A seasonal phenomenon, the clouds develop for a few weeks in the southern hemisphere spring, forming in the late morning and persisting into the afternoon. They were first recorded by E. C. Slipher in Chile in 1907, and they have been observed many times over the years, though the exact W pattern—particularly striking in Slipher's blue-filter photographs taken at Bloemfontein, South Africa, in 1954—has seldom been repeated.

The Curious Case of the Blue Clearings

Another interesting Martian phenomenon, noted as far back as 1909 in images taken through color filters by E. C. Slipher at Flagstaff and by Fernand Baldet (1885–1964) and Count Aymar de la Baume Pluvinel (1860–1938) at Pic du Midi Observatory in the French Pyrenees, involves the fact that familiar albedo markings always show up well in red and yellow light but (with the exception of the polar caps) usually disappear when blue, violet, and ultraviolet (UV) filters are used.[65] By way of explanation, it was long supposed that the Martian atmosphere must be acting as a screen to block out short-wavelength radiation in the same way that stratospheric ozone does on Earth—though on Mars, given the undetectability of oxygen in the Martian atmosphere, no one ever seriously suggested that ozone itself was responsible. The supposed screen was referred to as the "Blue Haze" or "Violet Layer" (something of a misnomer as there is no visible haze or layer). The importance of such a screen for the survival of life on the planet is obvious: if the haze existed, it would provide some protection against damaging UV radiation from the Sun.

However, there was something strange about the Blue Haze. As far back as 1926, E. C. Slipher began to suspect that it might not be permanent. Blue-filter images at that and the following opposition showed the dark areas plainly visible in the photographs. Later, upon reviewing several thousand blue-filter photographs taken over a number of oppositions, Slipher discovered that these "blue clearings" seemed to occur mainly around the dates of opposition, though why this would be the case was far from obvious. There was a great deal of discussion about all this, but finally, as astronomers at Pic du Midi showed, the reason the dark markings disappear in the usual blue-filter photographs is simply that the lighter yellowish areas of the surface darken sharply toward blue wavelengths. Why the blue clearings occur especially around opposition was only explained during the Viking era. This occurs because, owing to phase-angle effects of light-scattering by airborne dust, there is occasional enhancement of the low-contrast differences between light and dark areas in blue light.[66] Many of the blue clearings might have just been mere illusions, then, although more modern spacecraft observations have revealed that surface contrast can indeed be more detectable in blue wavelengths when the abundance of strongly blue-light-absorbing (iron-bearing) airborne dust is atypically low.

The Polar Caps and the Wave of Darkening Revisited

Though the existence of water vapor in the Martian atmosphere tantalized observers—and long eluded the spectroscope—there was never any real doubt that water did exist on Mars. An Irish physicist, George Johnstone Stoney (1826–1911), in the late 19th century, suggested that the perennial polar caps might be frozen carbon dioxide (dry ice), but though the idea appealed to Alfred Russel Wallace, it never really caught on.

Lowell had made much ado about the dusky band that hugged the polar caps in their retreat. He referred to a "badge of blue ribbon" and believed it to be produced by actual meltwater left behind by the retreating cap.[67] The existence of the "Lowell band" was doubted by Antoniadi, who thought it a mere contrast effect, but de Vaucouleurs, in 1939, found that though it was generally visible in winter when the cap was largest, the band was very prominent in the spring and thus could hardly be an illusion.[68] Its reality was also affirmed by Kuiper, using the 2.08-meter reflector at the McDonald Observatory in April 1950.

Kuiper showed that spectra of the permanent caps resembled that of ordinary snow, not frozen carbon dioxide. Moreover, radiometric observations such as those by Coblentz and Lampland seemed to show that even at the poles, Mars did not get cold enough for carbon dioxide to freeze out. In 1924, they found the temperature of the south polar region to be −50°C to −70°C (−58°F to −94°F). At the low atmospheric pressures on Mars, a far lower temperature of at least −100°C (−148°F) would be required for solid carbon dioxide to form a deposit on the surface (the later realization that the *seasonal* polar ice caps were indeed made of dry ice, especially so the residual north polar cap, would only come after the Mariner 4 flyby revealed that Mars polar surface and atmospheric temperatures are much lower than had been surmised earlier; see chapter 11). Thus, there could be little doubt of the permanent caps being water ice, but even so, they could contain only tiny amounts of water. From the rapidity of their melting, de Vaucouleurs estimated they might be only a few centimeters thick—a mere soufflé.

Was so little water sufficient to trigger the widespread germination of vegetation that Lowell had advanced as the explanation for the "wave of darkening"? Possibly this could be doubted, but there seemed to be no doubt about the wave of darkening itself. Georges Fournier wrote in 1926, "As the spring advances, the dark shading progressively encroaches from the [south] pole towards the equator across the seas. This advance, across wide expanses and along channels, takes place at a variable speed, but nearly always very rapidly; a few weeks only suffice to change the landscape completely."[69] Again, some of the dark markings seemed to change in a regular way with the seasons. Antoniadi claimed that Syrtis Major appeared broad and dark in winter, narrow and weak in summer, Pandorae Fretum darkened markedly each summer, and so on. Other changes were "secular"; that is, they did not appear to correlate with the seasons. Nepenthes-Thoth showed a dramatic broadening between 1911 and 1929. Solis Lacus, whose classical form had been an ellipse with its long axis lying east and west, underwent a marked transformation in 1926; now its long axis ran north and south. These changes were duly registered on photographs and maps. Regarding their explanation, Antoniadi was certain that they were due to "an invasion by the dark or greenish vegetation on to the lightly-shaded or rosy areas adjacent to the great dark areas."[70]

Seasonal color changes were also regularly alleged. Thus in 1924 Antoniadi and Baldet described complex changes spreading from the south polar regions northward. Antoniadi wrote, "Not only the green areas but also the greyish or blue surfaces, turned under my eyes to brown, lilac-brown or even carmine, while other green or bluish regions remained unaffected. . . . It was almost exactly the color of leaves which fall from trees in summer and autumn in our latitudes."[71] Antoniadi described these changes as a "brown wave." (With hindsight, they suggest that the Grand Lunette may have had more serious problems with chromatic aberration than Antoniadi acknowledged.)

Color observations with reflectors are of course more reliable. Barnard, using the 1.52-meter reflector at Mount Wilson in 1911, found the bright areas "feeble salmon," and dark markings such as Syrtis Major and Mare Tyrrhenum "light-grey."[72] In 1956, Kuiper and Lenham, observing during the Martian spring with the 2.08-meter reflector at McDonald Observatory, found the dark areas predominantly neutral gray, with no evidence of Antoniadi's color changes apart from perhaps a "touch of moss green" in some of the equatorial regions.[73]

The greenish colors, if real, naturally suggested chlorophyll, which looks green because it reflects light in the green part of the spectrum (while absorbing strongly in the infrared). If the dark areas on Mars contained chlorophyll-bearing plants—and even such primitive but hardy plants as lichens contain chlorophyll—then chlorophyll ought to show up in the spectrum of Mars. However, as V. M. Slipher had found back in the 1920s, this did not seem to be the case. Nonetheless, even into the late 1950s and early 1960s, the American astronomer William M. Sinton (1925–2004) argued that certain features he had observed in the near-infrared spectra of Mars could be produced by chlorophyll.[74]

Another argument against chlorophyll was that if it was present in the dark areas, the latter ought to look dark in infrared photographs. The opposite is true, mainly because darker surfaces are warmer (a result partly of their albedo and partly of their higher thermal inertia compared to lighter areas, as discussed in chapters 14 and 16). On the other hand, it was always perhaps provincial to expect Mars to be the abode of organisms with absorbing properties like those of green plants on Earth (and even on Earth plants with red chlorophylls exist). Why shouldn't the plants on Mars be red?

The Modern View of Mars Begins to Emerge

There was at least one astronomer, Dean B. McLaughlin (1901–65) of the University of Michigan, who found the whole picture of Mars during these years unconvincing. Not even a planetary astronomer—he specialized in stars—he brought a fresh outlook to the scene, and in a series of modern-sounding papers written between 1954 and 1956 he set forth a highly original reinterpretation of the nature of the Martian environment.[75]

McLaughlin knew of previous work by the French astronomer Bernard Lyot, who in the 1920s had written a foundational thesis on the polarization of the light of the Moon and planets. With a visual polarimeter that was a vastly refined version of the one William Henry Pickering had used in 1894 to try to determine whether the dark areas of Mars were seas, Lyot could ascertain the variation of light along one plane of polarization compared to another to one part in a thousand. Thus, he had a simple but powerful means of investigating the nature of the materials making up a surface, and he showed that the light from Mars was similarly polarized to that of the Moon and Mercury. It also closely resembled volcanic ash.

McLaughlin found this suggestive. If Mars was indeed covered with volcanic ash, this might explain the caret-shaped protrusions of a number of dark areas on Mars, such as the two-pronged Sinus Meridiani (Dawes Forked Bay). He further speculated that perhaps dark ash spewed from still-active volcanic vents at the tips of these features flared downwind to produce the characteristic pointed forms, and that seasonal variations in wind direction might redistribute the ash to change the shape and darkness of the Martian markings. In that case the well-documented changes including the wave of darkening, so long explained on the basis of the vegetation theory, might instead be aeolian features—changing patterns of windblown dust. Instead of dark ash being spread around, however, as McLaughlin supposed, it's the ubiquitous fine bright dust from the deserts that mainly causes the changes; even layers of bright dust only a few microns deposited from the atmosphere during a dust storm would be sufficient to increase the albedo of dark areas by a few tens of percent.[76] Perhaps not surprisingly, the most dramatic changes in albedo features recorded by observers of the planet have followed large regional or planet-encircling storms.

Soon after McLaughlin put forward his ideas, the Japanese astronomer Tsuneo Saheki (1916–96) of the Osaka Planetarium recalled a remarkable observation he had made on December 8, 1951. Examining Mars's tiny 5.3" disc through a 20-centimeter refractor, with a magnification of 400×, he had picked up the dark spot Tithonius Lacus just inside the limb. "Very soon afterwards," he wrote, "a very small and extremely brilliant spot became visible at the east end of this marking. At first I could not believe my eyes, because the appearance was so completely unexpected."[77] The brilliant spot remained visible for five minutes, with a brightness briefly surpassing that of the north polar cap and "twinkling like a star." It then began to fade; ten minutes after it had first appeared, it was "only a common white cloud near the limb"; a half hour later he could make out no trace of it.

Saheki's observation had been widely publicized at the time, and among the ideas offered to explain it was that the Martians had exploded an atomic bomb. (Recall that in 1951 memories of Hiroshima and Nagasaki were still fresh, particularly in Japan, and ever-more-powerful nuclear weapons were being tested at a feverish pace by the United States and USSR on remote atolls in the Pacific and on the Arctic island of Novaya Zemlya.) Saheki was at first inclined to believe he had observed nothing more than a reflection from ice on a mountainside, or perhaps a meteorite impact. However, after reading McLaughlin's ideas, he changed his mind and concluded that the most plausible explanation was a volcanic eruption. (In all likelihood, it was merely a reflection from clouds in the Noctis Labyrinthus canyon system or from the shield volcano Arsia Mons.)

The main weakness of McLaughlin's theory was its assumption that Mars was still volcanically active since, as pointed out by Kuiper, volcanic activity on the scale suggested was incompatible with the extreme dryness of the Martian atmosphere.[78] It still seems so today, but at least McLaughlin had had the kernel of an important idea, which, for the first time, steered Martian studies in a different direction from the venerable vegetation theory. Indeed, after witnessing the Great Dust Storm of 1956, Kuiper himself came to embrace a version of McLaughlin's theory. The obvious ability of winds to move dust on Mars led him to propose that the dark areas might be dust-covered lava fields, and that the wave of darkening was caused by the seasonal removal of the dust by wind currents.[79]

In 1958, the Russian astronomer Vsevolod V. Sharonov (1901–64) arrived independently at the same explanation. The air currents on Mars, he wrote, "vary from season to season, depositing dust at some times of the year and blowing it away at other times. Thus, for instance, the inherently dark surface may brighten at a definite time of the year as a result of settling of light-colored dust blown over from the desert areas."[80] The windblown dust theory was entirely vindicated on the basis of Mariner 9 imaging in 1972. The forks in which many classical albedo features end point along the directions in which dust is swept by prevailing winds, just as McLaughlin had surmised. Additionally, variations in the dark areas on both small and large scales are indeed produced by windblown dust.[81] McLaughlin thus stands apart as the individual who came closest to visualizing the real Mars in the pre-spacecraft era. It is a pity he did not live to see his ideas fully vindicated; he died in December 1965.

Mars on the Eve of the Spacecraft Era

During these years, in addition to the vegetation theory, other basic tenets of the Lowellian Mars were falling by the wayside. Lowell had identified the dark areas as dry seafloors. However, radar studies in the early 1960s indicated that at least some of them were elevated rather than low-lying areas. Also, the long-standing belief in a flat Mars, in which the scale of the highest mountains was no more than 1,200 meters and so resembled the mesas of northern Arizona, as Lowell had suggested, was called into question. Clyde Tombaugh (1906–97), a former assistant astronomer at Lowell Observatory and later professor at New Mexico State University, argued in 1961 that mountains on Mars would not be detectable from the shadows they cast at the terminator unless they were at least 8,500 meters tall, and more rounded features would have to be even taller. Incidentally, Tombaugh, along with Öpik and Ralph Belknap Baldwin (1912–2010), had independently (and reasonably) suggested around 1950 that there might be numerous impact craters on Mars.[82]

The tantalizing quest for Martian water vapor also finally came to fruition at the midwinter opposition of 1963. Two teams were successful. The first detection, near opposition in January 1963, was due to the French astronomer Audouin Dollfus (1924–2010), using a specially designed 50-centimeter telescope set up at the Jungfraujoch Scientific Station (elevation 3,500 meters) in the Swiss Alps.[83] The other, by the team of Hyron "Hap" Spinrad (1934–2015), Guido Münch (1921–2020), and Lewis D. Kaplan (1917–99), was carried out near the time Mars was in quadrature, and made use of a photographic emulsion especially sensitive to infrared radiation to record the spectrum of

Mars with the 2.54-meter reflector at Mount Wilson.[84] Spinrad, Münch, and Kaplan found that the average amount of precipitable water on Mars (defined as the equivalent thickness of liquid water if all the atmospheric water were condensed onto the surface) was only about 14 microns, compared with 1,000 microns in even the driest desert areas of Earth. In addition, they estimated that the partial pressure of carbon dioxide on Mars was 4.2 millibars and that the total atmospheric pressure at the surface could not be more than about 25 millibars. This was a drastic downward revision from the 85–90 millibars that de Vaucouleurs had adopted less than a decade earlier. Mars apparently had an even thinner and drier atmosphere than anyone had ever imagined—though this said nothing about the Martian soil, rocks, and underground, which could only be sampled and explored by future rovers on the surface. However, in November 1964, when Mariner 4 set out for Mars, that was a prospect so remote as to be almost beyond anyone's wildest dreams.

11

Marsniks and Flyby Mariners

The 1960s

Cold War Beginnings

The modern era of Mars spacecraft exploration began with the roar of a two-stage Atlas-Agena rocket—a converted Cold War missile—lifting off from Cape Canaveral Air Force Station's Launch Complex 12 on the morning of November 28, 1964. The date of launch was 305 years to the day since Christiaan Huygens had sketched Syrtis Major from the attic observatory in his father's house in the Binnenhof, The Hague (see figure 3.1). The Atlas-Agena carried the Mariner 4 spacecraft, humanity's first *successful* attempt to send a robotic probe to the Red Planet. The key word there is "successful," because Mariner 4 was far from humanity's first attempt at sending a spacecraft to Mars.[1] Indeed, starting about four years earlier, the USSR had failed on its first six attempts, and the United States had also had a misfire prior to the success of Mariner 4.

Despite its lack of any successes, the ambitiousness and speed of the early Soviet Mars exploration program is still impressive. As with the U.S. civilian space program, the Russians were beating swords into plowshares by repurposing some of their intercontinental ballistic missiles (ICBMs) as interplanetary spacecraft launchers. The Soviet system was based on the grandest and most tragic social experiment of a strange century: Soviet-style communism. Before it crumbled, it contributed to a surrealistic midcentury world in which two superpowers not only created giant rockets aimed at each other but adapted their ICBMs to more noble and enduring purposes.

The spiritual ancestors of these rockets were starry-eyed dreamers such as Lucian of Samosata (c. 125–after 180 CE), who wrote a very early account of a voyage to the Moon, and the 19th-century science fiction writer Jules Verne (1828–1905), who imagined a cannon powerful enough to fire a projectile from Earth to the Moon (and humans sturdy enough to survive the shock).

Those were the dreamers. There were also the thinkers. One was the "father of rocketry," the Russian Konstantin Tsiolkovsky (1857–1935). Born in Izhevsk, he was left permanently deaf after an attack of scarlet fever at age nine, and forced to leave school; henceforth, he was completely self-taught. In his early 20s he became a provincial schoolteacher, first in Borovsk and then, in 1892, at Kaluga, where he served as an instructor of physics at the Ladies Paris Local School, hardly a leading scientific center. There he began to think long and hard about rockets, which had been invented by the Chinese (who filled them with another of their inventions, gunpowder, and used them against the Mongols as early as the 13th century). "For a long time," Tsiolkovsky wrote, "I thought of the rocket as everybody else did—just as a means of diversion and of petty everyday uses. I do not remember exactly what prompted me to make calculations of its motions."[2] However he came to it, he managed to show that the way a rocket was propelled forward was simply a physical expression of Isaac Newton's third law of motion: any action produces an equal and opposite reaction. The hot gases escape at high speed through the nozzle (action), and as they rush back the rocket is propelled forward (reaction). The important point is that rocket propulsion is in no way dependent, like that of ordinary aircraft, on the presence of Earth's atmosphere. Provided one carries a supply of the necessary fuel for combustion, a rocket will work perfectly well in a complete vacuum—and in point of fact, it will work even better than in Earth's atmosphere, since in space the forward movement is unimpeded by air resistance.

Tsiolkovsky did not publish his first paper on the principles of rockets for space travel until 1902, a year before the Wright brothers' flight at Kitty Hawk. However, in this and subsequent papers he laid out clearly the basic principles of the rocket as a means of space travel. He showed the superiority of liquid fuels like kerosene (instead of gunpowder), calculated that the most efficient fuel of all would be hydrogen, recognized the advantages of multistage rockets, and even dreamed of solar power, interplanetary vehicles, and rotating space stations. His visionary writings captured the imaginations of others, such as Fridrikh Tsander (1887–1933), who worked incessantly during the 1920s across the Soviet Union to popularize the idea of spaceflight to the planets, and who never tired of exclaiming, "To Mars! To Mars! Onward to Mars!"[3] Though official recognition of Tsiolkovsky's contributions was late in coming, come it did. Shortly before his death 1934, Tsiolkovsky was made a hero of the Soviet Union. A meeting of the Soviet Academy of Sciences was held in his honor, and Josef Stalin (1878–1953) sent him a congratulatory telegram.

Tsiolkovsky was hardly a man of practical bent, and he never attempted to build a modern rocket himself. The first to do that was the American Robert Hutchings Goddard, who, as mentioned previously, had been inspired in his quest to discover a practical way of reaching Mars in 1899, after reading H. G. Wells's *War of the Worlds* as a sickly 16-year-old schoolboy. Working in complete ignorance of what Tsiolkovsky had been doing (in the West then, Russia was regarded as a backward country, and no one paid any attention to what was going on there), Goddard briefly worked at Princeton, but after becoming seriously ill with tuberculosis, he returned to his hometown of Worcester, Massachusetts, to convalesce. He now became obsessed with rockets and quickly worked out the main principles of rocket flight. By 1914 his health had improved sufficiently for him to take up a part-time position as an instructor and research fellow at Clark University. With support from the U.S. Army Signal Corps during World War I, he continued his rocket research, and in 1919 the Smithsonian Institution published his treatise *A Method of Reaching Extreme Altitudes*, which was mainly concerned with the idea of using rockets for upper atmospheric research. However, Goddard also mentioned the idea of sending "to the dark part of the new moon a sufficiently large amount of the most brilliant flash powder which, in being ignited on impact, would be plainly visible in a powerful telescope."[4] Predictably, he was viewed in some quarters as a crank, and in an editorial in the *New York Times* the following year, he was accused of ignorance of even the most elementary principles of physics in thinking a rocket could work in a vacuum!

Undaunted, Goddard pushed forward, though with greater resolve to avoid all publicity. Turning his attention from solid to liquid fuels, he successfully fired the first liquid-propellant rocket on March 16, 1926, on the farm of his Aunt Effie near Auburn, Massachusetts. The flight lasted 12½ seconds, and the rocket landed in a cabbage field 56 meters from its launch site. Later Goddard built more powerful rockets, and he introduced the use of two-axis control (gyroscopes and steerable thrust) to steer their flight. In the 1930s, with moral support from aviator Charles Lindbergh (1902–74) and financial support from the Guggenheim family, Goddard moved his base to a ranch near Roswell, New Mexico. However, his attempts to interest the U.S. Army in his experiments were rebuffed.

By then, others were entering the field. In Germany, interest was keen after 1923, when Hermann Oberth (1894–1989), a Romanian-born schoolteacher, published *Rakete zu den Planetenräumen* (The Rocket to Interplanetary Space). Through its influence, the Verein für Raumschiffahrt (Society for Space Travel) was organized in Germany, of which the leading members included Wernher von Braun (1912–77), who became enamored of the idea of interplanetary space flight after meeting Oberth in the mid-1920s. By the mid-1930s, despite the dire economic straits of the Soviet Union at the time, Tsander, Sergei P. Korolev (1907–66), Mikhail Tikhonravov (1900–1974), and others began building rockets at their own expense as members of the Group for the Investigation of Reactive Engines and Reactive Flight (GIRD). From this time onward, writes space historian Asif A. Siddiqi, "Korolev's whole life began to revolve around ideas of rocketry and astronautics, and there were many discussions during GIRD's early days of sending rockets into space and landing people on Mars. One of the more common inspiring phrases of the engineers was reportedly [Tsander's phrase]: 'To Mars! To Mars! Onward to Mars!'"[5]

As the Nazi Party gained power in Germany in the 1930s, the Luftwaffe embraced rockets with enthusiasm. Though he would always remain evasive about the matter, von Braun joined the Nazi Party and became an *Untersturmführer* (equivalent to a second lieutenant) in the Allgemeine SS. Thus credentialed, he took charge of the German army's rocket station at Peenemünde, on the Baltic, which, owing to labor shortages, employed slave labor in the Mittelwerk factory in the Harz Mountains. There, the world's first ballistic missile, the A-4, better known as the V-2 (short for *Vergeltungswaffe*, "retribution rocket"), was developed. These missiles were essentially weapons of terror, and in the final desperate stages of the war they rained down on London and the home counties as retaliation for Allied bombings against German cities. They did nothing to change the course of the war, but they caused the deaths of an estimated 9,000 civilians and military personnel. Far more died from the brutal slave labor and the concentration camp conditions at Mittelbau-Dora, where the factory workers were held as prisoners.

At the end of the war, the United States and the Soviet Union were the only two competing powers. With Germany's collapse, von Braun surrendered to the U.S. Army, and he brought over to the West (as part of Operation Paperclip) 120 of his engineers. A number of others, including Helmut Gröttrup (1916–81) and Fritz Karl Preikschat (1910–94), were captured by the Soviets. (After the Americans, with Apollo, beat the Soviets to the Moon, cosmonaut Alexei Leonov [1934–2019] confided to astronaut William A. Anders that in his opinion the reason for their success was, "Your Nazis were better than our Nazis.")[6]

Indeed, in getting to the Mittelwerk factory first, the United States was able to capture enough V-2 hardware to build approximately 80 of the missiles. However, the area being soon under Soviet control, the Soviets gained possession of the V-2 manufacturing facilities, where they reestablished V-2 production. Meanwhile, Stalin ordered the release of all the rocket experts—those whom he had not yet executed—from Soviet prisons. Among them was Korolev, who, because of his association with a general who had fallen out of favor with Stalin, had been rounded up in one of Stalin's purges. After narrowly escaping execution, he had languished in a gulag for several years. Valentin Glushko (1908–89), a leading designer of rocket engines, was another victim. Now Korolev became chief rocket designer at Special Design Bureau 1 (OKB-1),[7] an important position in the Soviet Ministry of Armaments, and he worked from blueprints from disassembled V-2s to design the R-1, R-2, and R-3 rockets.

Needless to say, the chief official interest in rockets at the time, in both the United States and the Soviet Union, was for military purposes, though both von Braun and Korolev were always much more interested in using them for space travel. To many youngsters of that era, interplanetary travel seemed just around the corner. Around 1950, von Braun and his colleagues wrote a series of articles, beautifully illustrated by the space artist Chesley Bonestell, for *Collier's* magazine. These articles showed the seeming feasibility of multistage spacefaring rockets, instrumented satellites, wheel-like space stations in orbit, landings on the Moon, and—most evocative of all—expeditions to Mars.

In a 1952 speech von Braun urged for the United States to build a "manned satellite to curb Russia's military ambitions," which by then had included success in exploding an atomic bomb. A year later von Braun published the English edition of *The Mars Project* (a slim volume translated from German, and based on a 1948 special issue of the magazine *Weltraumfahrt*), in which he pointed out that "the logistic requirements for a large elaborate expedition to Mars are no greater than those for a minor military operation extending over a limited theater of war."[8] (At the time, the United States was fighting a war on the Korean Peninsula.) His words fell on deaf ears. However, more modest plans were approved, and in July 1955, the United States announced plans to launch what it expected to be the first artificial satellite, Vanguard, into orbit around Earth.

In general, Americans at the time were rather complacent about their technological and military superiority. They had good reason to be, for even though the Soviets had exploded an atomic bomb in 1949 and a hydrogen bomb in 1953, their bombers at the time were utterly incapable of striking any target in the United States, while the Americans had by then completely encircled the Soviet Union with air bases. In an effort to achieve military parity with the United States, the Soviets began in 1954 to develop an ICBM capable of delivering a three-megaton nuclear charge. The result was the world's first ICBM, Korolev's R-7 (nicknamed Semyorka, "the seven"), a two-stage rocket with a maximum payload of 5.4 tons, sufficient to carry the Soviet Union's bulky atomic bomb from the Soviet launch facility in Baikonur (near Tyuratam, a village in the heart of the Kazakh Steppe) the 7,000-kilometer distance to the mainland United States. It was successfully tested in August 1957.

The success of the R-7 led the U.S. Army Ballistic Agency under the direction of von Braun to develop its own ICBMs, the Jupiter-A and Jupiter-C. These rockets could have been adapted for the purpose of launching a satellite. However, the military bureaucracy got in the way: the prerogative of a satellite launch belonged not to the army but to the Naval Research Laboratory, which was in charge of Vanguard and planning to attempt a launch in time for the International Geophysical Year (1957–58). The slow-moving and deliberative American schedule created an opening for the Soviets. An alert Korolev seized the initiative by putting a satellite atop the R-7, only just successfully tested, and lofting it into orbit on October 4, 1957—the date on which the Space Age began. The satellite was Sputnik 1, a round metallic ball with radio antennae that broadcast a radio signal—"bleep, bleep"—that could be heard as it went around Earth. Whereas the launch of a satellite had only two years earlier been dismissed as a "scientific amusement" without practical application, once it happened, it immediately took hold of the human imagination. It was widely covered in the American press and furnished the Soviets under their leader Nikita Khruschev (1894–1971) a tremendous opportunity to propagandize about the superiority of communism over capitalism.[9] In November, the Soviets followed up with another satellite, Sputnik 2, which carried the dog Laika on a terminal, one-way trip into space. Meanwhile, the American attempt to launch Vanguard on December 6, 1957, ended as a fireball on the launchpad at Cape Canaveral, Florida (the press called it "Flopnik"). Finally, on January 31, 1958, using a Juno booster (a modified Jupiter-C), the Americans managed to put Explorer 1 into orbit. Though smaller than either of the Sputniks, it was of greater scientific utility, carrying into orbit instruments designed by Iowa physicist James Van Allen (1914–2006), which established the existence of the radiation belts surrounding Earth that now bear his name.

For the first few years after Sputnik, it was generally conceded in the West that the Soviets had achieved a clear and perhaps even unassailable lead in space. In part, this apparent superiority was an illusion, created by the Soviet penchant for secrecy and propaganda. The official newspaper of the Communist Party was ironically called *Pravda* (Truth), and the newspaper of record of the Soviet Union *Isvestia* (News), prompting the saying "There's no truth in *Pravda* and no news in *Isvestia*." Siddiqi writes:

There is a tendency in the Western discourse on the Soviet space program to make repeated allusions to "the Soviets." It was always the generic "Soviets" who decided on a particular goal or the "Soviets" who launched a satellite, while in the United States, one could comfortably write about NASA or the Department of Defense. In the face of pervasive secrecy the inner workings of the program were as unknown as the secrets of the cosmos itself. It was as if there was a monolithic structure located in some far away place, an almost mythological quantity, which ran a program of gargantuan proportions. . . . What this myth did was to obscure a story of fallible people seeped in battles that were all too human.[10]

Thus, the "Soviets" during late 1958 and early 1959 succeeded spectacularly with the first probe past the Moon (Luna 1), the first probe to reach the surface of the Moon (Luna 2), and the first probe to swing around to the far side of the Moon and send back images of the Moon's hitherto-unseen face (Luna 3). The propaganda value was not lost on Khruschev, and Korolev—while remaining heavily involved in military work—now proposed an ambitious program of missions that included reaching outward into the solar system. Though later, of course, the emphasis would be on the race with the United States to achieve a piloted mission to the Moon, Korolev's original plans for the Soviet space program as set out in a June 1960 decree adopted by the Soviet Communist Party and government emphasized going not to the Moon but to Mars. According to Siddiqi, "Soviet space scientists consistently targeted Mars as the singular most important objective in plans to explore space. Piloted flight to Mars . . . figured prominently in the famous June 1960 decree on the Soviet space program; Korolev's draft of the decree includes mention of . . . sending cosmonauts around Mars and back to Earth again."[11]

But within a year of the June 1960 decree, Korolev had already begun to lose favor with Khruschev, who, as chairman of the Council of Ministers, was all powerful at the time. One of the reasons for this lack of favor was that, though it had been the world's first successful ICBM, the R-7 was not actually a very successful strategic missile. It used kerosene and liquid oxygen as propellants, making it well suited for space but not so good for use as an ICBM; in the best-case scenario, it took 8–10 hours to fuel, while if for some reason a launch were canceled, it took another 10 hours to empty on the pad. Increasingly, hypergolic propellants, such as nitrogen tetroxide and unsymmetrical dimethylhydrazine, were favored as propellants for Soviet ICBMs, since they could sit in their tanks for a long time. Since ICBMs were always the first priority of the all-powerful Soviet military—space came in a distant second—Korolev lost ground to rivals. Already in July 1961, a new decree was published scuttling Korolev's ambitious plans for space exploration and making clear that the emphasis was to be on goals of a defensive nature. Ironically, it was issued just after Korolev had succeeded in putting the first human into orbit, Yuri Gagarin (1934–68), in Vostok 1 in mid-April 1961, three weeks before American Alan Shepard's (1923–98) 15-minute suborbital flight. The exultation during Gagarin's subsequent procession in Moscow has been compared with the "victorious rapture of May 9, 1945, when Germany formally signed its terms of surrender."[12] In retrospect, this would be seen as Korolev's high-water mark.

Meanwhile, the Americans had been roused. Just two weeks after Gagarin's triumphant procession, President John F. Kennedy (1917–63)—whose young administration was then reeling from the Bay of Pigs fiasco in Cuba, and who was soliciting advice from his cabinet on ways to regain American prestige—decided to hitch his wagon to the star of space exploration. On May 25, 1961, in a stirring speech before a joint session of Congress, he proposed the goal "of putting a man on the Moon before this decade is out, and returning him safely to the Earth." The grandiose project (which the Soviets at the time did not believe possible) was to be coordinated through the National Aeronautics and Space Administration (NASA), founded on July 29, 1958, and supplanting the National Advisory Committee for Aeronautics (NACA). With NASA, a civilian organization, the American space program was separated from the country's military programs in a way that never happened in the Soviet Union. In what would become the Apollo program, "the effort was supported by a vast infrastructure spread across the United States, with hundreds of subcontractors and a management philosophy that was unparalleled in producing results."[13]

The Soviet program, meanwhile, would be hampered by unnecessary rivalries between different design bureaus, such as those that raged over propellant selection or booster designs, and a scarcity of funds owing to the persistent shortfalls of a poorly performing economy (especially in the agricultural sector). As NASA embarked on what novelist Tom Wolfe (1930–2018) called the era of "budgetless financing," Korolev's design bureau was starved for funds. It was also spread disastrously thin. Korolev's bureau was charged with development of the Vostok spacecraft for piloted missions; reconnaissance, communications, and scientific satellites; lunar, Venus, and Mars probes (eventually farmed out to another design bureau); and above all the massive new LOX/liquid hydrogen N1, which was to be the USSR's answer to the von Braun–designed Saturn V. Yet even so, the greatest part of the bureau's resources was dedicated to development of the long-range ballistic missiles favored by the Soviet Strategic Missile Forces. When Khruschev released the first 500 million rubles for the N1 program, Korolev's deputy chief designer, Leonid Voskresensky (1913–65), told Korolev that 10 times as much would be needed to achieve the goals of the program. Korolev cynically replied that if he ever asked the government for such enormous sums, the program would simply be canceled.[14]

Korolev was thus forced to cut corners on preflight testing and sometimes had to resort to downright sleight of hand to give the impression that the Soviets were still far ahead of the United States. Thus, the short-lived piloted Voskhod program of 1964–65 used a souped-up Vostok spacecraft to scoop the firsts of sending three cosmonauts into orbit and carrying out an extra-vehicular activity (EVA) ahead of NASA's Gemini program. The propaganda value of such achievements was, of course, of great importance to Soviet leadership, though Korolev's own highly competitive nature, his "almost pathological desire to be first—to beat the Americans at all cost,"[15] was also a factor.

Evident in all aspects of the Soviet space program, this competitiveness was nowhere more of a driving factor than in the conquest of interplanetary space. At least until mid-1963, when there was a major shift in thinking at OKB-1 toward a piloted Moon landing as a counter to Apollo, Mars remained the Soviets' priority. Already in 1960, the Soviets had attempted to reach the Red Planet with a four-stage Molniya ("lightning") rocket—a modified R-7 with enough thrust to hurl significant amounts of mass (over a ton) out as far as Mars. At a time when it was still a feat to send a spacecraft successfully to the Moon, Marsnik 1 and 2 (the name was a portmanteau of Mars and Sputnik) were launched on October 10 and 14, 1960, the first attempts by any nation to send spacecraft to another planet (figure 11.1).[16] For their time, they were unbelievably ambitious missions, planning high-resolution photography, infrared spectroscopy, and UV observations of the planet's surface and atmosphere as they flew past at close range. Failure was almost inevitable, and both suffered catastrophic Molniya third-stage malfunctions. And yet at this point the Soviets were fully *four years* ahead of NASA's first attempts to reach Mars. In their comprehensive history of the early Soviet planetary program, planetary scientists Wes Huntress and Mikhail Marov note the enormous potential global geopolitical impact of these missions, and others being attempted to Venus around the same time:

> Had these Mars spacecraft and the Venus spacecraft in February 1961 succeeded, then the world would have been treated to a spectacular planetary exploration

FIGURE 11.1 *Left*: Humanity's first attempt to launch a deep space mission to another planet was the launch of Soviet Marsnik 1 on October 10, 1960, from Baikonur on a Molniya R-7 rocket. Courtesy of OKB-1 and Gunter Krebs. *Right*: The main cylindrical body of the Marsnik 1, also known as Mars 1M, was about two meters tall. NASA.

coup in May 1961. The Venus probes would have arrived at their destination on May 11 and 19, and the Mars flybys would have occurred on May 13 and 15. Following only one month after Gagarin's orbital flight in April, the effect of these triumphs on the West would have exceeded even Sputnik.[17]

At the next Mars launch window, in fall 1962, the Soviets launched three more missions for Mars, known rather nondescriptly as Sputnik 22, 23, and 24.[18] (By then they had also attempted, unsuccessfully, to send a spacecraft to Venus.) Two of these Marsniks were to fly past the planet in an attempt to obtain the science lost from the Marsniks in 1960 (plus additional magnetic field and radiation environment data). The third was (incredibly, for the time) to attempt to reach the surface of the planet and obtain in situ atmospheric and surface measurements in the process.

In every way these missions bore Korolev's personal stamp, since his philosophy (largely necessitated by financial restraints) was "let's not work by stages, but let's assemble everything and try it. And at last, it will work."[19] Again, Molniya upper-stage malfunctions caused the loss of one of the flyby missions and the entry probe mission. The second flyby mission (Sputnik 23) managed to get off successfully, but communication with the spacecraft was lost about three months before the flyby date, due to attitude control problems with the propulsion system. Though Sputnik 23—also known as Mars 1 by the Soviets—probably made a very distant flyby of Mars in June 1963, it was unable to phone home to share the view (for a list of Mars missions, see appendix A).

To try to resolve the continuing launch and spacecraft propulsion system problems, Korolev and his team next launched a "test" mission of a new spacecraft design for Mars, in November 1963 (outside of a launch window). It, too, ended in launch vehicle failure. Another attempt was made in the late 1964 Mars launch window (the same window that would be used to launch NASA's first Mars mission attempts). This time the Molniya booster worked, and Zond 2 set out successfully on course toward Mars. Failure, however, continued to dog the Soviet quest for Mars. Communications with the spacecraft—always a weak point with the Soviets—failed abruptly a month or so after launch.

Korolev's health had never been good. During his detention in the gulag his jaw had been broken, and he had contracted a serious kidney ailment. An extreme workaholic, willing to put in 18 hours a day for extended periods when the situation called for it, he was under unimaginable stress, spread too thin and yet sticking his fingers into everything and trying to keep everything under his personal control. He had already suffered a first heart attack as far back as December 1960. His doctors repeatedly warned him to let up on himself, but without effect; he suffered a second heart attack in February 1964. Much of his stress had to do with fighting against Soviet mismanagement of the space program, mismanagement that by 1964–65 had become woefully apparent and was seriously affecting morale. Among his greatest disappointments was the automated lunar probe project, designed to achieve the first soft landing on the surface of the Moon. From January 1963 to December 1965, it suffered eleven consecutive failures. His health rapidly deteriorating, Korolev, in January 1966, was admitted to a Moscow hospital for what was supposed to be routine surgery; there were complications, and he died on the operating table, at the age of only 59. Only then was his name revealed to the wider public. He had been a unique figure, and in the end an irreplaceable one. Just in the area of the interplanetary program alone—a small part of all the activities Korolev supervised—the Soviet Union had launched a remarkable 30 different interplanetary spacecraft to the Moon, Mars, and Venus. However, apart from the early Luna probes, these efforts had been an almost complete failure. The prize for this six-year, multibillion-ruble effort had proved to be a handful of lunar impact basins with Russian names, none of which anyone can ever see from Earth.

After Korolev's death, the Soviet interplanetary program persevered, and it achieved a number of notable successes—including, ironically, the first lunar soft landing with Luna 9, later in the same month that Korolev died. But by the time the automated Lunokhods explored the Moon in the early 1970s, the lead had shifted decisively to the Americans.

Mariner 4—Success at Last!

In retrospect, it is apparent that Sputnik 1, more than any other development, shocked the West into recognizing that the communist system had apparently beaten it at its own game. Soviet leaders saw space exploration as a way of establishing the credibility of their system while at the same time developing weapons of mass destruction (powered by rockets, of course) that they could use, if necessary, against the perceived threat of imperialist capitalism. In getting elected president in 1960, John F. Kennedy suggested that there was a missile gap with the Soviets (this was not true) and that the United States needed to build up its own nuclear deterrent. In announcing the plan to land an American on the Moon before the decade was out, Kennedy was chiefly interested in increasing U.S. prestige rather than in exploring the Moon or other planets. In a 1963 tape released by the Kennedy Presidential Library in 2001, Kennedy actually says, "I'm not that interested in space. I think it's good, I think we ought to know about it. But we're talking about fantastic expenditures. . . . The only justification for it, in my opinion, is to do it in the time element I'm asking, in other words to win the race to the Moon."[20]

And yet those "fantastic expenditures" also provided an unprecedented and so far never-to-be repeated opportunity for scientists. U.S. Geological Survey (USGS) scientist Don Wilhelms has written, "Kennedy gave us a goal and purpose such as a nation rarely offers its citizens in peacetime."[21] Like the early Soviet robotic exploration program, NASA's early attempts at deep space missions (those beyond Earth orbit) relied on rockets adapted from previous

ICBM or other missile programs, standardized spacecraft designs, and incorporation of significant critical component redundancy within a spacecraft, as well as overall "block redundancy" of entire missions. This last principle explains why, for example, two identical, independent NASA missions were separately launched to Venus in 1962, Mariner 1 and Mariner 2, and two identical, independent NASA missions were separately launched to Mars in 1964, Mariner 3 and Mariner 4. Scientists saw a chance to hitch their instruments to rockets bound for other worlds, where they could finally learn whether generations of thinkers about the cosmos and humanity's place in it had been right in supposing there might be other Earthlike worlds in the universe.

The most likely places where answers to these perennial questions might be found were Venus and Mars. At the time, Venus was a cloud-shrouded enigma; it might be covered with oceans, dense jungles, or deserts—no one knew. Even its period of rotation, despite centuries of effort by telescopic observers, including Percival Lowell, was unknown. The Soviets made an early attempt to reach Venus in February 1961 with Venera 1 (it was actually the second of two Venus probes launched that month; the first, not publicized at the time and now designated Venera-1VA No. 1, failed to leave Earth orbit). Venera 1 got away successfully and seemed to be working well; however, owing to a malfunction of its orientation system, radio contact was lost. It is thought to have passed within 100,000 kilometers of Venus on May 19, but no useful data was obtained.

The first successful interplanetary spacecraft was NASA's Mariner 2, launched from Cape Canaveral on August 26, 1962 (Mariner 1 had blown up shortly after launch five weeks earlier because of a rocket malfunction, vindicating the block-redundant approach of the early U.S. deep space program). Its December 14 flyby within 34,773 kilometers of Venus was a complete success, and it showed that the planet was more like the medieval notion of hell than it was another Earth. Conditions on the surface were extreme, owing to runaway greenhouse warming produced by the massive carbon dioxide atmosphere. The surface temperatures reached 477°C (890°F), well above the melting point of lead. It seemed self-evident, at least at the time, that there was no possible role for life on this inhospitable planet (though recently the idea that microbes might exist in the Venusian clouds has been seriously entertained).[22]

That left Mars.

The first American attempt at Mars was Mariner 3, launched in early November 1964. It failed shortly after launch, with the root cause most likely being incomplete separation of the rocket's protective upper shroud or fairing shortly after the upper stage left Earth's atmosphere.[23] Fortunately, again following the block redundancy philosophy that the cash-strapped Soviets were never able to fully implement with their spacecraft, and reacting (quickly) to lessons learned from Mariner 3's failure, NASA launched Mariner 4, the second of the two spacecraft in NASA's "Mariner Mars 1964" program, three weeks later.

The timing of both launches fell within a so-called launch window to Mars. By way of explanation, in order to get to Mars or any other planet, a spacecraft has to be accelerated to escape velocity, 11 kilometers per second, or 40,000 kilometers per hour. At that point it untethers from Earth and becomes an independent body traveling in its own orbit around the Sun. However, since the fuel supply of a rocket is very limited, rather than traveling directly to a planet at constant full thrust it must be placed in what is referred to as a transfer orbit, which allows it to coast without using up fuel for most of the journey. To reach an inner planet, say Venus, the spacecraft must be slowed slightly relative to Earth; to reach Mars or one of the other outer planets, it has to be sped up.

The most energy-efficient trajectory between two planets is known as the Hohmann transfer ellipse, after the German engineer Walter Hohmann (1880–1945), who first described it in 1925. Though the Hohmann transfer ellipse is rather precisely defined, pinpointing a specific optimal day for launch,[24] the launch window itself is somewhat broader, with conditions satisfactory for launching a spacecraft to Mars occurring some two to three months before each opposition. Thus, Mariner 4 set out early in the launch window that opened up before the March 1965 opposition.

Although crude by today's standards, and resembling a Volkswagen-sized lampshade connected to a ceiling fan of solar panels, Mariner 4 represented the cutting edge of technology at the time and embodied NASA's original philosophy: spaceflight would be the test-bed for the latest technical advances in electronics, miniaturization, computers, and innovative design (figure 11.2). In those days, each new spacecraft was meant to be bigger and grander than the last. This philosophy dominated NASA missions until the 1990s, when, as described later, it was augmented by a "better, cheaper, faster" approach

FIGURE 11.2 Mariner 4 being weighed by technicians at JPL, November 1, 1963. The fiberglass shroud that protected the spacecraft during launch appears in the background. NASA/JPL.

designed to get more smaller probes off the ground by encouraging reliance on innovation related to miniaturization and maximizing the efficiency of critical resources like power, computing speed, and mass.

Mariner 4 carried a number of sensitive instruments to measure the magnetic fields, dust particles, and gas encountered on the way to Mars as well as to sample their presence in the Martian environment. The radio transmitter would be used not only to transmit data but to make a crucial scientific measurement that was impossible for Earthbound observers. As the spacecraft passed behind Mars (in terms of its line-of-sight direction from Earth) its radio signals would pass through the Martian atmosphere. Measurement of the resulting distortion of the radio signal would allow calculation of the pressure, temperature, and density of the atmosphere. In addition, Mariner 4 carried instruments to detect any magnetic field belonging to the planet. This was important because a strong dipole field—that is, one with a north and south magnetic pole, like Earth's—is thought to be generated by currents in a liquid iron core. A strong field of this type would suggest a sizable active core, still molten; a weak, patchy, or absent field would suggest a frozen, "dead" interior.

The most important experiment, however, was the Mariner 4 television camera system, which if everything worked according to plan would provide the first close-up images of anything beyond the Moon. (Given the predicted lack of visible features in Venus's cloud deck, Mariner 2 had not carried a camera system.) The investigator responsible was Robert B. Leighton at the California Institute of Technology (Caltech), the same person who had built the automatic guider for the Mount Wilson 1.52-meter reflector, which had produced such sensational color images of Mars.[25] The Leighton-designed television camera system consisted of a small reflecting telescope (focal length 30 centimeters), which focused an image onto the faceplate of a vidicon tube for scanning by an electron beam. The images were tiny, only 200 by 200 pixels in size, and with a dynamic range of just 6 bits (displaying only 2^6 or 64 possible data numbers; compare this to today's tens of megapixel, 64-bit imaging sensors). The camera could not be targeted on specific surface features such as possible canals. Instead, as Mariner 4 flew past the planet, the camera could do nothing more than snap a series of pictures covering a narrow strip passing across the disc of Mars. In the interest of clear reception, the output of the television camera was recorded on tape and had to be transmitted later to Earth at a much slower rate—only 8⅓ bits per seconds. Thus, it would take 10 hours for each image to be fully radioed back to Earth.

After a fairly uneventful 7½-month cruise out to Mars (the spacecraft studied solar magnetic fields and particles along the way), Mariner 4 reached its destination on July 14, 1965 (California time; July 15 GMT), passing within 9,800 kilometers (6,120 miles) of the planet's surface. This was actually a large-miss trajectory, selected to avoid any chance of the spacecraft crashing into the surface and contaminating the Martian soil with terrestrial microbes or organic molecules before it could be tested for native life-forms. As target practice, the mission proved to be a phenomenal success for NASA and its flagship deep space mission development and operations center, the Jet Propulsion Laboratory (JPL, a division of Caltech, in Pasadena). The process of successfully guiding a deep space probe to a flyby within less than three times the radius of Mars after traveling a distance of more than 400 million kilometers from Earth represented an accuracy to within 0.03 percent, an impressive and unprecedented deep space navigation achievement.

In general, the instruments performed to expectations and the mission was a resounding success. Mariner 4 discovered that Mars has no Earthlike magnetic field or radiation belts, and a beautifully choreographed pass of the spacecraft behind the limb of Mars as viewed from Earth allowed the radio signal to be monitored as it passed through the atmosphere (a radio "occultation"), revealing a low surface pressure of only 4.0 to 6.1 millibars, 0.4 to 0.7 percent of the surface pressure of Earth. When these results were combined with earlier ground-based work that had shown the partial pressure of carbon dioxide on the surface of Mars to be around 4 millibars, it became clear that the Martian atmosphere must be made up of something like 95 percent CO_2. Perhaps this meant that, rather than being water ice as had been long believed, the permanent polar caps were simply frozen carbon dioxide. (Though this idea was reasonable at the time, later spacecraft would show that this is not entirely the case, especially in the north.) Also, at such low pressures, liquid water, even in the relatively warmer equatorial regions, would not be stable on the surface.

Humans beings are highly visual creatures, and as such the results of the imaging investigation were, of course, the most eagerly awaited. The Leighton-designed camera system performed adequately if not brilliantly. The data it sent back consisted of bits of each picture in packets corresponding to the brightness of each pixel of an image. Completing their journey across interplanetary space, the bits were picked up in the giant dish of Goldstone's tracking station in the Mojave Desert, transmitted across California by teletype to JPL, and recorded onto a reel of ticker tape that engineers proceeded to cut into strips and pin to the walls. As noted, this was a slow process, and the engineer in charge of the tape recorder, Dick Grumm, grew impatient. Running out to a local art store and obtaining a pack of pastels, he produced his own free interpretation of the first image using burnt umber, Indian red, and yellow ocher, in which the edge of the planet began to emerge. This was a preview of Mariner 4's first-ever close-up image of Mars, which showed a section of the Amazonis desert, the same general area that Eugène Michel Antoniadi had rendered in his evocative impressionistic sketch with the Meudon refractor back in 1909.

The rest of the series (in gray scale, not pastel, and built up by computer, subjected to repeated operations to enhance contrast and to suppress electronic defects) covered a discontinuous swath of the planet's southern hemisphere, extending south and then eastward from Amazonis across Zephyria, Atlantis (a bright strip between the classical dark areas Cimmerium and Sirenum), Phaethontis, and Memnonia—enchanting, classical names. It took almost a week to transmit all the data back to Earth. There were 21 full images and one partial image, though the last three or four were featureless as they were exposed to the part of Mars that lay in darkness beyond the terminator. Minuscule, at only 200 × 200 pixels, and disfigured with coarse TV scanning lines, they were not great even by 1965 standards; a single photo with a good 35-millimeter camera contained about 25 times the amount of information as the entire Mariner 4 photo set. Nonetheless, they gave the first view of real topographic and geological features on Mars instead of the elusive broad shadings and shifting cloud forms that had been the domain of generations of telescopic observers. They stunningly revealed to Leighton and prominent geologists Robert P. Sharp (1911–2004) and Bruce C. Murray (1931–2013) of the imaging team a host of impact craters, some 300 in all, in all sizes, including one in Mare Sirenum measuring 120 kilometers (75 miles) across. It seemed as if Mariner 4 had simply photographed a strip of the Moon's surface, with a thin hazy atmosphere superimposed.

Recognizing the historical significance of these first-ever close-up images of Mars in the context of the late 19th- and early 20th-century debate about surface markings, Murray commented at a news conference in Washington, D.C., on July 29: "To the people working on the project, when we received Frame 7 [of the series], even in its initial [unprocessed] form, it was a very dramatic moment, because we began to recognize already many, many craters as well as some light to dark variations on the surface of the planet, so we knew we were going to have a very successful experiment when we reached this point."[26]

The craters were undoubtedly the big story of Mariner 4. Leighton, Murray, Sharp, and their colleagues wrote in their August 6, 1965, initial results paper in the journal *Science*, "The observed craters have rims rising to about 100 [meters] above the surrounding surface and depths of many hundred meters below the rims. Crater walls so far measured seem to slope at angles up to about 10°. The number of large craters present per unit area on the Martian

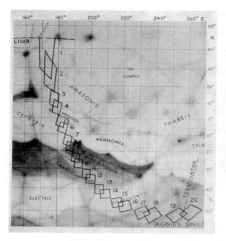

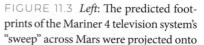

FIGURE 11.3 *Left*: The predicted footprints of the Mariner 4 television system's "sweep" across Mars were projected onto a canal-era globe of Mars. Courtesy of The Planetary Society. *Right*: Image #11, the most famous of the Mariner 4 series, showing detail in the dark region Mare Sirenum, including a heavily eroded 170-kilometer crater that now bears the name "Mariner." NASA/JPL.

surface and the size distribution of these craters resemble remarkably closely the lunar uplands." They briefly alluded to the failure of any canals to show up: "Although the line of flight crossed several 'canals' sketched from time to time on maps of Mars, no trace of these features was discernible. . . . No Earth-like features, such as mountain chains, great valleys, ocean basins, or continental plates were recognized."[27]

Reading such reports in the media, a 10-year-old boy (W. S.) who had just had his first view of Mars with a small telescope months before was crushed. No canals! It was like finding out there was no Santa Claus, though it was hardly the last time that reality would fail to meet romantic expectations. Keay Davidson has referred to the date July 14, 1965, as "the Bastille Day of exobiology. It revolutionized the young science in the hardest way possible, by humbling it in front of the world."[28]

In an editorial titled "The Dead Planet," the *New York Times* assured its readers that Mariner 4 had shown Mars to be lifeless. President Lyndon B. Johnson (1908–73), by then increasingly mired in the slowly unfolding disaster in Vietnam, also weighed in on the spacecraft results, telling a NASA group, "As a member of a generation that Orson Welles scared out of its wits, I must confess that I'm a little bit relieved that your photographs didn't show signs of life out there."[29]

In fact, these conclusions were clearly premature, but they made a tremendous impression at the time. Perhaps to soften the blow, or at the request of a peer reviewer from the traditional telescopic observation community, Leighton and his colleagues included a caveat in their rather sobering assessment. "It should be remembered," they wrote, "in this respect that the visibility of many Martian surface features, including the 'canals,' is variable with time."[30] Meanwhile JPL director William Hayward Pickering (1910–2004; not to be confused with astronomer William Henry Pickering, who had died in 1938) was even more sanguine, noting that of all the Television InfraRed Observation Satellite (TIROS) weather images, only one showed any evidence of intelligent life on Earth; so far we had taken only 20 pictures of Mars.[31]

This seemed at best a thin reed of hope at the time. For the most part, the revelation that Mars is full of craters brought about a shocked reappraisal not only among the general public but among many scientists, who, quite simply, had not expected them. (It is not quite clear just what they had expected; presumably, a Mars more like Earth than like the Moon.) In part this reaction shows the effect on readers, including some who became scientists, of generations of romantic writings about Mars, but it also says something about the state of planetary science at the time.[32]

Today, craters do not seem at all surprising on an alien world, but even as late as 1965, impact craters were only just beginning to be recognized as an important phenomenon shaping planetary surfaces. Arizona's Meteor Crater (a misnamed feature that should really be known as "meteorite crater" since "meteor" refers to a bolide burning up in the atmosphere and "meteorite" to one reaching the ground) had been recognized for a century. However, it was long mistakenly thought to be volcanic, a natural steam explosion "maar" rather than an impact crater. In any case it seemed to be almost unique, a freakish feature of no consequence. Of course, thousands of craters had been mapped on the surface of the Moon, but even they were still widely believed to be volcanic. Not until the late 1950s did geologist Eugene M. Shoemaker (1928–97), later head of the Astrogeology Branch of the USGS in Flagstaff, publish his youthful research on the Arizona crater, proving beyond any shadow of reasonable doubt that it was an impact crater. From there, he went on to argue persuasively that lunar craters were of the same type. Thus, the modern paradigm began to emerge, according to which both Earth and the Moon—and presumably other planets, like Mars, which after all was situated close to the asteroid belt—had been shelled, especially early in their history, by swarming shards of interplanetary debris.

Another development was the gradual discovery that Meteor Crater was far from unique; there were actually at least a few hundred such features on Earth. Though most craters had been effectively obliterated by the processes of erosion and plate tectonics, Canadian geologists began to recognize numerous ancient, glacier-eroded impact craters in Canada, usually visible on aerial photos as precisely circular lakes. The reason Canada proved to be an ideal hunting ground for such features was that it represented the several-billion-year-old core (craton) of the North American continent, a stable region that has been relatively spared disruption by mountain building, volcanism, or plate tectonic forces. Other cratons, including the Kaapvaal craton in South Africa and the Pilbara and Yilgarn cratons in Australia, were also good hunting grounds for ancient impact craters.

All this work in the early 1960s led to the recognition that ancient planetary surfaces in the solar system (including those on Earth) have accumulated many asteroid and meteorite hits. To put it the other way around, a very young surface, such as a lake-bed deposit or a lava flow that is, say, only 100 million years old, is likely to have been relatively spared. But if the surface is one to three billion years old like the Canadian shield, it will have many scattered craters, while if it is about four billion years old, like the ancient highlands of the Moon, it will show the scars of the heaviest era of bombardment, which accompanied planet formation itself. As on the Moon, such ancient surfaces will be saturated with craters, shoulder to shoulder, cheek by jowl.

Admittedly, in 1965, the significance of all this research had not yet been fully appreciated, though enough was already sinking in to lead the scientists charged with interpreting the Mariner 4 images to draw the main conclusion. Though destined to be regarded as sticks-in-the-mud by Mars dreamers everywhere, Leighton and his colleagues offered the following honest and forthright summary of their results: "We infer that the visible Martian surface is extremely old and that neither a dense atmosphere nor oceans have been present on the planet since the cratered surface was formed."[33] That meant that Mars's surface was likely at least two billion years old, and probably much older. Harvard astronomer Fred L. Whipple (1906–2004) came to an equally disappointing conclusion:

> The existence of a lunar-type cratered surface, even in only a 1-percent sample, has profound implications about the origin and evolution of Mars and further enhances the uniqueness of Earth within the Solar System. By analogy with the Moon, much of the heavily cratered surface of Mars must be very ancient, perhaps 2 to 5×10^9 years old. The remarkable state of preservation of such an ancient surface leads us to the inference that no atmosphere significantly denser than the present very thin one has characterized the planet since the surface was formed. Similarly, it is difficult to believe that free water in quantities sufficient to form streams or to fill oceans could have existed anywhere on Mars since that time. The presence of such amounts of water (and consequently atmosphere) would have caused severe erosion over the entire surface.[34]

The Armada of 1969: Hi-Fi Flyby Science

Mariner 4's achievement—bringing a TV camera and other instruments as far as any had ever been sent from Earth and successfully returning pictures and other data—was stunning and widely celebrated from an engineering and technology, as well as political, perspective. NASA had achieved on its second try what the Soviets had failed to do on six attempts, and with this success (and the earlier one of Mariner 2 to Venus), JPL had helped guarantee the continuation of the robotic planetary exploration program. Admittedly, Mars was refusing to cooperate with the history or mythology of Mars's public persona, and the discovery of craters on Mars had a dampening effect on some politicians and the general public for future missions to the Red Planet. The question was asked: why go so far to explore another Moon?

Before Mariner 4 arrived at Mars, NASA planners had already begun serious discussions of skipping the usual step-by-step approach by moving from an initial success straight to a larger, more complex *landed* mission featuring a gigantic robotic biological laboratory, known then as Voyager (not to be confused with the later mission to the outer solar system), which was to be launched on one of NASA's Saturn 1B boosters with a Centaur (liquid hydrogen propellant) upper stage. Ironically, Mariner 4's findings contributed significantly to the demise of the concept. Its measurements of the extreme thinness of the Martian atmosphere forced a major redesign of the landing system. Lightweight parachutes had to be replaced with heavier landing rockets, and the Saturn 1B and Centaur booster had to be replaced by the gigantic Saturn V, which was to be used for the Apollo Moon landings. But that sent the cost soaring above $2 billion just as the chances of finding life on Mars had seemingly dwindled to near zero. The idea of sending such a huge and costly biological laboratory to Mars at this moment seemed premature, to say the least. Ultimately it was the price tag that did the first Voyager mission in. The U.S. government's Budget Bureau (now known as the Office of Management and Budget) did not share NASA's enthusiasm for the project, and it was canceled. Instead funding was offered for the far more modest mission (see appendix H) of two flyby Mariners to be launched by Atlas-Centaur rockets during the launch window of winter 1969.[35]

Mariner 6 set out in February, and Mariner 7 in March, on trajectories leading to midsummer flybys. The Soviets not only matched that Red Planet bet but raised the ante with two launches of their own, designated Mars-69A and Mars-69B, both of which were designed to become the first spacecraft to orbit Mars. While they exploited new designs leveraging successful deep space technologies and subsystems that had enabled the USSR to finally achieve some remarkable robotic science successes at the Moon and Venus between 1966 and 1968, neither Mars-69 spacecraft ever got into Earth orbit, falling victim to failures of the Soviet's new Proton-K launch vehicle.[36]

The successful getaways of Mariner 6 and Mariner 7 marked the first time in the Mariner program that both spacecraft in the block-redundant mission pair launched without a hitch. This in itself was a significant success, as the dual-spacecraft missions were also the first to use NASA's new Atlas-Centaur launch vehicle combination. The launch of Mariner 6 almost began with a prelaunch catastrophe, however. On Valentine's Day 1969, about 10 days prior to launch, with Mariner 6 mounted atop the Atlas-Centaur booster, a faulty switch on the Atlas stage opened and began to release the propellant pressure that was helping support the rocket's massive structure. As the booster started to slowly crumple, two NASA engineers on the ground crew (actually, contract employees of General Dynamics) named Billy McClure (1924–

2009) and Jack Beverlin (1918–2013), recognizing the potential loss of the mission, rushed to the launchpad, closed the valves, and restarted the pumps, restoring pressure to the structure.[37] For their heroic actions in the face of a 12-story-tall stack of explosives potentially crashing down on them, the two men became the first recipients of the NASA Exceptional Bravery Medal.[38] Amazingly, compared to today's standards of risk aversion and seemingly endless investigations, Mariner 6 was quickly removed from the damaged Atlas and mounted atop a different Atlas-Centaur booster, with barely time to spare for the opening of the next Earth–Mars launch window.

Both Mariner 6 and Mariner 7 were laser focused on Mars flyby science, so no cruise observations were conducted on the way there. Both missions had an upgraded payload compared to Mariner 4, including instruments mounted on a two-axis scan platform that would let mission scientists track the planet all the way through each close flyby. The instruments included a visible-wavelength camera system with wide-angle and high-resolution lenses, ultraviolet and infrared spectrometers, an infrared temperature sensor, and the same kind of fly-behind-the-planet, radio-occultation-based atmospheric sounding strategy as Mariner 4.

Mariner 6 flew by Mars first, on July 31, 1969 (just 11 days after Neil Armstrong [1930–2012] and Buzz Aldrin landed on the Moon), at a closest-approach range of just 3,431 kilometers from the surface, just a fraction over one Mars radius away. Mariner 7 flew by five days later, on August 5, at an almost identical distance from the surface, but over a ground track slightly to the east. Because JPL flight controllers had a much higher radio bandwidth capability than with Mariner 4, they actually started acquiring images of the entire planet with both spacecraft from a much more distant range, starting about 50 hours from closest approach. As Mars grew bigger in the windshield, the resolution of the photographs taken during this "far encounter" phase of the mission (50 photographs by Mariner 6, and 93 photographs by Mariner 7) began to exceed the resolution of the best Earth-based telescope images. As the planet rotated, a variety of enigmatic features passed through the camera's field of view. Some were well known and expected, like the bright south polar cap and bluish clouds along the morning limb (the missions were the first to successfully use a color camera for space-based images of Mars). Other features revealed in the images, however, were harder to explain.

Once again, the camera team was led by Leighton and included Murray and Sharp, but the team had grown beyond the small Mariner 4 group. Additional collaborators with expertise in planetary astronomy and imaging included people like Bradford A. Smith (1931–2018), who would go on to lead the imaging team of the Voyager missions to the giant planets; Mert Davies (1917–2001), who would be involved in the successful mapping of numerous other worlds in subsequent missions; and Andy Young and Conway Leovy (1933–2011), who would use their terrestrial meteorologic and dynamical studies to make important discoveries about the atmospheres of Mars, Venus, and other worlds. The team also included—for the first time on a NASA mission—a dyed-in-the-wool biologist and geneticist. Caltech's Norm Horowitz (1915–2005) was the chief of JPL's biosciences section at the time, and was keenly interested in the question of life on Mars and the potential information about the habitability of the planet that the images and other data from Mariner 6 and Mariner 7 might provide. Horowitz would go on to lead one of the key biology experiments conducted by the Viking Landers, and he was among the first planetary science researchers (along with Carl Sagan) that could be considered an astrobiologist.

During the closest part of the flyby, Mariner 6 took 26 "near encounter" images with sampling as fine as 300 meters per pixel—unprecedented imaging at the time, which provided the capability to resolve and unambiguously identify features like impact craters only a few kilometers in size. Mariner 7 took 33 near encounter photos at a similar scale. The ground tracks for the highest-resolution images passed over the heavily cratered equatorial terrains to the west and the east of (and including) the classical dark surface feature called Sinus Meridiani. Meridiani was a useful benchmark for astronomers in trying to make the connection between telescopic observations and the new Mariner 6 and 7 images; unfortunately, as planetary scientists would later learn, bad luck and the vagaries of celestial mechanics meant that the high-resolution ground tracks of these flyby missions just happened to miss some of the planet's most spectacular volcanic, tectonic, and water-related surface features.

Nonetheless, Leighton and colleagues had a treasure trove of global-scale and high-resolution imaging data to work with compared to the comparatively sparse Mariner 4 data set, and they proceeded to interpret what they saw in a sprawling 18-page preliminary report published in *Science* magazine just two months after the flybys.[39] Among the most significant findings came from the fact that enough of the surface area of Mars had been covered at high resolution (about 20 percent) that the nature and distribution of impact craters could be accurately compared to the well-known cratered terrains on the Moon to assess similarities and differences in the surface histories of these terrestrial worlds.

And indeed, significant differences were found. As had been inferred from Mariner 4 images, there are ancient, heavily cratered terrains on Mars just as on the Moon. But Mariner 4 was unlucky in that those were the *only* terrains imaged well during that flyby. Mariners 6 and 7 flew over similar terrains, but they also flew over much less heavily cratered terrains, places where many of the craters were heavily eroded and perhaps even buried or completely eroded away by active surface processes. They also noticed a variety of large-scale "collapsed" features in places, "chaotic terrains" containing jumbled sets of irregular ridges, "featureless" areas of bright, flat, craterless landscapes unlike

FIGURE 11.4 Mosaics of multiple 1969 Mariner flyby images covering the south polar cap (Mariner 7, *top*), the classical dark region Sinus Meridiani (Mariner 7, *bottom left*), and the first example of "chaotic terrain" on Mars (Mariner 6, *bottom right*). NASA/JPL/Ted Stryk.

anywhere on the Moon, and quasi-linear alignments of small, dark impact craters at the same locations as some of the purported canal features identified from early telescopic observations.

The general lack of what Leighton and colleagues called "Earth-like forms," especially the apparent absence of obvious tectonic features, clearly weighed heavily on these scientists, who concluded that "the crust of Mars has not been subjected to the kinds of internal forces that have modified, and continue to modify, the surface of the earth."[40] They also rejected volcanic interpretations for features like the collapsed terrains because "the apparent absence of extensive volcanic terrains on the surface would seem to be a serious obstacle to such an interpretation."[41] Ironically, Mariner 7 had actually taken far encounter photos of the enormous (tectonically created) Valles Marineris canyon system, as well as Olympus Mons and other huge volcanic features in the Tharsis region, but the resolution of those more distant images was simply not high enough to allow an unambiguous geological interpretation. According to the Mariner 6 and 7 images, Mars was clearly not like the Moon; it was also clearly not like Earth. The idea that Mars has its own kind of distinctly Martian surface geological processes was born in the summer of 1969, and rightly persists to this day.

Along with the geologists, the atmospheric scientists and polar scientists were also having a field day with the Mariner 6 and 7 data sets. Early morning and north polar clouds and hazes could be mapped in the data, and the details of the composition and behavior of the south polar cap were being revealed for the first time. Infrared temperature and spectroscopic data revealed that the cap must be dominated by dry ice—solid carbon dioxide (CO_2)—rather than water ice. Indeed, Leighton and Murray had predicted back in 1966, based on simple thermal models and the Mariner 4 atmospheric data, that the seasonal polar caps of Mars would consist almost entirely of frozen CO_2.[42] Spectroscopy from Mariner 6 and 7 would confirm that hypothesis, and make "Leighton and Murray (1966)" a classic paper on all planetary scientists' required reading lists.[43]

Horowitz and colleagues jumped into the "life on Mars" debate wholeheartedly with the Mariner 6 and 7 results. The ancient nature of much of the surface and the lack of any obviously water-carved features seen in the images, the confirmation of the extremely low levels of water vapor in the atmosphere, and the apparent lack of significant amounts of water ice in the seasonal polar caps all pointed to an extremely desiccated planet. Even the reported evidence from the infrared spectrometers of small amounts—a few percent—of water (H_2O) or hydroxyl ions (OH) bound inside surface minerals still showed that Mars is overall a much, much drier world today than is Earth.[44] While leaving a little wiggle room for the possibility of future missions (like NASA's 1971 Mariner 9 orbiter, then being planned) and other discoveries changing their interpretations, the astrobiological view of Mars that emerged from the Mariner 6 and 7 flybys remained discouraging if not downright dismal:

The results thus reinforce the conclusion, drawn from Mariner 4 and ground-based observations, that scarcity of water is the most serious limiting factor for life on Mars. No terrestrial species known to us could live in the dry martian environment. If there is a permafrost layer near the surface, or if the small amount of atmospheric water vapor condenses as frost in favorable sites, it is conceivable that, by evolutionary adaptation, life as we know it could use this water and survive on the planet. In any case, the continued search for regions of water condensation on Mars will be an important task for the 1971 orbiter.[45]

12

A Martian Epic

Mariner 9

The Race to Mars Orbit

The successful results reported by the Mariner 6 and 7 science teams meant that Mariners 8 and 9, NASA's first attempts to orbit the Red Planet, were given the go ahead. But the Soviets were also in the hunt, their quest acquiring even greater piquancy given the humiliation of being beaten around the Moon (by Apollo 8 in December 1968) and to the surface of the Moon (by Apollo 11 in July 1969). So great was the humiliation that the Soviets engaged in one of the great deceptions of all time—claiming they had never planned to send cosmonauts to the Moon at all, but had only been interested in sending automated probes, like the successful Luna 16 of September 1970, which soft-landed in the Sea of Fertility and returned 105 grams of lunar soil to Earth, and Luna 17 that November, which delivered a remote-controlled rover known as Lunokhod 1 to Mare Imbrium. These successes were soon followed by other sample-return and Lunokhod missions. They were triumphs of a kind, but, of course, they were overshadowed by the ongoing series of Apollo human landings, which continued from Apollo 12 in November 1969 through Apollo 17 in December 1972.

There was no longer much prestige associated with returning again and again to the Moon, so the competition in space turned increasingly to the other planets.

As noted in the previous chapter, Mars had a been a particularly allur-ing target for Soviet mission planners going all the way back to Konstantin Tsiolkovsky, and—rather unbelievably from today's perspective—the Soviets, hoping to recoup the fortunes of their "dazed and confused" human space program in the aftermath of the spectacular Apollo 8 circumlunar mission, actually considered with some seriousness a series of missions to Mars on a breakneck schedule of completion, as follows:

Mars '73, a robotic vehicle to Mars for sample return, using the N1 booster.

Mars '75, a crewed mission to Mars to achieve orbit around the planet and return to Earth, using the N1F-V3 booster.

Mars '77, a crewed landing on Mars using an N1 with nuclear rocket engines.[1]

Of course, these were all pie-in-the-sky plans, and they were thrown into dis-array by the tendency of the gigantic N1 booster (designed by the late Sergei P. Korolev) to produce spectacular fireballs shortly after leaving the launchpad. But then, too, American plans for piloted missions to Mars, briefly mooted by NASA administrator Thomas O. Paine (1921–92) during the same period, were equally unrealistic and unfundable. Human spaceflight—for both the Soviet Union and the United States—remained bound to low Earth orbit after December 1972. And so it remains—as no one at that time would have predicted—almost 50 years later. Far from sending piloted missions to Mars, we have not even returned to the Moon.

The competition to reach the planets therefore devolved to automated probes, and curiously, the Soviets achieved a near monopoly on Venus, and the Americans on Mars.

In retrospect, the interest of the Soviets in the hellish planet Venus appears almost perverse. However, in some ways it seems to have been a product of inertia. When the Soviets launched their (unsuccessful) Venera 1 probe in February 1961, Venus was shrouded in mystery; even the rotation period was unknown. As Venus was an almost identical twin of Earth in size and mass, it was easy to imagine it a sister planet of our own. Speculation was rife, and ideas about what kind of planet might lurk beneath the clouds ranged from parched and arid deserts to lush tropical jungles like those on Earth during the Jurassic.[2]

When radio astronomers found evidence in the late 1950s that the surface of Venus was extremely hot, with an ovenlike temperature of perhaps several hundreds of degrees, there were few who realized the implications. One that did was Carl Sagan (1934–96), who in his 1960 doctoral dissertation had suggested not only that the temperatures were real but that they were caused by the "greenhouse effect"—the ability of carbon dioxide to trap solar heat.[3] He also argued that to produce such an intense greenhouse effect, the surface pressure of Venus's atmosphere was likely very great, and he placed a good-natured $100 wager with Gérard de Vaucouleurs, then at Harvard, that the surface pressure would prove to be 100 times that of Earth.[4]

In 1962, the plot began to thicken considerably. Radio astronomers discovered the rotation period of Venus—243 days, retrograde—while, during its December flyby, Mariner 2 detected microwave radiation from the surface, supporting Sagan's contention that it was a fiercely hot and inhospitable world. (Good-naturedly, de Vaucouleurs paid up on his wager, but he explained that his expertise was on Mars, not Venus.)

Soviet scientists did not at first accept these findings. They continued to explain away, as late as the mid-1960s, the Mariner 2 results by supposing its instruments to have measured the hot upper atmosphere and ionosphere rather than the actual surface, which they believed to be cool. As usual with the Soviet program, a large number of initial attempts to reach the planet ended in failure. Venera 1 was only one of a number of these. The first (partial) success was Venera 3 in February 1966. Its landing capsule apparently entered the atmosphere of the planet, but—not being designed to withstand either the high pressures or the temperatures it encountered in the lower atmosphere (which the Soviets did not believe in)—it failed to maintain contact during its downward plunge to the surface. Several better designed capsules followed: Veneras 4, 5, and 6 in 1967 (the same year the Americans succeeded with another successful flyby mission, Mariner 5), then Veneras 7 and 8 in 1970, which maintained communications for brief periods after reaching the surface. The Soviets' ultimate goal of capturing images and measuring the composition of soils at the surface itself was finally achieved by Veneras 9 to 14 in 1975. They managed to survive for about an hour before succumbing to the inhospitable conditions.

All these missions did was confirm the general impressions from Mariner 2. Venus has a crushing atmosphere consisting mostly of carbon dioxide, with a surface pressure of 90 bars and a temperature more closely approaching that of a blast furnace than an oven: 477°C (890°F). Deceptively alluring when it appears as the Evening or Morning Star, Venus in reality is an almost literal incarnation of the medieval vision of hell, with a surface even hotter than that of Mercury, which is the innermost to the Sun. The chief interest in Venus nowadays is in trying to understand how and when a planet of similar size and mass to Earth could have followed such a divergent evolutionary path, and in providing a vivid example of a runaway CO_2 greenhouse effect—a cautionary tale for us on Earth about the potential dangers of humans' continuing burning of fossil fuels.[5]

Despite having had success with the oppressive conditions on Venus, the Soviets continued to struggle with their Mars landing program. Presumably this had to do mainly with the fact that the descent process in the very dense Venusian atmosphere requires less sensitive control of the altitude than in the thin Martian atmosphere. Even before Korolev's death, all automated lunar and planetary programs had been transferred to the Lavochkin State Union Machine Building Plant, under the direction of its chief designer, Georgy N. Babakin (1914–71). He was still in charge in 1971, when, after a thorough redesign of the Mars orbiters following the failed Mars-69 attempts, they seemed to be closing in on success, and even raised the bar higher than it had been by those earlier missions by enhancing two of their spacecraft with vehicles that would be deployed from the orbiters and attempt to land on the planet for the first time. Sadly, Babakin himself did not live to see it; he died, at age 57, just before the latest flotilla of Soviet spacecraft set out for the Red Planet. He was succeeded as chief designer of Lavochkin by Sergei Kryukov (1918–2005).

In fact, five missions attempted to reach Mars during the very favorable 1971 Earth–Mars launch window, which opened in May of that year. (Since 1971 was a perihelic opposition year, the transit time for spacecraft to reach Mars was shorter than usual.) There were two American and three Soviet: the Mariner 8 and Mariner 9 orbiters from the United States, and the M71-S ("S" for Sputnik) orbiter and Mars 2 and Mars 3 orbiter/lander missions from the USSR.

The campaign didn't begin well. NASA's Mariner 8 spacecraft was the first to launch, on an Atlas-Centaur booster on May 9, 1971, but the Centaur upper-stage guidance system failed only about four minutes after launch. The upper stage and spacecraft tumbled out of control and eventually separated, and both crashed down into the Atlantic north of Puerto Rico (perhaps one day they will be recovered by Robert Ballard or another deep-sea explorer).[6] Two days later, the Soviet's M71-S mission launched from the Baikonur Cosmodrome spaceport in southern Kazakhstan on a four-stage Proton-K booster. After the spacecraft successfully reached low Earth orbit, the Proton's fourth stage failed to ignite on time due to a software error, and the mission ended when the spacecraft came tumbling back to Earth two days later.[7]

Prospects improved dramatically in the second half of May 1971, however, with the successful launches of the Soviet Mars 2 (May 19) and Mars 3 (May 28) and the NASA Mariner 9 (May 30) missions on perfect Mars-bound trajectories. The Soviet missions would attempt to orbit *and* land on Mars, but they were somewhat hobbled from the start because the earlier failed M71-S mission had been intended to serve as a reconnaissance and relay communications orbiter for those landers. Without M71-S getting there ahead of time, both lander missions faced significantly increased risks. As it turned out, all three missions would face existential threats imposed on them by conditions on Mars itself.

The conditions were largely unprepared for, though they could have been foreseen—and in fact, they were foreseen. Just two months before the flotilla of Mars spacecraft set out across interplanetary space, Charles F. "Chick" Capen (1926–86) published an article that contained an interesting prediction. Capen had gained expertise in imaging the planets under Clyde Tombaugh at New Mexico State University, and by 1971 had been hired by William A. Baum (1924–2012) at Lowell Observatory as an observer with the International Planetary Patrol (IPP), a ground-based network of observatories established in 1969 for the purpose of monitoring and making continuous observations of the planets, including of the clouds, dust storms, and climate changes on Mars. (Apart from Lowell, the other observatories were Mauna Kea in Hawai'i, Mount Stromlo and Perth in Australia, the Republic Observatory in South Africa, and the Magdalena Peak Station of New Mexico State University.) Writing about the dust clouds ("yellow clouds"), Capen wrote in February 1971:

> Though yellow clouds have been recorded in all Martian seasons, the largest outbreaks seem to occur during Martian perihelic oppositions, when the insolation [exposure to the Sun's rays] is greatest on the planet, and the thermal

equator is far south of the geometric equator. . . . If a bright yellow cloud again develops in the Hellespontus region [as was the case in 1956] . . . it will likely do so after opposition. A vast atmospheric disturbance could interfere with . . . the first Mariner orbiter spacecraft mission, which is planned to begin reconnaissance of the planet in November.[8]

As three spacecraft headed toward Mars, a network of professional and amateur astronomers conducted detailed campaigns of telescopic observations designed to help mission planners understand what to expect when they got there, and to put the hoped-for orbital and lander measurements into a seasonal context. What the astronomers discovered in late September 1971, however, was that Mars was rapidly being engulfed in one of the planet's famous but only occasional (and unpredictable) great dust storms.[9] During the southern hemisphere summer months, solar energy is the most intense because Mars is at its closest point to the Sun in its elliptical orbit. Strong solar heating drives strong winds (just as on Earth), which can pick up tiny particles of dust and transport them across the planet. Similar dust storms fueled by strong temperature contrasts (sometimes called haboobs, Arabic for "blasting or drifting") can occur in the Sahara, the Middle East, the American desert Southwest, and elsewhere, and they can lift large amounts of dust into the atmosphere. While they can attain impressive heights, dust lifting on Earth by such storms is limited to 10,000 meters or so by our thick atmosphere, short supply of super-fine-grained dust, relatively high humidity, and topographic obstacles. On Mars, however, dust-lifting events can sometimes exhibit positive feedback effects and lift enormous amounts of abundant fine-grained (micron-sized, comparable to smoke particles) dust high into the atmosphere. What causes these storms to only sometimes grow to planet-encircling and even global scale is not entirely clear, though significant insights into the mechanisms for Mars dust storms were to come from Mariner 9's data. The effect, however, is much like the effects of a large volcanic explosion on Earth, injecting high concentrations of aerosols into the upper atmosphere and, as established by later Mars spacecraft, raising the mid- to upper-atmospheric temperature by some 30°C. At the same time, by blocking incoming sunlight, the dust lowers the average surface temperature. Thus, dust storms lead to major short-term changes in local climatic conditions on Mars.

Actually, the understanding of how Martian dust storms develop was still at a fairly rudimentary level in 1971, when the only large-scale planet-encircling dust storm that had been observed and photographed in detail had been that of 1956. The 1971 event seems to have been heralded by the development of a regional dust storm over Hellas in July, but it was only in September that a really significant event got underway, in the same region. On September 21, Alan W. Heath, an English amateur astronomer using only a 30-centimeter reflector, noticed that Hellas appeared bright and also traced a faint tendril of cloud running from northwest Hellas to the desert area known as Aeria. A night later, the cloud was photographed from the Republic Observatory in Johannesburg, South Africa. Astronomers at New Mexico State University Observatory, Lowell Observatory, and elsewhere then proceeded to follow its increasingly rapid expansion to the west. Within a week of the appearance of the streaklike core, the storm had grown to stretch from the east edge of Hellas across Noachis, and was encroaching on Syrtis Major to the north. Observers at the University of Arizona detected a new cloud in Eos, a feature later found to be part of the great canyon system on Mars. Markings still far away from the visible clouds, including Syrtis Major and Mare Cimmerium, were starting to fade, presumably owing to dust injected into the atmosphere by the storm, and within 16–17 days of the storm's onset a continuous belt of dust clouds extended through the mid-southern latitudes.[10] A curious feature, one that had also been observed in 1956, was the appearance of so-called anomalous dark markings, very dark areas where the storm's powerful winds completely scoured away the mantling of fine dust to reveal darker terrains below.

Within three weeks the entire planet was totally obscured, including the polar caps, making this the first global dust storm ever observed. Highly inconvenient! All three of the 1971 orbiter missions, and especially the two Soviet landers, were flying right into a maelstrom, the detailed environmental conditions of which were essentially unknown. On November 10, personnel at mission control at the Jet Propulsion Laboratory (JPL) in Pasadena, California, switched on Mariner 9's cameras as the spacecraft drew within 800,000 kilometers of the planet. The Martian globe remained hopelessly obscured, with the first images showing no detail whatsoever except for the bright south polar cap at the bottom and four mysterious dark spots near the equator.

Two days later, five leading personalities associated with the lore and scientific exploration of the Red Planet—*New York Times* journalist Walter Sullivan (1918–96), science fiction writers Ray Bradbury and Arthur C. Clarke, and scientists Bruce C. Murray and Carl Sagan—gathered for an event on the Caltech campus to discuss what they expected to learn from Mariner 9 once the dust cleared. Inevitably, there were significant references to the Mars of romance, and to the (diminished) prospects of finding life on the planet, which everyone agreed was something that would be determined not from an orbiting spacecraft but only from biological assays carried out on the surface of the planet—and probably not even then. Among the most interesting remarks were those of Murray, a member of the Mariner 9 television team and by all odds the least susceptible to Martian romance of any of the five participants:

> My own personal view is that we are so captive to Edgar Rice Burroughs and [Percival] Lowell that the observations are going to have to beat us over the head and tell us the answer in spite of ourselves. I think the observations will have to become so unambiguous and so compelling that finally we are forced to recognize the real Mars. Now, Mariner 9 is going to return more than five thousand pictures, to carry out nearly one hundred radio occultations, and to acquire vast amounts of spectral and radiometric data. It will be an enormous step upward. It will provide the observational stick to beat us with and to help us recognize the answers.[11]

Several of the other participants stressed the importance of romance in first stimulating interest in science. Thus, as Bradbury put it, "I think it's part of the nature of man to start with romance and to build with reality."[12] There is no doubt that this is true, and so Burroughs and Lowell, despite the fact that their writings contain a fair amount of nonsense, had done their part. Murray, however, framed things more broadly. Despite the Cold War overtones of the present race to Mars, he invoked higher ideals of human curiosity and the thirst for knowledge for its own sake. His words reflected an American view, though he also attributed the same motivations to the Soviets as they pursued their lunar and planetary programs:

We as a people, as a nation, are spending our money on an uneconomic endeavor. We will not recover a product. We won't get military benefits. Instead, we're doing something that really has cultural value. That Mariner spacecraft up there is a cultural edifice dedicated by this country to an idea, to the idea of exploration, to learning about something we don't know.

The extent to which a people will do this is a measure of their optimism and imagination. I don't think we have to be looking for life on Mars to justify it any more than we had to justify exploring the polar areas on the grounds that there might be great economic benefits. The fact that we as a people have advanced far enough to explore another planet is something of which we should be very proud. . . . In my mind, it balances many of the negative things we have to live with—Vietnam, smog, bureaucracy—all these and other things that we don't like. I feel it is a real privilege to be able to do things that couldn't ever be done before.[13]

Noble aspirations! But first the dust had to clear.

At the moment, there was nothing to be done about the dust, but on November 14, a 15-minute, 23-second main engine braking burn successfully allowed Mariner 9 to slip into a stable orbit, with an inclination of 65°, around Mars—becoming, in the process, the first spacecraft to ever orbit another planet. The last of the five launches in the 1971 window became the first to reach its destination. Mars 2 successfully attained Mars orbit just over two weeks later, and Mars 3 also reached orbit, albeit in a much larger elliptical path than planned because of an only partially successful braking burn, about five days after that. The Soviets had lost the race to Mars orbit, but they still aimed to steal the show by becoming the first to successfully land there.

Sadly, they came close, but fell just short. The Mars 2 lander was released in almost the right direction toward Mars just four and a half hours before the orbiter arrived at Mars. But in rocket science, and especially when attempting to land on Mars, *almost* is not good enough. The lander came in at too steep an angle, causing the descent system, which consisted of the now-familiar combination of aerodynamic braking, parachutes, and retrorockets that the Soviets had already tested on Venus, to malfunction. The lander crashed at high speed (before parachute deployment) on the surface.[14] The Mars 3 lander was also released just before its orbiter got to Mars, and it appears to have

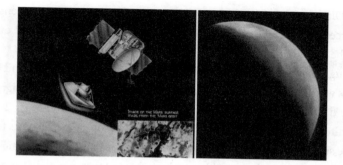

FIGURE 12.1 Artwork and an example orbital image from the Soviet Mars 3 mission. While the lander was not successful, the orbiter was able to study the planet for about eight months after arriving in orbit in late 1971. Courtesy of the Russian S. A. Lavochkin Institute (*both*) and Ted Stryk (*right*).

encountered the Martian atmosphere at the correct angle and speed. This time the descent system worked as planned (figure 12.1). Telemetry received in real time by the Mars 3 orbiter indicated that on December 2, 1971, the lander successfully touched down on the surface near 45°S and 158°W (in the desert area Phaethontis). (The Soviets, by the way, had responsibly taken the precaution of surface sterilizing the probe by gaseous sterilization, radiation, and heat, to avoid contaminating Mars with terrestrial microbes.) Once on the surface, the Mars 3 lander appears to have ejected its protective cover and opened its flowerlike petals to right itself and deploy its instruments and a pennant containing the hammer-and-sickle insignia of the Soviet Union. Its television cameras were switched on. Then, about 90 seconds after landing, data transmission began. It was little more than an audible hiccup from a dying spacecraft, however. After only 20 seconds of radio contact, the transmission suddenly ended, and the lander was never heard from again. The 20 seconds of data received were mostly noise, and unfortunately did not provide any usable information about Mars itself.[15] There has been much speculation that the raging global dust storm somehow contributed to the lander's failure to survive more than a few minutes on the surface, but of course there's no way to really know. An interesting postscript to the mission is that the Mars 3 lander and its heat shield and parachute may have been identified by amateur Russian space enthusiasts in high-resolution NASA images from the Mars Reconnaissance Orbiter taken near the expected latitude and longitude in

2013.[16] Ultimately, despite the impressive entry, descent, and landing engineering success of Mars 3, the Soviet efforts to become the first to land on Mars would be regarded as failures.

Though the lander was thus a tantalizing near miss, the Mars 2 and Mars 3 orbiters were somewhat more successful, and did manage to secure significant science data sets.[17] Plagued by communication problems, the Mars 2 orbiter was limited to frequent atmospheric radio occultation measurements to characterize surface pressure and temperature variations and the properties of the planet's upper atmosphere and ionosphere. Mars 3 was somewhat more productive, returning about 60 color images, the first images of another planet ever obtained from a Soviet interplanetary probe.[18] Unfortunately, since they were taken while the dust storm was still abating, the quality was rather low, and many of them were rather bland and uninformative. However, useful information was obtained from other instruments, including measurements of the temperature at various points on the Martian surface. The coldest point proved to be the north polar cap, where the temperature was −110°C; elsewhere, the values ranged from −93°C to +13°C, depending on the latitude and time of day. Other useful data included measures of topographic variations on the surface, determinations of the composition of the atmosphere, and verification of the fact that Mars has no strong magnetic field. (If there was a magnetic field, it was at least 4,000 times weaker than that of Earth.) Finally, radio tracking of both Soviet orbiters allowed the Martian gravity field to be mapped, and included the discovery of significant differences in the gravity field from place to place that were likely indicative of large, lunarlike impact basins.

And yet, hardly anyone remembers Mars 2 or Mars 3. The mission that everyone remembers is Mariner 9, one of the most successful interplanetary missions of all time and the one that forever shifted our paradigm about Mars. In contrast to the pre-spacecraft era, when Mars seemed in so many ways Earthlike, or the flyby Mariner era, when it seemed more or less to be another Moon, Mariner 9 showed Mars to be Mars—it was itself, itself alone.

In contrast to the Soviet orbiters, which had been preprogrammed to carry out their imaging sequences automatically and consequently could not wait out the dust storm, Mariner 9's television cameras were initially used rather sparingly—not to save energy, as is sometimes claimed (the spacecraft was powered by solar panels), but because there was so little to see. Instead of just taking rote frame after rote frame of nearly featureless expanses, the space-craft adopted more flexible targeting procedures, which included imaging the two Martian satellites, Phobos and Deimos (see chapter 20). Because of the built-in flexibility in its program objectives, Mariner 9 was able to outlast the global dust storm (and the two Soviet orbiters). Also, in contrast to the secrecy that had shrouded every Soviet space mission, the Mariner 9 mission was conducted openly, with frequent press/media/science conference updates, rapid publication of scientific results in the peer-reviewed literature, and a strong NASA-supported program of public outreach built in to the mission's culture.

After the initial mid-November 1971 engine burn, additional smaller engine burns settled Mariner 9 into a 12-hour elliptical orbit, inclined about 64° from the Martian equator, and coming within about 1,650 kilometers of the surface at each orbit's closest approach. The dust very gradually subsided. By mid-December, the enigmatic dusky spots in Tharsis had begun to show their true nature (figure 12.2). Their tops, rising above the dust, proved to be calderas atop enormous shield volcanoes. Planetary scientist and space artist William K. Hartmann, then a young member of the Mariner 9 science team, has recalled how after the initial encounter everything was completely obscured by dust except for the four mysterious dark spots:

One day, Carl Sagan came running down to the science team room from the upstairs "computer" room that had been receiving images from the Goldstone tracking antenna, waving a polaroid photo of the TV screen (the mode of initial transfer of Mariner 9 images to the science team!). The photo revealed a pretty clear volcanic caldera in a summit protruding from the clouds. That was the day that everyone realized the dark spots were enormous shield volcanoes.[19]

The grandest of the shield volcanoes—coinciding in position with Giovanni Virginio Schiaparelli's Nix Olympica (the Snows of Olympus)—was christened Olympus Mons. As is now known, it measures 624 kilometers at its base (about the size of Arizona), has an 80-kilometer-wide caldera at the summit, and rises some 25 kilometers (82,000 feet!) above the surrounding plains. The other dusky spots corresponded to the shield volcanoes now known as Ascraeus Mons, Pavonis Mons, and Arsia Mons, from north to

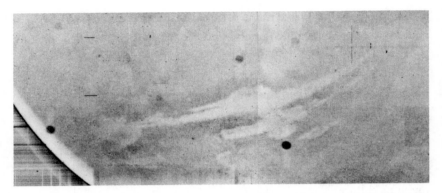

FIGURE 12.2 One of the earliest photos (from December 1971) of the enormous Martian canyon system eventually named Valles Marineris (after the Mariner 9 mission), still partially filled with the remnant clouds and hazes from the abating global dust storm. NASA/JPL.

south. A Mariner 9 oblique view taken on December 17 provided a first dramatic glimpse of the enormous (and still partially dust-filled) canyon of Valles Marineris, which extends along the equator for 4,000 kilometers, a full quarter of the way around the planet's circumference (and roughly the distance between New York and Los Angeles). The same image also showed an associated branching latticework of canyons, Noctis Labyrinthis.

While waiting out the dust, mission scientists also used the onboard ultraviolet and infrared spectrometers to learn as much as possible about the famous Martian dust storms, and especially how they influence the atmosphere and surface. Infrared observations are highly sensitive to the abundance, composition, and particle size of dust grains suspended in the atmosphere. When these observations were combined with information on the atmospheric pressure and temperature profiles from radio occultation measurements and imaging data of high-altitude clouds and hazes along the planet's limb, a more complete understanding of the dust's effects was obtained. Among the more puzzling discoveries, at least initially, was that the planet's surface temperature appeared to *drop* significantly during the large-scale dust storm. Once the dust eventually cleared, mission researchers ended up with an outstanding before-and-after data set that helped provide new insights into the surface temperature puzzle and the role that dust and aerosols play in *any* planetary atmosphere, including our own.

Sagan, already well on his way to becoming the most recognizable planetary scientist of his generation, did much of his best work at this time, in collaboration with his first graduate student, planetary atmospheres expert Jim Pollack (1938–94).[20] Together, Pollack and Sagan attempted to account for the unexpected tendency of the Martian surface to cool during dust storms. Armed with in situ data from missions to both Mars and Venus (including, eventually, detailed surface temperature history records from the Viking Landers later in the 1970s), as well as terrestrial data taken during enormous volcanic explosions that inject huge amounts of dust into our stratosphere, these researchers were able to develop models that explained not only CO_2 greenhouse warming, but also the surface cooling (and upper atmosphere warming) caused by suspended dust in a planetary atmosphere. The latter models were eventually used by Sagan and colleagues to predict that a so-called nuclear winter—massive cooling of Earth's surface—would be the likely result of a global nuclear war between the superpowers.[21] In more recent years, of course, worries about nuclear winter have subsided somewhat, and another effect first recognized from the study of Venus—runaway greenhouse warming, in Earth's case owing to the massive injection of carbon dioxide into the atmosphere through the burning of fossil fuels—has caused concern and fears of another impending mass extinction event.[22] Clearly, understanding our neighbor planets, Venus and Mars, has proved to be anything but academic.

But to return to 1971–72: when the dust finally cleared completely in early 1972, Mars was a long way from Earth, hardly more than a pinpoint in amateur-sized telescopes but looming vast and magnificent beneath the spacecraft's television cameras. Only now, at last, could Mariner 9 begin its primary mission objective of systematically mapping the entire surface of the planet. Hartmann was engaged at the time in converting his lunar crater chronometry to Mars, having already used his crater-counting system to make a correct prediction that the lunar lava plains were about 3.6 billion years old. He now counted craters on various regions of Mars and achieved a second success, with a conclusion rather prescient for the time: that the youngest volcanic regions, such as lava plains in Tharsis, the slopes of Olympus Mons, and so forth, were only a few hundred million years old. This conclusion was controversial at first, because Mariners 4, 6, and 7 had left the impression that Mars was somewhat moonlike, and being smaller than Earth it would have cooled off rapidly, so that volcanism would have ended in the planet's earliest years. As we will see, Hartmann's Mariner 9 result was later supported by actual rocks from Mars (Martian meteorites).

Meanwhile, as the mission progressed and more and more images of Martian regions accumulated at JPL, two globes were prepared. JPL's darkroom prepared images at a specific scale, and then they were glued onto the globes, giving members of the science team the very first look at the global patterns that emerged. The globes were set up not in the science team area but rather in one of the more public conference areas, Von Karman Hall. Hartmann says:

> Even the individual photos, as they came down from the upstairs imaging screen areas to the lower floor science team area, were wonders in themselves, as you never knew but that you might be the first person in human history to see some amazing scene on another planet. But I'm something of a back-off-and-look-at-the-big-picture guy, so there was also the excitement of how all the individual regions fit together into the global picture.[23]

All this, of course, could be done today in a matter of hours by simply plastering digital images onto computer-generated globes. It may not have been as fast, but the anticipation was certainly much more exciting the way it was done in 1972.

Over the course of more than 600 more orbits and 10 more months of operations, Mariner 9 completed not only its original mission objectives (monitoring changes in the Martian surface and atmosphere over time) but also those of the failed-at-launch Mariner 8 (mapping at least 70 percent of the Martian surface).

The list of Mars science and technical "firsts" achieved by the operations and science teams of Mariner 9 is impressive, as were many of the prescient follow-on hypotheses that the mission's discoveries helped the planetary science community formulate. It was hardly an exaggeration to call it, as associate administrator for space science at NASA John E. Naugle (1923–2013) did in the aftermath of the mission, "by all odds the most productive planetary mission that has ever been flown."[24] Arguably the two Voyager missions, launched in 1977 on their Grand Tour of the giant planets, would contest with Mariner 9 for the title of the "most productive planetary mission." Still, Mariner 9 was—

and will remain—among the most productive missions ever flown to Mars. The pre–Mariner 9 era was that of the Old Mars; the post–Mariner 9 era that of the New Mars. The former had been largely that of myth and conjecture; the latter was that in which the reality of the Martian landscape began to emerge.

A new Mars required new names. The classic Schiaparellian nomenclature had been inspired by the albedo features, which suggested, back in the day, Mediterranean lands and seas. Mariner 9 showed that the albedo features were not as significant as they looked. The underlying topography did not always defer to the fineness of the dust spread over it like a tablecloth.

Even before the dust cleared, the first features to be revealed were the Tharsis shield volcanoes—so towering they had poked up even above the pall. (They appeared dark by contrast with the surrounding bright dust swirling around them.) These were among the most striking topographical features on the planet. Schiaparelli had ventured a lucky guess—it was no more than that—in naming Mars's most imposing shield volcano Nix Olympica (the Snows of Olympus), which he saw as a tiny whitish patch in November 1879. In the new era, it would become Olympus Mons. With its summit towering 25 kilometers above the surrounding plains, Olympus Mons is the tallest mountain the solar system; but as it measures 624 kilometers wide at the base, the slope is very gentle, averaging only 5° on the volcano's flanks. (By comparison, the largest shield volcano on Earth, Mauna Loa, measures only 120 kilometers at the base, and the summit rises 9 kilometers above the ocean floor.)

The other three Tharsis shield volcanoes poking above the pall before the dust began to clear were Arsia Mons, Pavonis Mons, and Ascraeus Mons, which correspond with the classical "oases" Ascraeus Lacus, Pavonis Lacus, and Arsia Silva. They are spaced about 700 kilometers apart and aligned southwest to northeast along the crest of the great rise known as the Tharsis bulge; their summits reach some 17 kilometers above the surrounding plains.

As the dust cleared, Mariner 9 revealed in detail landscapes that had hitherto registered only as distant blurs in the telescope, making geologists on the science team rather like modern Adams walking through the Garden of Eden and needing to confer names on the species of animals they encountered. If, as in the Ray Bradbury story "Dark They Were, and Golden-Eyed," there had once been Martians, they would undoubtedly have had proud Martian names for these places. But we didn't know the Martian names; we didn't even have anything like the Native American names the colonists gave to the rivers and countries they settled after the *Mayflower* arrived and pushed the original inhabitants themselves aside. Bradbury imagines his character Mr. Bittering glance up from the garden to the Martian mountains:

> He thought of the proud old Martian names that had once been on those peaks. Earthmen, dropping from the sky, had gazed upon hills, rivers, Martian seas left nameless in spite of names. Once Martians had built cities, named cities; climbed mountains, named mountains; sailed seas, named seas. Mountains melted, seas drained, cities tumbled. In spite of this, the Earthmen had felt a silent guilt at putting new names to these ancient hills and valleys.
>
> Nevertheless, man lives by symbol and label. The names were given.[25]

The new names were chosen by scientists. The much-revered Caltech geomorphologist Robert P. Sharp spent a good part of the Mariner 9 mission brainstorming neutral terms for the newly revealed features of Mars. Following lessons of the early years of geology, Sharp was determined not to use names that might hint at origins—as some early 19th-century geologists committed to establishing Noah's flood might term something "sedimentary" terrain when what they were documenting might turn out to be a lava flow layer. Many of these new terms were uncontroversial: planum for low plain, planitia for high plain, terra for large, rugged areas on Mars corresponding in many cases with the classical desert regions. Other terms were more controversial, and though not everyone would agree with his assessment, Hartmann, then a young scientist on the geology team, thought they caused more problems than they avoided. "Sharp introduced all those deliberately meaningless, now somewhat annoying-to-me, terms," Hartmann says,

> like "fretted terrain," "chaotic terrain," "hummocky," "corrugated," "speckled," "polka-dotted." Okay, some of those I made up in a moment of sarcastic weakness to make the point. Their problem was that the early generation of planetary geologists were trained to present Mariner 9 and later photos in those terms, and only then discuss an interpretation of what it meant. I don't know how many ten-minute conference talks I attended where a young speaker spent eight minutes translating the photos and their clear features in to neutral words, and

then ran out of time to report anything meaningful or experiential (my term) about Mars and its history. The concern with terminology began to obstruct the understanding.[26]

The International Astronomical Union (IAU) was the ultimate custodian of official nomenclature for Mars. At the 1967 IAU assembly in Prague, an ad hoc committee for Martian nomenclature under the chairmanship of Gerard Peter Kuiper was appointed, and it laid down guidelines for the naming of craters whose existence had been revealed by Mariner 4: "The most prominent of these (craters) may be named after deceased scientists and members of other suitable professions."[27] In anticipation of results from the 1971 orbiters (both Soviet and American), a working group for Mars nomenclature under de Vaucouleurs's chairmanship met at the IAU General Assembly in Brighton, making proposals for an appropriate system of names for prominent Martian features.[28] Drawing from the atlases and maps produced on the basis of the Mariner 9 results, and adopting many of Sharp's suggestions, the working group, now headed by Mariner 9 science team member Brad Smith, approved the elaborate propositions for Martian nomenclature (which have been used ever since) at the IAU General Assembly in Sydney in 1973.[29] (See appendix E on the nomenclature of Mars.) For better or worse they will be with us for as long as humans concern themselves with Mars—or perhaps until Martians of the future rename them themselves.

O Brave New World!

Mariner 9's first mapping cycle took place predominantly over the southern hemisphere—the region between 25°S and 65°S. The most prominent features seen in this cycle included the gigantic impact basins Hellas and Argyre. That this is ancient, heavily cratered terrain had been known since the flyby Mariners; but some completely unexpected features emerged as well. Mariner 9 imaged networks of channels and tributaries that looked for all the world like runoff channels and dry riverbeds, and that seemed to imply that conditions on Mars were once very different from what they are today—so that liquid water could run on the surface.

The next mapping cycle included regions as far north as latitude 25°N, and revealed the enormous canyon system known as Valles Marineris—the Grand Canyon of Mars, as it is sometimes called, though it utterly dwarfs its terrestrial counterpart. Valles Marineris extends along the equator for 4,000 kilometers, a quarter of the entire circumference of the planet. Its origin lies close to the summit of the Tharsis bulge, at Syria Planum, where it consists of a series of short, deep gashes intersecting at all angles. These features have been formed by faulting and resemble grabens. This labyrinthine system has been aptly named Noctis Labyrinthus, the Labyrinth of Night. (The reason for the name is that Eugène Michel Antoniadi, in 1909, had named a dusky patch seen in this location Noctis Lacus, the Lake of Night.) In the middle section of Valles Marineris, the canyons become more continuous and run roughly in parallel as three main branches, known as Ophir, Candor, and Melas Chasmata, which are separated by intervening ridges. The combined width across all three canyons reaches 700 kilometers, and the depth, in places, is as much as 7 kilometers. These canyons connect with Coprates Chasma, which runs eastward and joins Eos Chasma, before merging with an area of jumbled and enmeshed ridges, cracks, and plains known as "chaos terrain." This particular example of chaos terrain is known as Margaritifer chaos, and is associated with several ancient dry riverbeds, of which we will say more presently.

We do not propose to go into great detail about Martian geology here; it is a vast subject that has been ably treated elsewhere.[30] A few generalizations, however, may be useful. First, with few exceptions, the boundaries of the classical albedo features do not show a good correlation with Martian topography (figure 12.3). Among the exceptions are the impact basins Hellas and Argyre, which had been well known from Earth as large, bright circular features. On the other hand, the most prominent albedo feature, Syrtis Major, proved not to be a depressed area (dry seabed) as formerly assumed. Instead it proved to be a low-relief shield volcano, and has been renamed Syrtis Major Planum. Another dark area, Mare Acidalium, sprawls across a large area in the northern hemisphere, and is referred to as Acidalia Planitia, and so on. (As would be established by the later spacecraft mission Mars Global Surveyor, Syrtis Major, Aurorae Sinus, Mare Sirenum, Mare Acidalium, and similar classical dark areas show up as having high calcium-rich pyroxene content, which identifies

FIGURE 12.3   The first global map of Mars made from an orbiting spacecraft. Mariner 9 took more than 7,000 images of Mars and revealed that the planet had both ancient cratered terrain and more modern tectonized and eroded areas. NASA/JPL/USGS/Phil Stooke.

them as basalts similar in composition to the oceanic crust on Earth and the maria on the Moon.)

The albedo markings merely overlay the more significant topographic units. In fact, the difference between light and dark areas, which seems so important in the telescope, reflects nothing more than fundamental differences in dust cover or dust-grain sizes. The dark areas are generally covered by less dust and contain a wider range of particles, including coarser ones, while the light areas are generally covered only with finer-grained dust (known also as "drift" or "fines" among Martian geologists). It has become evident, from subsequent spacecraft results, that the dark areas are either relatively dust-free coarser-grained fines (like sand dunes), or surface exposures of bedrock (often interpreted as consisting of igneous rocks such as basalts), while the bright areas are typically bedrock covered in deposits of fine weathered dust, in some places as a relatively thin veneer, in others, such as the Tharsis region, as a very thick blanket. The seasonal/decadal changes in albedo markings are due to removals and incursions of dust. Indeed, laboratory and field studies have shown that the removal of only a few-micron-thick cover of dust can turn a bright region into a dark region, and vice versa. Thus, as became apparent with Mariner 9 but would not be fully worked out until the later Viking Orbiter and Mars Global Surveyor missions, at least within certain limits albedo markings on Mars appear to be ephemeral. As summed up by British Astronomical Association (BAA) Mars Section director and Mars dust storm analyst Richard J. McKim,

> There is an intimate link between the albedo changes on the Martian surface and yellow cloud [dust storm] activity. The removal of a dusty veneer of high albedo surface deposits by Martian winds can quickly expose darker underlying bedrock. This is visible microscopically [at a small scale] in the "tailed craters" imaged by [Mariner 9 and] Viking and Mars global Surveyor, and macroscopically in changes in the classical albedo patterns. Changes visible telescopically can occur in as little as a day.[31]

Mariner 9 showed that the earlier flyby Mariners had passed mainly over areas of the southern hemisphere, where most of the prominent and thus historically important-looking dark markings are located. Thus, they sampled mostly heavily cratered terrains, and in doing so, they missed many of Mars's most intriguing features. Mars to the earlier flyby Mariners appeared blander and more lunarlike than it really is.

Broadly speaking—that is, looking at Mars from a "back-off-and-look-at-the-big-picture" point of view—the most important generalization about the Martian surface is the difference between the northern and southern halves of the planet, the great crustal dichotomy. South of a circle inclined roughly 35° to the equator, the surface consists mostly of ancient, heavily cratered terrain. North of this circle lie relatively smooth plains and volcanic features of obviously much younger age. The boundary between the two regions is formed by a gentle, irregular scarp and low, knobby hills. On average the southern highlands are some 2.1 kilometers higher in elevation than the northern lowlands.

Even a cursory glance at a map or globe of Mars shows different areas that vary widely in cratering densities. This allows the relative ages of surface units to be worked out. The most heavily cratered areas are obviously the most ancient. The oldest surviving areas are classified as being of Noachian age (so called after Noachis Terra, the prototypical surface of this type). Younger areas are called Hesperian, after Hesperia Planum, and Amazonian, after Amazonis Planitia, respectively. (A summary is given in table 12.1, but note that, though we can certainly tell the youngest from the oldest surfaces, the dates are highly uncertain, especially for the beginning of the Amazonian. The main problem in crater chronometry of Mars is the uncertainty in the current rate of crater formation, which is estimated by shifting the lunar cratering rate to Mars, which is closer to the asteroid belt. But there is a range of estimates for the ratio of these two rates.)[32]

The southern highlands, which include most of the subequatorial parts of the planet as well as a rather wide tongue extending northward beyond Terra Sabaea and Meridani Terra, are packed cheek by jowl with craters; at least superficially, the view looks very much like that of the highlands of the Moon. Though a debate between volcanic and impact formation of the lunar craters raged for more than a century, as we have seen it was all but settled in the 1960s in favor of the impact theory (though a few dissenters remained). Indeed, the impact process has now been worked out in considerable detail. Whenever an object—say, a small asteroid—strikes the surface of a planet, it produces two interacting shock waves. The first engulfs the asteroid, vaporizes

TABLE 12.1 The Geological Periods of Mars

Era name	Description	Age, billions of years ago[a]
Pre-Noachian	The interval from the accretion and differentiation of the planet to the formation of the Hellas impact basin. The great crustal dichotomy is thought to have formed during this time.	4.5 to 4.1
Noachian	Formation of the oldest extant surfaces of Mars. Heavily scarred with large impact craters. The Tharsis bulge is thought to have formed during this period, along with river valley networks caused by extensive erosion by liquid water. Large lakes or oceans may have been present. Named after Noachis Terra.	4.1 to 3.7
Hesperian	Formation of extensive lava plains. Olympus Mons probably began to form. Catastrophic releases of water carved extensive outflow channels around Chryse Planitia and elsewhere, and ephemeral lakes or seas may have formed in the northern lowlands. Named after Hesperia Planum.	3.7 to 3.0
Amazonian	These regions have relatively few meteorite impact craters but are otherwise quite varied. Characteristic features of this period include lava flows, glacial/periglacial activity, and minor releases of liquid water. Named after Amazonis Planitia.	3.0 to present

Source: W. K. Hartmann and G. Neukum, "Cratering Chronology and the Evolution of Mars," Space Sci. Rev. 96 (2001): 165–194.

[a] Ages, which were initially defined based on Mariner 9 data, are approximate and generally rather uncertain. They have been estimated from studies of impact crater distributions. For example, the date of the Hesperian/Amazonian boundary is particularly uncertain and could range from 3.0 to 1.5 billion years ago.

it, and melts rock at the immediate point of impact. This part of the process absorbs a relatively small fraction of the energy of impact; the much greater share goes into producing a second shock wave, traveling radially away from the point of impact, excavating the crater, and throwing a rim of disintegrated material around it known as an ejecta blanket.

Not surprisingly, craters of different diameter vary in form. On the Moon, very small craters are simply bowl-shaped pits that have a fairly constant depth-to-diameter ratio of about 0.15 to 0.20. Larger craters are more complex. Violent rebound of the floor from the shock of impact gives rise to a central peak or an interior "ring" of peaks. In addition, many of the larger craters have terraced outer walls caused by landslides, the slumping in of rim materials toward the center of the crater. Partly because of this partial filling in with wall material, more complex craters become shallower with increasing diameter.

Already by the time of the flyby Mariners, it was obvious that Martian craters were generally flatter and "softer" in outline than their lunar counterparts. There was also a relative paucity of smaller pits on Mars. This was clear evidence that on Mars, unlike the Moon, considerable weathering has occurred over time. The most interesting difference involves certain craters with diameters of over 5 kilometers. On both the Moon and Mars, smaller craters have ejecta blankets laid down with ballistic trajectories, but on Mars some craters larger than about 5 kilometers in diameter, especially those in mid- to polar latitudes, are surrounded with ejecta blankets resembling mudflows. Such forms suggest that subsurface permafrost or ice was melted during impact, to produce a slurry of fluidized water and rock that was released by the energy of impact and that flowed across the surface (some planetary scientists even call these "splosh craters"). The fact that the mudflows are generally seen only around craters above a certain size and north or south of specific latitudes suggest there is some kind of transition, presumably corresponding to the minimum depth needing to be excavated in order to release subsurface water ice. Moreover, since near-equatorial features are generally crisp and well defined, while craters of the same size appear sloshy at higher latitudes, the distribution of these fluidized ejecta deposits suggests that subsurface ice must be hidden at greater depths (or absent) near the warmer equator and exposed nearer to the surface in the colder polar latitudes. The existence of significant amounts of water ice buried just beneath the surface of Mars, especially at high latitudes, would be confirmed and elaborated by later spacecraft.

Among Mariner 9's most important achievements was a detailed analysis of the composition of the polar caps. After Mariners 6 and 7, some scientists, including most notably Bruce C. Murray, bleakly announced that the polar caps were entirely frozen CO_2. However, though carbon dioxide freezes out onto the polar caps, especially onto the south polar cap during the long and bitter southern hemisphere winter, Mariner 9 demonstrated the existence of water ice in both the northern and southern caps. (The amount and purity of this ice would not be apparent until radar sounders on the European Mars Express and the NASA Mars Reconnaissance Orbiter spacecraft confirmed the existence of relatively clean ice in both polar caps extending to a depth of several kilometers below the surface.) Together, the volume of ice present in the Martian north and south polar ice caps is now estimated to be similar to that in the Greenland ice cap on Earth.

As had been noted long ago, the seasonal behavior of the two caps is markedly different, and it was evident that the difference between the northern and southern seasons was responsible. Recall that because of Mars's eccentric orbit, the southern polar summers are short and warmer, and the southern polar winters are long and colder; by contrast, the northern polar summers are long and milder (by Martian standards), its winters short and moderate. Around 95 percent of the atmosphere of Mars consists of carbon dioxide, and the amount of CO_2 gas changes with the seasons. The deep cold southern polar winter removes CO_2 gas from the atmosphere by freezing it directly onto the south polar cap. As temperatures drop below −123°C, CO_2 freezes out as frost, snow, or ice onto the permanent water ice cap, causing the atmospheric pressure all over the planet to drop by 25 to 30 percent. As the cap warms during the summer, the CO_2 sublimes back into the atmosphere. As it does so, the south polar cap "remnant"—the permanent water-ice cap—is revealed, as well as abundant internal layers of ice and dust that form spiral-shaped troughs cutting through the water ice (figure 12.4). Similar, though more modest (partly because of the shorter winter), effects occur at the north polar cap. The layered terrains within and surrounding both caps likely contain, rather analogously to those found in the ice sheets on Earth, a detailed record of past climates on Mars, just waiting to be explored.

Another way of thinking about all this is to imagine the top layer of one polar cap migrating to the opposite polar cap—not by a system of water-filled canals, as Percival Lowell once imagined, but by air. Half a Martian year later, an opposite migration occurs. This migration takes place almost with clockwork regularity because the circulation of Mars's atmosphere is rather straightforward compared to Earth's, mainly because Mars has no oceans. The

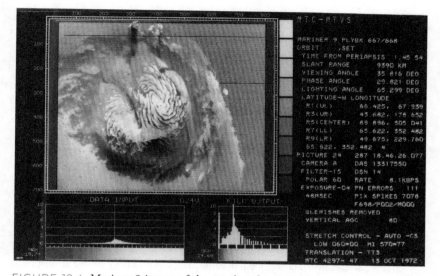

FIGURE 12.4 Mariner 9 image of the north polar cap, taken near the end of the mission. This image was taken about a half Martian month after summer solstice, at which point the cap had reached its minimal extent. The cap is about 1,000 kilometers across; the interior dark markings are frost-free sun-facing slopes. Thick stacks of flat sedimentary sheets of dust and water-ice underlie the cap. The image is centered at 89°N, 200°W. NASA/JPL.

circulation at low latitudes is dominated by Hadley cell motion (named for George Hadley [1685–1768], who first described it on Earth; figure 12.5). Air warmed at the equator flows one part north and one part south, making it about as far as 30°N and 30°S latitude, where it cools, sinks, and then flows back toward the equator at the surface. The direction of flow would be along a direct north and south line were it not disrupted by the planet's rotation (through the Coriolis effect) and surface landforms (such as the Tharsis volcanoes). Because of these effects, in actuality northern hemisphere winds generally blow from the northeast, in the southern, from the southeast, patterns that are broadly similar to Earth's strong westerly polar vortices, the subtropical westerlies, and so forth.

At high latitudes, polar air masses dominate. A series of high and low pressure areas sweep around the planet from west to east, and where they interact with Hadley cell movements, sharp weather fronts and storms can arise. These storms, however, tend to be much less violent than those on Earth because of

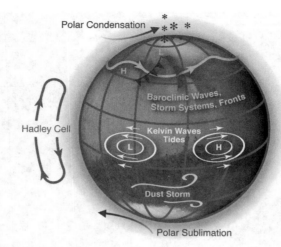

Polar Condensation

Hadley Cell

Baroclinic Waves, Storm Systems, Fronts

Kelvin Waves Tides

Dust Storm

Polar Sublimation

FIGURE 12.5 Because it lacks global oceans, Mars has relatively simple atmospheric flow patterns compared to Earth's. On a planetary scale the flow is dominated by the Hadley cell. Locally, however, the flow can be more complex and includes air that sublimates off of and moves away from the summer polar cap; air that condenses onto the winter polar cap; and regional waves, fronts, and individual storm systems that can create ice clouds and lift dust to form small-scale to planet-encircling dust storms. NASA Mars Climate Modeling Center.

the thinness of the Martian air, the lower temperatures, and the fact that water vapor—which can transport a lot of energy as it changes phase—is minimal.

Global circulation patterns as enhanced by local effects in certain regions can lead to dust-raising through saltation or by dust devil activity. For example, according to Richard J. McKim, "In southern summer, a large temperature gradient exists between the freshly exposed surroundings of the shrunken S. polar cap, and the cap remnant itself. The evaporation of the cap also generates large pressure gradients in the atmosphere, and atmospheric tides will have maximum dust-raising potential, especially in places where the Sun is overhead."[33] These gradients may be enhanced in locations such as the sloping plains between the northwest rim of Hellas and the Noachis highlands, where both the 1956 and 1971 storms developed; the sloping plains to the west, south, and southeast of Claritas Fossae; and the low-lying Isidis Planitia to the east of Syrtis Major. These local dust clouds expand slowly during the first few days, then more rapidly as new centers of activity develop and old ones coalesce.[34] In the northern hemisphere, large-scale dust activity as observed telescopically occurs disproportionately in Tempe and Chryse, which lie adjacent to the large neighboring dark area of Mare Acidalium; according to McKim, "the mare has a much higher thermal inertia [than the adjacent brighter regions], and this thermal contrast will act in enhancing regional winds and lead to dust-raising given the appropriate forcing seasonal conditions."[35]

In contrast to Earth, where the blue of oceans is the predominant theme of the surface, on Mars everything is the color of dust. The apparent blues reported by visual observers—the blues that were not really blues, as Ray Bradbury put it—were indeed, as had been long suspected, illusions.[36] Mars is indeed the Red Planet. Its color consists of nothing more than various shades of rust-red, produced by the ubiquitous presence of oxidized dust, in turn generated by the weathering of basaltic rocks on the planet. Just as the Inuit are said to have 50 different words for "snow," future Martians are almost certain to have as many words for "red."

The classical desert areas simply consist of finer, brighter, and more oxidized—thus more intensely rust-red—particles. (Presumably, the greater degree of oxidation is due to the smaller size of the dust grains, since smaller grains have a higher surface-to-volume ratio and thus are more exposed to weathering factors.) The presence of dust is attested to by not only the shades but also the shapes of the dark areas—for instance, the classical albedo areas Sinus Meridiani (Dawes Forked Bay) and Margaritifer Sinus, which roughly overlie the heavily cratered Meridiani Terra and Margaritifer Terra, end in forks, as do many others. This gives them a decidedly windswept look, and windswept they certainly are. In general, as was first inferred from Mariner 9 imagery by Sagan and Pollack, the features point along the directions that dust is swept by the current prevailing winds, and in a few cases broad dusky wind-blown streaks do correspond with some of the classical "canals."[37] At a smaller scale, aeolian features, including streaky or "tailed" craters, were visualized in many Mariner 9 images, as were wind-eroded hills known as yardangs, sand dunes, and wind-carved ripples.

The widespread presence of dunes certainly underscores Mars's desert reputation. A vast field of transverse dunes was discovered in Mariner 9 images in the region between latitudes 75°N and 85°N, where they form a dusky collar (long known as the Lowell band) around the north polar cap.

Many craters here are completely filled with dunes. Dunes are present in many other parts of the planet, too. However, where sand is presumably scarcer than in the circumpolar region, the characteristic forms of the dunes are crescent shapes known as barchans. (Dean B. McLaughlin's theories from the 1950s, minus active volcanism, proved to be remarkably prescient; of all the theorists of the pre-spacecraft era, he came perhaps closest to guessing the truth about Mars.)

Adding to the long list of its accomplishments, Mariner 9 also provided the first accurate measures of Martian wind speeds. During the most extreme Martian storms, such as that of 1971, wind speeds were inferred to reach up to about 180 kilometers per hour—and yet, because of the thinness of the Martian atmosphere, they would feel to us like gentle breezes of only about 21 kilometers per hour. The force of those winds, though extremely weak, is able to mobilize very fine materials to produce dust devils and dust storms, but the actual amount of material transported is rather small. Nevertheless, because Mars is so dry, the dust once airborne is able to remain aloft far longer than on Earth, where rain soon washes it away. Thus a patina of light-colored (fine) dust is spread far and wide across the surface, producing temporary albedo changes on both large scales (visible from Earth) and small scales (as in the reorientation of windblown bright and dark streaks in craters).[38] However, these changes are typically rather cosmetic: it has been estimated that even during a major dust storm, the dust forms a layer that would be only a few micrometers thick if deposited with uniform thickness all across the planet between latitudes 58°N and 58°S—considerably less than the breadth of a human hair.

It may seem surprising that, despite the fact that light-colored dust is blown so readily about the surface, the whole planet has escaped becoming a uniform brightness. There are several reasons for this. After a dust storm, the deposited veneer is sometimes cleared after another Martian season or two by subsequent "normal" prevailing winds. Also, it is hypothesized that wind erosion could be producing dust mainly in certain areas of Mars, such as the Syrtis Major plateau, where there is a higher abundance of exposed rocks or outcrops. Slow physical and chemical erosion of rocky surfaces produces smaller rocky particles that are sifted by the wind, with heavier (silt- and sand-sized) particles bounding and jumping across the surface through a process known as saltation, creating (via tiny impacts) and stirring up finer particles that can become windborne in the atmosphere. But since only the finer materials are able to be borne aloft on the thin atmosphere, they can be scattered to great distances and accumulate selectively in certain classically bright areas such as Tharsis, Arabia, and Elysium. Subsequent detailed radar and thermal studies of these areas showed that they appear to be covered in a relatively thick (up to centimeters or more in places) layer of fine dust particles. Since there are fewer larger particles exposed in these areas to trigger as much saltation, once dust settles in them it cannot be easily removed (at least under current average wind and climatic conditions) but remains permanently in place in these "dust sinks."

The overall yearly pattern is for dust to be scoured off regions like the Syrtis Major plateau and deposited in dust sinks like Elysium. Immediately after a large dust storm, the dark areas appear lighter because they are wearing a thin coating of lighter dust. Hence the dark marking Syrtis Major always appears fainter after a major dust storm. However, ongoing scouring and lifting action by the wind soon removes this dust veneer to re-expose the underlying darker surface, and after a season or two it produces a "revival" of the previous darkness. This, then, is the real basis of the famed "wave of darkening," the increase in contrast between the dark areas and surrounding ocher areas that proceeds apace toward the equator every Martian spring and reverses every autumn, with the changes in the prevailing winds. It would more properly be called not a "wave of darkening" but a "wave of brightening," as was first realized by IPP photographer Chick Capen during the early post–Mariner 9 period.

Perhaps the most surprising of Mariner 9's many discoveries—and certainly the most unexpected, after the discouraging flyby Mariner results—is the presence of a variety of kinds of dry river valleys on Mars. Many of these are referred to as "outflow channels" and are associated with chaos terrains, of which more than a couple dozen examples exist in the southern highland regions. The largest of these are found in the region south of Chryse Planitia, in the Oxia Palus quadrangle, and along the crustal dichotomy. They called to mind no immediately obvious analogs on Earth, though post–Mariner 9 researchers have called attention to their similarity to the anastomosing channels of the Channeled Scablands of eastern Washington State in the United States, formed by cataclysmic floods during the Last Glacial Maximum.[39]

The channels are typically only one to two kilometers wide, and for the most part they are not very long. Even including their tributary systems, the networks seldom go on for more than a few hundred kilometers, though the longest of them, Ma'adim Vallis, is 700 kilometers long, 20 kilometers wide, and 2 kilometers deep, and thus significantly longer than the Grand Canyon on Earth.[40]

The outflow channels, dry river valleys, and smaller valley networks also found in the ancient highlands were surprising because they showed that sometime in the past history of Mars, liquid water was able to flow on the surface. This was not a universal view, however. Even after Mariner 9 and the Vikings, and until the missions of the late 1990s, there was a fraction of the Mars community that believed that the canyons and other fluvial features could not have been produced by liquid water. Instead, other possible fluids or flow processes were proposed. Also, at a time when most of the community was focused on liquid water, researchers like Baerbel Lucchitta from the U.S. Geological Survey (USGS) in Flagstaff importantly recognized that ice and glacial action had played a ubiquitous role in sculpting a number of classes of features on Mars. Hers was pioneering work that was to be abundantly confirmed by later missions.

Since the outflow channels, dry river valleys, and smaller valley networks exist almost entirely in the ancient, heavily cratered southern highlands of the planet, the era of liquid water flow must have been similarly ancient. At present, the temperatures and atmospheric pressures on Mars are too low for liquid water to exist stably, and thus the water molecule exists only as ice on or just under the surface and as vapor in the atmosphere. Rarely, liquid water could briefly make an appearance, since present Martian conditions are relatively close to the so-called triple point of water—the point at which water can exist in stable equilibrium as liquid, solid, and vapor. On Mars, the triple point lies at 0°C and 6.1 millibars, with the latter number also being the average atmospheric pressure on Mars.[41]

So if liquid water cannot exist on Mars at present, at least under usual conditions, there must have been a time in the past when Mars had a thicker atmosphere, providing the necessary pressure and temperature for it to flow on the surface. Based partly on Mariner 9 results, Carl Sagan and his colleagues speculated that this could even have been true in the relatively recent past.[42] It was well-known that because of the gravitational pull of the Sun and planets on Mars, the inclination of its axis oscillates over a period of about 50,000 years, between 35° and only 14° (that its present inclination of 24° is nearly the same as Earth's is a mere coincidence). With the changing axial tilt, the theory was proposed that the polar caps, which were still thought to be mostly carbon dioxide, might completely disappear, releasing CO_2 and significantly thickening the atmosphere—perhaps enough to cause, through greenhouse warming, a sufficient increase in the temperature of the planet to allow liquid water to flow onto and across the surface. It all seemed very reasonable at the time, though obviously, as Mariner 9 itself showed, if the polar caps are largely ordinary water ice (and, as shown later, not particularly voluminous), the idea breaks down. As described in later chapters, the existence of a thicker atmosphere and a warmer Mars continues to be assumed, but much earlier in the history of the planet—presumably within the first billion years after it formed. The quest for Martian water, meanwhile, goes on.

Seeing Mars with New Eyes

Mariner 9 so utterly revolutionized our views of Mars that it is rather difficult to remember what things were like prior to its arrival in Martian orbit, and just how little we knew. We were still largely in the throes of the fantasies of Lowell, Wells, and Burroughs, imagining a Mars with canals as late as 1950, when Arthur C. Clarke became the first science fiction writer to break with the convention in the *Sands of Mars*.[43] The scientists who interpreted the results from Mariner 9, which sent its final data in October 1972, and the rest of us who simply followed along with keen interest, saw Mars, with wonder and awe, for the first time as it really is.

Carl Sagan, perhaps the most prominent of the Mariner 9 scientists, would in 1973, just after Mariner 9 completed its mission, write: "In all the history of mankind, there will be only one generation that will be the first to explore the Solar System, one generation for which, in childhood, the planets are distant and indistinct disks moving through the night sky, and for which, in old age, the planets are places, diverse new worlds in the course of exploration."[44] Other planets remained to be explored; Mars, however, had already become a "diverse new world in the course of exploration."

As remarked earlier, one of us (W.S.) was in grade school when, in March 1965—only months before Mariner 4's flyby—he first looked at Mars with a small telescope. Though it was just possible to make out that there were lighter and darker areas, the view was still highly stimulating to the imagination, and it remained at least theoretically possible that that small shimmering disc was the world Lowell had dreamed up, the world of a dying civilization and canals. Six years later as a senior in high school he was still observing Mars as Mariner 9 entered orbit around the planet and revealed the shield volcanoes, the Martian Grand Canyon, river valleys, and more. Martian explorers who lived through that time can truly say: we have heard the chimes at midnight.

Troubled as the present time is, with myriad problems besetting our globe (of which the greatest include climate change), no one can ever take away from us that we were the first generations to explore the solar system up close, the first to explore Mars *from Mars*. For every generation that comes after, Mars will be the Mars of Mariner 9, the real Mars that humans first discovered and interpreted and explored in 1971–72.

And all science fiction writers who invoke Mars will have to invoke the names and places that were conjured into existence by the IAU working group for Mars nomenclature, after that never-to-be-forgotten year of discovery. In his remarkable book *Red Mars*, Kim Stanley Robinson brings the scenery of Mars, as viewed from Phobos, to life as has no one else:

During perigee Mars filled most of the sky, as if they flew over it in a high jet. The depth of Valles Marineris was perceptible, the height of the four big volcanoes obvious; their broad peaks appeared over the horizon well before the surrounding countryside came into view. There were craters everywhere on the surface. Their round interiors were a vivid sandy orange, a slightly lighter color than the surrounding countryside. Dust, presumably. The short rugged curved mountain ranges were darker than the surrounding countryside, a rust color broken by black shadows. But both the light and dark colors were just a shade away from the omnipresent rusty-orangish-red, which was the color of every peak, crater, canyon, dune, and even the curved slice of the dust-filled atmosphere, visible high above the bright curve of the planet. Red Mars! It was transfixing, mesmerizing. Everyone felt it.[45]

This is brilliant prose; it familiarizes and domesticates what Mariner 9 found strange and new, and in it, Mars is no longer a world to be discovered, but a place and—almost—a second home.

13

Vikings Invade the Red Planet

1976–1980

A Frontal Attack on One of the Great Questions in Human History

One of the longest journeys in human history has been the search for our place in the cosmos. From the sky people of Aboriginal Dreamtime to Percival Lowell's dying Martians, the idea of extraterrestrial life has been stitched into the fabric of human imagination since the evolution of thought. But only in the past few decades has the human species succeeded in crossing interplanetary space in order to gain a better understanding of our planetary neighbors like Mars and begun to look seriously and systematically for evidence of life elsewhere in the solar system. And only in the past few decades has our race begun to comprehend the true and profound vastness of space.

From a cosmic perspective, even our host galaxy, the Milky Way, is not particularly special: it is but one of several hundred billion to perhaps two trillion galaxies, according to the latest estimates. While our Sun is certainly special to us, it is not in any particular way special from a cosmic perspective: it is an average star among some 200 billion others in the Milky Way. And, as we now know, our Sun is not the only star to have planets. "We find that we inhabit an insignificant planet of a hum-drum star," Carl Sagan wrote, "lost in a galaxy tucked away in some forgotten corner of a universe in which there are far more galaxies than people."[1]

The burning question that Sagan sought to answer until his untimely death in 1996 is: "Are we alone?" He died not knowing the answer; we still do not know. Yet, from the cosmic perspective, it is hard to believe that Earth's thin blue line of atmosphere could embrace the only biota in the universe. The famed Harvard astronomer Harlow Shapley (1885–1972) once observed that the idea that life is not special to Earth is unsettling to many people, but only because we have always been "bedeviled by a natural and persisting sense of anthropomorphism. Corrections to our vanities are provided by modern science, but we still suffer relapses and return to believing that we are somehow important and supremely powerful and understanding. Of course we are not."[2]

Born in Depression-era Brooklyn in 1934, Sagan had already reached this conclusion by the time he was eight. "It seemed absolutely certain to me," he recalled long afterward, "that if the stars were like the Sun, there must be planets around them. And they must have life on them."[3] At the same time, Sagan savored the novels of Edgar Rice Burroughs (being especially taken with the phrase "the hurtling moons of Barsoom"), and by the time he was in high school he had graduated to the works of Arthur C. Clarke. In one of Clarke's books, *Interplanetary Flight*, published in 1951, he read: "The challenge of the great spaces between the worlds is a stupendous one; but if we fail to meet it, the story of our race will be drawing to its close. Humanity will have turned its back upon the still-untrodden heights and will be descending again the long slope that stretches, across a thousand million years of time, down to the shores of the primeval sea."[4]

In the same year that *Interplanetary Flight* appeared, Sagan began undergraduate studies as a physics major at the University of Chicago. It was a demanding and highly rigorous curriculum, and Sagan recalled in 1996, "I was a physics student in a department orbiting around Enrico Fermi [1901–54]; I discovered what true mathematical elegance is from Subrahmanyan Chandrasekhar [1910–85]."[5] He got himself qualified to use the small telescope in the observatory on the top of Ryerson Hall, though he seldom observed. (Neither then nor later was he ever much of an observer.) Among the subjects outside physics that he was most interested in, however, was evolutionary biology, and his determination to straddle disciplines led him to introduce himself to chemist Harold C. Urey (1893–1985), who in turn, fatefully, introduced him to Stanley Miller (1930–2007), who began graduate studies at Chicago the same year Sagan arrived as an undergraduate.

Urey had won the 1934 Nobel Prize in Chemistry for his discovery of deuterium (a form of hydrogen that has a neutron added to the atom's nucleus), but by the early 1950s, he had become passionately interested in the origin of the Moon and planets, and in his seminar spoke about the possibilities of the synthesis of organic molecules in Earth's primordial atmosphere. Miller was in attendance, and he left the seminar inspired. In September 1952 he approached Urey with the request that he might test the organic synthesis hypothesis in the laboratory.

Urey is said to have at first regarded the experiment Miller proposed as too doubtful for a doctoral thesis, but he eventually came around and helped Miller with the experimental design, in which a flask filled with hydrogen, methane, and ammonia gases—then believed to be the constituents of the primordial atmosphere of Earth—was exposed to an electric spark, simulating lightning. Earth's primitive ocean was represented by water vapor circulated through the glass apparatus and recondensed. Any molecules produced by the reaction fell into a trap at the bottom of the apparatus, which prevented them from being destroyed by the next spark. Within a few days, the liquid in the trap at the bottom of the apparatus became pink; after another week it turned deep red. Miller noted that these colors, as well as a yellow-brown material, were due to the presence of solid organic compounds generated in the experiment. When the liquid was analyzed, it was found to contain at least two amino acids, glycine and alanine, which are fundamental to living organisms.[6]

Sagan attended Miller's colloquium in which the now-famous experiment was described, and he left convinced that the natural emergence of living organisms from a primordial soup was not only possible but unavoidable. No miracles or supernatural intervention were needed. What's more, nothing in Miller's experiment was unique to Earth, and so it seemed obvious to Sagan life could have formed in this way not only on Earth but on Mars and other worlds.

Inspired, Sagan took one of Urey's undergraduate courses and wrote an honors essay on the origin of life. He later admitted that it was very naïve, and riddled with misconceptions, but he learned from his mistakes, and after completing his BA in physics and his MA a year later, he embarked on graduate studies under Urey's great rival in the attempt to learn about the origin of the Moon and planets, Gerard Peter Kuiper.[7] During the four years of Sagan's graduate studies (1956–60), the chance of finding life on Mars looked rather promising, as most popular books of the time suggested. Astronomers still echoed the view that Giovanni Virginio Schiaparelli had enunciated long ago: "The planet is not a desert of arid rocks. *It lives*."[8] Typical were the views of Mount Wilson astronomer Robert S. Richardson, who wrote in his book *Exploring Mars*:

> The fact that large, distinctly marked dark areas exist on Mars indicates that these surfaces cannot be dead, for no surface without the power to regenerate itself could withstand the continual inpouring of dust for ages. Hence the only reasonable explanation of the dark areas would seem to be that they consist of some living substance which stubbornly refuses to be obliterated by defying the sand drifts and feeding upon the dust itself.[9]

As we have seen, even Kuiper favored the idea that the dark areas were tracts of vegetation, until, with the Great Dust Storm of 1956, he changed his views on the matter and instead accepted that the changes in the albedo regions could well be an effect—as Dean B. McLaughlin had argued—of windblown dust. Sagan, just before starting his graduate studies that autumn, accepted an invitation from Kuiper to observe Mars's 1956 opposition from McDonald Observatory, then operated under contract with the University of Chicago, in West Texas. Sagan had been planning a trip to Europe, but he wrote to Kuiper to accept the invitation to McDonald, saying that while Europe would forever be at a constant distance from him, Mars would not. Unfortunately, the weather was lousy both in Texas and on Mars, and Sagan left with stiffened resolve "that the way to learn about the planets is not by peering through this ocean of air but to go there"—by spacecraft.[10]

After completing his dissertation under Kuiper (titled "Physical Studies of the Planets," it was unusually speculative and broad in scope for a dissertation, and included consideration of possible lunar organics and life, organic molecules in the atmosphere of Jupiter, and most importantly a cogent explanation of the unexpectedly intense microwave radiation on Venus), Sagan won a fellowship to the University of California at Berkeley. There, in 1961, he published "The Planet Venus," a literary tour de force for the top-drawer journal *Science*;[11] and "The Question of Life on Mars" for the National Academy of Science and the National Research Council. His views on Mars at the time were as follows:

> The evidence taken as a whole is suggestive of life on Mars. In particular, the response to the availability of water vapor is just what is expected on a planet which is now relatively arid, but which once probably had much more surface water. The limited evidence we have is directly relevant only to the presence of microorganisms; there are no valid data for or against the existence of larger organisms and motile animals. . . . The ancient and exciting question of the possible existence of life on Mars will probably be answered in the next decade.[12]

His views had by now diverged greatly from those of his mentor. Thus, in 1967, Kuiper, then director of the Lunar and Planetary Laboratory in Tucson, flatly declared in a speech to the Arizona Academy of Science: "Mariner 4 seems to have done what . . . careful observers of the past half century were unable to do, namely, to destroy in the public mind the myth of the canals of Mars and all that it implied. This indicates, if such were necessary, that even reports by scientists may at times be found to be premature or foolish and that no subject is so well established that continued and more careful scientific investigation is superfluous."[13] Sagan, by contrast, tried to keep the canal controversy alive by arguing—mistakenly, as we now know—that "at least some of the finer lines found by Mariner 4—as we interpret the evidence—turn out to be ridges comparable to the oceanic ridges and sea mounts that lace ocean bottoms on earth."[14]

Even after Mariners 6 and 7 seemed to show that Mars was, in the words of one team member, "a dull landscape, as dead as a dodo,"[15] Sagan still persisted, and refused to discountenance even the existence of higher organisms. "Intelligent life on the Earth," he announced, "would be entirely undetectable by photography in reflected sunlight unless about 100-meter (330-foot) resolution was achieved, at which point the urban and agricultural geometry of our technological civilization would become strikingly evident."[16]

Mariner 9 did have that resolution, and though its images did not show evidence of urban and agricultural geometry, they did show volcanic formations and evidence for water—dry riverbeds and channels carved by flash floods, as Sagan and others interpreted them—and so the possibility of life, seemingly a dead-letter issue after the flyby Mariners, returned to the Red Planet. Harold Masursky (1922–90), head of the Mariner 9 imaging team, enthused, "Our image of a cratered, static Mars was replaced by one of a dynamic and fascinatingly complex world that called out for further exploration."[17]

Sagan (who, after an unsuccessful bid for tenure at Harvard, had been appointed to the faculty at Cornell University, where he would remain the rest of his life) was one of the most prominent members of the Mariner 9 imaging team. As ever, he was quixotic about the possibility of life on Mars, just as Lowell had once been about his civilized race of dying Martians. He was also, like Lowell, an unrivaled master of the English language, highly skillful at wooing the public and communicating his own enthusiasms—which, in contrast to President Johnson's relief that the Mariner 4 images did not show evidence of life on Mars—largely lined up with the public's own willingness to entertain the fascinating possibility. Nevertheless, the Sagan whose books (and also, increasingly, television appearances) appealed so widely to the public was not always the Sagan his colleagues encountered at Cornell or the Jet Propulsion Laboratory (JPL). "He wore his Optimist hat in public, when he talked blithely about the possibility (however remote) of 'advanced life forms' on Mars," Sagan's biographer Keay Davidson recounts, "but in his conversations with colleagues, he often donned his Skeptic hat." At times he was a master at wearing both hats simultaneously. "Look," he'd argue, "here is a niche in which you can imagine life. And don't you have to know more about that niche, or look directly for life, before you exclude it?" The skeptic side had, for instance, led him to reject a paper arguing that intelligent life might exist not only on Mars but elsewhere in the solar system—a position he had himself often argued in favor of—on the grounds that, recalling Eugène Michel Antoniadi's observations demolishing the canal illusion, much of the evidence "may be psychological rather than astronomical in origin."[18] Also, with his first graduate student, Jim Pollack, Sagan clearly recognized that the tantalizing albedo changes that for so long seemed evidence of the existence of vegetation on Mars were merely due, as McLaughlin had suggested, to the shifting outlines produced by windblown dust. These examples of the skeptic side were, however, often offset by the sudden reappearance of the optimist.

Thus, in summer 1975, as the two Viking missions to Mars were being readied for launch, Sagan was back in the major key, declaring: "We are only now beginning an adequate reconnaissance of our neighboring world. There is no question that astonishments and delights await us."[19]

Born from Apollo

The Vikings were complicated—and expensive—missions, and as might be expected they had had a long and complicated history. The successes of the U.S. Mariner 4 flyby and Mariner 9 orbiter missions to Mars in the 1960s and early 1970s were achieved against the backdrop of the even more renowned Apollo test flights and human landing missions to the Moon. With victory declared over the Soviets, and government (and even to some extent, media and public) interest in additional crewed lunar missions waning, Apollo mission managers and—more importantly—commercial rocket system contractors like Boeing, Grumman, and Rockwell eagerly commenced the search for alternate potential uses for the Saturn 1B and Saturn V launch vehicles. NASA had purchased enough Saturn V rockets, for example, to conduct missions through a hoped-for Apollo 20 lunar expedition. The Nixon administration, however, grew sour on the price tag for additional lunar missions after Apollo 17, and it canceled what would have been Apollo 18, 19, and 20 in favor of starting a new low–Earth orbit space transportation program called the Space Shuttle.[20]

As early as 1964, even before the successful Mariner 4 flyby of Mars, NASA was starting to think about additional longer-term uses for the Saturn rocket boosters. That summer, the Apollo Extension Program (AEP) was started, becoming the Apollo Applications Program (AAP) in 1965. AEP and then AAP searched for ways to utilize additional heavy-lift boosters for robotic and crewed science missions beyond the Apollo Moon landings. Some fascinating human mission concepts were explored, like flybys of Venus and Mars and a series of launches to develop a long-term lunar base.[21] While most were never implemented, some crewed missions—like the four Skylab launches in 1973 and the Apollo-Soyuz Test Project in 1975—were launched using Apollo boosters.

The AAP also explored potential uncrewed (robotic) deep space missions that could be enabled using Apollo hardware. These included an Outer Solar System Grand Tour mission concept that would take advantage of a once-every-175-year planetary alignment to enable spacecraft to rapidly travel from Jupiter to Saturn to Uranus to Neptune using gravity assist trajectories. This mission would have used one or more Saturn V launches to send two to four flyby spacecraft and atmospheric entry probes to the giant planets in the late 1970s to late 1980s. Ultimately eschewed by Congress because of the high cost, the mission concept eventually evolved into what became the Mariner 11 and 12 giant planet flyby missions, renamed Voyagers 1 and 2, launched on smaller Titan-Centaur boosters.[22]

The Voyager missions that we have come to know and love were not the first NASA Voyagers, however. Another set of robotic mission concepts that were studied in general by NASA starting as early as 1960 and specifically by the AAP starting around 1966 was called the Voyager Mars program.[23] Early mission concepts in this program included orbiters to Venus and Mars based on modifications to the Mariner probe design, and a landed capsule for Mars based on the Apollo command module and designed to use the heavy-lift capabilities of the Saturn launch vehicles. As it became clear that the focus of the Saturn rockets would be on getting the Apollo astronauts to and from the Moon, however, congressional support for funding to support a parallel robotic Saturn program to other planets quickly dwindled. An additional factor that ultimately led to the cancellation of the Voyager Mars program came from the results streaming in from Mariner 4 in 1965: the Martian atmosphere was much thinner than had been thought, meaning that more mass, propulsion, and technical complexity would be needed to slow down the descent module as it approached the Martian surface. Facing tight budget constraints and social and political turmoil at home and abroad, Congress eventually canceled the Voyager Mars program in 1967, making it the first major space science program cancellation in the history of NASA.[24]

All was not lost, however, as years of engineering effort and significant amounts of work had been performed by engineers at JPL and other NASA centers and industry partners in the particularly thorny problem of Mars EDL: entry, descent, and landing. EDL at Mars is difficult because the atmosphere is too thin to allow landers to slow down to soft-landing velocities using only parachutes, and thus more complex systems would have to be developed. Among the most studied were the kinds of retrorocket propulsion systems and methods that were being used to successfully soft-land the Surveyor spacecraft on the Moon. In addition, the Mariner probes of the 1960s provided new environmental and scientific information about the Red Planet. Combined, the Mariner results and the precursor technical work from the Voyager Mars program would ultimately pave the way for Viking: a new, scaled-back but still ambitious program to orbit and land on Mars.

Vikings from Virginia

After the cancellation of the Voyager Mars program and despite the successes of the Mariner flyby missions and especially Mariner 9, NASA was faced with exactly zero new deep space robotic missions on the drawing board. This created an opportunity vacuum for potential new players in the agency's deep space exploration business. One organization that quickly stepped up to the challenge was NASA's Langley Research Center, in Hampton, Virginia, which had established itself as one of the premiere NASA centers for challenging problems in aerodynamics, the study of the interactions between air and solid bodies moving through it. Langley engineers knew that EDL at Mars would be extremely difficult, but they had made major advances in aerodynamics that enabled the successes of the Apollo reentry capsules as well as cutting-edge supersonic and hypersonic aircraft. The Mars problem was just a special case where—potentially—the same fluid dynamics theories would apply, just under a different gravity, atmospheric pressure, and atmospheric composition scenario. How hard could it be?

Indeed, Langley engineers had started to try to tackle the Mars EDL problem as early as 1964, in collaboration with science and mission design experts from JPL.[25] They came up with a Mars lander concept that would fit within the Atlas-Centaur launch capabilities and would use parachutes and retrorockets to soft-land, and pitched it to NASA Headquarters. With the cancellation of the Voyager Mars program coming shortly thereafter, the Langley-JPL Mars lander proposal presented an attractive potential fallback for continuing exploration of Mars.

In late 1967, facing the effective cancellation of NASA's planetary exploration program as well as potentially losing the race with the Soviets to be the first to land on Mars, NASA administrator James Webb (1906–92), for whom the James Webb Space Telescope is named, took the case directly to Congress and the Johnson administration to enable a still-ambitious but scaled-down version of the Voyager Mars program to move forward. Launches could continue to use smaller Atlas-class rockets (which cost 10 times less than the Saturn V), and spacecraft could continue to take advantage of the Mariner legacy and other work done by NASA centers like JPL and Langley. Webb was persuasive. Thus, Congress approved the 1971 Mars orbiter that eventually became Mariner 9, as well as a new two-orbiter, two-lander mission tentatively called Titan Mars 1973 because it would launch on two Air Force Titan III rockets. Langley would lead this ambitious new mission, which would become Viking, with JPL support primarily for the orbiters (figure 13.1).

Significant NASA budget cuts initiated by the Nixon administration starting in 1969 (after the success of Apollo 11), as well as booming cost estimates for the mission, caused immediate problems for the Viking program, and led to a two-year delay of the planned 1973 launch.[26] Still, the mission survived potential cancellation during what was a rather austere time for NASA, and by 1972 work on the spacecraft—two orbiters and two landers managed by Langley and built by a combination of Martin Marietta Aerospace, JPL, Langley, and a bevy of other contractors—began in earnest. Launches were scheduled for the Mars launch window in summer 1975.

Seeing Mars Again with New Eyes

Viking 1 was launched from the Cape Canaveral Air Force launch complex on August 20, 1975; Viking 2 followed about three weeks later, on September 9. Each mission's Titan IIIE rocket and Centaur upper stage successfully injected a merged orbiter-lander spacecraft into a perfect trajectory to Mars that would enable its arrival on June 19 and August 2, 1976, respectively. Flawless main engine burns on those dates allowed both missions to be captured into elliptical Mars orbits, with each spacecraft passing within about 320 kilometers of the surface at closest approach. Viking 1 orbited across the equator and

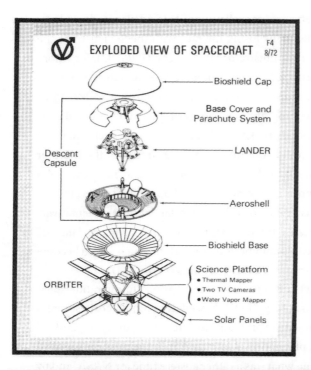

EXPLODED VIEW OF SPACECRAFT F4 8/72

Bioshield Cap

Base Cover and Parachute System

LANDER

Descent Capsule

Aeroshell

Bioshield Base

Science Platform
• Thermal Mapper
• Two TV Cameras
• Water Vapor Mapper

ORBITER

Solar Panels

FIGURE 13.1 Drawing of the stacked-up configuration of the Viking Orbiter and Lander (*left*), and one of the Viking Landers being lowered into a thermal vacuum test chamber at the Martin Marietta Aerospace facility in Denver, circa 1974. NASA.

midlatitudes (inclination = 39°), while Viking 2 was placed into a more polarlike orbit (final inclination around 80°).

The Viking missions became the world's fifth and sixth Mars orbiters (after Mariner 9, Mars 2, Mars 3, and Mars 5), but after six failed Soviet landing attempts dating back to 1962 (see chapter 11), they still had the chance to become the first successful landers on the Red Planet. Indeed, the orbit of Viking 1 was intentionally set to provide maximum earliest-possible coverage and resolution over equatorial and midlatitude regions that previous data from Mariner 9 and other earlier orbiters had indicated might be the safest potential landing sites. In contrast, the more polar orbit of Viking 2 provided enhanced mapping capability (as the planet rotated underneath) as well as the potential for a landing at higher latitudes, where ice and other volatiles might be more abundant (figure 13.2).

Shortly after Viking Orbiter 1 arrived at Mars, new images of the surface began streaming in to a team of planetary scientists gathered at JPL whose job it was to identify a safe (and, hopefully, interesting) place for Viking Lander 1 to attempt to set down. Pressure was on to choose a landing site as soon as possible while the orbiter was still functioning well (the lifetime of a space mission is always a major concern). However, the job was made doubly hard by two related and conspiring effects: the outstanding resolution of the images, and the unexpected clarity of the Martian atmosphere.

As we have seen, when Mariner 9 arrived at Mars in 1971 the planet was in the midst of a massive, planet-encircling dust storm, the likes of which hadn't been seen in a decade and a half.[27] Frustrated but patient imaging scientists on the Mariner 9 camera team discovered that it took many months for the dust to clear enough to begin to see the surface below. As the dust cleared, their patience was rewarded by spectacular images of towering volcanoes, deep canyons, ancient river valleys, and other geological wonders. What the Mariner 9 imaging team didn't fully understand, however, but what the Viking Orbiter imaging team quickly discovered, was that the airborne dust from the 1971 storm had never fully cleared out of the atmosphere, even by the end of the Mariner 9 mission in late 1972. In contrast, when Viking Orbiter 1 arrived at Mars in June 1976, there had been no large dust storms since the one in 1971, and thus the atmosphere was much clearer. While there was still some low level of dust suspended in the Martian atmosphere (as future missions

FIGURE 13.2 Mars global imaging mosaic constructed from Viking Orbiter 1 and Viking Orbiter 2 mapping data. This high-resolution digital image model is in a simple cylindrical projection and runs from the north pole (*top*) to south pole, centered on 0° longitude. The pixel scale of approximately one kilometer per pixel at the equator enabled significant new geological studies of the planet. NASA/JPL/USGS.

would show there always is), the image quality was correspondingly much higher.

The Viking Orbiter cameras were assisted not just by the increased atmospheric clarity but by their own significantly increased capabilities to resolve features on the surface of Mars.[28] The Mariner 9 cameras had what imaging scientists call a ground sampling distance (or what optics experts call the instantaneous field of view, IFOV) ranging from about 1,000 meters per pixel down to about 100 meters per pixel, depending on how far from the planet the spacecraft was in its elliptical orbit. This was, of course, unprecedented resolution compared to previous flyby missions or telescopic observations, and it allowed geological features like small craters, faults, or valleys ranging from around 500 meters to 5 kilometers across to be clearly resolved (in remote sensing, features need to be at least 3 to 5 pixels across in order for their shapes to be clearly resolved, and thus the "resolution" of an imaging system is really something like 3 to 5 times the IFOV). Because of Mariner 9's elliptical orbit, however, only a small fraction of the planet was imaged at those highest resolutions.

The Viking Orbiter cameras, which were more advanced and able to view the surface from up to five times closer than the Mariner 9 cameras, were able to resolve features as small as around 30 to 50 meters across. Because of their longevity and dual-mission approach (Viking Orbiters 1 and 2 would operate for a combined total of about 2,200 orbits, compared to Mariner 9's approximately 700), they were able to map most of the surface at this unprecedented resolution. To the geologists on the Viking Orbiter imaging team, having two or three times better resolution compared to Mariner 9, through a clearer atmosphere and consistently across the entire planet, represented a revolutionary breakthrough in their ability to interpret the history of the Red Planet. The Mars of Mariner 9 was not the Mars of the Viking Orbiters. They were truly seeing Mars with new eyes.

Over the lifetimes of Viking Orbiters 1 and 2 (the first operated until August 1980, and the second until July 1978), they acquired more than 32,400 and 15,600 images of Mars, respectively. This enormous (for the time) deep space imaging data set presented unprecedented challenges to the data transmission, computational analysis, and image interpretation pipelines of the day. At the same time, the images were rich with fundamental new discoveries about the Red Planet. Mosaics were assembled (many times by hand, from taped-together photographic prints) from groups of adjacent images to provide broader, regional context with which to associate individual areas seen at higher resolution than ever before. Geologists found themselves in a candy store of exposed geological landforms and processes covering an entire planet devoid of vegetation cover (plate 7). Enormous Hawaiian-style shield volcanoes with long lava flows and gaping tectonic canyons, discovered in lower-resolution Mariner 9 data earlier in the decade, were suddenly revealed in a scale and clarity comparable to the high-altitude aerial photos that the researchers and their students used to guide their fieldwork back on Earth.

One of the often-forgotten (on Earth) major geological processes that alters planetary surfaces—impact cratering—was revealed to be a major force of change on Mars. Impact craters on Mars span the size range from enormous, ancient crustal basins that likely hark back to the earliest times of Martian history, some four billion years ago or more, to the tiniest and likely relatively young holes in the ground, barely visible at the limit of the imaging resolution. Contrary to the impression gleaned from the earlier Mariner 4 flyby, however, only certain parts of the planet—mostly in the southern hemisphere—are heavily cratered like the surface of the Moon and thus likely to be as ancient. Other places show many fewer craters, or intermediate numbers, attesting to the role that impact craters would play in the coming decades of geological studies of Mars in determining the *relative ages* of different parts of the surface.[29] Just like on the Moon, regions with fewer craters had to have been either more recently resurfaced or more eroded down compared to regions with more craters, providing clues about the history of volcanic lava or ash eruptions, or the history of potential erosive agents like wind, glaciers, or even rivers and streams in wearing down and erasing craters and other landforms.

In the polar regions, imaging from the Viking Orbiters revealed fine details of narrow alternating layers of bright and dark materials that had been first recognized in Mariner 9 and even Mariner 7 images. These were especially visible in cross section along the edge of the polar caps. Color and thermal data revealed that the brighter layers are relatively clean water ice stored in the permanent year-round polar cap (as opposed to the seasonal carbon dioxide ice cap, which only covers the polar surface during the winters), and that the darker layers, which are also quite reddish, are either dusty ice or perhaps layers primarily composed of dust and soil. The alternating layers, like the alternating dusty and clean layers seen in terrestrial ice cores from Greenland or Antarctica but on a much larger scale, implied that there have been periodic, almost rhythmic changes in the Martian climate over long periods of time. Other researchers have since concluded that Mars undergoes periodic changes in its orbital properties (like its tilt, or its orbital eccentricity) similar to but perhaps more extreme than Earth's. On Earth, these so-called Milankovitch cycles are thought to be a major factor that drives periods of glaciation or warmer interglacial periods in our planet's climate.[30] On Mars, similar cycles could have had even larger planetary climate effects over the course of the planet's history, resulting in a record of major advances and retreats of the polar caps within the enigmatic polar layered terrains.

While Viking Orbiter images revealed enormous landslides and huge fields of dark sand dunes that attest to the erosive power of gravity and wind on Mars (just like back on Earth), perhaps the most surprising geological wonders revealed from the extensive global coverage and high resolution of the new images were the various Earthlike styles of valley networks dotting the most ancient (heavily cratered) terrains. Some of the valley networks show intricate interconnected branches of smaller tributaries that grow into larger and larger valleys, as in major river basin catchments on Earth that are formed from rainfall. Others show more stubby ends with little evidence of tributaries, similar to some valley networks that form in terrestrial regions with high subsurface groundwater flow rates. Still other valleys suddenly appear in the landscape at the locations of severely jumbled and chaotic terrain, as if a great slump occurred and released an enormous flood of groundwater all at once. Indeed, some of the landscape in those areas appears to have been subjected to scouring and erosion on a scale that would dwarf the largest floods known to have occurred on Earth.

Overall, the view of Mars that emerged from the Viking Orbiters abundantly confirmed that liquid (presumed by many to have been water)—flowing over the surface, flowing underground, and stored in underground aquifers—had played a major role in shaping the geology. The fact that most of the putatively water-carved features were found in the most ancient, heavily cratered parts of the planet (including most of the southern hemisphere) further implied that liquid water might have been there long ago, perhaps more than three to four billion years ago, when Mars was young. But for how long? And if the liquid was water, how much water was there? Viking Orbiter images couldn't provide answers to those questions, but the fact that many ancient southern hemisphere craters are heavily eroded but not completely eroded away implies that the planet wasn't covered by a long-lived ocean.

At least in the south. Viking Orbiter images and geological mapping revealed a different situation for much of the northern hemisphere. In general, the north is smoother and topographically lower than the south—this came to be known as the "hemispheric dichotomy" by Viking-era researchers. But why is the north lower and smoother? Some researchers advocated that a large ocean filled the northern basin early in Mars history, and ocean sediments buried the impact craters that had once been exposed there. Others hypothesized that the lowlands were the result of an early giant impact that gouged a deep hole in the planet's northern hemisphere, and that windblown dust and other impact sediments have since been burying preexisting terrain there. The "Did Mars have a global ocean?" debate was not ultimately resolvable with the Viking Orbiter data, but various members of the Mars community would continue to advocate nonetheless for their respective positions for another few decades, until better images from Mars Global Surveyor (chapter 14) and later missions would begin to provide additional clarity.

The Viking Orbiters also carried instruments designed to study the planet's atmosphere and other characteristics. These included the Mars Atmospheric Water Detector (MAWD), which determined that the current atmosphere holds only tiny amounts of water vapor. Indeed, the amount of water in the Mars atmosphere today is equivalent to a global layer only about 10 microns deep—thinner than a human hair—if it could all be turned to liquid and spread around the planet. For comparison, if all of Earth's atmospheric water were turned to liquid and spread around the planet, it would form a layer more than 5,000 times as deep. Mars is a dry world, and its atmosphere is drier by orders of magnitude than the driest desert air on Earth. And yet, there in the Viking images was stunning evidence for ancient features, presumably carved by liquid water.

If so much water was clearly there once, where did it all go?

Site Selection

While there would be decades to try to tease out the geological and atmospheric story of Mars being told in the high-resolution images, in summer and fall 1976 the clock was ticking for Viking Orbiter scientists and engineers to find safe places to try to deploy the lander that each orbiter carried. The team had actually been engaged for years in a process of site selection for Viking Landers 1 and 2, using Mariner 9 and other data sets to try to identify places that would be smooth at the scale of the landers (comparable in size to a small sports car).[31] That process had actually started way back in 1969, when the relatively newly formed Viking Project Science Steering Group started thinking about future landing sites using data from the Mariner 4 flyby several years earlier. Ultimately, neither the Mariner 4 nor the Mariner 6 and 7 flyby images proved adequate in terms of resolution or coverage for the team to make well-informed selections.

Instead, it was the new Mars revealed by Mariner 9 orbital images that would provide the Viking team with much of the data needed to make reasonable guesses at the best sites to land. Thus, once the data started streaming in from Mariner 9, the Viking team's landing site working group began to investigate the images that were shared by the Mariner 9 imaging team. The Viking group developed a set of five engineering and science criteria that it would use to rank potential sites.[32] These included assessments of (1) the "landability" of various places based on local terrain statistics like size and number of craters and topographic slopes; (2) the diversity of feature types (like craters, ridges, and boulders) that could represent either hazards or opportunities for the landers; (3) the absolute topographic elevation of potential sites, which would affect the performance of the landers' parachutes and other landing systems like radar; (4) the geological nature and history of various sites, as windows into fundamental Martian and planetary processes; and (5) the seasonal variability of winds, clouds, storms, and surface markings, indicators of possible conditions at times of future landings.

In summer 1972, after Mariner 9 had finished its primary mapping mission, and following many internal meetings and much discussion, the Viking team requested that the Mariner 9 imaging focus on 35 potential landing sites where those five criteria could be assessed. The Viking team ranked the

35 sites according to a best-guess scoring of the criteria (with landing safety weighted much higher than science value). Then the Mariner 9 team set about trying to photograph as many of these sites as possible at high resolution, while still achieving its own extended-mission science goals with the orbiter's instruments.

Potential Viking landing site images kept coming in from Mariner 9 until the mission ended in late October 1972. But during the rest of the year, and into the winter and spring of 1973, the Viking team analyzed and debated intensely about which sites to rank highest. Finally, in May 1973, NASA announced that Viking Lander 1 would be directed to land in a relatively flat region called Chryse Planitia near 20°N and 34°W and near the mouth of what appeared to be an ancient river valley. If that landing was successful, the Viking Lander 2 would be sent to land in a region called Cydonia, closer to the north pole, near 73°N and 350°W. (The old Schiaparellian names added the luster of romance to these dusty places; Chryse was an island rich in gold, and Cydonia an ancient seaport in Crete said to have been founded by King Minos.) If the Viking Lander 1 failed, however, the second lander would be sent to an alternate near-equatorial landing site.

There was significant disagreement and uncertainty among the science and engineering communities about the choice of landing sites. Using Mariner 9 images plus available Earth-based radar data, the Viking team had to extrapolate the behavior of Mars seen at a few-kilometers resolution down to a best guess of what that surface would look like at a few-meters resolution, the scale of the landers themselves. This caused a lot of unease on the team, especially among the geologists who had dealt with imaging and fieldwork on Earth, and even more so among those who had been tasked with extrapolating Lunar Orbiter–level imaging resolution down to the meter scale needed for the successful landings of the Surveyor landers and Apollo astronauts. Such estimates had often been wrong for Apollo, and most pilots who flew the Lunar Module—most famously, Neil Armstrong on Apollo 11—had to make real-time adjustments to the flight path to avoid small but hazardous obstacles that had not been visible in the available orbital images or other data. For the Viking Landers, of course, no such real-time piloting and obstacle avoidance was possible. These kinds of concerns had provided strong justifications for beefing up the capabilities and resolution of the Viking Orbiter imaging system relative to Mariner 9's, and resulted in the first phase of the Viking Orbiter 1 mission being devoted to the process of verifying those landing site choices using higher-resolution images and other instrument measurements.

Site Certification

Once Viking Orbiter 1 was safely in Mars orbit, the team quickly transitioned from the process of site selection to a new process, that of site certification. In site certification, orbital images and other data are rapidly acquired about the finalist landing locations identified in the earlier site selection process. The higher resolution of the images and increased clarity of the atmosphere relative to the Mariner 9 data yielded stunning new assessments of the finalist candidate sites as countless newly identified small craters, boulder fields, and steep slopes suddenly came crisply into view where previously there had only been relatively bland, safe-looking terrain. The preferred location in Chryse, for example, appeared to be near the hoped-for mouth of a channel where the team might encounter sedimentary rocks, but upon closer inspection that channel turned out to be a deeply incised and etched riverbed with enormous numbers of small craters and boulders down to the limit of the imaging resolution. As the first images of Chryse came down around June 22, 1976, U.S. Geological Survey (USGS) geologist Mike Carr, the leader of the imaging team, was quoted as saying, "We were just astounded—both a mixture of elation and shock."[33] Because the preferred site was likely full of potential obstacles on par with the scale of the lander, the landing site decision would have to be reassessed.

But the team couldn't take long to decide whether to stick with Chryse, the so-called A-1 site, switch to the A-2 backup site, or choose another site altogether, because the second spacecraft, Viking Orbiter 2, was beginning its approach to the planet. Once it got close enough to Mars (starting around July 26), communications between Earth and both spacecraft would become more complicated, and the cadence of imaging and other observations from the first orbiter would have to be slowed down. Plus, NASA and Viking project management had hoped to make a big public relations splash by landing Viking Lander 1 on Mars on July 4, 1976, the bicentennial anniversary of the American Declaration of Independence.

Team members worked frantically through late June hoping to identify a safe haven within the large elliptical surface region in Chryse that had been dubbed site A-1.[34] Unfortunately, the complexity and apparent danger of the prime A-1 site that was revealed by the Viking Orbiter images of Chryse meant that the hoped-for landing date would need to be delayed beyond July 4. But could they even identify a safe site to deploy the lander before July 26? Nearly 400 people—scientists, engineers, and students from JPL, the USGS, and other universities and laboratories around the country—engaged in the sometimes-tedious work of assembling mosaics from the Viking Orbiter images and then cataloging and counting the thousands of craters, large boulders, ridges, and other potential geological hazards (and targets) that those images revealed. Under the intense pressure of media scrutiny and a looming countdown clock, many team members were being asked to make snap interpretations and judgments—instant science.

By early July several landing site options had been identified near the prime A-1 site based partly on additional photographic analysis, but also on new Earth-based radar data that provided unique information on the centimeter-scale roughness of the potential landing sites.[35] Ultimately, the team voted and project management decided on a site known as A1-NW, a few hundred kilometers downstream from the original A-1 site (near 22°N, 48°W), which seemed to be safer in the photography but still potentially dangerous according to the radar data. Rather than land within the boulder-laden channel, they surmised, perhaps it would be better to land on the outwash plain coming out of the channel, even if it was rocky.[36] After all, the engineers had designed the lander to be capable of landing on small rocks and operating even on slopes of over 20°. Just how big the rocks would turn out to be at A1-NW (or any other potential site, for that matter) was unknown. Rolling the dice, engineers executed orbit trim maneuvers on the spacecraft on July 8 and then again on July 16 to shift its orbit to a better position for the A1-NW landing attempt. That attempt would be made on July 20, 1976—exactly seven years after Neil Armstrong and Buzz Aldrin landed on the Moon in Apollo 11.

Touchdown in Chryse

At 5:12 a.m. Pacific daylight time on that Tuesday in July, Viking Lander 1 became the first spacecraft to successfully land on Mars and live to phone home about it. The first photographs had been preprogrammed to be taken starting just 15 minutes after landing, and they were to be relayed through the orbiter and then directly to Earth. The photos would thus be "live from Mars" except for the one-way travel time at the speed of light, which was about 20 minutes when Viking Lander 1 set down. Millions of people (including many budding space nerds on summer vacation from school) watched live network television special reports that morning that showed the images slowly coming in, column by column, from left to right. It was like watching paint drip down a wall. . . . Was that a rock? A tiny sand dune? But the reward came once the painstakingly slow downlink enabled the entire image to be assembled—there it was, the lander's footpad, on terra firma!

The elation was tangible, not just for the thousands of people who had been directly involved in the success of the Viking missions thus far, but for fans of space exploration young and old alike, as well as the nation. President Gerald R. Ford (1913–2006) proclaimed July 20, 1976, "Space Exploration Day" in America, and the landing was the lead story on all three major television networks' nightly news broadcasts. While the Soviets had tried and failed six times to land a robotic probe on Mars, the United States had succeeded on its first attempt, a fact not lost on a world still in the midst of the Cold War and a country still reeling from the relatively recent resignation of a president.

With the uncertainty of landing behind it, the Viking Lander 1 team quickly set to work characterizing the landscape and environment of the spacecraft's permanent new home (figure 13.3). The lander (like the lander for Viking 2) was equipped with a payload of scientific instruments consisting of two "facsimile" cameras (essentially fax machines that acquired images by elevation and azimuth scanning of small numbers of digital sensors, or "pixels," across a scene); three sets of biology experiments to search for evidence of life in the soil by assessing levels of metabolism, growth, and photosynthesis; a mass spectrometer to determine the composition of the atmosphere as well as of any gases given off by the soil samples; an X-ray spectrometer to measure the chemistry of the soil; a seismometer to search for evidence of marsquakes

FIGURE 13.3 The first panoramic views downlinked from Viking Lander 1 in Chryse Planitia (*top*), and from Viking Lander 2 in Utopia Planitia. Both landers ended up slightly tilted, causing the curved shape of the distant horizon in these 300° views. The Chryse site has more topography (low hills and impact crater rims) than the flatter Utopia site. NASA/JPL.

(although it never deployed properly after landing and thus never worked); and a weather station designed to monitor temperature, pressure, and wind speed over the lifetime of the mission. The lander also deployed a robotic arm and sample scooping/sieving tool that could dig small trenches and deliver soil samples to the instruments inside the lander's body. Several small but powerful magnets mounted on the sampling arm provided a way to assess the magnetic properties of the soil.[37]

Life on Mars?

From the first, the biology experiments commanded the lion's share of the attention. This was understandable, as Mars had long been of compelling interest precisely because it had always been the most promising extraterrestrial "abode of life" in the solar system. Now, for the first time, humanity was searching in situ for life on that other world. The question of how to detect organisms on Mars, should they exist, had been carefully considered by the biologists on the Viking team. Rather than just asking whether there was life on Mars, they also wondered whether, if there is life on Mars, it would be of different origin from terrestrial life. Though some scientists had also speculated that extraterrestrial life might be silicon- rather than carbon-based like life on Earth, this was seen as improbable. Carbon is unique in its ability to combine in myriad ways with other elements (especially hydrogen, nitrogen, oxygen, phosphorus, and sulfur), allowing strange and wonderful creatures to develop. Any life on Mars would almost surely be made up of organic molecules, like those of terrestrial organisms.

The scientists also had to agree on the definition of life itself. Sagan, for one, did not like the traditional scientific definition of life as something that consumes energy, moves, and reproduces. "My automobile eats and breathes and metabolizes and moves," he insisted. "Crystals grow and even reproduce."[38] For him the key was genetics. Automobiles might mimic some properties of life, Sagan said, but they did not carry a genetic code or evolve; they rust (there was certainly abundant evidence of *rust* on Mars!). Unfortunately, no Viking experiment would be able to look at the genetic attributes of any suspected Martian life-forms found, nor would the experiments last long enough to see whether that life would evolve. However, even genetics was the exclusive province of carbon. As Norm Horowitz pointed out, "Carbon is so superior for the building of complex molecules that the possibility of forming genetic systems with other elements has never seriously been considered. . . . This does not imply, of course, that the genetic systems of extraterrestrial species must be chemically identical with our own, only that they must be built out of carbon compounds."[39]

In the end, Horowitz and the other designers of the biological experiments equipped each lander with a $55 million (about $250 million today) biology lab capable of detecting a wide spectrum of microscopic life within the area reachable by the sampling arm. The biologists designed three experiments:

1. **Pyrolytic Release experiment.** The sampling arm scoops up a tiny sample (about 0.1 gram) of Martian soil and places it inside the test chamber of the biology lab. Carbon dioxide and carbon monoxide labeled with the radioactive isotope carbon-14 are then admitted, the mixture is incubated under a sunlamp for several days, and everything is heated to break down (pyrolyze) any organic compounds present. Finally, hydrogen gas is admitted into the chamber to sweep the pyrolosis products into a gas chromatograph and mass spectrograph capable of detecting carbon-14 and other organic molecules. Since any organisms present should carry out metabolic processes during which they will assimilate carbon-14 from the gas in the chamber, detection of carbon-14 would be considered a positive result. However, this in and of itself would not be entirely conclusive, since a first peak of radioactivity might conceivably be due to chemical processes not involving living organisms, and so, in order to rule out this latter possibility, other samples (controls) are sterilized by heating before the carbon source is admitted.

2. **Labeled Release experiment.** A sample of Martian material is placed into the chamber, and a moist nutrient material containing carbon-14 is added. Any Martian organisms present will metabolize the nutrient material and release carbon-14-labeled gas, which is then registered by the detector.

3. **Gas Exchange experiment (popularly known as the "chicken soup" experiment).** At the beginning of the experiment, the atmosphere within the chamber consists of carbon dioxide and the inert gases helium and

krypton. A nutrient-rich material and water vapor are then added to a Martian soil sample. On suddenly finding themselves in a water- and nutrient-rich environment, the Martian organisms are expected to respond with a vigorous spurt of metabolism, leading to the sudden build-up of gases in the chamber.

How likely was it that Viking would find life on Mars? The bets ran the gamut. Ever the skeptic, Bruce C. Murray said the chances were zilch. Horowitz, who designed the Pyrolytic Release experiment, put the odds at "not quite zero." Harold P. Klein (1921–2001), head of Viking's biology team, gave it a 1 in 50 chance. Sagan played the Pascal's wager card: there was either life or no life, so he guessed 50 percent. Team member Clark R. Chapman guessed that the odds were much less.

The results, alas, were equivocal. The Pyrolytic Release experiment showed two peaks, and at first was felt to be weakly positive; however, later attempts to duplicate the effect were unsuccessful. The Labeled Release experiment showed an immediate, rather startling, rise in the level of carbon-14 activity after the nutrient was introduced into the chamber. This result seemed to attest to a positive reaction, so much so that the experiment team immediately rushed out and ordered a bottle of champagne. The "chicken soup" experiment also produced dramatic and unexpected results: when the samples were humidified, there was a sudden burst of oxygen—something that had never occurred in earlier tests with terrestrial samples. However, there was a very weak response when the nutrient material was added.

The exact meaning of the results was not immediately obvious, and generated a great deal of discussion. In the end, the consensus was that no indigenous organic compounds had been found by the Viking Landers.[40] The fine-grained materials that were sampled and delivered to the organic chemistry instruments were found to be highly reactive, however, releasing oxygen when a watery solution was added to them, oxidizing organic nutrients that had been brought inside the instruments, and "fixing" (converting from a gas to a solid) atmospheric CO_2. While some mission scientists at the time were inclined to believe that these results were at least potentially indicative of the presence of active microorganisms within the soils of Mars, most in the scientific community, then and since, concluded that the experiments had revealed the presence of some extremely reactive but nonbiological compound(s) in the soil mimicking some of the kinds of reactions that Earthlike biology would produce. Hydrogen peroxide or other similar reactive molecules were implicated as possible culprits, though no definitive identification was made during the mission. Indeed, in experiments with the Earthbound test versions of the instruments, no terrestrial samples were found that reproduced all the attributes of the Viking Lander samples. Whatever was responsible, it appeared to be uniquely Martian.

Not until more than 30 years later, during the Phoenix lander mission to the north polar region of Mars, would scientists learn definitively that a compound called *perchlorate*—not uniquely Martian but ubiquitous in Mars soils and rarely occurring naturally on Earth—was the most likely culprit for the highly reactive results recorded in the Viking Lander biology experiments. Since perchlorate also breaks down organic molecules, its presence in the soils could be at least partially responsible for the nondetection of organics in the Viking biology experiments.

The Mission Continues: Life After Biology

Over the course of 2,245 sols of operation (more than 6.3 years, until November 11, 1982), the Viking Lander 1 team conducted a comprehensive examination of the surface and shallow subsurface around the landing site on Mars, and it collected a rich meteorologic data set documenting daily and seasonal patterns of weather on the Red Planet. While many preliminary science results were announced rather quickly during periodic press conferences held at JPL and other NASA facilities during the mission, it took many more years to plumb the depths of the photographic, compositional, atmospheric, and biological cornucopia of data generated by the mission.[41]

The cameras on Viking Lander 1 revealed Chryse to be a gently rolling but generally flat plain. Several low hills in the distance mark the eroded rims of small, ancient impact craters. The site is littered with rocky fragments, from meter-sized boulders to small pebbles down to the limits of the resolution of the cameras. Indeed, the lander set down perilously close to one of the largest rocks around, a two-meter-wide boulder nicknamed "Big Joe." Viking imaging team member and Washington University (Saint Louis) planetary scientist Ray Arvidson notes that the rock was originally nicknamed "Big Bertha," but that the name was changed to something that didn't invoke a World War I German artillery weapon. Had one of the lander footpads set down on Big Joe, the extreme tilt would have likely ended the mission in catastrophe. Large rocks like this were likely excavated during impact cratering events, both local and potentially far flung. Several exposures of what look bedrock outcrop can be seen poking out in areas where the loose rock abundance is lower. Permeating the whole scene, though, are loose, bright-red sediments—soil made out of sand and dust—mixed with small rocky fragments in some places, accumulated into small wind-formed dunes or "drifts" in other places. Close in to the lander, where the retrorockets heavily scoured the surface during the lander's terminal descent, the loose soil was blown away, revealing the sediments below to be relatively coherent and perhaps even cemented. Mission geologists began to refer to these sediments as "indurated," a word that derives from the Latin for "made hard."

Should the loose or indurated dust and sand on the surface of Mars really be called "soil"? Going back to the early beginnings of soil science in the 1870s, geologists had known that the formation of soil on Earth is intimately tied to the presence of life. Earth soils are partially created by chemical reactions involving water and organic molecules from both living and dead organisms (predominantly microorganisms), as well as by the physical breakdown of rock fragments from burrowing, tunneling, or other activities typical of subterranean organisms, both microscopic and macroscopic. But does the very use of the word "soil" require the presence of living things? Planetary scientists ran up against similar issues before and during the Apollo missions, when, first, landers like the Surveyors, and then the astronauts themselves, found that much of the surface of the Moon is covered by a layer of loose, unconsolidated dust, sand, and pebbles. Was that lunar "soil"? Surely life was not involved in its creation.

Many planetary scientists fall back to another term, "regolith" (from the Latin for blanket + rock), to refer in a general way to the layer of unconsolidated material covering bedrock. Regolith is a great term to use when talking about the thick, jumbled-up layer of rocks and debris created on an ancient planetary surface like the Moon or Mars, where countless impact cratering events over billions of years have shattered the bedrock and created a layer of rocky debris many tens to even hundreds of meters or more deep. But the use of the term "fine-grained regolith" is arguably too cumbersome and certainly jargonistic, when in fact, plain old "soil" will do. Indeed, the term "soil" has become commonly used in planetary science to describe the fine-grained, porous, uppermost layers of regolith, despite the more purist viewpoint that the term applies only to terrestrial material formed by or in the presence of life.[42]

While the Viking Lander instruments didn't include the kinds of spectrometers that had been used on previous flyby and orbiter missions to help identify surface minerals, their identical dual cameras were able to use color filters (in red, green, and blue, as well as three filters in the infrared just beyond the limit of human red vision) to compare the colors of the rocks and soils on Mars to a variety of different kinds of rocks and soils measured in the same colors in laboratories on Earth.[43]

The lander team's color imaging campaign got off to a rather rocky start, however, as the first color images released to the public and the media, based on not yet properly calibrated data, showed the reddish Martian landscape under a relatively bluish Earthly sky. Ray Arvidson tells the story of atmospheric scientist and Viking imaging team member Jim Pollack getting excited by these early views, because Rayleigh scattering—the process that makes Earth's sky blue because of our relatively thick atmosphere—was not at all expected to be significant in the ultrathin Martian air. Unfortunately, while familiar looking, the blue sky of Mars was actually a gross misinterpretation.[44]

After the initial miscue, subsequent press releases used images created with the lander's color calibration targets to balance the colors more properly. When this was done, the sky came out in a much more expected reddish-brown hue. The original blue-sky photos were replaced with the properly calibrated reddish-brown sky ones, but inevitably some in the fringe media (and some still in the modern Internet's conspiracy world) called foul, claiming that NASA was covering up the fact that the landing was a hoax and that the mission was actually operating from an isolated desert on Earth. Despite the silliness of such ideas given the significant evidence to the contrary, fringe conspiracies about Mars seem to have an uncanny ability to take hold and survive—for decades if not centuries. The same silliness would again come to the fore when the old "Face on Mars" debate was reignited during the early days of the Mars Global Surveyor mapping mission in 1999.

Once a reliable calibration was established, imaging scientists like Arvidson and others found the best match to the colors of the soils of Mars to be a reddish, fine-grained sedimentary material on Earth called palagonite. Palagonite is a weathering product formed in places like Hawai'i and Iceland when water chemically breaks down iron-bearing volcanic lava or ash. As the original volcanic rock weathers, the iron in it becomes oxidized, changing color from black to brownish to reddish. Essentially, Mars is red because it's slowly rusting. Support for this idea was also found in the color properties of the dark rocks and rock fragments imaged by the lander color filters, which showed a good match to the kinds of iron-rich igneous rocks (what geologists call basalts) that are typical of the kinds of volcanic materials erupted from Earth's mid-ocean ridges. However, most of the rocks, indeed most everything in the entire scene around the lander, were covered or coated with the super fine-grained (smoke-sized) dust that is picked up easily by the wind and blown around the whole planet during the famous Mars dust storms.

In addition to facilitating these kinds of mineral inferences from the cameras, the lander could use its X-ray spectrometer to measure the chemical elements in the dust and sand directly. The team would use the lander's robotic arm to scoop up and sift some loose materials off the surface and dump them into an inlet on the lander deck that led to the spectrometer. The abundances of key elements from the periodic table—iron, silicon, calcium, sodium, aluminum, magnesium, chlorine, sulfur, and others—provided diagnostic ways for geochemists on the team to compare the composition of the soil and dust to that of Earth, the Moon, and elsewhere.

One key Viking team leader in the development of the lander's X-ray spectrometer was Benton C. "Ben" Clark, who served as the chief scientist of space exploration systems at Lockheed Martin outside of Denver, where the instruments had been built (as have, indeed, many other NASA robotic spacecraft and instruments since). Clark has been deeply involved in many Mars (and other deep space) missions since Viking, including work on the chemical and mineral analysis instruments on the Spirit, Opportunity, Curiosity, and Perseverance rovers. For most of his career he's managed to carve out a role as a leading space scientist working for a major aerospace company, certainly not a traditional role or environment for an academic interested in discovering the details of the chemistry and mineralogy of other worlds. Ben's work has been critical to establishing a successful interface between scientists and engineers working on space missions, however, and his role as a scientist embedded in an engineering organization is a great example of a possible alternative career path for planetary scientists, presaging the transdisciplinary and entre-preneurial space science and engineering job market that is becoming more widespread today.

Ben Clark led the team doing the first detailed chemical analysis of the reddish soils of Mars. What they found was that the composition was pretty uniform no matter where they sampled it from around the lander, and that it matched a mixture of clays, salts (mostly sulfur- and magnesium-based salts rather than sodium-based ones), and unweathered rock fragments once again typical of the kinds of iron-rich basalts found on Earth and other terrestrial planets. Some of the more indurated soils showed elevated abundances of chlorine and sulfur, suggesting that chloride and/or sulfate salts might be the "cements" responsible for those soils being crusty or cloddy. A fraction of all of the soils sampled also stuck to the magnets on the rover's arm, which was also consistent with basalts since those kinds of volcanic rocks also contain tiny amounts of magnetic minerals like magnetite.

Unfortunately, the mission wasn't able to measure the elemental chemistry of rocks larger than small pebbles, however, because of a limit on the size of the inlet to the X-ray instrument. Thus, deeper discoveries about the kinds of (likely basaltic) bedrock materials that the soil and dust originally came from would have to wait more than 20 years for the first mission to be able to truly measure the chemistry of rocks on Mars, Mars Pathfinder.

Viking Lander 1 operated for more than three Mars years, with the end of the mission occurring in November 1982, when ground controllers acci-dentally sent a command that caused the lander's communications antenna to be unable to establish a lock on Earth. Because of the sheer longevity of the mission, in situ monitoring of many local dust storms was achieved, and in addition, two major (planet-encircling) storms were observed. (These are designated in the literature by *1977a* and *1977b*; the occurrence of two planet-encircling storms during the same year was and still is unprecedented.)

Among the findings, the large storms in 1977 not only affected lander sky opacity measurements but also produced a sudden increase in the daily pressure range. The first major storm was preceded by several local storms, and it began over Thaumasia Fossae on February 15, 1977, before quickly spreading to cover the entire southern hemisphere in thick dust storms. The second major storm began on May 27, 1977, in the Valles Marineris region; it was more extensive than the earlier major storm, and the dust took longer to settle. Viking Landers 1 and 2 (the latter was by then also monitoring events from the surface) both detected suspended dust from these storms.

In the absence of the transmission of any data after November 1982, it's impossible to know how long Viking Lander 1 remained operational on Mars, perhaps in "safe mode," desperately awaiting a phone call from home with instructions on how to proceed. Meanwhile, in 1980, about two years before the mission ended, a much more human tragedy struck, when Thomas A. "Tim" Mutch, leader of the Viking Lander imaging team and one of the science administration leaders at NASA Headquarters, died while mountain climbing in the Himalayas. In honor of Tim's important role in the success of the mission, and to cement his legacy forever in the future annals of Mars geography, NASA named the Viking Lander 1 landing site Mutch Memorial Station.

Despite the unfortunate human error that led to the loss of the mission, and despite the lack of compelling evidence for extant life-forms or even organic molecules from indigenous extinct life-forms, Viking Lander 1 had still collected a rich and unprecedented data set of geological, geochemical, and meteorologic data from Chryse Planitia. The mission had also added another laurel to NASA's growing list of robotic space mission accomplishments: first to land safely on Mars.

Utopia!

To return to 1976: once the final landing site had been selected for Viking Lander 1, and before that successful landing and even before the Mars orbit insertion of the second orbiter, the Viking Orbiter team had started work on the same kind of intense and controversial site certification activity needed to choose a final site for Viking Lander 2. Just like at Chryse, the first Viking

Orbiter images of the B-1 primary site in Cydonia revealed extensive boulder outcrops, ridges, and numerous small craters that could have been major obstacles to a successful landing.[45] Battle worn from the experience of recently choosing a new A-1 landing site for Viking Lander 1, the team hunkered down and started collecting and interpreting images and other data about alternate candidate sites for the second lander, focusing on the strong scientific desire to land much farther north than the first lander, in a latitude band that might be more favorable for the occurrence of surface water (ice or frost) or shallow ground ice. Besides just images, the site selection team was also able to use new measurements from the Viking Orbiter MAWD to try to find a site that had high relative water vapor abundance, thought to be important for the life-detection experiments on Viking Lander 2.[46]

Even as Viking Orbiter 2 decelerated successfully into orbit on August 7, the imaging team had still not found a good site for its lander. Indeed, team members scoured a number of possible regions that had looked promising in the Mariner 9 images, and seriously debated the merits of about a half dozen of them until, finally, on August 30, project manager Jim Martin (1920–2002) decided (based on a consensus, but certainly not unanimity, of the imaging team) that the second lander would be sent to a relatively smooth-looking region called Utopia Planitia, near 48° north latitude.[47] It was a gutsy call because the decision was based only on imaging data; high-resolution Earth-based radar wasn't available for that area. Whether there were lander-sized boulders awaiting the team in Utopia would have to be discovered in real time; landing date was set for September 3.

Happily, Viking Lander 2 set down successfully on Mars just before 4:00 p.m. Pacific daylight time, and after some heart-stopping tension related to an earlier temporary loss of attitude control on Viking Orbiter 2, was able to relay its first two pictures back to Earth before local sunset, just like Viking Lander 1 had done some six weeks before. The second lander carried an identical scientific payload as the first, but the landing site, about 26° farther north in latitude (roughly the latitude difference between Havana and Paris), turned out to have a very different landscape and environment.

Over the course of a substantially shorter mission than Viking Lander 1 (Viking Lander 2 operated for 1,278 sols, about two Mars years), the Viking Lander 2 team conducted a comprehensive examination of the surface and shallow subsurface at the more northerly landing site on Mars (plate 8). It was a heady time for NASA and JPL—operating the first two successful landers on the surface of Mars at the same time—as well as for the dozens of newly minted Mars surface scientists from NASA and other agencies and universities around the world who were working tirelessly to plan for and then interpret the results coming back from the Red Planet.

The terrain around Viking Lander 2 in Utopia Planitia turned out to be flatter than Viking Lander 1's site in Chryse, but with similar numbers of rocks. Just like in Chryse, the rocks were likely excavated from nearby impact craters (like the large, 100-kilometer-wide crater called Mie about 200 kilometers to the west of the landing site) that were then exposed and weathered by billions of years of erosion. And just like in Chryse, many rocks in Utopia are embedded within a sort of matrix of crusty to cloddy duricrust: finer-grained, cohesive mixtures of pebbles, cobbles, sand, and dust. The rocks are so well distributed across the landscape around the lander that exposed regions that could uniquely be considered bedrock are extremely rare. Also rare are wind-blown dunes or small drifts of dust; perhaps the numerous rocks provide a natural obstacle to organized movement of sand and dust across this terrain.

A unique aspect to the geology around Viking Lander 2 compared to Chryse is the appearance of "polygonal terrain" at the more northerly Utopia landing site. Polygonal terrain, also called "patterned ground," is characterized by relatively flat surfaces that are segmented into numerous typically four-sided, five-sided, or six-sided individual polygonal areas that each range from a few meters to a few hundred meters across.[48] Shallow trenches occur between the polygons, and the edges of some of the polygons might show slightly raised rims. This kind of terrain is also found in higher-latitude and polar regions on Earth because it is diagnostic of the presence of ground ice in the shallow subsurface. During daily and seasonal cycles of temperature changes (and, on Earth at least, of freezing and thawing), the ice expands and contracts and perhaps even desiccates entirely, moving the enclosed volume of soil and rocks in slow, rhythmic patterns that lead to the formation of polygons. The size of the polygons is often correlated with the amount and/or depth of the ground ice (larger polygons can mean more and/or deeper ice). Was there freezing and thawing of ice in the subsurface in this part of Mars long ago? Did ancient glaciers and their meltwaters pass across this terrain? Viking observations (even trenches dug into the shallow subsurface using the robotic arm) couldn't get deep enough to tell, but there is certainly ice still existing in the shallow subsurface in Utopia.[49] The polygonal terrain provides indirect evidence; more direct evidence would come from orbital detection of shallow ice at those latitudes from NASA's later Mars Odyssey, which entered Mars orbit in 2001 (and is still operational at the time of writing, March 2021), and from the exposure of ground ice by robotic arm trenching during the later NASA Phoenix lander mission that same year, which explored Martian polygonal terrain of similarly high latitude. Thin, bright deposits of water and CO_2 ices (depending on the season), formed by snowfall or direct condensation from the atmosphere, can exist right on the surface at Utopia as well, as discovered by analysis of Viking Lander 2 images of frost deposits (only about a half to a few millimeters thick) that formed on the surface during the winter.[50]

The elemental chemistry of the fine-grained soils at the Viking Lander 2 site is remarkably similar to that at the Viking Lander 1 site, more than 6,500 kilometers away. This is one piece of evidence that the windblown sands and dust of Mars have been widely transported across the surface and effectively homogenized during billions of years of dust storms and the slow migration of sand dunes. Similarly, results from the Viking Lander 2 biology experiments largely mimicked those from Viking Lander 1—again, most likely because the main materials being ingested into the instruments were the fine-grained, globally distributed dust and soil. This was a disappointment to some on the Viking team who had hoped that the choice of a far northerly landing site would mean a "wetter" environment with a greater likelihood of harboring extant life of some kind. While there was certainly evidence for thin layers of frost and more ground ice in Utopia than at Chryse, atmospheric measurements from both the landers and the orbiters revealed that the Martian environment today is still orders of magnitude colder and drier than even the driest, coldest deserts on Earth. The Viking Landers discovered that Mars makes Antarctica seem like a tropical playground in comparison.

The higher-latitude landing site in Utopia meant colder temperatures, especially at night, when it would routinely dip down to −110°C (−165°F) or colder during the winter season at the lander's location. This meant that more of the lander's radioactively generated electricity had to be used to recharge the batteries and run the heaters to keep the inside of the lander to a temperature above about −40°C (−40°F), not too cold for sensitive instruments and electronics. But the deeper battery charge and discharge cycles at Utopia compared to Chryse eventually took their toll, and the Viking Lander 2's batteries ultimately failed to generate enough electricity to keep the interior warm. Last communication with the lander was on April 11, 1980, which was nearly 1,200 sols longer than mission planners had hoped for when the landers' original missions were envisioned, each intended to last a mere 90 sols (Mars days). (Recall that a sol = 24 hours, 39.6 minutes, the length of the mean solar day on Mars.)

Just as for the landing site in Chryse, NASA and Viking mission members sought to honor a departed colleague by naming the landing site in their honor. Ultimately, in 2001, the Utopia landing site was named the Soffen Memorial Station, after longtime Viking project scientist and then NASA Headquarters director of life sciences Gerald A. Soffen (1926–2000). Back in 1977, during the early days of Viking, Soffen had appeared on the popular Leonard Nimoy–hosted television show *In Search of. . .*, during which Soffen said of Viking, "We have started what will become an adventure of mankind in searching for not only the lower forms of life but also the search for intelligent life. This is one of the milestones in the course of human destiny to find cousins."[51]

Lessons Learned

Many aspects of the Viking results are still being modeled, interpreted, and debated today, more than four decades after the orbiters and landers arrived. Still, the overall legacy of Viking is of a mission of dramatically successful firsts: first (and second) to successfully land on the surface; first to map the surface at high-enough resolution to enable detailed global geological and climatic interpretations; first to search for life on another world. If the solar system is ever (in the far, far future) home to an interplanetary park system,[52] then the Viking landing sites in Chryse and Utopia are sure to be among the most commonly visited attractions. The Viking results surpassed those of Mariner 9 in every way. It may be regarded both as a capstone on the first era of reconnaissance of Mars by spacecraft, and as the cornerstone for the planning of later explorations.

Viking was also an important mission for NASA in that it cemented the agency's dedication to mounting *robotic* deep space missions with close-knit, intermingled teams of scientists, engineers, and managers, all working toward a common set of specific and focused goals. The team dynamics during the landing site selection process in particular would become a model for future NASA Mars lander efforts. During Viking, scientists were able to realize and internalize the fact that new information on potential landing hazards might not allow them to land exactly where they wanted. At the same time, engineers and managers were able to realize and internalize that scientists usually didn't want to land where it was simply "easiest" to do so. The fact that there was an inverse correlation between what appeared to be scientifically interesting places to land and the perceived safety of such potential landing sites, and that teams never really had all the information that they would want to make the best possible decision between science potential and landing safety, was a lesson that would continue to crop up in Mars exploration during the decades ahead.

Another important legacy of Viking had to do with the old parable of putting all your eggs into one basket. The main goal of the lander missions was to determine whether life existed on Mars. The most sensitive and logical biological experiments possible to deploy onto a deep space mission using early 1970s technology (and early 1970s knowledge about Mars) were devised, and those experiments were successfully performed at both landing sites. The fact that they came up ambiguously negative—with unanticipated atmospheric and/or geochemical effects being most likely responsible for the potentially lifelike signals detected during the experiments—caused some in the broader planetary science community to wonder in retrospect whether Viking should have spent more effort characterizing the basic chemical, mineral, and atmospheric environment of the planet before attempting to search for Earthlike life. Perhaps a more measured approach would have led to different kinds of instruments or experiments being devised to search for life. It is a specious argument, however, as the scientists, engineers, and managers who designed, built, and operated the Viking Landers were only using the best available data and inferences at the time to formulate the best possible approach to tackling a critical but fundamentally difficult question: is there life on Mars? Arguably, an inadvertent legacy of the negative Viking Lander biology results would be that the next several Mars surface missions would focus on geological and compositional/mineralogical goals rather than biological ones. It would take more than 30 years to mount the next Mars surface mission that could actively search for extant life on Mars.

14

A Sedimentary Planet

Mars Rocks the World

Despite the fact that the Viking missions were successful in so many ways, one of their lingering ironies is that their results created both enormous enthusiasm and enormous disappointment among various segments of the planetary science community. Most astrobiologists, for example, when faced with the reality of the overwhelming harshness of the current Martian surface environment (extreme cold, extreme unfiltered solar ultraviolet [UV] radiation reaching the surface, likely presence of an aggressive oxidizing agent in the soil), had to substantially reassess the strategy of continuing to search for evidence of extant life using methods like those employed in the Viking Lander biology experiments. Indeed, the biology experiment results—especially the failure to detect indigenous organic molecules at a level above the threshold of the mass spectrograph's sensitivity of a few parts per billion in the soils examined on the plains of Chryse and Utopia— produced a level of skepticism about "life on Mars" that would infuse the broader community, as well as some at NASA Headquarters, for decades. In some ways the situation was similar to that which followed the flyby Mariners, which had, at least as some scientists saw it, shown Mars to be little more than a dead and faraway Moon.

Not until the mid-1990s did NASA again dive into getting back to the surface of Mars, partially spurred by claims—announced with great fanfare on August 7, 1996, by David McKay (1936–2013) of NASA's Johnson Space Center—that researchers had found compelling evidence of life on Mars, ironically in a stone found on Earth. Thus began the curious and still not completely settled saga of possible extraterrestrial fossilized life-forms in a potato-sized meteorite called ALH84001. (This small rock was found by the U.S. Antarctic Meteorite Recovery Team in 1984 in the Allan Hills region of Antarctica, which is why it's called ALH84001. This motivates a special note about acronyms as we dive deeper into the modern age of Mars exploration: while for the scientists and engineers involved in space research they are often held in almost as much affection as close members of the family, for the general reader acronyms are apt to seem a mere jumble of confusing letters. For convenience, all the space-related acronyms discussed in this book are listed in appendix B.)

The astonishing idea that some meteorites found on Earth might consist of material from the Moon, asteroids, and even Mars began circulating as far back as the 1970s, but the first direct evidence of Martian origin for a meteorite did not occur until 1983, when noble gases locked within the glassy nodules of a meteorite from Antarctica were compared to the noble gases found in the Martian atmosphere by instruments on the Viking 1 and 2 landers, and found to have the same profile. More Martian meteorites have since been identified, forming what is known as the SNC group of samples, where the acronym SNC refers to the first letters of the three type specimens after which they are named: the S for shergottite, named after a meteorite that fell in 1865 at Bihar, India, near the town of Shergotty; the N for nakhlites, after the type specimen that fell in Nakhla, Egypt, in 1911; and the C for a single meteorite that fell in Chassigny, France, in 1915.[1] When the shergottite ALH84001 was found, it was just the 10th Martian meteorite known, and it was and remains the oldest known of the rocks from Mars. Of course, many others—now more than 100—have been found since. These include one in particular that has defined a whole new class of Martian meteorites. Named Northwest Africa 7034, it was discovered in the Sahara Desert in 2011 and is the second-oldest Mars meteorite yet found—and, like ALH84001, it dates back to the earliest epoch of Mars history. The baseball-sized sample weighs less than a pound, and has a very dark, volcanic appearance that has led meteorite dealers and researchers to nickname it "Black Beauty." Stunningly, analysis appears to show that Black Beauty might be pieces of rock that were formed by cooling lava *on* the surface of Mars, and that still harbor some of the water that once flowed there long ago.[2]

All SNCs are so-called achondrites, a class of stony meteorites that consist of material similar to terrestrial basalts or other igneous rocks and that have been geologically reprocessed or remelted within their original parent planet or asteroid. The Martian achondrites differ distinctly from other achondrites, however, and ALH84001 stood out dramatically from the hundreds of other meteorites found on the Antarctic ice cap that year. Later analysis gave its age as 3.9 to 4.1 billion years, or several hundred million years younger than the original crust of Mars. It followed a remarkable odyssey leading from its original formation near the surface of Mars to the Antarctic ice cap on Earth.[3] Some 16 million years ago it was blasted into space by an asteroid impact on Mars, and it hurled across the inner solar system, possibly crossing Earth's orbit many times, until it finally entered Earth's atmosphere. About two kilograms of this "lamp of earthly flame" (to quote a phrase from Percy Shelley's "Epipsychidion") survived the blazing journey. It then sat another 13,000 years in frigid isolation on the Antarctic ice cap until researcher Roberta Score chanced across it. After finding it, Score recalled that it was one of the greenest meteorites she'd ever seen. "I've always thought that rock was weird," she later recounted.[4]

The claim that this charred stone from outer space might contain fossils (figure 14.1) was based on microscopic analysis of strange structures mimicking some Earthlike biological forms (albeit 10 times smaller than the smallest terrestrial bacteria or viruses), lurking within its green crystals and carbon deposits and along fractures found in unlikely association with magnetite and iron sulfide minerals, an assemblage of compounds typical of decaying organic matter on Earth. Along with the announcement at Johnson Space Center, a paper on the findings was published in the top-drawer journal *Science*.[5] It had been reviewed and approved for publication by none other than Carl Sagan, who was ill with myelodysplasia and had just returned home after a bone marrow transplant. Sagan had made a comment about Martian meteorites two years before, saying that there had been no reports of microbes found in them—"so far."[6] He died in December 1996 before the debate about

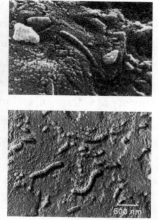

FIGURE 14.1 The potato-sized Martian meteorite ALH84001 (*left*), discovered on top of an Antarctic ice field in 1984, and vaulted into fame by the 1996 scientific claim that it contained fossilized evidence of Martian microbes. These high-resolution transmission electron microscope images are of a cast from a chip of the meteorite, showing the outlines of tubular structures that resemble in form and size (approximately a micrometer) some terrestrial bacteria. The resemblance in morphology does not, however, make the identification definitive. NASA/JSC.

the extraordinary claims regarding ALH84001 could be resolved. Indeed, they continue to be subject to debate, with McKay and his team continuing to argue in favor of the biogenic properties of the meteorite, while others, echoing Sagan's paraphrasing of Occam's razor that "extraordinary claims require extraordinary evidence," have been more reticent.[7] But at least the environment in which the carbonates found in the rock matrix formed has been better clarified; according to some research, they were deposited from a gradually evaporating, subsurface aqueous body, likely a shallow aquifer meters to tens of meters below the surface.[8]

Beyond the esoteric scientific debate, the Martian meteorite announcement electrified both the public and the astrobiology community at the time, and it gave a much-needed jolt to planners of additional Mars missions, who had already been devising future mission plans beginning with an initial set of engineering and technology demonstrations that would lay the groundwork for potential future networks of landers or potential future rovers. The geologists and atmospheric scientists, in contrast to the astrobiologists, had hardly needed any further incentivization. They had already been giddy over the Viking results, and especially the global mapping results from the Viking Orbiters, which finally brought most of the surface into relatively sharp view. These results had made abundantly clear that Mars was *not* just the Moon with a thin atmosphere (as had been suspected by some researchers after the Mariner 4, 6, and 7 missions). Instead it was a world with a substantial and complex history of volcanism, tectonism, impacts, and erosion, as well as a significant history of global climate variations and an ongoing rich and complex cycle of weather and surface/atmosphere volatile exchange, right up to the present day.

A Decade of Near Silence

Even though Mariner 9 had revealed a new Mars back in 1971–72 and Viking had renewed the planet again in 1976–82, it was clear that there was still much to be learned, and that the most Earthlike neighboring world in the solar system deserved a much closer look. A new generation of scientists eagerly began to train so as to take up the torch, as the great first generation of mission scientists, who had worked out so many of the important first-order relationships in the treasure trove of Mariner and Viking data, began to retire or pass away. Unfortunately, this new generation had to wait for a long time in the wings, as the 1980s proved to be a spectacularly lean decade for Mars exploration.

The only missions launched from Earth to Mars then were the Soviet Union's Phobos 1 and Phobos 2 spacecraft.[9] Each mission consisted of an orbiter and a lander based on an enhanced Venera spacecraft design but modified to optimize the orbiters for the Mars environment and the landers—"hoppers," really— for attempts to land in the low-gravity environment of the innermost Martian moon, Phobos. Phobos 1 was launched on July 7, 1988, but—continuing the history of heartbreak that has characterized the entirety of Soviet Mars exploration—communication was lost with the spacecraft during cruise, only a few months later, due to an erroneous command sent by ground controllers. Phobos 2, launched on July 12, 1988, successfully went into Martian orbit on January 29, 1989, and it took a significant number of images and infrared spectra of the equatorial region of Mars before contact was lost in late March, just before the landing on Phobos was to occur. While the mission was only partially successful, an important legacy of Phobos 2 was the first detection of interesting iron-bearing silicates and even water- or OH- (hydroxyl molecule) bearing clay minerals, by the French team operating the mission's imaging spectrometer (ISM).[10] ISM was the first instrument to get visible to near-infrared spectroscopic data of the surface from Mars orbit, and it could thus detect small variations in mineralogy at scales 50 to 100 times smaller than the best available Earth-based telescopic data at the time. Similarly, visible and thermal infrared imaging from Phobos 2 provided new images documenting the geology of Mars and Phobos, as well as new data on the infrared (thermal) properties of the Martian surface. Even though Phobos 2 only survived a few months at Mars, the data sets and results gleaned from that mission would go on to influence the design and operations of many future Mars orbital investigations, on both Russian and U.S. spacecraft.

Through the late 1970s and early 1980s, the U.S. planetary science community was having little success in getting a new mission to Mars approved. Partly this was a result of the residual disappointment from the ambiguous to negative Viking Lander life-detection findings, but more so (and as is often the case in high-profile government-funded activities), it was the result of politics. In his first inaugural address, in January 1981, U.S. president Ronald Reagan (1911–2004) said, "government is not the solution to our problem; government *is* the problem," giving the cue to his administration to start work in earnest to fulfill his campaign promise to cut taxes and shrink the size of government, until—in the words of a later anti-tax activist—it could be "reduced to the size where it could be dragged into the bathroom and drowned in the bathtub."[11] At the same time that the Reagan administration was vastly increasing expenditures for a military buildup to bring the "Evil Empire" of the Soviet Union to its knees, cuts to so-called discretionary spending like the funds that paid for NASA missions or other nondefense federal agencies were unleashed in draconian fashion. Led by Reagan's chief of the Office of Management and Budget (OMB), David Stockman, the administration proposed to cut NASA's planetary exploration budget substantially, instead wanting to focus the agency almost exclusively on flying the Space Shuttle and, eventually, starting to build a space station.[12]

Though much of the American population may have become blasé about space by then, there was significant support from a hardly negligible subset of enthusiasts who had been enthralled by highly productive robotic missions like Viking and the twin Voyagers to the giant planets in the 1970s. Recognizing this, three insiders and leaders in the U.S. planetary science community—the omnipresent Sagan, who had achieved a level of celebrity seldom achieved by scientists through his hit Public Broadcasting Service series *Cosmos* and appearances on *The Tonight Show* with Johnny Carson; Jet Propulsion Laboratory (JPL) lab director and planetary scientist Bruce C. Murray; and JPL engineer and project manager Louis Friedman—mobilized a new public advocacy group called The Planetary Society that attempted to take the message of broad public support for NASA's robotic exploration program directly to Congress. For a time The Planetary Society was the fastest-growing advocacy organization in America, with more than 100,000 members. Postcard- and letter-writing campaigns and phone calls to Congress demonstrated to lawmakers the significant public support for NASA's robotic exploration program, and—along with NASA's own internal cost-cutting measures—helped ensure continued funding for future missions.[13]

Taking advantage of this level of public support and leveraging the discoveries continuing to emerge from the longer-term analysis of the Viking data, the U.S. planetary science community, via influential policy groups like the Solar System Exploration Committee (an independent community-led group of scientists, engineers, and managers assembled by the National Academies of Science and Engineering to advise Congress about future NASA priorities), began to advocate for a new, so-called Planetary Observer class of missions, one of which would return to Mars orbit and survey the planet in new and higher-resolution ways, but using Earth-orbiting satellite technologies to help lower costs.[14] The first Observer-class mission, approved by OMB's Stockman in 1983, was initially called the Mars Geoscience/Climatology Orbiter but was soon renamed Mars Observer. It would be designed to take data from a low polar orbit that would enable long-duration global observations. The mission was predicted to cost just over $800 million (around $2 billion today), and to fit within the context of NASA's overall program concept, which then involved launching all spacecraft from NASA's workhorse crew and cargo delivery system, the Space Shuttle. Other planetary missions to be launched by the shuttle would include the Magellan orbiter to Venus, the Galileo orbiter to Jupiter, and the Hubble Space Telescope.

Space Shuttle

Things are always clearer in hindsight, and it is now rather easy to see that the robotic space program, including the robotic Mars program, was ultimately to be held hostage to decisions that dated back to the start of the post-Apollo era. In order to understand the 1980s malaise in the Mars program, it is necessarily to consider, if only briefly, what was happening at NASA more generally.

By the time the last Apollo landed and returned from the Moon in December 1972, with public interest already on the wane and Vietnam and economic issues displacing the Space Program as a national priority, NASA had searched, under its administrator Thomas O. Paine, for new, exciting, and nationally important missions in order to justify its continued existence and funding. Under Paine's leadership, a human mission to Mars, a space station in Earth orbit, and a Space Shuttle to transport people and materials to the space station and to move objects around in space were considered. At the same time, however, funding cuts began reducing NASA personnel by 1,000 employees each year, from an all-time high of nearly 34,000 in 1969. Continuing budgetary pressures forced NASA to abandon its more ambitious Mars and space station goals and to settle for the Space Transportation System, of which the Space Shuttle was the crew and cargo delivery vehicle.

Boston College sociologist Diane Vaughan has teased apart the chain of decision-making and trade-offs that NASA, in a desperate effort to secure funding, was forced to make in its design for the Space Transportation System. From the beginning, the program became inextricably linked to the U.S. Department of Defense, so military goals were added to NASA's original goal of excellence in space. Development costs were also slashed, leading to a final design that was far from NASA's original concept, so that "instead of a Cadillac, NASA got a camel."[15] Despite its backpedaling, NASA continued to argue hard in congressional hearings for the compromised shuttle design, especially for its supposed cost-effectiveness. A flawed analysis using data furnished by contractors—an analysis later described by experts as "pure fantasy"—indicated that it would eventually become revenue neutral and self-supporting. This was based on the wildly optimistic assumption that it would be able to provide a launch rate of more than 30 flights a year (considered a conservative estimate in 1971!).

There were further problems with this analysis, as the actual costs of development rose in unpredictable ways: the launch turnaround time necessary to break even was never achieved, simply because methods of mass production could never successfully be applied to the launching of such a complex, human-rated spacecraft. In addition, design compromises seriously affected missions scheduled to be launched by the shuttle. For instance, the Hubble Space Telescope (HST), originally proposed as far back as 1962, and intended to incorporate a 3-meter telescope, was canceled in 1974, then revived but with a scaled-back 2.4-meter telescope (the size was the largest that could fit inside the shuttle's cargo bay, which had been designed to accommodate the largest missile in use at the time). Ridden with cost overruns (as was the shuttle, charged with launching it), HST had its scheduled launch dates pushed back from 1979 to 1983. The first shuttle orbital mission, with a crew of John Young (1930–2018) and Roger L. Crippen, was completed in April 1981, but continuing delays pushed the HST launch back to October 1986. Then, in January 1986, disaster struck with the explosion of the Space Shuttle Challenger shortly after launch and the loss of a seven-person crew.

The sequence of events that had caused this national tragedy led to rapid reappraisal from the top down. In a prescient *Scientific American* article published in January 1986 and available on newsstands even before the Challenger launch, James Van Allen (1914–2006), the legendary University of Iowa scientist whose instruments on board America's first successful satellite, Explorer 1, in 1958 had detected the radiation belts that were eventually named after him, took issue with NASA's emphasis on the development of human spaceflight and military directions and lamented achievements delayed or lost because of NASA's (mis)direction.[16] Among other things, he pointed out that NASA's emphasis on the Space Shuttle had deterred the space agency's development of a mission to Halley's Comet. Five spacecraft would visit Halley's Comet in 1986, as it returned after 76 years to the inner solar system: two were Russian, two Japanese, and one European. American participation was notable by its absence.

The Rogers Commission, which was set up to investigate the causes of the Challenger disaster, eventually settled on the failure, on a cold January morning, of field joints containing rubber O-ring seals installed between each fuel segment of the Morton Thiokol–built solid-fuel rockets. It turned out that these O-ring seals had never been tested in such cold weather! Delays caused by this commission and the inevitable retreat from the overambitious frenzy of shuttle flights caused a significant reorganization of planned launches, including that of Mars Observer. In general, there was a great deal of soul-searching within NASA about the use of expensive and risky human-rated launch vehicles like the shuttle, and a renewed search for reliable expendable rockets such as those that had been the workhorses of NASA in the pre-shuttle era. Several missions, however, including Magellan, Galileo, and the HST, were too far along in development or too massive to find a feasible alternative to their destinations. In the end, they were successfully launched by shuttles in April 1989, October 1989, and April 1990, respectively. They were all to be highly successful missions, and at the time of writing, HST continues to be functional, more than 30 years after launch.[17]

Mars Observer was luckier than its bulkier counterparts; it was small enough to skip out of the shuttle's backlog and find an alternative ticket to Mars on a late 1970s vintage rocket, the Titan III. That was to be the extent of its luck, however.

Mars Observer

With a mass of only a metric ton and with a bus comparable in size to many commercial communications satellites, Mars Observer (figure 14.2) was small enough to be accommodated by an old-fashioned launch vehicle. Nevertheless, despite being relatively compact, it was still an example of what was called a "Christmas tree" spacecraft, one loaded with a large number and variety of scientific instruments to try to achieve an impressive range of scientific goals related to the planet's geology, topography, mineralogy, internal structure, magnetic field, and atmosphere.[18] It was a lot to ask of one mission, but similar "all eggs in one basket" objectives had been or were being asked of other major NASA missions (known as flagship-class missions) like Viking, Voyager, Magellan, Galileo, and the nascent Cassini mission to Saturn.

NASA had solicited instruments for the Shuttle-launched Mars Observer via a competitive proposal process back in 1985, and since many of those instruments were already under development, the team opted to use a subset of those instruments on the reorganized 1992 mission. Ultimately, the

FIGURE 14.2 *Left*: Artist's rendering of the Mars Observer spacecraft. *Right*: An MOC narrow-angle photo of Mars acquired on July 27, 1993, about 28 days before the planned orbital insertion. This "quarter Mars" view (with north up and tilted at about the one o'clock position) prominently features Syrtis Major and the Hellas basin. This is one of just five MOC images that essentially represent the entire science data set of the short-lived Mars Observer mission. NASA/JPL and NASA/JPL/MSSS.

spacecraft's payload included both high-resolution and wide-angle imaging, an infrared spectrometer, a magnetometer, a laser-ranging topography instrument, a gamma ray spectrometer, and an infrared atmospheric sounding instrument.[19] These instruments were tested between 1987 and 1992, in preparation for a September 1992 launch, 17 years after Viking. Like almost every other large NASA mission of that era, the team had to struggle mightily with cost overruns, descopes of some of the instruments and goals, and a range of other engineering challenges.

In addition, since April of that year, NASA had been led by a new administrator, Dan Goldin, "a self-described cold warrior who had earned his space management stripes working on the Strategic Defense Initiative (Star Wars) and spy satellite projects for a variety of agencies such as the National Reconnaissance Office."[20] Goldin was appointed by President George H. W. Bush with an explicit mandate to reform what the White House had concluded was an excessively bureaucratic and technologically stagnant space agency. He would significantly rechart the direction of NASA later, as we shall see, but obviously Mars Observer was too far along—and he was too new in the job—for him to deserve either credit or blame for the results, and in any case he was then thoroughly absorbed in bureaucratic infighting with Congress over the future of the International Space Station.

The Mars Observer team had to deal not only with the broader disarray within NASA in the post-Challenger era but also with intense pressure to control costs, which, according to then NASA Ames Research Center director and soon-to-be NASA-wide "Mars Czar" G. Scott Hubbard, "drove the cost of the modification to a level comparable with building a new spacecraft from scratch."[21] More than that, though, the team also faced some bizarre naturally caused challenges, including the need to clean the spacecraft on the launchpad because of dust and other contamination related to preparations for Hurricane Andrew, which passed across Cape Canaveral in August 1992. Nevertheless, the launch went smoothly, and the spacecraft was injected by the old-but-reliable Titan into a Mars encounter trajectory on September 25, 1992. The Christmas tree was on its way!

For deep space missions, the period between launch and the ultimate planetary flyby, orbit insertion, or landing is called interplanetary cruise. Cruise can be short, from just a few days for some missions to the Moon, or it can be extremely long, like the 9 years it took the New Horizons mission to get to Pluto, or the 12 years, punctuated by frenetic flybys of giant planets, of Voyager 2 out to Neptune. During cruise, the instruments and spacecraft are checked out and calibrated, and the operations team learns how to "fly" the spacecraft in its native environment. Cruise is generally a quiet time of low activity as the team prepares for whatever Big Event awaits at the end. For Mars Observer, the Big Event would be the firing of a powerful onboard rocket engine just before arriving at Mars, which would slow down the spacecraft and allow it to be captured into orbit around the planet. Everything was on track to fire that engine on August 24, 1993, and to begin a historic Mars year of orbital exploration and discovery.

As part of the planned end-of-cruise activities, about three days prior to orbital insertion, the Mars Observer team commanded the main engine to perform a small trajectory correction maneuver to position the spacecraft properly for the main orbit insertion burn. It was the first time that the engine had been used since launch, and, unbeknown to the team, there was apparently a small leak in the propulsion system that had allowed propellant to accumulate in the fuel lines over time. When the engine was ignited, the propellant in the lines appears to have exploded, rupturing the propulsion system, forcing the spacecraft into a crippling spin, and cutting off communications with Earth.[22] Ground controllers made frantic attempts to regain contact with the spacecraft, to no avail. On August 24, Mars Observer sailed past the Red Planet on a sad and silent flyby mission that dashed the hopes of planetary scientists and space enthusiasts around the world. The mission was lost; the Christmas tree had gone up in flames, at least metaphorically. To quote Hubbard once again, "Mars Observer represented an ultimately unsuccessful experiment under . . . stringent cost reduction."[23]

Better, Faster, Cheaper?

The loss of Mars Observer in summer 1993 was a huge shock and setback to NASA's Mars exploration efforts, and to the scientists and engineers who were planning to devote some of their careers to the mission. All the Martian eggs had been put into that one basket: there were no other orbiters, landers, or rovers waiting in the queue to follow up on whatever Mars Observer would discover. NASA's administrator, Dan Goldin, who had only been on the job about 18 months, was faced with an interplanetary political dilemma: more detailed global studies of Mars were needed to address the big questions remaining after Viking, but NASA's budget couldn't stomach another billion-dollar-scale flagship mission to replace Mars Observer.

Goldin was trained as a mechanical engineer and had spent a 25-year career as an aerospace engineer and propulsion expert at the aerospace giant TRW, rising through management to eventually become vice president and general manager before being appointed as NASA administrator in 1992. With a reputation for focusing on innovation and disruption as well as a personality that has been characterized as "at times brilliant, at times abusive, and always mercurial,"[24] Goldin set out to change the culture of NASA by reducing the agency's heavy reliance on large ("Christmas tree") flagship missions and instead embarking on a series of competitively selected, smaller missions that would spread out the inherent risk. Adopting the mantra of "better, faster, cheaper," Goldin would help create a new class of smaller NASA missions called the Discovery series. Even so, he faced a certain amount of pushback, or at least alarmed resignation. Mars geology expert Ken Edgett, who was just beginning his career in the early 1990s, captured a concern that many in the community had about Goldin's disruptive efforts: "I worried, at the time, that he was Gorbachev—that to save NASA, he'd end up accidentally killing NASA."[25] As we shall see, the new philosophy would have its ups and downs. It would be rewarded with the spectacular success of Pathfinder in 1997; would experience significant growing pains with two failed missions in 1999; and in the end would establish itself as a key part of NASA's planetary exploration program, with future Discovery small missions including the InSight lander, followed by the similar but Mars-specific "Mars Scout" small mission program that would eventually fly the Phoenix and MAVEN missions.

All that had yet to play out, however. In the meantime, the hiatus in Mars missions between the end of Viking in the early 1980s and the start of new orbital observations in the late 1990s was a challenging time for the Mars research community. Cuts to funding in the 1980s and then the failure of Mars Observer meant that jobs and research grants were scarce, and often mostly won by more senior scientists who had deeply established research records and were much more familiar with the details of the available Mariner and Viking data sets. Debates uninformed by new data dragged on among the community of researchers, who were picking over earlier data sets for new clues that might support their hypotheses about a past Mars ocean, or the possibility that there were meteorites from Mars, or the timing and magnitude of past planetwide climate changes.

Necessity was to prove the mother of invention, however, as some Mars researchers took a step back and returned to the telescope. Despite the much, much lower spatial resolution achievable from ground-based telescopes (which could study "pixels" the size of Arizona at best, compared to the football field–sized resolution of the then-best-available spacecraft data), it was still possible to observe the planet, its atmosphere, and its seasonal polar caps in unique wavelengths that had not yet been studied by spacecraft, or for which no firm space-based measurement plans had been developed. Thus, some modern Martians returned to their astronomical roots and studied the planet using optical, infrared, and radio telescopes. For example, the 305-meter Arecibo radio telescope in Puerto Rico, the largest single-aperture telescope in the world from 1963 until 2016, was used to probe the physical properties of dusty surface regions, while optical and infrared telescopes were used to search for evidence of specific kinds of surface minerals or ices. A group of Mars researchers even began to use the HST to observe the planet's atmospheric composition and cloud patterns (plate 9) in the ultraviolet (a wavelength region that cannot be studied from Earth's surface), and to take advantage of the fact that HST could image the planet in visible and infrared wavelengths at resolutions not too much worse than from a wide-angle camera in Martian orbit.[26] These ground- and space-based observations during the 1980s and 1990s not only helped bridge the research gap between successful missions to Mars,[27] they also helped train a new generation of planetary scientists in the design, fabrication, and operation of complex telescope-mounted cameras and spectrometers that would, eventually, be made more rugged, miniaturized, and modernized for use on orbiters, landers, and rovers sent to Mars and other deep space destinations. Some of those 1980s and 1990s telescope observers even went on to apply this experience as members of future Mars mission teams.

In addition, there was a breakthrough at NASA. Even before the first of the Discovery missions was launched, Goldin saw the chance to both demonstrate his developing new philosophy and, at the same time, recover the science lost with Mars Observer. He and others asked, why not break up the payload of Mars Observer into three smaller and cheaper missions, whose launches would be spread over several launch windows and thus work to incrementally recover the lost science from Mars Observer while accommodating themselves to NASA's perennially tight budget profile? The idea resonated not only with NASA management and congressional budget makers, but also with the Mars science community. As hard as researchers had been trying to make do with Earth-based telescope observations or reanalysis of older data, they were more than ready for the kind of ringside seat in orbit around Mars that had been lacking since Viking Orbiter 1 shut down due to depletion of its attitude control fuel way back in August 1980. Edgett recalls, "I remember an anecdote that Goldin also said he wanted the instrument Principal Investigators to not have to wait their entire careers for results from their instruments; that you should be able to go from proposal to results in a few years' time ... Mars is very amenable to this."[28] Spreading the eggs among many smaller baskets seemed like a prudent and expeditious path forward at the time, and it still does.

Mars Global Surveyor

The first of the smaller egg baskets was to be called Mars Global Surveyor, or just MGS. It was designed to quickly proceed to launch by using flight-spare or closely copied versions of the imaging, infrared spectroscopy, topography, and magnetic field instruments that had flown on Mars Observer. The speed at which MGS was built, tested, and launched was impressive and unprecedented for NASA in that era. The heavy reliance on flight-spare hardware and lessons learned from Mars Observer enabled Goldin's "better, faster, cheaper" vision for Mars exploration to be realized with the launch of MGS on November 7, 1996, only about three years after the failure of Mars Observer. Not since the lunar missions of the 1960s had such a major NASA effort gone from concept to launch in such a short period of time. A new era of Mars exploration was underway.

In addition to the smaller spacecraft and slimmed-down suite of instruments incorporated in the MGS mission, NASA celestial mechanics and spacecraft systems engineers used another innovative trick to reduce costs: aerobraking. First demonstrated by Earth orbital missions in the early 1990s, aerobraking is the use of a planet's upper atmosphere to help reduce a spacecraft's velocity and lower its orbital altitude. In 1993, JPL flight system engineers had exploited the super-dense nature of the atmosphere of Venus to aerobrake the Magellan spacecraft from its initial elliptical radar-mapping orbit into a new, nearly circular gravity-mapping orbit. Conceivably, the engineers knew that the same approach could be used for Mars, but that planet's much thinner atmosphere meant that a spacecraft would have to dip much deeper into the atmosphere at periapsis than would a mission attempting to lower its orbit at Earth or Venus. The idea for MGS was to use a hybrid approach: first, fire a smaller main rocket engine at the right time at Mars to let the spacecraft be captured into a large, 45-hour-long elliptical orbit; and second, use the spacecraft's solar panels as "wings" to provide friction against the upper atmosphere during closest approach (figure 14.3), removing energy from the orbit and slowly lowering the height of the "top" of the ellipse.

The idea of aerobraking at Mars worked great on paper and in computer simulations, but in actual practice it proved much more difficult. Unbeknownst to mission engineers, the hinge of one of the MGS solar panels (which had

FIGURE 14.3 Artist's rendering of the Mars Global Surveyor spacecraft skimming through the upper atmosphere during the aerobraking phase of the mission. NASA/JPL.

been outfitted with extra flaps on their ends to increase drag) had been damaged during or shortly after launch. Friction from the initial aerobraking skims through the upper atmosphere began to stress and excessively flex that hinge, threatening to damage—and potentially completely tear off—the panel. After only a dozen aerobraking passes, the orbit of the spacecraft had to be raised up out of the upper atmosphere before the panels were permanently damaged, while engineers figured out a plan B. (There always has to be a plan B.) Aerobraking was on hold.

Within a month or so, a solution was found that would enable aerobraking during periapsis passes to resume without overstressing the solar panels, by essentially reversing the orientation of the panels relative to the atmospheric friction and to their originally intended configuration.[29] By delaying MGS's entry into a circular mapping orbit by about an Earth year (about half a Mars year), the team could fly the spacecraft in its intended polar orbit but with the orbital motion going from south to north over the dayside of the planet, rather than north to south. MGS aerobraking was originally supposed to end around March 1998, but instead, the more cautious, less vigorous, "backward" aerobraking strategy ended up taking until March 1999 to get the spacecraft into its final, nearly circular, two-hour polar mapping orbit. Despite the delay, *MGS* was the first mission to demonstrate a slower but "cheaper" way, in terms of fuel needs and thus spacecraft mass, to get into a desired orbit around Mars. Aerobraking maneuvers like this would come to be used to get all future NASA Mars orbiter missions (at least as of 2021), as well as several other international Mars orbiter missions, into their final mapping orbits.

Old MOC, New MOC

The improvement in imaging resolution between the Viking Orbiter cameras and the MGS high-resolution camera provided a dramatic increase in the ability to uniquely interpret the geological processes that have occurred on the Red Planet over time. In typical Viking Orbiter images, features from about five kilometers to a few hundred meters across could be clearly resolved (a small fraction of the planet was also imaged at a resolution of just a few tens of meters, when the spacecraft were closest to Mars in their elliptical orbits). In such images at typical Viking Orbiter resolutions, it would be possible to identify a big region like Central Park in New York City, but it wouldn't be easy to identify an individual football stadium like the Rose Bowl. The MGS high-resolution camera, in contrast, could resolve features as small as 5 to 50 meters across, resulting in images of about 10 to 100 times higher resolution than Viking. If MGS was imaging Earth, not only individual football stadiums would be identifiable, but image analysts could probably even tell how many buses were parked in the stadium parking lots. To photogeologists, that factor of 10 to 100 improvement in image resolution was profound: features diagnostic of specific geological processes and events—like from the actions of rivers, glaciers, or individual volcanic eruptions—began to come much more sharply into focus. The improvement in resolution was comparable to or even greater than that between Eugène Michel Antoniadi's views of Mars with the Meudon refractor and those of Galileo with his small telescope in Padua.

The high-resolution camera on MGS was called the Mars Orbiter Camera, narrow-angle camera (MOC-NA, or just MOC for short), to distinguish it from the Mars Orbiter Camera, wide-angle camera (or MOC-WA), which was a separate device, a pair of cameras (one with a red filter, the other with a blue filter) used to take relatively low-resolution, global weather monitoring images over the course of the mission. The MOC was originally designed, built, tested, and operated in flight for the Mars Observer mission by a team of scientists and engineers led by planetary geologist Michael Malin, who won the competition to lead the MOC investigation in 1986, when he was a professor at Arizona State University (ASU).

Malin was trained as a geologist at Caltech by some of the most prestigious Earth and planetary scientists in the field (including Mariners 4, 6, 7, and 9 imaging team members Robert P. Sharp and Bruce C. Murray). Another key leader of the MOC development effort was Caltech/JPL imaging systems engineer G. Edward "Ed" Danielson (1939–2005), who had played a prominent role in the development and testing of the imaging systems on Voyager, HST, and other missions. Malin and Danielson attracted a young team of talented scientists and engineers to help get the innovative MOC design to the launchpad, and they augmented the development team with a cadre of seasoned Mariner 9 and Viking Orbiter imaging scientists.[30]

Generally disaffected with the life of an academic, Malin, who was awarded a MacArthur "Genius Grant" in 1987, left ASU partway through the development and testing of the MOC for Mars Observer, using his hefty cash award to found a small business devoted to building high-quality, reliable, low-cost, science-worthy imaging systems for NASA deep space missions. Malin Space Science Systems Inc. (MSSS), in San Diego, California, became and still is a small business run by a nimble and passionate group of engineers and scientists who have been able to propose imaging systems to NASA for much lower cost than the "traditional" camera vendors among the large NASA centers and aerospace companies. MOC was the first big NASA instrument operated by MSSS, and it represented somewhat of a gamble for NASA. Could a small business with no prior direct experience with NASA deep space hardware build and operate reliable, high-quality instruments? Could a much cheaper imaging system deliver the needed science?

It turned out that the answer to both questions was yes. But while Malin and team were able to deliver the MOC to the required specifications and operate it successfully during cruise for Mars Observer, the failure of the mission just three days shy of entering Mars orbit meant that the true potential of that imaging system was never actually realized. Fortunately, when NASA came looking for relatively inexpensive and rapid-development instruments for the MGS mission's attempt to resurrect much of the science lost from Mars Observer, Malin and team had enough spare parts to quickly build a near copy of the MOC.

Between 1997 and 2006 the MOC for MGS ended up returning almost 100,000 high-resolution images of Mars, and those images ended up once again revolutionizing our understanding of the Red Planet.[31] Because the MOC observed such a tiny part of the surface in each image, even nearly 100,000 MOC images ultimately ended up covering only a few percent of the entire surface. Thus, each image was carefully targeted to a specific region where Viking and previous images (or, later in the MGS mission, where previous MOC images or data from the newly arrived Mars Odyssey orbiter's thermal infrared imaging system [THEMIS] camera) had indicated that there was the potential to reveal new and interesting insights about the planet's history. That targeting was often done by Malin himself, by Edgett (who had joined MSSS in 1998 after obtaining his PhD in geology from ASU), or by other MOC team members or MSSS staff who worked closely with them. For nearly a decade, Malin, Edgett, and the MOC science and operations team controlled the highest-resolution imaging system ever used to study Mars. Based on the images that they helped collect, they certainly took amazing advantage of their unique situation. Edgett describes himself and the team as feeling like they had "a responsibility to collect the best data for the Mars science community to address a broad array of questions and hypotheses about the Red Planet."[32]

Layers, Layers, Everywhere

It is hard to underestimate the effect that an increase in resolution of 10 to 100 times that of previous images had on our understanding of the past history of our neighboring planet Mars. Perhaps one way to sum it up is using the words of Malin and Edgett themselves: "The Mars revealed by MOC is very different from the planet we anticipated finding and that most of the scientific literature describes. . . . The pervasive layering of the crust, to depths of ten kilometers or more, attests to a place that is most definitely not the Moon with an atmosphere."[33]

The stunning revelation apparent in the high-resolution MOC images was that the surface of Mars is "pervasively layered," with many layers occurring in repeated, almost rhythmic patterns (figure 14.4). "Layered" is a kind of a holy grail word for geologists, because the presence of a layer in the geological record means that *something* in the climate, the volcanic record, the impact record, or elsewhere has changed over time, and the different layers preserve evidence of those changes. Even the break between one layer and the next says that a change occurred. This basic geological principle is in fact why layered landscapes like the Grand Canyon have proven so critical to untangling the geological history of our own planet. Layered rocks are like the pages of a book, and they can be read by trained geologists to gain amazing insights about the past.

MOC found that layers on Mars occur not only in volcanic lava flows, representing multiple episodes of eruptions, but also in sedimentary rocks.[34] On Earth, sedimentary rock layers represent places where wind, water, or glaciers have created accumulations of continental, ocean floor, stream bed, or delta deposits downstream (or downwind) in low-elevation basins on land or in

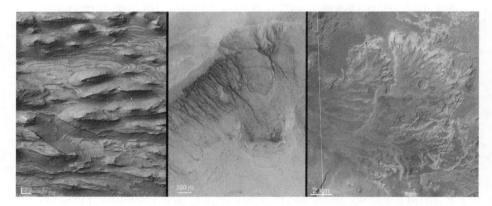

FIGURE 14.4 *Left*: An early Mars Global Surveyor MOC photo of a 1.5 × 2.9 kilometer area in far southwestern Candor Chasma, showing extensive layering of sedimentary rocks deep within the Valles Marineris canyon system. *Center*: MOC photo showing gullies on the walls of an impact crater near 39.0°S, 166.1°W. The resolution of the view here is about 1.5 meters (5 feet) per pixel, among the highest resolution views of the surface obtained by MOC. *Right*: MOC photo of the ancient, preserved delta in Eberswalde crater. The field of view is about 15 kilometers wide, and what were once meandering deltaic river channels have been preserved here in "inverted relief" after billions of years of erosion. NASA/JPL/MSSS.

the sea. One key aspect of such layering is that it takes time—sometimes not much time but sometimes significant amounts of geological time—for layers to be deposited, depending on the way those layers were formed. On Earth, a layer of coarse, bouldery rubble and debris on a lake bed can form quickly during a violent landslide, while a thick layer of fine-grained mudstone can take millennia to build up on some parts of the ocean floor. On Earth, fossils often provide key markers in time with which to date the relative ages of different layers, and even in some cases the durations over which they formed. On Mars, the task of estimating ages and durations is harder, because images from orbit wouldn't provide adequate resolution to identify fossils, and even from the ground there's no evidence (at least not yet) of a fossil record in Martian history.

Still, the presence of pervasive layers seen in MOC images implied a much more active and complex geological history of Mars than previously thought. Were the layers formed on land by wind (so-called aeolian deposits, after Aeolus, the Greek god of the wind), impact events, or volcanic eruptions? Or were they formed underwater, via subaqueous processes? Or by some mixture of those possibilities? Mariner 9 and Viking Orbiter images had shown that most of the planet's fluvial features (landforms created through the action of liquid, presumably water) occur in the most ancient terrains, implying that Mars had a warmer and wetter (perhaps more Earthlike) climate early in its history. The presence of pervasive and deep sediments preserved in many places could thus mean that the period of possibly more Earthlike conditions was much longer than previously thought. Indeed, Malin and Edgett acknowledged that some features in the landscape could have been formed or modified over an enormous span of Mars history: "Not every process operated early in Martian history and then subsequently with less or no vigor."[35]

Many astrobiologists were also excited by these discoveries because they implied that if life had developed on early Mars (a big if!), then perhaps it might have had time to become more complex (like Earth life), or at least to evolve into specific, perhaps subsurface niches where conditions might have remained habitable after the planet's surface climate changed. These kinds of ideas are, of course, highly speculative, and determinations of the absolute ages of Mars surface features and the durations of different geological eras in the planet's past (both estimated from counting impact craters in high-resolution photos) have large uncertainties. Still, much of science and exploration is speculative; the key is to develop *testable* hypotheses that can turn speculation into theory.

Deltas and Gullies

One of the most enigmatic and diagnostic landforms discovered in the MOC high-resolution images was a fan-shaped layered deposit on the floor of a crater called Eberswalde that has been interpreted to be the eroded remains of a lithified (turned to stone) delta (figure 14.4).[36] On Earth, deltas like those at the end of the Mississippi or Mekong Rivers are places where enormous amounts of sediment are gently dumped into shallow lakes or bays. As the river channels meander, the sediments are spread out into a fan-shaped deposit. The discovery of evidence of similar kinds of deposition, including meandering rivers, implies persistent, long-term surface water flow at one time on Mars. As well, features like this (dozens more have since been discovered on Mars in high-resolution imaging, though Eberswalde is among the best preserved) imply that long-lived lakes also occurred on Mars, perhaps many of them occupying large impact craters. This discovery created another big stir in the astrobiology community because on Earth deltas and their basins are often environments where fossils and organic molecules transported downstream are well preserved and protected from the elements, buried in shallow delta sediments. Eberswalde would become a high-ranking candidate for a future NASA rover mission, and the Curiosity rover would end up landing in and exploring the sediments of another former ancient lake in a crater called Gale. And in 2021, NASA's 2020-launched rover, Perseverance, has begun to explore another extremely well-preserved delta in a former ancient lake called Jezero, while the European Space Agency's (ESA) Rosalind Franklin rover is scheduled to launch in 2022 to eventually explore a region not far from another delta located at the terminus of Coogoon Vallis, at the eastern end of the rover's landing ellipse.[37]

Another spectacular class of landforms first identified in high-resolution MOC images were gullies: small channels on steep slopes that suggest groundwater seepage and surface runoff.[38] The channels, typically only one to two kilometers long and a few tens of meters wide (figure 14.4), were observed primarily in the southern highlands, poleward of 30° latitude, and on poleward-facing slopes. Malin and Edgett reported finding about 150 gullies within the more than 20,000 MOC high-resolution photos taken in the first few years of the MGS mission (eventually, tens of thousands would be discovered). The typical characteristics of the channels (a headlike source alcove, primary and secondary channels, and a wider debris apron near the terminus) have terrestrial analogs in Icelandic hillside gullies and similar systems.[39] They suggested to Malin and Edgett "fluid-mobilized mass-movement processes" analogous to gullies on Earth. Some are straight, some have sinuous or meandering secondary channels, and most begin on slopes a few hundred meters below the surrounding surface. Furthermore, even though they occur in some of the most ancient terrains on the planet, the gullies appeared relatively young, showing no "cross-cutting" relationships with other features, or superimposed small impact craters, in MOC images. In a follow-on study in 2006, Malin and colleagues reported the stunning evidence that some gully activity is still ongoing today, with new, brighter deposits having formed at two gully sites within the span of about three to five years between MOC images.[40] While the attributes of these new features (like their diversion around obstacles) suggested "emplacement aided by a fluid with the properties of liquid water," the research team acknowledged that alternative hypotheses—like dry, granular dust flows—could also be valid.[41] Indeed, imaging from the High Resolution Imaging Science Experiment (HiRISE) camera on the Mars Reconnaissance Orbiter (MRO), which entered Mars orbit in 2006, has provided significant new insights and hypotheses about gullies, especially focused on whether the action of liquid water is required to explain them.[42] Still, many puzzles about gullies and other channel-like features on Mars remain controversial or unexplained more than 20 years after their initial discovery. While these features' origin might not turn out to be ultimately related to liquid water, Malin and colleagues concluded with some rather philosophical speculations about just such a possibility:

> As with many discoveries, the possibility that liquid water may be coming to the surface of Mars today poses many questions: Where is the water coming from? How is it being maintained in liquid form given the present and most likely past

environments? How widespread is the water? Can it be used as a resource in further Mars exploration? Finally, has it acted as an agent to promote or sustain a Martian biosphere?[43]

Topo Mapping

The discoveries enabled by MGS weren't just limited to features seen in amazing images, however. Each of the spacecraft's other scientific instruments also helped enable fundamental new advances in our understanding of the Red Planet. The Mars Orbiter Laser Altimeter (MOLA), for example, bounced hundreds of millions of laser shots off the surface and used the travel time for the light reflected back to the instrument to map the topography of the planet to a level of detail and resolution that was better, at the time, than our knowledge of the global topography of Earth (figure A5.2).[44] As first discovered from Mariner 9 and Viking Orbiter atmospheric sounding and stereo imaging, that topography turns out to be pretty dramatic (plate 10). Despite its smaller size, Mars has about 50 percent more topographic relief between the highest highs and the lowest lows than Earth's topography. And, strangely, almost the entire northern hemisphere sits about five kilometers lower in altitude than the southern hemisphere, suggesting that an enormous impact event or perhaps other global-scale internal geological processes may have created this hemispheric dichotomy in elevations.[45] MOLA data enabled a new zero elevation reference, or topographic datum, to be established for Mars. Since Mars has no "sea level," the zero point of Mars elevations had previously been set at the elevation corresponding to 6.1 millibar atmospheric pressure, the triple point of water at 0°C. In 2001, however, the zero elevation datum was redefined as being tied to the average radius of the planet, since that had now been determined highly accurately from MOLA mapping.[46] MOLA also took the study of altimetry beyond this global scale, however, to allow researchers to zoom in on the altimetry of specific regional and local features, like volcanoes, tectonic features, water-carved channels, impact craters, and the polar caps. Estimates of surface roughness and small-scale surface slopes from the MOLA data have also been used to help determine the safety and traversability of all proposed Mars surface mission landing sites since the late 1990s.

Magnetic Stripes

While Mars does not currently have a planetary-scale magnetic field like Earth does, the MGS Magnetometer/Electron Reflectometer (MAG/ER) instrument discovered that the ancient crust of Mars has preserved evidence that the planet once did have a strong magnetic field, probably when the core was still partially molten (as Earth's core is today). Over geological time, smaller planets cool off faster than larger planets. Mars has apparently cooled off so much that its once-molten core is now solidified, shutting off the planet's magnetic dynamo and exposing the atmosphere and surface to the solar wind and other harsh radiation. Still, ancient volcanic rocks and minerals that were created when the field was active have preserved the imprints of that strong magnetism in a way that was detected and mapped from orbit by MGS.[47] The most striking discovery was a series of lineations or "stripes" of alternating (adjacent strongly positive then strongly negative) magnetic field intensities, primarily within the ancient southern highlands,[48] but especially focused between about 270° and 90°W longitude (plate 11). The stripes are reminiscent of the magnetic polarity reversal stripes discovered on Earth's seafloor in the 1960s, which is one of the key observations supporting the theory of plate tectonics on our planet. Did early Mars release its interior heat and form new basaltic crust by lava erupting from ridges between crustal plates as Earth still does today? Do the alternating stripes reveal that the early Mars core dynamo underwent magnetic polarity reversals as Earth's core dynamo still does occasionally?

Exactly when the magnetic field of Mars died off is the subject of considerable ongoing study and debate, and is of course a big concern to astrobiologists and others interested in the potential habitability of Mars. Did the planet's magnetic field protect the surface from harmful radiation (like Earth's magnetic field still does) for long enough to let life take hold there? The late MGS magnetometer principal investigator and NASA planetary physicist Mario Acuña (1940–2009) and colleagues were sanguine in their summary of the importance of this discovery: "A complete understanding of the crustal magnetic record remains as one of the most significant challenges in Martian geophysical research, one with great potential for understanding not only Mars' evolution but also many aspects of that of the terrestrial planets, asteroids, and the Moon."[49]

Infrared Vision

The first infrared spectroscopy observations of Mars had been made during the Mariner 6 and 7 flybys and the Mariner 9 orbiter mission and revealed evidence for interesting variations in atmospheric gases and surface ices and rocks/minerals. The short durations and relatively low resolutions of those investigations compelled additional higher-resolution follow-up observations, however, which was the job of the Thermal Emission Spectrometer (TES) instrument on MGS. TES could measure the infrared radiation coming from a region only about three by six kilometers in size on the surface (comparable to about a third the length of Manhattan Island), and the instrument would sweep that field of view across the surface as the spacecraft orbited, slowly building up global maps of that infrared energy. The TES team then used the data to measure the temperature and other physical properties of the surface, as well as to detect the unique spectroscopic "fingerprints" of certain kinds of rocks and minerals.[50]

TES data enabled the confirmation that, as suspected from decades of earlier ground-based telescopic observations and Mariner mission results, the dark regions on the surface of Mars are dominated by mineral compositions consistent with the presence of so-called mafic (magnesium- and iron-rich) igneous rock. Indeed, most volcanic rocks detected by TES fall into the same kinds of categories as volcanic rocks on Earth, the Moon, and even Venus. These volcanic rocks are typified by basalt, an igneous rock that is common on Earth in volcanic eruptions or intrusions in Hawai'i, in Iceland, and along the mid-ocean seafloor ridges, and that is made from a mixture of minerals like olivine, pyroxene, plagioclase, and magnetite—all of which could be fingerprinted from TES data in Mars's classical dark regions. Furthermore, a major discovery enabled by TES data was that a compositional dichotomy mirrors the topographic dichotomy between the planet's hemispheres: the southern highlands are dominated by lower-silicon igneous rocks like typical basalt on Earth, and the northern lowlands have been interpreted as chemically altered (weathered) basalt or as a higher-silicon igneous composition more like that known terrestrially as basaltic andesite (plate 12).[51] If the higher-silicon compositional interpretation is correct, it might indicate an increased level of crustal recycling in the northern hemisphere crust, perhaps akin to (but not necessarily as extensive as) the crustal recycling processes that enhance silicon at Earth's convergent plate boundaries (indeed, andesite is named after the Andes mountains in South America, which formed along one of these convergent plate boundaries). While detailed analysis and interpretation of the MGS infrared data is still ongoing, one result is clear: Mars has continent-sized compositional provinces not recognized prior to TES.

Interestingly, TES researchers found only rare examples of the mineral quartz (SiO_2) on Mars.[52] Quartz is a major component of the continental crust of Earth, and is thought to form partly as a result of plate tectonics and the remelting (and separation of silicon) from more basaltic oceanic crust. The dearth of quartz on Mars is consistent with the lack of overall geological and geophysical evidence for plate tectonic–like crustal recycling on that world. Unlike Earth, Mars today appears to be a one-plate planet, although some interpretations of the composition of the northern lowlands and the presence of the puzzling magnetic stripes discussed above raise the possibility of nascent multiplate tectonics early in the Red Planet's history.

The bright, reddish regions of the planet are different in composition from the dark regions, as they are covered in fine particulate dust and are thought from telescopic data and other subsequent shorter-wavelength studies to consist of oxidized (rusted) iron-bearing minerals formed from the physical or chemical weathering of the dark igneous basalts. In the infrared, TES spectra also displayed evidence of a few percent carbonate in the dust, which points to minor aqueous (by liquid water) alteration of precursor igneous rocks in the presence of a CO_2 atmosphere. In general, however, TES results suggested that aqueous alteration has not been extensive throughout the Martian crust. As TES principal investigator and ASU planetary scientist Phil Christensen and his colleagues concluded in their 2008 review paper: "Thus, while Mars has a large inventory of water, it is likely that this water has remained frozen throughout much of Mars history, resulting in very little chemical weathering. Occasionally, this water was melted and released to form channels, lakes, and groundwater."[53] This is a view still held by many Mars researchers today, one that is generally inconsistent with alternative hypotheses that postulate early Mars as having an Earthlike environment and lakes, seas, or even oceans persisting over geologic time scales. Although many deposits of water-formed minerals have been identified from visible to near-infrared observations acquired by missions that arrived after MGS, they appear to represent only a very small fraction of the crust, perhaps consistent with only relatively short-term, occasional "warmer and wetter" climatic conditions.

TES data also allowed the thermal inertia of the surface to be globally mapped at fine spatial scales. Thermal inertia characterizes how a surface heats up or cools down over time; for example, a high thermal inertia surface like a boulder will stay warmer well into the night and take a longer time to heat up in the morning than a low thermal inertia surface like a fine-grained sandy or dusty region, which cools down and heats up quickly. TES thermal inertia data thus provided a way to estimate either the rock abundance of a region, or the nature of the finer-grained soil or regolith deposits in places with few or no rocks.[54] Importantly, TES data nailed down the idea that most of the dark, low-albedo surfaces on Mars—many of which have been observed and monitored telescopically for centuries—are not bare rock exposures but rather, more often than not, regions covered in windblown sand. Even the famous low-albedo surface of Syrtis Major (now known to be a shield volcano) is mostly sandy regolith rather than bare rock (although higher-resolution thermal infrared measurements from the Mars Odyssey THEMIS instrument have since revealed small exposures of bare rock there and elsewhere). Dark dunes found all over Mars appear to be made of fragments of mafic igneous rock, primarily basalt, pulverized into sand by billions of years of erosion. Mars is a basaltic planet, and in that regard it is similar in many ways to the other terrestrial planets in our solar system.

A Dusty Link to the Past

The unprecedented surveillance provided by MGS from its arrival in Mars orbit in 1999 until the end of its mission in 2006 added a great deal to our understanding of dust storms, and among its discoveries were some real surprises.

The large planet-encircling dust storms had long been known to be sporadic and unpredictable. Thus, Viking Lander 1 saw planet-encircling storms during the first Martian year—in fact, two!—as well as again during the fourth year but nothing larger than a regional storm during the second year and nothing larger than local storms during the third year. The planet-encircling storms—so rare in the historical record going back to the 19th century, but occurring with unusual frequency between the 1971 arrival of Mariner 9 and the 1982 end of the Viking mission, with six storms in all—then petered out through the end of the century, with only one more (questionable) planet-encircling storm occurring in 1994.

Since the planet-encircling dust storms were most frequently identified with sources in the southern hemisphere, where the increased solar insolation was associated with the warmer (if shorter) summer, it seemed reasonable to assume that a combination of warming and local topography might lead to increased local winds and thus to increased dust further warming the atmosphere, and so on through a series of positive feedback loops producing ever-larger storms. MGS, however, showed that though this is certainly true in some cases, many storms follow a more complex history of development. Remarkably, many of the great southern summer storms start out tacking along tracks of seasonally repeating winter dust storms originating in the northern hemisphere.

As noted in chapter 12, Mars's atmospheric circulation, though simpler than Earth's because the planet lacks oceans, is similar in that it too can also be modeled in terms of Hadley cells: warm air rises near the equator and moves toward the poles, sinks as it cools, then flows back toward the equator at ground level. In addition to this basic scheme, Mars's rotation also produces through the Coriolis effect deflection of these moving air masses to create a circulating pattern of strong winds (somewhat similar to Earth's), including polar vortices in high latitudes. On Mars, the polar vortices are associated with the autumn and winter clouds of the polar hoods and sometimes give rise to circumpolar cyclones (similar to those observed in Earth's polar regions), which sometimes travel down through the lower latitudes (plate 9).

MGS was lucky in that, after a hiatus going back at least to 1994 and perhaps all the way back to the end of the Viking mission in 1982, a new global dust storm broke out in June and July 2001 (plates 9 and 13). Several bands of dust originating in northern hemisphere circumpolar cyclone activity were transported in a southeasterly direction through Chryse and Acidalia and across the equator, possibly augmented by dust in transit, and then they proceeded to strip fine bright dust from Tharsis. (Tharsis, rather surprisingly, appeared darker after the storm.) Some of this dust was then injected into Solis Planum, changing the shape of Solis Lacus (which was now elongated to the northwest as it had been in 1926–29). The dust also increased the opacity of the region southeast of Valles Marineris. Then, spread by the strong westerly winds of the southern subtropical jet, the dust rapidly swept from west to east in longitude to encircle the planet, leaving albedo changes in its wake, including a deposit of thin bright dust over an area of a million square kilometers on the western side of Syrtis Major. Though much of this coating was abruptly blown away in the direction of Libya and Isidis Planitia (an old impact basin and another important dust sink) in January 2002, not all of Syrtis Major was cleared in the event; large areas to the west and north of Syrtis were still visible as brightish patches even in Earth-based telescopes well into 2003.[55]

Analysis of the 2001 dust storm event suggested that there were two seasonal windows, at L_s = 210–230° and L_s = 310–350°, in which northern circumpolar cyclones develop into large regional and planet-encircling storms. One was in the northern hemisphere autumn, the other in the late winter.[56] The next significant dust storm, the large regional storm of October 2005, also developed from cyclones in the large dark area Mare Acidalium, north of Chryse (in what has been described as the "Acidalium storm factory").[57] As observed from Earth, this appeared as a typical southern hemisphere midsummer dust storm moving northward and from Chryse to involve Valles Marineris and Solis Planum, and eventually reaching across Noachis Terra.

Though the MGS mission ended before the next planet-encircling dust storm, which began in June 2007, by then MGS was no longer alone in Martian orbit. The torch of keeping the dusty Red Planet under surveillance had now been passed to NASA's Mars Odyssey, the ESA's Mars Express, and NASA's Mars Reconnaissance Orbiter, all of which reached Mars orbit by November 2006, when contact with MGS was lost.

A Global Legacy

As had happened several times during previous missions, the discoveries of MGS yet again established a new Mars, far different and more interesting than the one that came before. As aptly described by Malin and Edgett in the conclusion to their landmark paper summarizing the MGS high-resolution imaging results:

> These attributes strongly suggest that Mars as we see it today is not what it has been like over all of its history. Just as deposits and landforms in Europe and North America strained interpretations in the nineteenth century before the realization that ice ages had occurred, so too is it likely that many Martian landforms are relics from environments other than those found on Mars today.
>
> If there is one idea to take from this report, it is that Mars is substantially more complicated, and its geology more complex, than anyone had previously thought. . . . Nothing can be taken for granted when considering the "new Mars."[58]

MGS showed that deep, pervasive layers of sedimentary rock and familiar patterns of deposition and erosion point to a rich and complex history of aeolian, impact, volcanic, and fluvial processes on the surface and within the crust of Mars that rival the nature and complexity of similar processes that have shaped the crust of our own world. The mission revealed the remnants of a once-strong global magnetic field preserved in igneous rocks and minerals on the surface. MGS data allowed the similarities in the chemistry and mineralogy of those rocks to igneous rocks on Earth to provide us with a context in which to compare Mars to all of the terrestrial planets.

The results from MGS cemented the inference—first gleaned from Mariner 9 and Viking Orbiter data—that the planet experienced at least one, but perhaps more, profound climate change episodes that saw the surface environment go from more clement, more Earthlike conditions, to colder, dryer, harsher conditions like those on the surface today. Perhaps most importantly, MGS reopened the door, slammed shut on astrobiologists by the Viking Lander biology results, for the possibility of long-lived, persistent, and habitable climate conditions having existed on the Martian surface or in the shallow subsurface for a significant fraction of the planet's early history. And the discovery of abundant sedimentary rocks on Mars also increases the prospects that a fossil record of those early habitable conditions—either physical fossils or, perhaps more likely, chemical, isotopic, or textural trace evidence—might still be preserved and accessible on the surface. How long was the surface of Mars habitable for life as we know it? Is the subsurface potentially habitable still? MGS data alone could not be used to answer such questions, further reinforcing the idea that additional measurements needed to be made—first, to complete the original scientific goals of the Mars Observer mission, to build on these results using new orbital missions and measurements, and then, eventually, to get back down to the surface and pick up where the Viking Landers left off.

15

Baby Steps

Back to the Surface with Pathfinder

Discovery

The launch delays of the Magellan and Galileo missions because of the Challenger disaster in 1986, the cancellation (for budgetary reasons) of NASA's Comet Rendezvous / Asteroid Flyby flagship-class mission in 1992, and the failure of Mars Observer in 1993 pointed out the brittleness of NASA's planetary exploration program. Flagships could produce spectacular science (indeed, Voyager and Viking had), but they represented rare mission opportunities (maybe one per decade), were vulnerable to budget overruns and thus extra congressional and administration scrutiny, and were generally high-science-risk endeavors designed to achieve many different and sometimes competing objectives in a single mission. Partly in response to these concerns, a series of National Academy of Sciences panels and NASA and planetary science community workshops from the mid-1970s to the late 1980s looked critically at the issue of *balance* within NASA's robotic mission portfolio. That is, was too much emphasis being placed on large, rare flagship-class missions, to the detriment of both the science and workforce training required to sustain the community? Many thought so, and one recommendation of the workshops was the realization that an exploration strategy that included a larger fraction of smaller, more focused, more frequently launched planetary missions (like the Explorer small mission program in the space physics community) was needed to balance NASA's portfolio.

In addition, studies by the National Academy of Sciences' Space Science Board (SSB) provided guidance to NASA about ways to focus future Mars exploration efforts on the longer-term goal of a robotic sample return mission, which had been studied and considered since well before Viking. Specifically, in a 1976 letter to NASA administrator James Fletcher (1919–91), the SSB recommended that "[to] better define the nature and state of Martian materials for intelligent selection for sample return, it is essential that precursor investigations explore the diversity of Martian terrains that are apparent on both global and local scales. To this end, measurements at single points . . . should be carried out as well as intensive local investigations of areas 10–100 [kilometers] in extent."[1]

One of the community advisory groups that was formed in the late 1970s, partially in response to the SSB's advice, was called the Mars Science Working Group (MSWG). Tasked by NASA to provide recommendations on future Mars mission science goals post-Viking, the MSWG performed a variety of studies that embraced the idea of exploring the diversity of Martian terrains. For example, in a foundational 1991 report,[2] the MSWG advocated strongly for a landed network strategy that would build on the orbital reconnaissance that was intended to be achieved by the Mars Observer mission. A network approach was judged to be especially important for science goals related to meteorology, seismology, and compositional "ground truth" for orbital remote sensing interpretations. Just such a network-based program of planned surface missions, called the Mars Environmental Survey, or MESUR, had actually been studied in detail starting in the late 1980s by NASA's Ames Research Center.[3] The Ames MESUR concept would deploy up to 16 small, low-cost lander stations widely across the planet during the 1999, 2001, and 2003 launch opportunities. Part of the way costs would be kept low would be by using a novel "rough landing" system based not on retrorockets like the Viking Landers, but on airbags— then a relatively new technology. To demonstrate the viability of the lower-cost airbag-assisted landing concept, an initial technology demonstration landing mission, dubbed MESUR Pathfinder, was envisioned as the first station.

While individual mission costs were predicted to be relatively low, estimates for the full MESUR network were predicted to exceed $1 billion, which was too much for NASA to take on during the relatively lean budget times of the early 1990s. Indeed, budget pressure was part of the motivation leading the then-new NASA administrator Dan Goldin to create the new Discovery line of smaller, robotic solar system exploration missions chosen through a competitive selection process. Rather than delay the start of the Discovery program while waiting for a first round of community-led proposals to be written, submitted, and then reviewed, however, Goldin seeded the program by directing the choice of the first two missions. One of them would be based on the already-conceived idea of MESUR Pathfinder, but reassigned to NASA's Jet Propulsion Lab (JPL) at Caltech and renamed Mars Pathfinder. The other would be the Near Earth Asteroid Rendezvous (NEAR) mission, designed to rendezvous with and study the Mars-crossing asteroid 433 Eros during one of its near-Earth approaches, and led by the Johns Hopkins University Applied Physics Lab (APL).

Goldin's idea was to demonstrate, via these first two directed missions, that the agency's newly minted motto, "better, faster, cheaper," could be used to achieve complex but relatively focused science objectives for a diverse range of mission architectures. NEAR would be the first spacecraft to orbit a small solar system body, and would thus need to develop innovate ways to maintain orbit around a low-gravity world and properly point its instruments at an object that was, as yet, hardly more than a starlike point of light. Mars Pathfinder would attempt to deploy the first rover on another world since the Soviet Lunokhods of the early 1970s, and the first ever on Mars. It would also need to demonstrate that landing on Mars does not require the budget (or complexity) of a flagship-class mission, and that important new Mars science could be done with a small mission team. Not so subtly, NASA was also challenging the engineers and management of APL and JPL to do things in a more streamlined way, to innovate with new technologies, and to accept additional mission risk in trade for less launch mass and fewer mission objectives, all of which translated into substantially lower mission cost. One goal of the early Discovery program was to keep those mission costs (not including launch vehicle) below about $250 million, or somewhere between three and five times less than a typical flagship mission budget. It was a high bar, and many in the community were skeptical that these small missions would succeed. Skepticism was particularly high at JPL, which had a recent history dominated by flagship-class missions, and a culture that was convinced that flagship-class budgets were required to dramatically reduce risk and guarantee mission success.

Chutzpah

It was perhaps easy to predict that scientists, engineers, and managers who would categorize themselves as "risk averse" generally steered away from involvement in Mars Pathfinder at JPL. The project faced the daunting challenges of designing, building, and testing a lander and a rover, as well as the cruise stage to carry them to Mars and the equipment needed to land safely on the surface. However, it was operating on a shoestring budget compared to most others at JPL. The team also had to adhere to an aggressive build and test schedule (less than three years) in order to have the spacecraft ready and on the rocket before the opening of a three-week launch window in late 1996. Luckily, a number of eager young engineers and scientists were willing to work on a more risky and cutting-edge small project, and just as fortunately a few seasoned veterans had the experience and chutzpah to help guide the effort.[4] Still, it was a small team for a NASA planetary mission (with only around 150 JPLers directly involved in day-to-day work on the rover and lander and other equipment at least part time, plus additional support from a small group of external team members and contractors).[5] In many ways the team operated "below the radar" at JPL, needing to cut corners on cost, streamline new assembly and test processes, and react quickly to problems within a parent organization that regarded most of those approaches as anathema.

Some of the work for Pathfinder was done at other institutions as well, in collaboration with (and in most cases under funding contracts from) JPL. For example, the lander's camera system, playfully called IMP, for Imager for Mars Pathfinder, was designed, built, and tested by a team based at the University of Arizona. Scientists from the Max Planck Institute in Germany and the University of Chicago led the development of the rover's elemental chemistry instrument, called the Alpha Proton X-Ray Spectrometer, or APXS. Scientists and engineers from Cornell and Arizona State Universities played significant roles in the development of a wind sock experiment within the lander's meteorology package, which was part of a facility instrument suit called the Atmospheric Structure Instrument/Meteorology package (ASI/MET). And shortly before launch, a group of about 20 additional scientists from universities and government labs around the world were added to the team in a competitively selected proposal process.[6] Most of the team from JPL and the other institutions had never been involved in a small, low-cost, fast-paced mission like Pathfinder. The ensuing bonding experience of trial and error, and essentially the process of making up new ways to design, build, test, and operate a deep space mission as the team went along, made for a relatively tight-knit group.

One aspect of the Pathfinder mission that helped mitigate some of the risk inherent in a rapidly developed and low-cost project was the acknowledgment by NASA Headquarters and other stakeholders that Pathfinder was actually a technology demonstration mission designed to feed-forward to future missions. As a result, the ability to do cutting-edge science was effectively relegated to "bonus" status. One way that NASA and other space agencies formalize these kinds of strategies is by defining what it will take to achieve "mission success." In the case of Pathfinder, mission success was deemed to include a successful landing on Mars, the successful acquisition and downlink of a panorama of the landing site from the lander, and the successful deployment of the rover onto the surface (along with the attainment of appropriate photographic documentation from the lander and rover). Anything beyond that, like elemental chemistry results for surface soils and rocks, or detailed meteorologic data, or mineralogical inferences gleaned from the lander camera's dozen color filters, or any other science results culled from the mission data, would be icing on the cake. Setting the bar this way put most of the pressure on the engineers, especially those designing the rover and the landing system. Because of the tight nature of the relatively small team, however, almost everyone felt that the ability to do some great science during the mission was also a critical part of mission success.

Grab Bag

Because of the failure of Mars Observer and the fact that Mars Global Surveyor hadn't yet gone into orbit, planning during the early 1990s for where to land Pathfinder in summer 1997 was mostly based on the highest-resolution available Viking Orbiter imaging of Mars. The group that organized the landing site selection process, led by Pathfinder project scientist and JPL planetary geologist Matt Golombek, faced a similar problem to that faced by the Viking Lander team 20 years earlier. They had to balance a variety of constraints to find the optimal solution to the problem.[7] On the one hand, the group needed to pick a safe place to land. In the absence of any other constraints, this requirement could easily drive the team to a flat, boulder-free, low-wind, low-dust, parking lot kind of landing site. Indeed, most of the pressure on the site selection process for the Viking Landers was similarly focused on these kinds of landing site safety issues. On the other hand, the Pathfinder team also wanted to give significant weight to the *science potential* of the chosen landing site, especially to the potential to augment the science returned from the Viking Landers in new and different ways.

Safety constraints for the Pathfinder landing system came in a variety of different flavors, all of which had to be incorporated into the final recipe, and into the final assessment of the uncertainty in the precise location of the safest place to land (captured in the derived "error ellipse" of the predicted landing site; figure 15.1). The entry system, for example, would rely partially on parachutes (like Viking's), and thus the landing site had to be *below* a certain elevation to ensure that there would be enough atmosphere (and enough time) to make the parachutes effective. That elevation proved to be below the average elevation of the surface (the topographical datum or zero elevation) and is analogous to landing below sea level on Earth. Obviously, Pathfinder wouldn't be able to land on any of the planet's large volcanoes or other mountain peaks.

Another engineering safety constraint was that because the lander and rover used solar panels to generate their electricity, the landing site had to be at latitudes where the Sun traveled high enough in the sky to shine at optimal angles on the solar panels. This turned out to limit the landing site to a zone between about 10° and 20° north latitude—in other words, the northern hemisphere tropics—given the season during which the mission would land and operate. (For comparison, if this constraint were applied to Earth, the landing would have had to have been in Central America, sub-Saharan Africa, or Southeast Asia.) A similar latitude constraint was to land in a zone where Earth would also be as high as possible in the Martian sky during the mis-

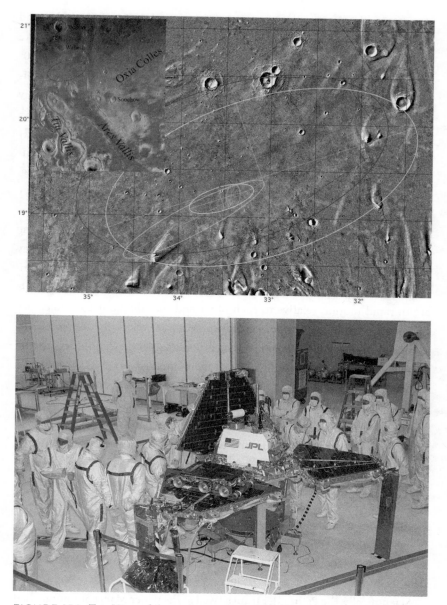

FIGURE 15.1 *Top*: View of the various estimated Mars Pathfinder landing system's "error ellipses" among the streamlined islands near the mouth of the Ares Valles out-flow channel, as shown on an image from Viking Orbiter. The yellow *X* is thee actual landing site. *Bottom*: Engineers inspect the Mars Pathfinder lander and folded-down Sojourner rover, prior to closing the lander's solar panel–covered petals. NASA/JPL.

sion, to enable efficient communications from Earth to Mars ("uplink") and back again ("downlink"). From Mars, Earth is a Morning or Evening Star (like Venus or Mercury are from Earth) that never gets much more than about 40° from the Sun in the sky. Nonetheless, the optimal latitude for Earth communications during the mission turned out to be around 25°N, which was roughly consistent with the solar panel latitude restriction.

Finally, mission navigators needed to identify a relatively flat, relatively rock-free, east-west-aligned elliptical area about 70 kilometers wide (north-south) and 200 kilometers long (east-west) in which to target the landing, because that was roughly the accuracy of the smallest-sized region that they could guarantee NASA they could guide the spacecraft to after the eight-month cruise to Mars and passage through the planet's atmosphere. A landing uncertainty of 44,000 square kilometers might seem like pretty poor accuracy at first, but if the situation were reversed it would be like shooting a bullet from Mars and trying to hit a target in Denmark. Given the fact that the planets are in rapid motion relative to each other and that the nature of trajectory disturbances from winds and weather on Mars are relatively unpredictable, the accuracy with which mission planners could target the Pathfinder landing site turned out to be both unprecedented and impressive.

Between fall 1993 and spring 1994, Golombek and the rest of the landing site selection group identified three broad areas on Mars where these engineering safety constraints could be met: Chryse (where Viking Lander 1 had set down), Amazonis, and Isidis-Elysium. During this time they also solicited the input of the global Mars science community via a landing site selection workshop held in Houston.[8] Dozens of potential landing sites were pitched by scientists from around the world during the workshop, with several more being added by the JPL-led site selection team itself. Overall, 40 different potential landing ellipses were put forth for consideration by the Pathfinder team; of those, about 10 ended up making the cut in terms of the final engineering and landing safety constraints.[9]

Scientifically, the Pathfinder team as well as the wider Mars community had agreed to classify the scientific potential of the sites under consideration according to their likely rock types: (1) "grab bag" sites, located near the mouths of large outflow channels, where a wide variety of individual rock types might be available for study; (2) large, uniform, relatively dust-free regions of unknown or uncertain rock types, where potentially new and different kinds of compositions could be identified compared to the Viking Lander sites; and (3) large, uniform, relatively dust-free regions of known or suspected rock types (like basaltic lava flows), where the details of presumably "typical" surface materials could be confirmed in ways that had not been possible for Viking. In the end, the Pathfinder site selection team felt that grab bag sites would have the highest scientific potential, enabling measurements of both ancient and more modern Mars materials all within the relatively small auditorium-sized area around which they expected to be able to maneuver the rover.

At a key meeting in June 1994, the science team voted to advocate for one particular ellipse that Golombek had championed above the rest as the most highly desired landing site. That ellipse was located in a region called Ares Vallis (near 19.5°N, 32.8°W), in what looked like a fan-shaped deposit of sediments near the mouth of a large catastrophic outflow channel. The site was predicted to have relatively smooth slopes based on imaging at Viking Orbiter resolution (under 50 meters per pixel), but models of thermal data from the orbiters suggested that there would be a fairly high rock abundance, with perhaps 18–25 percent of the surface covered by small rocks and less than 1 percent covered by boulders larger than about 50 centimeter tall, which would represent the most significant obstacles to the lander.[10] The region around the landing site also exhibited streamlined islands indicating that significant amounts of fluid (presumably water) had violently flowed through this region much earlier in Martian history—perhaps, some geologists believed, during cataclysmic flooding events similar to those that had formed the Channeled Scablands of eastern Washington—and likely contained both ancient crustal materials and more recent volcanic materials transported downstream by those flows.[11]

The Ares Vallis site had to undergo additional scrutiny and site certification, including Earth-based radar studies like those that had been used to certify the Viking landing sites. Other sites selected as potential backups, as well as the two Viking Lander sites themselves, were also studied in detail for comparison to Ares Vallis. Ultimately, the Ares site was found to be both safe and potentially the most scientifically compelling, and the Pathfinder team recommended it as the landing site in November 1995. In March 1996, NASA Headquarters—the arbiter of such profound decisions for NASA-led missions—concurred with the Pathfinder team and officially designated Ares Vallis as the intended Pathfinder landing site. The die was cast, the gauntlet set: Pathfinder would attempt to land in the mouth of an ancient catastrophic outflow channel and thus become the first mission to land on Mars and directly measure the composition of ancient rocks. As the mission got off to a picture-perfect Delta II rocket launch from Cape Canaveral on December 4, 1996, spirits were high, but no one was really sure whether such a low-cost, high-risk gamble would pay off.

Entry, Descent, and Landing

Though one might guess as much from the rather dismal record of unsuccessful missions to Mars, it is not quite intuitively obvious why Mars is one of the most difficult places on which to land in the entire solar system. It doesn't have as much gravitational pull as Venus or Earth, so an incoming spacecraft doesn't approach with quite as much velocity. However, both of those planets have relatively thick atmospheres that allow heat shields and parachutes to help significantly slow a spacecraft down as it heads toward the surface. The Moon and small asteroids have less gravity than Mars, but no atmospheres at all. Mars's atmosphere, though thin, is not negligible, and it causes significant frictional heating as a spacecraft comes screaming in—meteorlike—at supersonic velocities. The result is that heat shields and parachutes alone can't provide enough of a slowdown to do the job, so it is necessary to go to a significant amount of extra trouble (and expend extra energy) in order to get a spacecraft to land on Mars gracefully. Less-than-graceful landings (known informally in the business as "lithobraking") had been performed previously, including by the half-dozen Soviet attempts to land on the Red Planet.

For the Viking Landers, that additional energy came from the powered descent thrust of substantial retrorocket engines. Viking was the first to demonstrate that the combined parachute and retrorocket approach could work on Mars, but those landers were large and heavy (and expensive) spacecraft compared to the slimmed-down, higher-tech Mars Pathfinder lander and its rover, so going with a Viking-style landing system would be engineering overkill. It also wouldn't be consistent with the "cheaper" part of "better, faster, cheaper."

The team had to find another way to land a smaller spacecraft on Mars for less than 10 percent the cost of the Vikings.

By the mid-1990s it had become commonplace for new cars to be outfitted with airbags, at least for the driver and front passenger. After the sacrifice of countless numbers of crash test dummies, there could no longer be any doubt that airbags saved lives in many kinds of automobile collisions, especially the kinds of head-on collisions at typical highway speeds that were causing so much tragic loss of life. Engineers at NASA's Ames Research Center and at JPL, including Pathfinder chief engineer Rob Manning—who also led the mission's entry, descent, and landing (EDL) team—must have had this fact in the back of their minds when they dreamed up, tested, and modified the Rube Goldberg–like technique that would be used to try to land Pathfinder safely on Mars (figure 15.2). Fortunately, NASA (and, earlier, the Soviets) had learned much about the use of heat shields and supersonic parachutes to slow down space vehicles attempting to land on Mars. And fortunately, the mass of the Mars Pathfinder entry capsule (which housed the lander, rover, parachute, and other EDL systems) was only a little more than half the mass of the Viking Lander entry capsule. This meant that even though Pathfinder's parachute had to be smaller (about 12.5 meters in diameter compared to Viking's 16-meter chute) and the heat shield was also correspondingly smaller, they could still exert about the same level of braking force. Still, the heat shield and parachute wouldn't be enough to bring even the lighter Pathfinder lander to a gentle stop on Mars; just as on Viking, an additional thrust from retrorockets would be needed to slow down even more. So the team designed three solid rocket boosters (like tiny versions of the Space Shuttle solid rocket boosters) that would add about a ton of braking force during their brief two-second ignition, and a Kevlar tether system that would lower the lander away from the backshell that housed the retrorockets. The rockets fired about 100 meters above the surface, bringing the entire tethered lander to a hover only about 10 meters above the ground (figure 15.2).[12] All of this happened fully autonomously, because ground controllers on Earth were simply too far away to "joystick" a process that took the lander from space to the ground in just a few minutes. An onboard radar system was constantly pinging the surface to tell the computer the vehicle's height off the ground, but because of the uncertainties in those height estimates and the time it takes for the computer

FIGURE 15.2 *Top*: Engineers inspect the ability of the Mars Pathfinder mission's five-meter-tall airbag design to withstand possible punctures from rocks in the "Mars Yard" at JPL. *Left*: Artist's concept of the airbags being dropped from the mission's backshell/retrorocket descent system. *Right*: Artist's concept of the Mars Pathfinder lander and the laser-printer-sized Sojourner rover on Mars. NASA/JPL.

to process the results, the engineers didn't dare try to fire the retrorockets any closer to the surface.

So now, in theory, the whole folded-up lander is just hovering there, a few stories above the surface. It still needs to actually land! Cutting the tether and allowing the entire tetrahedrally-shaped structure to just drop from a height of about 10 meters would allow it to hit the ground at speeds of up to 100 kilometers per hour, depending on how much horizontal velocity remained after the retrorockets removed all the vertical velocity from the system. A crash at this speed would kill the mission. This is where the idea of airbags came in . . . a system that can deploy rapidly in a car, for example, and protect extremely sensitive cargo (us) during high-speed collisions. So the NASA engineers devised a system to rapidly (in just one-third of a second) inflate a cluster of a few dozen spherical airbags surrounding the lander while it was suspended on its tether, but before the retrorockets fired. Once the rockets did their job, then the computer would command the tether to be cut, and the airbag-encased lander would just fall to the ground.

And bounce. And bounce, and bounce, and bounce. . . . Some simulations predicted that the airbags—which were not vented and designed to deflate like car airbags—could initially bounce most of the way back up to the height from which they were dropped, and could bounce dozens of more times, with hundreds of meters of horizontal travel between the early bounces, until they eventually came to rest. It seemed like a pretty crazy way to land hundreds of millions of dollars of sensitive electronic and optical equipment, but it was also elegant in that the team found a defensible way to demonstrate EDL on a shoestring budget compared to past missions (scrimping not just on cost, but also on mass and volume).

The big day was set (on purpose, to maximize public interest) to be July 4, 1997, a festive Friday on a holiday weekend. Landing would happen around 10:00 a.m. local time at JPL in Pasadena, and around 3:00 a.m. local time in Ares Vallis on Mars. After powering the encapsulated lander and rover "entry vehicle" for seven months, the cruise stage with its solar panels was jettisoned about 30 minutes before landing, at a distance of about 8,500 kilometers from the surface. The entry vehicle, including its Viking-derived protective heat shield, continued traveling at more than 7.5 kilometers per second (more than 16,700 miles per hour) toward Mars, and it finally started to "feel" the planet's thin upper atmosphere about 125 kilometers above the surface, only around four to five minutes before landing. During the next few minutes the heat shield absorbed a huge amount of thermal energy (estimated at more than 100 megawatts),[13] as atmospheric friction slowed the entry vehicle down to only around 400 meters per second (900 miles per hour). During parts of this descent, the deceleration forces on the vehicle exceeded 20 times Earth gravity. Just two minutes before landing the parachute was deployed, followed by the separation of the heat shield about 114 seconds before landing, then the lowering of the lander/rover package down the tether starting about 20 seconds later. About half a minute before landing, the radar started detecting the surface, and by just around 10 seconds before landing, when the radar indicated a height of about 350 meters, the computer commanded the airbags to inflate. Four seconds later, just 100 meters from the ground, the retrorockets fired, and then just 3.8 seconds before landing the tether was cut, letting the airbags free-fall to the surface from a height of about 21.5 meters.

All these events, including the complex, orchestrated firing of 41 different pyrotechnic devices (fitting for the Fourth of July) had of course to happen automatically by onboard computer control, given the 10 minutes and 40 seconds it took for radio waves to travel the more than 190 million kilometers from Mars at the time of landing. However, a clever series of communication semaphores or "beeps" at various frequencies let JPL ground controllers—and an anxiously waiting world watching live on CNN or over a relatively new communication medium called "the Internet"—get some basic confirmation that all these steps were happening properly.[14] Or at least, had hopefully happened properly. . . . As the beeps came in on schedule, step by step, mission scientists, engineers, and managers noted them nervously but professionally on various computer screens. The final beep was a long time in coming, however. Postlanding telemetry reconstruction revealed that the airbags bounced at least 15 times, up to 12 meters high, without rupturing.[15] During that harrowing period of bouncing and rolling, however, the signal from the internal radio transmitter couldn't be detected, and so the team had to wait. After a minute or so that seemed like a century, Rob Manning, the main console narrator for the landing events for the world, made an announcement: "EDL comm reports that a signal is barely visible. . . . That's a very good sign everybody!"[16]

The room, and space fans watching from around the world, erupted in raucous cheers of delight.

It would take another four predawn hours on Mars for the airbags to deflate and be retracted, and for the tetrahedral lander petals to unfold, revealing the solar panels that would power both the lander and the rover during the mission. As the Sun rose on sol 1, Mars Pathfinder switched from its survival batteries to solar power to begin its historic mission.

Mission Success

The day after landing, NASA administrator Dan Goldin announced that the Mars Pathfinder lander would be officially named the Carl Sagan Memorial Station. The noted astronomer and planetary scientist, educator, Mariner 9 and Viking Lander team member, science popularizer, and philosopher had passed away shortly after Pathfinder had launched in December 1996. Aptly, Goldin made the announcement in front of a cheering crowd of space enthusiasts attending Planetfest '97 in Pasadena, an event organized by The Planetary Society, the public space interest group that Sagan had founded with Bruce C. Murray and Lou Friedman back in 1980 in order to promote the kinds of continuing missions of robotic exploration exemplified by Pathfinder.

The Pathfinder team had demonstrated that a lower-cost, fundamentally different (and thus higher-risk) landing system could effectively deliver a modest-sized lander and rover to the surface of Mars. Much of that technology and know-how would directly transfer to future Mars surface missions. Once on the surface, however, could the team also demonstrate that it was possible to do important new Mars science with a similarly lower-cost set of instruments?

Organizing that effort fell to Matt Golombek. He had been involved with the Viking Orbiter missions early in his career, and so was familiar with some of the stresses of Mars spacecraft operations and the need to sometimes generate "instant science" for media and public consumption. Ebullient and outgoing, Golombek quickly became the well-known scientific voice of Pathfinder during the mission's frequent media briefings. His enthusiasm proved infectious to the rest of the team, and he worked hard to not only closely integrate the scientists with the engineers, but also to involve early-career people in key roles in press events and in the preparation of peer-reviewed team publications.

Another key Pathfinder scientist was the leader of the lander's camera team, Peter Smith. Based at the University of Arizona, Smith was among the earliest of NASA's major mission principal investigators to design, build, and test a camera system at a university rather than at a traditional NASA center or aerospace company. As such, Smith's team, like the JPL team, was a mix of students, early-career researchers, and a few seasoned veterans who helped guide the way. Smith, whose own character was often quite mischievous in the many press briefings in which he participated during the mission, had (as mentioned earlier) called his camera the Imager for Mars Pathfinder, or IMP for short, which those who knew him regarded as eminently fitting. The IMP team was immediately put to the test shortly after landing, when the camera was commanded to take a small mosaic known as the Mission Success panorama (figure 15.3), intended to capture the still-folded-up rover on the lander

FIGURE 15.3 Mars Pathfinder's Mission Success panorama, which graced the front pages of many newspapers shortly after the successful July 4, 1997, landing. The Rock Garden is in the foreground, and the Twin Peaks can be seen on the horizon. NASA/ JPL/Doug Ellison.

in the foreground, as well as a swath of whatever features Ares Vallis had in store looking out toward the horizon.

Fittingly, the Mission Success panorama became an instant success, appearing above the fold in huge numbers of newspapers on July 5, and garnering then-record numbers of "hits" on NASA's Mars Pathfinder website. Smith waxed poetic at the NASA press conference showcasing the first views from Pathfinder's new home: "The eyes of the camera are our eyes, and in that sense, we are all on Mars. We are there together. You might say that the people of Earth are the soul of this robot. So for the first presentation of images, forget about the scientific and engineering aspects. They're very important, but open your imagination to the experience and beauty of the landscape of Mars."[17]

Not only could the world clearly see that the rover had been safely delivered to Mars and was soon to deploy to the surface, but the scene beyond the lander was mesmerizing, including a field of dark boulders and a beautifully composed pair of hills on the horizon dubbed "Twin Peaks," partly after an American mystery horror drama television series that had appeared for two seasons in the early 1990s and won something of a cult following. It was a landscape both familiar but also eerily alien, especially considering that the temperatures were like those in Antarctica and the thin oxygenless air was unbreathable. Indeed, the photo became such an icon of space exploration success that it was later turned into a $3 U.S. Postal Service stamp to commemorate the mission.

Meanwhile, the rover, named Sojourner, was waiting to be deployed. The project manager for the rover was Donna Shirley, a native of Oklahoma who became interested in building aircraft at age 10, became fascinated with Mars and space exploration at 12 when she read Arthur C. Clarke's 1951 classic *The Sands of Mars*, abandoned high school home economics class in favor of mechanical drawing, and obtained her pilot's license at 16. During the summer of Pathfinder, she recalled a passage from the book that had inspired her long-ago interest, in which an astronaut, Gibson, has landed on Mars and looks out across a landscape—including crimson hills—not unlike that across which Pathfinder was looking now:

He was on Mars. He had reached what to ancient man had been a moving red light among the stars, what to the men of only a century ago had been a mysterious and utterly unattainable world—and what was now the frontier of the human race. . . . But he knew he was still a stranger: he had really seen less than a thousand millionth of the whole surface of Mars. Beyond . . . the crimson hills, over the edge of the . . . plain—all the rest of this world was a mystery.[18]

At the time Shirley joined JPL in 1966, she was the only woman among 2,000 engineers at JPL with an engineering degree. At JPL, she served as mission analyst and later program manager for the Mariner 10 mission to Venus and Mercury, then as a manager for the Mars Exploration Program, in which she worked on the heat shield for a vehicle to enter Mars's atmosphere, and finally on the Pathfinder rover.

Shirley wanted to get a rover to hitch a ride to Mars with Pathfinder, and was allocated a budget of $25 million to do so. She later recounted the troubled early days of the rover's development and in particular the skepticism with which the whole project was viewed by Mars Pathfinder project manager Tony Spear:

> No one at JPL gave our rover project much of a chance, least of all Tony. We had a budget of $25 million at a time when billion dollar projects were the norm. . . . Tony's assignment was to build a spacecraft that could land safely on Mars at a fraction of the cost of a regular mission. To Tony, initially, the rover was a parasite and my team and I were an annoyance.
>
> Tony believed it was hard enough to design and construct a mission to Mars for under $200 million when the Viking missions had cost over $3 billion in 1997 dollars. He didn't want the rover taking up space, consuming power, and requiring its own communications hardware partly because, like most everybody else at JPL, he didn't believe the rover would really work.[19]

Not only would the rover "work," it would also prove to be a public relations stroke of genius. Public interest in it had been seeded early when The Planetary Society was engaged by NASA to involve school students in an essay competition to choose a name for it. The winner was 12-year-old Valerie Ambroise from Connecticut, who suggested Sojourner in honor of the 19th-century African American abolitionist and women's rights activist Sojourner Truth. Valerie's essay deftly ended, "Sojourner will travel around Mars bringing back the truth."[20] Once on Mars, Sojourner demonstrated just how susceptible humans are to anthropomorphic projections.[21] It had all the charm of a favorite pet, proved to be a singularly charismatic television performer, and arguably came closer to superstardom than any other piece of space hardware up to its time.

As soon as Pathfinder was safely on Mars, Shirley and her team's attention quickly turned to the unfolding and deployment of the Sojourner rover. Like the lander itself, the rover was folded up during cruise and landing, latched to the inside of one of the lander's flower-like petals. The first step in the process was thus to command the rover to "stand up" to its full height of about 30 centimeters by firing some pyrotechnic cable cutters that freed the vehicle from its vertical restraints. A set of spring-loaded latches and pins that further restrained the rover and its stowed radio antenna were then deactivated by moving the rover a short distance backward on the petal. Finally, two semirigid carpetlike ramps were unstowed and rolled from the petal onto the surface in front of and behind the rover, to provide a pathway to egress across the deflated, bunched-up airbags.

With something as novel as the deployment of a rover onto the surface of Mars, there were bound to be a few hiccups. First, the lander and Sojourner had trouble communicating. The problem was diagnosed as being caused by a temperature difference between the modem in the rover and the modem in the lander. The other problem, though not unforeseen, was that one of the landing airbags had crumpled against the petal the rover was to slide down. This forced mission controllers to carry out an extra maneuver. Just before sunset on sol 1, they sent a command to Pathfinder to lift one of its petals 45°. With the petal tilted, the lander was reoriented, the airbag further retracted, and the petal re-extended. Although it took longer than expected to deploy the rover, this wasn't really a problem, because there really wasn't any deadline for getting the rover off the lander. Just before the fast-sinking Earth set below the horizon of Ares Vallis, moments ahead of the Sun, severing the communications link with ground control, Pathfinder captured an image showing that the maneuver had succeeded. Though part of the airbag still dropped over the petal, there was now room enough to allow the rover to roll down the ramp.

The radio link between the lander and rover was restored on sol 2. Now Sojourner backed down the ramp to begin its epic trek across the Ares Vallis floodplain.[22] A short movie made from time-lapse IMP photos showing the rover's egress delighted the world and put the final exclamation mark on mission success.[23] NASA had a rover on Mars!

The rover's maximum speed was only 46 meters per hour, about the speed of a tortoise's gait. Built-in gyroscopes equipped it with a primitive vestibular system to keep it from toppling over as it explored nearby boulders. Five laser beams helped it "feel" its way, in blindman's buff fashion, across the surface. Sojourner's progress could be traced in the thin tracks it left in the rust-red Martian soil, and was recorded by the cameras of the fabric-draped lander back at Sagan Memorial Station.

As Sojourner began its exploration, the skies overhead presented an intense, ever-changing panorama. Despite the thinness of the planet's atmosphere, it had been known, since the Viking Landers, that the Martian skies were surprisingly bright, like the skies over Los Angeles on a smoggy day, as the Viking scientists were wont to quip. The reason for the brightness, it was realized, was the presence of micron-sized dust particles, seeded by the planet's dust storms and extending to altitudes of up to 30 kilometers above the surface. By sol 16, predawn images from Ares Vallis were also showing thin wisps of water-ice cirrus clouds in violet skies. The clouds, 16 kilometers high, floated across the landscape from the northeast, wafted along by winds at 24 kilometers per hour. Another intriguing image showed tiny Deimos suspended in the sky, the first time a Martian moon had ever been captured from the surface of the planet.

The Martian sunrises were grand Homeric stirrings: glorious bursts of color and light in a white sky tinged with the faintest hint of blue. They were lilac rather than (as the sunrises of the Aegean described by Homer) rosy-fingered affairs. The Sun, as seen from Mars, was noticeably smaller than as seen from Earth, which was only to be expected given the planet's one-and-one-half-times greater distance from our shared star. As the Sun climbed higher above the horizon during the day, the wispy cirrus burned off. By noon the clouds were gone without a trace, and the dusty skies had assumed a murky brownish-orange—strange skies on a strange world.

Sunsets, too, were events of breathtaking beauty as darkness and repose settled once more on the lonely plains of Ares Vallis. The feeble glow of twilight faded to dark and muddy brown; the brown to violet; the violet to eerie blue. (The blue color, by the way, was produced by the same process of forward-scattering of blue light by fine suspended dust particles that can, rarely, cause a bluish-tinged Moon or Sun on Earth.) Then followed the chilly Martian night, in which (from Pathfinder's location in the northern hemisphere of Mars) the stars would have been seen to wheel not around Polaris as do the stars of Earth but around a point lying about halfway between the bright star Deneb in Cygnus and the fourth-magnitude star Erakis in the constellation Cepheus, since this is the point in the sky toward which the north pole of the Martian axis points. The night sky view would be dominated by the potato-shaped and fast-flying moon Phobos, about a third as large as the Moon seen from Earth, racing across the sky, rising in the west and setting in the east twice during a single night and exhibiting twice every night what Isaac Asimov imagined would be an "interplay of light and shadow . . . a fascinating display of kaleidoscopic changes that will never exhaust the fancy."[24] Meanwhile, Deimos, much farther away and about a 20th the brightness of Phobos, would lumber imperceptibly across the sky in the opposite direction.

Most interesting to us, however, might be the sight of the bright blue-tinted Earth and its nearby companion the Moon, running now ahead and now behind the Sun and joining Mercury and Venus in the ever-shifting lineup of Mars's Morning and Evening Stars. With the naked eye it would be splendid, shining as bright as magnitude −2.5, about as bright as Mars or Jupiter at their best as seen from Earth (though, oddly enough, Earth would still be outshone by Venus and its brilliantly reflecting clouds). Through a small telescope, it would show a gamut of phases. Pathfinder's cameras weren't able to observe Earth in the brilliant predawn twilight; it would be another seven years until the Spirit rover (chapter 17) would finally show us a view from Mars of Tennyson's "fairest of the evening stars."

83 Sols in the Sun

The playbook for the Mars Pathfinder mission was to demonstrate a successful landing, to operate the lander and its instruments for at least 30 sols, to deploy the rover successfully to the surface, and to operate the rover for at least two weeks and drive it across a few tens of meters to study different targets around the lander. In all those respects, Pathfinder was phenomenally successful. Indeed, the mission significantly surpassed all those operational expectations while making important new science discoveries and providing NASA—thanks in large part to the winning personality of Sojourner—with arguably its biggest public relations success in decades. Pathfinder was, at the time, the largest Internet event in history, and it garnered front page headlines for a full week. No other planetary mission had that kind of attention before, and in terms of attracting the largest fraction of the currently available global media audience, it is arguably still the most popular planetary mission ever.

The IMP team's imaging priorities early in the mission focused on collecting black-and-white stereo panoramas around the lander so that a digital terrain model (a 3-D representation of the topography of the surface) could be generated to help the rover drivers avoid obstacles and other hazards (figure 15.4). Then the team focused on building up color coverage of the landing site to help determine some details about the mineralogy and degree of dust cover of the local rocks and soils. Especially important was to try to quickly identify relatively clean and interesting-looking rocks for the rover to drive up to and deploy its elemental chemistry instrument (the APXS) onto. One of

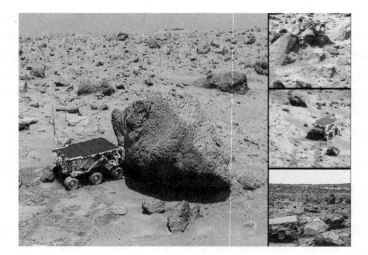

FIGURE 15.4 Mars Pathfinder IMP photos of the Sojourner rover driving and taking measurements at the landing site: measuring the chemistry of the boulder called "Yogi" (*left*), perching precariously on the rock called "Wedge" (*top right*), driving over "Mermaid Dune" (*middle right*), and measuring the chemistry of the rock called "Moe" (*bottom right*). NASA/JPL/University of Arizona.

the unique contributions to Mars science that the mission had the potential to make was measuring the chemistry of rocks on Mars. The Viking Landers had done a great job of measuring the chemistry of loose soils and dust, but it was not possible to get rocks into the Viking elemental chemistry instrument. How are the rocks and soils related on Mars? Is there a diversity of rock types? Are they comparable to some kinds of Earth rocks? The ability of the rover to drive to and deploy the APXS on various rocks around the lander would be key to answering such questions.

"Rover driver" was a new job description for a very specialized activity at JPL. Unlike the earlier Soviet Lunokhod rovers, which were relatively close to home on the Moon, Sojourner was too far away to be real-time "joysticked" by remote control drivers because the one-way light-time delay was more than 10 minutes. Instead, a digital model of the topography around the rover was built up inside a computer at JPL from stereo images taken by the IMP on the lander, and then the rover drivers would "drive" a computer-simulated version of Sojourner inside the computerized terrain. When they were convinced that they could find a safe and conservative path in the computerized terrain to get the rover from desired point A to point B, they would then package up the driving commands used in the simulation, and send them to the real rover on Mars to perform on the next sol. It was like playing one of the most expensive slow-motion video games in history, with the goal to keep the rover on relatively safe terrain within the line of sight of the lander. It was all enabled by significant advances that NASA had made in microcomputer hardware miniaturization and programming technology, and especially by new "smart" software that enabled a limited amount of onboard decision-making by the rover itself. In many ways, Sojourner was the first artificial intelligence on Mars.

The IMP team would use the knowledge of where Sojourner was expected (commanded) to be at the completion of that sol's planned drive and blindly point the camera to that location to take an "end of sol" stereo image of where the rover was supposed to have ended up. On most sols, the rover went where it was commanded to go and thus appeared in the end of sol IMP images. On some sols, however, there was no rover in the end of sol IMP image (because some unforeseen or underestimated obstacle had caused the rover to veer off its intended course), and so the next sol would sometimes have to spent playing a sort of "find the rover" hunt using the camera. Despite such occasional hiccups, the use of high-fidelity stereo images to guide a rover through virtual and then real geological terrain was generally highly successful on Pathfinder, and the methodology would ultimately be adopted in very similar ways for future NASA rover missions.

The geological interpretation of the landing site gleaned from the IMP panoramas was consistent with the Viking Orbiter–based hypothesis that Ares Vallis had been carved by a large catastrophic outpouring of water early in Martian history.[25] In particular, the landing site itself contains piles of tabular and partially rounded boulders (the nearest pile was dubbed the Rock Garden by the team), some of which are perched atop one another, some of which appear partially cemented together (possibly a kind of typical streambed rock called a conglomerate), and others of which are tilted in the direction that the water flowed (a characteristic like tipped-over dominoes known as "imbrication" to geologists). Based on numerous terrestrial analogies, all these observations are consistent with the area being part of the depositional plain of a catastrophic flood. Even the Twin Peaks, interpreted to be part of a streamlined island that was formed in the flooding event, seemed to fit the model. The rims of several nearby impact craters were seen in both the panoramas and in high-resolution Viking Orbiter images of the area; ejecta from these craters was tagged as the likely source of many of the small, dark, angular rocks seen around the lander. In addition to flooding and impacts, the continuing erosive and depositional action of the wind is also obvious at the site, as the team noted numerous rock surfaces that appeared to be scoured and grooved by sand, fine-grained wind tails behind many rocks (similar to the much larger impact craters with windblown tails first noted in Mariner 9 images), and numerous "drifts" of materials with similar color properties as atmospheric dust. Several small, darker sand dunes were also studied, including one behind the Rock Garden that the rover was commanded to drive right through.

Sojourner's APXS measurements of six bright soil and dust deposits at the Pathfinder landing site showed a composition very similar to the chemistry of the soils at both Viking Lander sites, further cementing the idea that bright, reddish, fine-grained materials have been globally distributed and homogenized across the planet by billions of years of dust storms and blowing sand.[26] Magnets on the Viking Landers had discovered that the bright soils and dust on Mars are also magnetic, a result used to refine a new magnetic properties experiment for Mars Pathfinder that would involve a series of magnets that included some that were much less strong than those used on Viking. Over time during the Pathfinder mission, the atmospheric dust settled on and stuck to several of the magnets of different strengths mounted on the lander, showing that almost all the dust grains are not pure magnetic minerals but are instead composites of silicates mixed with just a small percentage of strongly magnetic minerals, most likely one or more iron minerals like maghemite (γ-Fe_2O_3), a kind of iron oxide that forms from the aqueous (water-related) weathering of other magnetic minerals like magnetite that occur in typical terrestrial volcanic rocks.[27] The possibility that at least the magnetic part of the globally distributed bright soils and dust might have originated from chemical weathering of rocks by water was an exciting hypothesis that would be further tested by follow-on magnetic properties experiments on the later Spirit and Opportunity rovers and the Phoenix lander.

The rover's measurement of eight different rocks near the lander provided even more interesting new results (figure 15.5). As predicted by previous telescopic measurements, by some of the Viking Lander results, and by rover images showing the pitted, frothy, or vesiculated nature of most of these rocks, the composition of the rocks is generally similar to basalt, the kind of low-silica volcanic rocks typical of volcanic flows coming from Earth's upper mantle. However, all the rocks had generally higher silicon than basalts on Earth, and some rocks (including the very first rock measured by Sojourner, on sol 3, called "Barnacle Bill") showed a higher silicon abundance that is more like a type of terrestrial volcanic rock called andesite, rather than basalt.[28] On Earth, andesitic rocks are usually a clue that the crust where those rocks are found has been differentiated (melted and recrystallized, perhaps many times, a process that segregates and concentrates heavier iron- and magnesium-bearing rocks from lighter silicon-bearing rocks), and the source of that processing is usually plate tectonics. There's no direct geological evidence for plate tectonics on Mars, but the Sojourner rock chemistry results provided the first evidence that perhaps the Martian crust had a more complex evolution than previously thought, potentially involving reprocessing of melt rocks in underground

FIGURE 15.5 Photos taken from the Mars Pathfinder rover Sojourner. *Top row, left to right*: View of the Pathfinder lander; sandblasted surface of the rock "Moe"; wheel-dug trench near the "Mermaid" dune. *Bottom row*: View approaching the rock "Yogi"; pitted surface of the rock "Chimp"; cobbles, pebbles, granules, and wind tails in "Cabbage Patch." NASA/JPL/University of Arizona.

magma chambers. Also interestingly, there is apparently no simple, direct way to form the typical bright soil composition seen at the landing site (and across Mars) from weathering of the local rocks, so some aspects of the chemistry of the Pathfinder rocks might represent specific local, rather than global, conditions and processes.

Since the lander carried a meteorologic station, the mission also provided the first new set of daily pressure, temperature, and wind measurements from the surface of Mars since the end of the Viking Lander 1 mission in 1982, helping refine models of the seasonal and longer-term behavior of the current Martian climate. The weather reports from the lander's ASI/MET station revealed a dramatic temperature difference between ground level and the meteorology mast, about 1.5 meters above the surface. Thus, at the hottest point in the daytime, a person standing at Ares Vallis would have felt the ground beneath her feet at a comfortable 18°C (65°F), but the top of her head would have been a decidedly uncomfortable 0°C (15°F). This was hardly surprising, and is similar (if much less extreme, because of the planet's much thinner atmosphere) to what one experiences walking on the sidewalk or a beach, barefoot, on a hot summer day on Earth. The air is hot, but the ground is much hotter, and one could easily burn one's feet. At night, the situation was reversed: the surface temperature dropped to −90°C (−130°F), though the air above, having a greater heat capacity than the ground, only dropped to −19°C (−2°F). Of course, the temperatures changed as the mission went on, and one of the popular highlights of the whole Pathfinder mission was to report each day the latest weather reports coming in from Mars. Even the IMP team got in on the atmospheric science game, taking images of the Sun through a special filter that allowed the dustiness of the atmosphere to be assessed over time, as well as taking time-lapse photos of numerous dust devils passing near the lander during the mission.

Final Resting Place

With an endless supply of sunlight to power the solar panels, the mission faced as its most tangible resource limitation the lander's and rover's lithium-based rechargeable batteries. Mission planners knew from the Viking Lander experience that nighttime temperatures would be extremely low, and so battery-powered heaters had to work hard every night to keep the internal temperatures of the vehicles above about −40°C (−40°F), lest the sensitive electronics components break in the extreme cold. Spaceflight-qualified batteries can only go through so many repeated charge-discharge cycles, however, before they lose their ability to hold adequate charge (and early 1990s-era batteries had much worse performance than today's).

After the lander had lasted for nearly three times its expected lifetime, radio contact started to become intermittent by late September 1997. One hypothesis is that after repeated charging and discharging, the batteries started becoming unable to fully recharge. Vital communications equipment inside the lander began to intermittently fail, and despite repeated attempts by JPL and NASA's Deep Space Network operators for many weeks, no new signals from the vehicle were received after October 7. The rover was still working fine when last heard from, but because of its very low power transmitter it could only communicate with the lander, not with Earth. Interestingly, the rover's onboard software was designed to react if it didn't hear from the lander within a certain period of time (assuming, for example, that its radio line of sight to the lander might be getting blocked by a rock) by driving around the vicinity of the lander and trying again. Thus, once the lander batteries had died, if the rover was still operational it could very well have continued to move around, continually trying to reestablish contact with the lander, sol after sol for who knows how long (one gets a kind of sad mental image of a lonely bear cub circling her dead mother, plaintively crying out for help . . .). Someday, then, when people visit the Mars Pathfinder landing site, we might find out how the rover's mission really ended.

16

Mineral Mappers

When I go out of the house for a walk, uncertain as yet whither I will bend my steps, and submit myself to my instinct to decide for me, I find, strange and whimsical as it may seem, that I finally and inevitably settle southwest, toward some particular wood or meadow or deserted pasture or hill in that direction. My needle is slow to settle—varies a few degrees and does not always point due southwest, it is true, and it has good authority for this variation, but it always settles between west and south-southwest. The future lies that way to me, and the earth seems more unexhausted and richer on that side.

—Henry David Thoreau, "Walking," Concord Lyceum lecture, 1851

New Martian Frontiers

Mars is the new frontier, akin to the West for explorers of North America in the 19th century, to Greenland and Labrador for Norse sailors a thousand years ago, to the new and impossibly distant tropical islands for navigators of the Polynesian diaspora two thousand years before that. As noted in chapter 10, there are good reasons for associating the frontier (the West that is Mars) with Arizona in particular, whose landscapes—so close to the observatory on "Mars Hill" that Percival Lowell founded especially for the study of the Red Planet—arguably evoke those of Mars more than do any others on Earth. Lowell saw in the colors of the Painted Desert as seen from the San Francisco Peaks analogs to those that enchanted him in the eyepiece of his telescope. As more was learned about Mars, the analogy became more apt, rather than less, so that when Eugene M. Shoemaker decided to move the astrogeology branch of the U.S. Geological Survey (USGS) from Menlo Park, California, to Flagstaff, he did so specifically in order to be able to facilitate doing field work on terrestrial geological features that presented analogs to extraterrestrial ones.

Though it wasn't planned that way, the long hiatus between the first generation of Mars spacecraft (the flyby Mariners, Mariner 9, and the Vikings) and the next, which included some devastating disappointments but also the triumphs of Pathfinder and Mars Global Surveyor, meant that many scientists and engineers involved in the earlier missions had retired or, in a few cases— perhaps most notably Tim Mutch and Carl Sagan—passed away. In the interval between these two generations, a tremendous leap forward in technology, including the development of charge-coupled devices (CCDs) for imaging and miniaturized electronics and computers, meant that it was possible to pack much more varied and sophisticated instrumentation for remote sensing and imaging Mars than had been possible before.

This achievement was largely made on the engineering side. However, on the science side, there were similar advances. The early missions, especially Mariner 9, had been rather like Lewis and Clark's small expedition to "explore the course of the Missouri River," which as proposed in 1804 by President Thomas Jefferson, "in reality . . . would have no other view than the advancement of the geography."[1] In the end, that mission was to go far beyond what Jefferson at first envisaged for it, and reveal for the first time the vastness and scope of the American West. Mariner 9 was the Lewis and Clark mission to Mars, and it showed the great Tharsis shield volcanoes, the canyons of Valles Marineris, and the dry river valleys. By contrast, the succession of spacecraft that began to set out for Mars 20 years after the two Vikings landed on the planet were comparable to the subsequent waves of surveyors, prospectors, botanists, geologists, corporate scouts, and others who set out to count, map, classify, and assess the topography, flora, fauna, history, and, yes, exploitable resource potential of the new American frontier.

In contrast to the mere inventory-taking and rough impressions of a first reconnaissance, this later phase of Mars exploration would involve more mature methods (including geophysics, geodesy, geology, hydrology, meteorology, stratigraphy, mineralogy, soil science, and, should it prove to be relevant, biology), with a different end in view: attaining a level of knowledge approaching that of our own Earth.

Clearly, the nature of the science that would be done during the new more intensive and thorough phase of exploration at which we have arrived was bound to be qualitatively different from that of the previous ones. Up to this point, the story of Mars exploration has been the story of a relatively few individuals in the earlier telescopic era, heroic figures like Schiaparelli and Lowell and Barnard and Antoniadi, and also a relatively few in the early space-craft era, like Leighton and Murray and Sharp. Their careers took place when much of what was previously thought to be known about Mars was wrong and the rest uncharted, when the blank areas on our maps might well have been labeled with "here be dragons!" In the early spacecraft era—and especially with Mariner 9 and the Vikings—most of the blank areas were at least roughly filled in, and later Mars scientists (and would-be scientists) were in the position of latecomers hoping there might still be some blank areas left to chart. Indeed, the Mars scientists who interpreted the results from the flyby Mariners, Mariner 9, and the Vikings were fortunate in being the ones who filled in the blank areas on Mars and enjoyed the satisfaction of primogeniture in exploration that the ancient Latin poet Horace described in one of his Epistles (1.19.21–22): "I was the first to plant free footsteps on a virgin soil, / I walked not where others trod."

The 20-year gap between Viking and later close-up Mars missions meant that when Mars work "geared up again in the 1990s," some of the tracks had been lost, and first-generation work, even fundamental work, had to be "rediscovered," and in some cases was published as new.[2] Some of this was inevitable given the way that information storage had changed in the interval (with journals, for instance, largely going from paper to online). But it also reflected, at a more fundamental level, the dramatic way in which the entire enterprise of Mars research was being transformed, as the science became more and more multidisciplinary, and as the volume of literature dealing with this science increased exponentially. The proliferation of new instruments, the sheer duration of the missions, the overlap of one mission with another, the enormous amount of data being received, all meant that interpretations could only be undertaken by an ever-vaster team of highly trained specialists. Mars scientists of the earlier generation were rather in the (enviable) position of bees ranging happily from flower to flower gathering nectar and converting it to honey and wax. More recent ones have perhaps necessarily been more like spiders, confined within a shorter radius, indeed preferring a corner. It has become impossible for any individual to even remotely keep up with everything.

Already during the Viking era, this had been somewhat true. Thus, the passionate and energetic Cornell University Mars researcher Steve Squyres, who during the heyday of the Viking missions in 1977 was a geology student at Cornell, recalled:

> A few weeks into the semester I figured I'd better start thinking about what I was going to do for my term paper, so I asked for a key to the "Mars Room," where all of the new pictures from the Viking orbiters were being kept. I found the Mars Room in Clark Hall, behind the Space Sciences building. It was a deserted and disorderly place, more like a warehouse than a scientific data archive. A few of the pictures that had been taken during the earliest part of the mission were in glossy blue three-ring binders, arranged in chronological order on gray-painted steel shelves. Most, though, were on long rolls of photographic paper, stacked on the floor or still in their shipping cartons. My idea had been to spend fifteen or twenty minutes flipping through pictures, hoping to find inspiration for a term paper topic. Instead, I was in that room for four hours, racing through the pictures, stunned. I understood almost nothing that I saw, of course, that was the beauty of it. *Nobody* understood most of this stuff. . . . I walked out of that room knowing exactly what I wanted to do with the rest of my life.[3]

Squyres would, as the principal investigator behind the Mars Exploration Rover missions, and working with a team of thousands, eventually realize the dream he dreamed that day in the Mars Room in Clark Hall. He and they would homestead the Columbia Hills region of Gusev Crater with Spirit, and Endeavour Crater in Meridiani Terra with Opportunity. Many of his generation, and still more that he and they have trained since, would go on to rove among the flanks of Mount Sharp with Curiosity, to dig for ice near the north pole with Phoenix, to hunt for marsquakes with InSight, and to chronicle the details of the Red Planet's geology, meteorology, and climatology from a half-dozen more high-tech, long-lived orbiters helping to explore the new frontier.

We now raise the curtain on an era of Mars research in which individual spacecraft have transmitted more than 300 terabits of data back to Earth, and in which several missions, productive at that level, and continuing to be productive for years and even decades, operate simultaneously and—to consider the imaging capabilities alone—have achieved such a dramatic increase of resolution that our knowledge of many areas on Mars has indeed become comparable to that of similarly sized regions on Earth. The exponential increase of information means that researchers now have so much to wade through that they are apt to become more and more specialized. This means—in Mars science as anywhere else—that it is harder and harder to see the bigger picture. It has simply become too big.

Spectacular Success Followed by Two Steps Back

The Mars Pathfinder mission, with its dramatic airbag landing and its doughty little anthropomorphic rover Sojourner venturing out across the rock- and sand-strewn soil of Ares Vallis, injected tremendous public enthusiasm into the Mars program. At least as important, though perhaps less familiar to the wider public, was the successful development and launch of the Mars Global Surveyor (MGS) orbiter, which cemented and emboldened the concept of "better, faster, cheaper" within the NASA robotic mission culture. Indeed, the success of rapidly and relatively cheaply designing, launching, and eventually operating those missions provided impetus to an expanded NASA Mars exploration effort within a dedicated new program called Mars Surveyor. With Pathfinder and MGS setting the pace, the goal of Mars Surveyor was to launch a low-cost orbital and landed mission to Mars at every subsequent opportunity. The first new mission in that program, designed to take advantage of the late 1998/early 1999 launch window to Mars, was called Mars Surveyor '98. Led by the Jet Propulsion Laboratory (JPL), Mars Surveyor '98 would consist of an orbiter and a separately launched lander designed to study the planet's current climate, climate history, and abundance and behavior of volatiles (like water ice and CO_2 ice) in great detail. An open proposal competition in 1995 helped determine the payload—the scientific instruments and investigations—that the orbiter and lander would carry.

The Mars Surveyor '98 orbiter, subsequently renamed Mars Climate Orbiter (MCO), carried a competitively selected dual-camera imaging system called the Mars Color Imager (MARCI), led by Mike Malin and built by many of the same folks at Malin Space Science Systems Inc. (MSSS) in San Diego, who had built and operated the Mars Orbiter Camera (MOC) on MGS.[4] MARCI was designed to augment the capabilities of MOC by imaging the entire surface in the ultraviolet, visible, and shortwave near infrared through color filters spanning 15 different wavelengths. With multiple lenses providing both wide-angle global coverage and medium-angle imaging down to one kilometer per pixel, MARCI was designed to observe Martian atmospheric processes daily at the global scale, studying details of the interaction of the atmosphere with the surface at a variety of scales in both space and time, and examining surface features characteristic of the evolution of the Martian climate over time. In addition to the cameras, MCO also carried one of the two remaining Mars Observer instruments that had not flown on MGS, an atmospheric science instrument called the Pressure Modulated Infrared Radiometer, or PMIRR.[5] Built by JPL, PMIRR was designed to map atmospheric temperatures, pressures, water vapor, and dust abundances in 3-D from the surface to about 80 kilometers altitude. MCO's instrument suite was poised to make the orbiter similar to one of Earth's weather satellites, the most similar to be deployed around Mars. Importantly, MCO would also serve as a communications relay satellite for the Mars Surveyor '98 lander and subsequent surface missions.

It was a spectacularly promising and ambitious mission, but alas, as so often in the history of Mars exploration, success proved elusive. After a flawless launch on December 11, 1998, the mission seemed—from the outside—to be having an uneventful cruise to Mars. However, behind the scenes, trouble was brewing. MCO was unlike any deep space spacecraft that the mission navigators from JPL and Lockheed Martin (the spacecraft's manufacturer) had ever flown, mostly because it was "asymmetrical." That is, it had a big three-panel solar array deployed on just one side of the spacecraft, as opposed to the symmetrical solar array "wings" used on previous JPL missions (figure 16.1). Thus, as the spacecraft cruised through interplanetary space, torques from solar radiation pressure (the tiny amount of force exerted by photons coming from the Sun) tried to spin the spacecraft preferentially in one direction, rather than randomly averaging out. As a result, the spacecraft's attitude control system—a set of "momentum wheels" that could use their changing rate and direction of spin to help control the angular orientation—had to preferentially compensate for solar torque in a consistent opposite direction. Such wheels can only spin so fast, and when they reach their maximum spin rate ("saturation"), the spacecraft's thrusters are used to compensate for the

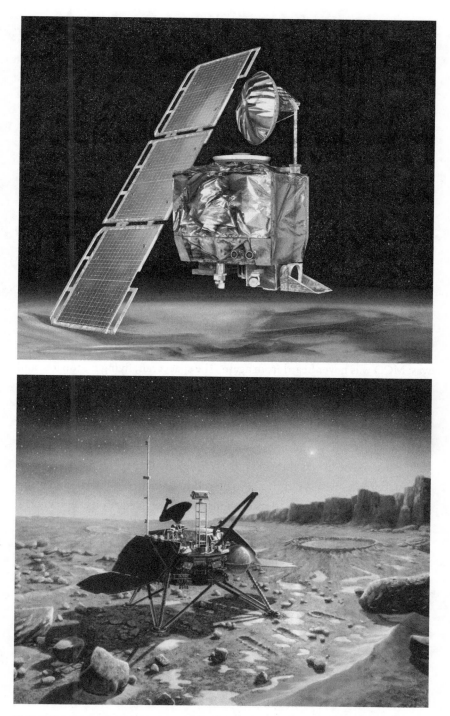

FIGURE 16.1 Artists' depictions of the Mars Climate Orbiter (*top*) and Mars Polar Lander. Both missions failed upon arrival at Mars in 1999. NASA/JPL.

counterspin that would otherwise occur when needing to slow down ("desaturate") the wheels. The need for so many of these angular momentum desaturation events (more than 10 times as often as was originally expected by the operations navigation team) resulted in small errors being introduced in the trajectory estimate over the course of the nine-month journey.[6]

By the time of the planned Mars insertion on September 23, 1999, when MCO was configured to fire its main engine to slow down and be captured into Mars orbit, the spacecraft trajectory was approximately 170 kilometers closer to Mars than originally planned. Right after the main engine was ignited, the planned trajectory was supposed to have taken the spacecraft behind Mars as viewed from Earth, and thus out of radio contact for most of the main orbital capture event. The 16-minute main engine burn began on time, but mission controllers at JPL in Pasadena and at Lockheed Martin in Denver were puzzled when the orbiter went out of radio contact behind Mars a full 49 seconds earlier than planned. That might not seem like much of an error, but in the space navigation business, 49 seconds might as well be an eternity. In fact, it was: MCO was never heard from again. The spacecraft, flying far too low and deep in the Martian air, either was destroyed in the atmosphere or reentered heliocentric space after leaving Mars's atmosphere. The mission was lost.

The subsequent investigation of the "mishap" by a blue-ribbon panel of engineers, managers, and mission navigators revealed that the root cause of the failure of MCO was human error.[7] While calculating the parameters for the final orbital tweaks ("trajectory correction maneuvers") needed to properly align the spacecraft for Mars orbit insertion, the navigation experts at JPL and Lockheed Martin were trading data files back and forth to establish the amount of thruster firing needed to achieve the new orbital parameters. However, the JPL team assumed that the numbers in the files were in metric units (kilograms and meters), while the Lockheed Martin team assumed that the numbers were in the more customary (for U.S. engineering companies) imperial units (pounds and feet). While the differences in the values were small, the effect on the position of the spacecraft near Mars was significant. Instead of hitting an aim point hundreds of kilometers above the surface of Mars, where the atmosphere is extremely thin, MCO ended up swinging behind Mars a little less than 60 kilometers above the surface, where the atmosphere is much thicker. At that altitude, and with the spacecraft traveling at such high speed, the demise of MCO must have produced a spectacular fireball in the Red Planet's night sky.

Rather than becoming the first Martian weather satellite, MCO had the unfortunate legacy of being mocked on late-night television and the Internet,[8] as well as being an object lesson (still widely taught in project management and systems engineering courses) in the potential implications of poor communications among teams of people working on complex projects. Why didn't mission navigators notice the trajectory errors earlier? In addition to the units conversion error, the subsequent investigation identified numerous additional "contributing causes" to the mission's failure, including a too-lean management structure; insufficient staffing, training, and testing; inattention to key documentation and telemetry details; and the inability of people working in the trenches to alert top project decision makers to their potential concerns (or the inability of those decision makers to listen).[9] While it is perhaps not such a bad thing to remind the world that even rocket scientists are still human, NASA's reputation suffered significant damage because of the failure of the mission. It was not the end of the bad news for Mars Surveyor '98, however: like the orbiter, the lander would suffer a catastrophic loss.

The scientific instruments for the Mars Surveyor '98 lander were competitively selected in 1997 from suites of instruments that had to be designed to work together to solve pressing questions across a wide range of Mars science disciplines. NASA called these "integrated payloads," and multiple teams of U.S. and international planetary scientists vied for their proposed integrated payload to be selected to fly. Among those that competed head-to-head as principal investigators were seasoned Viking and now MGS mission planetary scientist Ray Arvidson; Larry Soderblom, another venerable geologist and planetary science researcher at the USGS who had been heavily involved in Mariner, Viking, and Voyager (among other missions); and the less experienced but no less capable Steve Squyres from Cornell. Arvidson, Soderblom, and Squyres would all lose the Mars Surveyor '98 lander payload competition, but as a result of the process they would decide to pool their resources to become an unbeatable team competing for future Mars surface mission proposals.

The winning proposal for the Mars Surveyor '98 lander was an integrated imaging, chemistry, and meteorology package led by planetary geophysicist Dave Paige of University of California, Los Angeles, and called the

Mars Volatiles and Climate Surveyor (MVACS). Paige's winning strategy was designed to exploit a fairly rare opportunity during the 1998/99 launch window to send a lander to the high southern latitude polar regions for less energy (thus less launch mass and a smaller, lower-cost rocket) than during typical every-26-month launch window opportunities to Mars. He and the team targeted a landing site in Planum Australe, near 76°S latitude and on polar layered terrain not far from the planet's south polar cap (figure 16.1). From that location, the lander would have a front-row seat to witness the major changes in surface temperature, pressure, and ice cover (water and CO_2 ice) that characterize the seasonal cycles on Mars. Other main goals of the proposal included searching for evidence related to ancient climates and more recent periodic climate change, analyzing the soil for physically and chemically bound carbon dioxide and water, and studying the geology, topography, and weather of the landing site.[10]

The Mars Surveyor '98 lander was subsequently renamed Mars Polar Lander (MPL), and its MVACS instrument suite included a descent-imaging camera and microphone from MSSS, and an atmospheric structure laser instrument provided to NASA by the Russian Space Research Institute for Space Science (the successor to the Soviet space agency, the Soviet Union having ceased to exist in December 1991). MPL also carried a technology demonstration component, a pair of high-velocity impact probes called Deep Space 2 A and B (Deep Space 1 was an earlier-launched asteroid and comet flyby technology demonstration mission). The plan was for the 2.4-kilogram (5.3-pound) Deep Space 2 probes to be released from the cruise stage just before MPL arrived at Mars, so that they could decelerate via atmospheric friction within their protective aeroshells, independently impact the surface (with a force of about 60,000 g's), and relay their data on surface properties back to the MGS satellite passing overhead.

Like MCO, MPL was designed, built, and operated on a shoestring "better, faster, cheaper" budget profile. A particularly novel approach for a deep space mission was for NASA to entrust a substantial fraction of the MPL mission operations responsibility to Paige's university-led team, helping to keep costs lower compared to those at NASA centers. After a picture-perfect launch on January 3, 1999, and a relatively uneventful 11-month cruise to Mars, MPL was set for a landing attempt on December 3. After the shocking loss of MCO just a few months earlier, hopes among the MPL team were high that this mission could redeem the entire Mars Surveyor '98 program. But again, fate did not so decree. A preprogrammed command sequence had been loaded on board to perform the initial descent protected by the heat shield, followed by the heat shield separation, parachute deployment and deceleration, and retrorocket firing needed for a soft landing. As small thrusters properly oriented the spacecraft for atmospheric entry just prior to cruise stage separation, signals from the spacecraft indicated that all systems were go. As the heat shield began to ablate, flight controllers at JPL knew that they would lose the radio signal and then not hear from the lander again until it was safe on the ground. Alas, those "safe on Mars" signals never came. MPL had apparently crashed, and the mission was lost. Simultaneously, the Deep Space 2 probes also entered the atmosphere on independent trajectories from each other and from MPL, but neither of them was ever heard from again either.

Another blue-ribbon panel conducted another mishap investigation for NASA,[11] trying to figure out what had gone wrong in the absence of any actual telemetry from the lander or probes during the descent. The panel's sleuthing suggested that an undetected software bug in the lander's leg deployment system was the likely cause of the mission failure: the shock from the deployment of the legs some 40 meters above the surface during the last part of the descent seems to have been interpreted by the software as the actual *thump* of the legs touching down on the surface, leading to the issuance of a premature command to the engines to shut off. Now in free fall, the lander would have quickly plunged to its death from that height. The most likely reason for the failure of both Deep Space 2 probes was never identified. Similar to the findings for the failure of MCO, the board noted several contributing factors to the loss of the mission: a shortage of staffing, deficiencies in the review process, and the project's inability to conduct a robust program of "verification and validation" of the flight hardware.

Had the limits of "better, faster, cheaper" been reached? At around $160 million each (in 1999 dollars, including launch; see appendix H), MCO and MPL cost much less than any other Mars missions previously attempted, and the low price tag meant necessarily lower staffing, oversight, and test levels—factors that can dramatically increase risk and that were all implicated in both mission failures. The MPL Failure Review Board refused to abandon the idea of continuing to try to push traditional mission development and operations costs lower, however, cautioning that while the budget ax had perhaps cut too deeply on Mars Surveyor '98, the prior success of missions like Pathfinder and MGS as well as likely trends toward more efficient, lower-mass, lower-cost technologies for many key spacecraft components, subsystems, and payloads gave cause to remain sanguine about the potential for future missions. As stated in the conclusion to the review board's executive summary: "One lesson that should not be learned is to reject out of hand all the management and implementation approaches used by these projects to operate within constraints that, in hindsight, were not realistic. A more appropriate point of departure would be to evaluate the approaches, and improve, modify, or augment them in response to implementing the Recommendations contained herein."[12]

Indeed, one specific lesson and recommendation that NASA's Mars Exploration Program took to heart on all subsequent missions to the planet's surface was to establish continuous communications between future landers and either Earth or the Mars-orbiting relay satellites (or both) during the entry, descent, and landing process, so that if a mission did fail in the future, teams would have the forensic data needed to properly diagnose the problem and get it right the next time.

Sadly, NASA's two highly visible Mars mission failures weren't the only casualties of the late 1998 launch window to Mars. The Japanese Institute of Space and Aeronautical Science (ISAS) had launched its first Mars mission during that window as well. The Nozomi ("wish" or "hope" in Japanese) orbiter was a mission designed to study the interaction of the Red Planet's upper atmosphere with the solar wind and to test new technologies for future ISAS planetary missions.[13] Problems with the spacecraft's thruster system, and then eventually with its electrical system, prevented the would-be orbiter from ever doing more than a distant Mars flyby in 2003, however, bringing frustration and disappointment to citizens of yet another nation that had experienced the challenges of safely reaching Mars. The Nozomi failure meant that as of the 1998 launch window, only around a third of the world's attempted missions to Mars had succeeded.

2001: A Mars Odyssey Begins

The next launch window to Mars would open in spring 2001. Prior to the severe setback of the Mars Surveyor '98 failures, NASA had been planning to send another orbiter and another lander to Mars in that opportunity as part of a mission called Mars Surveyor '01. The Mars Surveyor '01 surface mission was originally intended to deploy a long-range sample-caching rover, chosen via a competitive proposal process, to traverse more than 10 kilometers across some of the planet's most ancient terrains. However, severe budgetary issues at NASA as well as some technological hurdles related to the rover forced the mission to be descoped to an MPL-like lander that would instead deploy a short-range rover very similar to Sojourner. With the failure of MPL occurring so deep into the development of that nearly identical Mars Surveyor '01 lander, NASA ended up canceling the Mars Surveyor '01 lander mission and going back to the drawing board to rethink the best way to proceed with future surface missions (see chapter 17). The partially completed Mars Surveyor '01 lander was put into storage at JPL, from where it would eventually emerge from the ashes to be reborn as the Phoenix Mars lander mission in 2008. The team whose proposal had been selected to lead those Mars Surveyor '01 rover (then lander) missions was led by Squyres and included former competitors Arvidson and Soderblom. Squyres and his team had to swallow the bitter pill of spending several years working on a hard-won mission that was ultimately canceled, and thus not really winning anything at all.[14]

Fortunately, and because the failure of MCO was human, rather than technological, NASA decided to press on with the orbital mission in 2001, naming that vehicle Mars Odyssey, in a nod to the Arthur C. Clarke book and Stanley Kubrick blockbuster film *2001: A Space Odyssey*. The goals of Odyssey (as it became known, for short) were much more focused on the Martian surface because the mission planners had presumed that MCO would be successfully focusing on the atmosphere, weather, and climate. Specifically, major objectives for Odyssey included globally mapping the geology, elemental composition, and mineralogy of the surface; searching for concentrations of hydrogen (presumably as ground ice) in the shallow subsurface; and characterizing the radiation environment around Mars to assess radiation risks to future human explorers.

Once again, NASA conducted an international competition to select the specific instruments that would enable the orbiter to achieve its mission goals. (The countless hours and grueling work needed to compete for and win a place for one's instrument on a spacecraft betrays the less glamorous, but unavoidable, side of a scientific career in the age of Big Science.) Ultimately selected in 1997 were the Thermal Infrared Imaging System (THEMIS), led by Arizona State University (ASU) in collaboration with Raytheon Inc. and MSSS, which could study the surface at high resolution in both visible and infrared (heat) wavelengths; a combined Gamma Ray Spectrometer (GRS) and Neutron Spectrometer (NS) package that was a collaboration among the University of Arizona, Los Alamos National Laboratory, and the Russian Space Research Institute for Space Science, which could measure the elemental chemistry and hydrogen content of the surface (GRS was yet another reflight of an instrument that had been on the ill-fated Mars Observer mission four years earlier); and a radiation sensor led by NASA's Johnson Space Center. In addition, Odyssey would carry a new high-speed communications system called Electra that could potentially be used to relay signals from future surface missions.

Odyssey was launched from Cape Canaveral, Florida, on April 7, 2001, and spent some 200 days cruising to Mars—among the quickest trips possible because of a very favorable Earth-Mars alignment for the 2001 launch window. With new team communication procedures in place because of the MCO mishap two years earlier, Odyssey successfully fired its main engine at the proper altitude and entered into an elliptical orbit around Mars on October 24, 2001 (figure 16.2). Using the kinds of aerobraking methods pioneered by the MGS team, the orbiter spent the next three months skimming through the planet's upper atmosphere, slowing down a bit more at each close pass, and eventually trimming the orbit into a nearly perfect circular path over the poles of the planet, about 400 kilometers above the surface. Odyssey circled the planet every 2 hours, and Mars rotated below the spacecraft once every 24 hours, 37 minutes, 22 seconds, allowing the orbiter to collect strips of imaging and compositional data across the entire planet. A similar circular polar orbit had also been used for the mapping mission of MGS, which was still operating successfully, with a highly complementary set of instruments, as Odyssey arrived.

Since getting the spacecraft into its circular mapping orbit in early 2002, the Odyssey mission team has collected a stunning global data set that once again began creating a "new Mars" paradigm shift in our understanding of the Red Planet. Among the most important discoveries enabled by the mission is the realization, based primarily on globally mapped GRS and NS data, that there are significant supplies of ground ice in the shallow subsurface at mid- to high latitudes in both the northern and southern hemispheres.

How can underground ice be detected and mapped from orbit? Cosmic rays from galactic and extragalactic sources (like black holes, supernova explosions, and neutron star mergers), as well as high-energy particles from our own Sun, are constantly bombarding the surfaces (and atmospheres) of solar system bodies. When these high-energy particles collide with the nuclei of chemical elements within the uppermost meter or so of the surface, they can create showers of neutrons with a range of energies, which themselves can collide with other nearby nuclei to create showers of gamma rays of element-dependent energies. Specially designed, sensitive spectrometers like the GRS and NS instruments on Odyssey can detect and measure the energies of the neutrons and gamma rays that end up escaping from the uppermost surface and heading back out into space.[15] The spectra of the gamma rays' energy, detected by instruments like GRS, act like fingerprints, revealing the abundances of large numbers of chemical elements in the surface below. Similarly, the energies of neutrons detected by NS also act like a fingerprinting tool, with the greatest sensitivity existing for hydrogen: places where the neutrons are significantly "slowed" on a planetary surface tend to be places with the most hydrogen (because the nucleus of a hydrogen atom is comparable in mass to the neutron colliding with it). And although there are complexities, the simplest interpretation of places with the most hydrogen is that those are also the places with the most near-surface water (on frigid Mars, presumably as water ice).

It took several years for Odyssey to build up enough coverage and signal levels for the team to start publishing the first global maps of shallow subsurface ground ice on Mars (figure 16.2).[16] The maps revealed extensive deposits of ice buried beneath a thin layer of soil or dust, as well as, perhaps, a concentration of nearly pure ground ice poleward of about 45° latitude in both the north and south.[17] The ice appears to be patchy, however, and is perhaps closer to the surface and/or greater in abundance in some places like far northern Tharsis and Acidalia Planitia, and deeper and/or lesser in abundance in other places, like Utopia Planitia. The latter, recall, was the location of the Viking Lander 2 mission, and the relatively lower abundance of ground ice detected by Odyssey in that area is likely among the explanations for ground ice not being discovered by Viking, despite the fact that the lander was sent to that high latitude site ostensibly to maximize the chances of detecting ice or other volatiles. Conversely, these Odyssey ground ice maps were used to choose

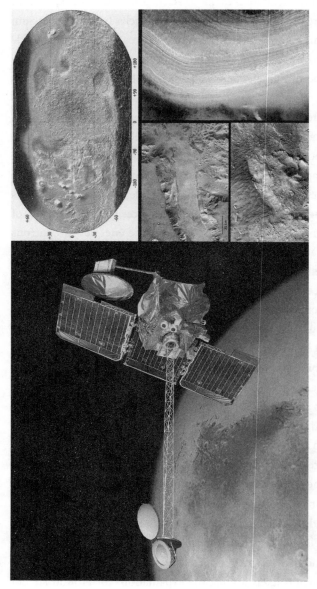

FIGURE 16.2 *Left:* Artist's conception of the NASA Mars Odyssey orbiter. *Top right:* Global map of the inferred water ice abundance in the uppermost meter of the Martian surface from Odyssey GRS/NS data (bluer colors are greater abundances). *Middle, top to bottom:* Odyssey THEMIS false-color infrared mosaic of mineral variations in Ganges Chasma (olivine-rich basalt is purple, basalt is yellow/orange, dust covered regions are bluer); Odyssey THEMIS-derived thermal inertia image of a 30-kilometer-wide region near 26.5°N, 357°W, showing a range of lower-inertia sand dunes (bluer) to higher-inertia bedrock (redder). *Lower right corner:* Odyssey THEMIS visible-wavelength false-color image of finely layered south polar deposits (the scene is just under 20 kilometers across). NASA/JPL/ASU/UA/LANL.

the landing site region for the NASA Phoenix lander's 2008 mission, which would confirm the validity of the Odyssey results directly when its retrorocket engines cleared away a loose layer of dust and soil to reveal previously buried deposits of nearly pure-white water ice.

Odyssey data also revealed a second, lower-abundance (perhaps 10 percent or less) concentration of near-surface hydrogen at equatorial and midlatitudes on Mars.[18] Near-surface water-ice deposits would find it difficult to persist in equatorial regions because summertime temperatures get above freezing and drive water ice out of the regolith and into the atmosphere. Instead the near-equatorial hydrogen is likely to be tied up in minerals containing water or hydroxyl (OH), like hydrated sulfates. The Odyssey team's discovery of hydrated minerals in the equatorial regions of Mars lent support to similar conclusions derived from the earlier Mariner 6 and 7 missions, as well as from the Soviet Phobos 2 mission, whose limited infrared spectrometer investigations found hydrated minerals and potentially clays on the surface of Mars. The Odyssey findings also provided a preview of (and part of the impetus for) the confirmation and extensive mapping of clays and other hydrated minerals that would be achieved by the infrared spectrometer on the Mars Reconnaissance Orbiter, as described below.

Mapping of gamma ray energies by Odyssey's GRS instrument proved similarly successful and revealing, significantly expanding on the limited gamma ray data set obtained by the Soviet Mars 5 orbiter nearly 30 years earlier. GRS maps of chemical elements like silicon, iron, calcium, aluminum, potassium, thorium, chlorine, and others showed that there are detectable variations in these elements from region to region, providing ways to directly assess the planet's average composition as well as the origin, evolution, and diversity of volcanic and impact processes that alter the composition of the surface of Mars. For example, maps of the planet's iron abundance show that Mars overall is more enriched in oxidized iron (by 4–5 percent) than is Earth, a finding that is consistent with its nickname the "Red Planet" and that suggests fundamental differences in the mantles of these two planets that could be related to formation conditions or the action of subsequent planetary-scale processes (like mantle convection or, for Earth, plate tectonics).[19] GRS data also appears to show that Mars has a higher relative abundance of moderately volatile elements (elements like potassium and chlorine that participate in more chemical and mineralogical reactions than others), consistent with the planet forming in an environment farther from the Sun, where there was likely a higher relative percentage of volatiles than closer in.[20] Generally, even though they cover much larger areas, the Odyssey measurements of chemical abundances from orbit also closely match the abundances determined locally at the Viking and other landing sites on Mars, providing a sort of "ground truth" verification that allows the orbital results to be trusted for locations where landers and rovers have not yet gone.

Another Odyssey instrument, THEMIS, returned pictures of the surface through multiple color filters at both visible and thermal infrared (heat) wavelengths, with one part of the instrument designed to map the geology and visible color properties of the surface at tens of meters scale,[21] and the other part designed to map the mineralogy, temperatures, thermal inertia, and other thermal properties of the surface and atmosphere at hundreds of meters scale (figure 16.2).[22] The thermal infrared capabilities of THEMIS (which expand on the infrared spectroscopic capabilities of the MGS Thermal Emission Spectrometer, TES) enabled most of that instrument's new discoveries about Mars during the mission, including the result that there do not appear to be any active "hot spots," areas of even a few degrees above average local temperatures, on Mars today. Nonetheless, THEMIS data does reveal a wide range of differences in surface thermal properties on Mars, some correlated with differences in mineral content from region to region, others with differences in the amount of rockiness of the surface, and still others with the degree to which the surface is covered or buried by loose dust, sand, and soil.[23]

For example, while THEMIS infrared data confirms that the composition of the surface of Mars is dominated by volcanic minerals similar to terrestrial basalts, it also reveals a much richer diversity of igneous minerals than previously realized, ranging from so-called ultramafic (very low silicon and high iron and magnesium) basaltic rocks rich in the mineral olivine, to basaltic rocks very similar in composition to those in Earth's average upper mantle, through dacite (very high silicon) compositions seen in specific small volcanic cones or lava flows.[24] On Earth, similar compositional diversity is often due to igneous differentiation (for example by addition of water in subducting slabs leading to preferential melting of silicon-rich minerals), much of which is driven by the action of plate tectonics. THEMIS data shows that Mars also appears to exhibit igneous differentiation processes, but they are driven by other forces—

impact melting, ascent of mantle plumes, or fractional crystallization (sinking of hot rock from deep in the mantle to concentrate heavier iron-rich minerals in magma chambers)—because Mars does not appear to have had an extensive history of plate tectonics. A particularly exciting discovery enabled by the ability of THEMIS to obtain infrared images during the nighttime on Mars is that bedrock outcrops have been mapped globally down to the scale of a typical football field: after the Sun sets, the rocks and outcrops hold their heat well into the night, glowing brightly in infrared images. THEMIS has also been a sensitive detector of ice, both at and just beneath the surface.

Since exposure to dangerous levels of radiation is a key concern related to astronaut safety during long space mission flights, especially for missions that leave the protective confines of Earth's magnetic field, one of the goals of Odyssey's Martian Radiation Experiment (MARIE) was to provide data on the average and peak radiation levels that astronauts would be exposed to both during a typical cruise to Mars and then once in Mars orbit. MARIE found that, not surprisingly, radiation levels during the trip to Mars and then in Mars orbit were two to three times greater than in low Earth orbit. However, occasional (and potentially lethal) doses of radiation up to a hundred times higher were observed in association with solar flares or other high-energy solar-charged particle events on the way out to Mars. MARIE data reinforces the idea that interplanetary travelers making long-duration flights across the solar system (whether robotic or, eventually, human) will have to take radiation shielding and other protective measures extremely seriously.[25]

With the loss of contact with the MGS spacecraft in 2006, Odyssey assumed the mantle of being the longest-running Mars mission in history, and it is still going strong as of this writing (late 2020). The mission's supply of orbit maintenance thruster fuel and battery lifetime is predicted to potentially allow it to operate well into the mid-2020s, continuing to collect imaging and other data sets and to provide relay satellite capabilities for future landers and rovers on the surface below.

Europe Goes to Mars, Express

Until 2003, only the American, Soviet, and Japanese space agencies had attempted to launch missions to Mars (and collectively achieved success rates well under the chance of a coin flip). In June 2003, however, a new player entered the game, when the European Space Agency's (ESA) Mars Express mission was launched on a Soyuz rocket from the Baikonur launch facility, on the steppes of southern Kazakhstan. ESA, which was established in 1975 as a membership-based consortium of 10 European member states (now expanded to 22),[26] provides an organizational scheme that enables spacefaring European nations to pool their resources and thus conduct larger-scale space science missions than any individual member state could conduct alone. Mars Express consisted of an orbiter, known simply as the Mars Express Orbiter, and a lander named Beagle 2, named after the 19th-century HMS *Beagle*, which took naturalist Charles Darwin on his initial voyage of scientific and evolutionary exploration. The scientific objectives of the Mars Express Orbiter included high-resolution stereo imaging for topography and geology, mapping of surface mineralogy and atmospheric composition, geophysical studies of the structure of the subsurface, and studies of the interaction of the atmosphere and the interplanetary medium. Beagle 2 was designed to demonstrate European-developed Mars landing technologies and had scientific objectives that included characterizing the landing site's geology, composition, and weather.

Mars Express arrived at Mars on Christmas Day 2003, performing its orbital insertion main engine burn about six days after releasing Beagle 2 for its landing attempt (figure 16.3). The lander was confirmed to have separated successfully and been placed on the proper ballistic trajectory to reach its intended location, a broad elliptical region within Isidis Planitia thought to be relatively free of landing hazards. However, communications were not possible once the lander began its preprogrammed series of entry, descent, and landing events (involving a heat shield, parachute, and large protective airbags), so the ESA control center team in Leicester, England, had to wait patiently for Beagle 2 to phone home once it was safely on Mars. Sadly, that call never came; like so many attempted Mars landings before, Beagle 2's mission had ended in failure. Various hypotheses to explain the loss of the mission in the absence of any actual telemetry were developed by the subsequent investigative review board. However, it wasn't until February 2007, when NASA's Mars Reconnaissance Orbiter (MRO) was able to spot Beagle 2 in a high-resolution image, that the most likely cause of the failure was identified. The MRO images showed that Beagle 2 had

FIGURE 16.3 Artist's depictions of the European Space Agency's Mars Express Orbiter (*left*), which successfully went into Mars orbit in 2003, and the NASA Mars Reconnaissance Orbiter, which successfully went into Mars orbit in 2006. NASA and ESA.

apparently landed safely, but incomplete deployment of its solar panels had apparently blocked the lander's radio antenna from properly communicating with either Earth or any of the orbiters above. So close, but yet so far!

Meanwhile, the Mars Express Orbiter began its mapping mission from its highly elliptical polar orbit (passing within about 300 kilometers of the surface every seven hours) in early 2004. The orbiter's truly international science instrument payload included two different spectrometers from France: one for visible and infrared studies, called Observatoire pour la Minéralogie, l'Eau, les Glaces et l'Activité (OMEGA); and the other for ultraviolet (UV) studies of the atmosphere, called SPICAM, an acronym for Spectroscopy for the Investigation of the Characteristics of the Atmosphere of Mars (and capturing the historical importance of observations of the star Spica in searching for a potential atmosphere on the Moon). Italy provided another atmospheric spectrometer for the mission, called the Planetary Fourier Spectrometer (PFS), as well as a subsurface radar sounding instrument called the Mars Advanced Radar for Subsurface and Ionosphere Sounding (MARSIS). Germany contributed a powerful stereo and topography imager called the High Resolution Stereo Camera (HRSC); and Sweden provided an instrument to study the solar wind's interaction with the upper atmosphere, called the Analyzer of Space Plasmas and Energetic Atoms (ASPERA). All these instruments would return outstanding data sets, and a special telecommunications system built in the United Kingdom would demonstrate that the orbiter can successfully serve as a relay satellite for landers and rovers down on the surface.

The Mars Express Orbiter continues to be functional as of late 2020. Some of the most outstanding discoveries so far have come out of the OMEGA data. Building on the earlier results from spectrometers on the Mariners 6 and 7 and Phobos 2 missions, OMEGA has shown the existence of widespread but relatively small (a few to tens of kilometers, typically) outcrops of hydrated clays, hydrated sulfates, and iron oxides across the planet (plate 14). A particularly interesting finding is an impressive accumulation of the hydrated sulfate mineral gypsum ($CaSO_4 \cdot 2H_2O$) within the high-latitude polar dunes of Olympia Planitia, in huge kilometer-high mounds of magnesium sulfates in the Valles Marineris canyons, and in outcrops with high concentrations of different aluminum-, iron-, and magnesium-rich phyllosilicate minerals (claylike minerals made of sheets of those elements plus silicon, oxygen, and hydroxyl, often with water molecules trapped between the sheets) in an ancient, heavily eroded part of Mars known as Mawrth Vallis.[27] Analysis of the data globally has resulted in new hypotheses about the history of water on Mars: the oldest exposed rocks had clay minerals from chemical reactions with water; the next oldest had only sulfates, left behind by evaporating water; and the younger rocks (actually, not *that* young, at 2.5 billion years old!) had only iron oxide rusts, signifying very little exposure to water.

Another impressive result, from the mission's HRSC imaging data, has been the accumulation of an enormous number of small-scale digital terrain models (3-D computer representations of a landscape) created from images taken by the orbiter's stereo camera system (plate 14). In terrestrial geology and remote sensing, such models are critical for mapping and modeling of the terrain, especially for hydrologists and geomorphologists reading the history of water, ice, and lavas via landforms. Having the ability to drape other data sets—like mineral maps—on top of a topographic model at resolutions of around 100 meters provides a powerful way to synthesize not only the Mars Express data sets, but also those from other Mars orbiter missions.

Another interesting—albeit contentious—result reported by the Mars Express Orbiter team was the detection in the PFS data of small amounts (around 10 parts per billion, or 0.000001 percent) of methane gas (CH_4) in the atmosphere of Mars.[28] Although the presence of methane was suspected from previous telescopic observations as a very minor or "trace" component of the atmosphere, the existence of any methane at all is nonetheless noteworthy because individual molecules of methane are very quickly (on a timescale of just a few centuries) broken apart in the atmosphere by ultraviolet sunlight. Thus, the detection of any methane must indicate that there are one or more sources resupplying it to the atmosphere. On Earth, methane has about 2,000 times the relative abundance and millions of times the absolute abundance as methane on Mars, with more than 90 percent of it being generated by biological sources. While there are also geochemical (nonbiological) ways to create methane and release it into the atmosphere, the possibility of methane being a potential biosignature indicating methane-producing (subsurface) life on Mars makes its scientific and public interest far outweigh its tiny abundance.[29]

The contentious aspect of the discovery arose partly because of the inherent difficulty in detecting such tiny amounts of methane within a (relatively) much thicker atmosphere of other Martian gas molecules, inconsistencies between abundances and variations derived from telescopic and the orbital spacecraft data sets, and (for the telescopic observations) interference from much-larger quantities of methane and other gases in Earth's atmosphere. Still, a flurry of media coverage surrounding the possible detection once again generated enormous public interest in the possibility of life on Mars, with planetary scientists trying to temper—without utterly squelching—that interest by recalling Sagan's reminder that "extraordinary claims require extraordinary evidence." In the case of the methane detection reported by the Mars Express Orbiter team, additional evidence to try to decide the question one way or the other would not come for nearly a decade, first when NASA's Curiosity Mars rover deployed sensors able to detect methane on the ground, and then a few years later, when ESA's Trace Gas Orbiter arrived to perform high-sensitivity atmospheric observations. As we shall see later, the results from those missions have also been controversial.

Mars Reconnaissance Orbiter

One more orbiter was sent to Mars in the first decade of the 21st century, NASA's Mars Reconnaissance Orbiter (MRO), launched in summer 2005. It was another legacy of the failed Mars Observer of 1993, and of NASA's new "better, faster, cheaper" strategy of mission development in which the science investigations lost with Mars Observer were more frugally and cautiously split among three smaller, less-costly future orbiters: MGS, MCO, and Odyssey. Though MGS and Odyssey were successful, the failure of MCO left a still-glaring hole in the Mars science that needed to be done regarding atmospheric dust, cloud, and water vapor monitoring. Long-term atmospheric monitoring and high-resolution surface imaging continued to be provided by MGS, but by the middle of the first decade of the 2000s, the spacecraft was more than a decade old, and so the continuity of its unique monitoring of the planet was at serious risk of being interrupted.

As an indication of NASA's increased confidence in its ability to successfully carry out Mars missions, MRO represented something of a departure from the "better, faster, cheaper" strategy. With a price tag of $720 million (plus launch, in 2005 dollars; see appendix H), it was not a low-cost, high-risk mission. Instead, NASA hoped that spending additional money to increase the reliability, redundancy, and robustness of the spacecraft would assure a longer lifetime so that the orbiter could serve as a key scientific and infrastructure resource at Mars for many years to come. The mission had a complement of six competitively selected scientific instruments,[30] making the orbiter a much closer descendent of Mars Observer than of the leaner, more recent MGS and Odyssey spacecraft. And MRO carried the last of the failed Mars Observer scientific instruments, the Mars Climate Sounder (MCS) instrument, which was designed to monitor the planet's atmospheric pressure, temperature, and dust abundance as a function of altitude. MCS was a miniaturized, modernized version of the PMIRR instrument, which was originally launched 13 years earlier on Mars Observer, and then again 7 years earlier on Mars Climate Orbiter. The long-suffering JPL-led MCS team, led by planetary atmospheric scientist Dan McCleese, was truly hoping that the third time would be the charm. Other instruments on MRO included the highest-resolution camera (and telescope) yet flown in deep space (the University of Arizona's High Resolution Imaging Science Experiment, or HiRISE, capable of imaging the surface at 30 centimeters per pixel); a medium-resolution mapping Context Camera (CTX) and wide-angle Mars Color Imager (MARCI, a redesigned version of the MARCI on the MCO mission), both from Malin Space Science Systems Inc.; a visible to near-infrared mineral mapping instrument called the Compact Reconnaissance Imaging Spectrometer for Mars (CRISM), from the Johns Hopkins University Applied Physics Lab (APL); and the Shallow Radar (SHARAD) instrument developed by the Italian Space Agency that was designed to map layered materials (rock, ice, and perhaps groundwater) at fine resolution in the uppermost few hundred meters of the subsurface.[31] One of the key features of this payload was that the high-resolution imagers (HiRISE and SHARAD) were nested with each other and with the spectrometer (CRISM), enabling the collection of compositional data paired with detailed geological imaging context.

Transmitting unprecedented amounts of deep space data from all these instruments required a large (three-meter-diameter) radio antenna and the Electra ultrahigh frequency (UHF) radio system, which could enable large data volumes to be sent back to Earth, both from the onboard instruments and from landers and rovers on the surface below, relayed through the orbiter. Indeed, MRO to date has transmitted more than 300 terabits of data back to Earth, more than all previous interplanetary (nonlunar) missions combined.[32] This massive stream of data, from a diverse set of highly capable science instruments, has enabled a plethora of new discoveries about the Red Planet since MRO went into Mars orbit in March 2006 and aerobraked to its final mapping orbit in September of that year (figure 16.3).

Perhaps the most dramatic discoveries have come from the ability—once again—to see a new Mars at a higher imaging resolution than ever before. While the improvement in resolution from the typical Viking Orbiter imaging scale to the more than 50 times smaller pixel scale of the MOC on MGS was dramatic, the increase in detail and geological interpretability between the pixel scale of the MOC and the five times smaller still pixel scale of HiRISE was just as much so.[33] In particular, small-scale details have come streaming forth from more than seven Mars years of HiRISE imaging covering volcanic, sedimentary, and polar layers, outcrop exposures, individual lava flows, landslides, small impact craters, sand dunes of all shapes and sizes, and even rovers and landers seen as if from an airplane passing overhead. Even after more than 15 Earth years of photography, HiRISE has only observed a few percent of the surface of Mars because its field of view is so small and its spatial resolution is so high. However, the University of Arizona and JPL team responsible for collecting HiRISE images, led by planetary scientist Alfred McEwen, have done a masterful job of *targeting* precisely those pieces of Martian real estate that contain the most revealing highlights of the planet's rich geological history.[34]

An Active Planet

Among the most important HiRISE discoveries so far is that Mars has a much higher degree of current surface activity than previously imagined. "Science nuggets"[35] that have come out of this unprecedented high-resolution imaging of the Red Planet have included a campaign to quickly get high-resolution photos of small impact craters that have recently formed on Mars and that were initially detected by MRO's CTX camera or other previous imaging systems (figure 16.4). Hundreds of such new craters have now been identified,[36]

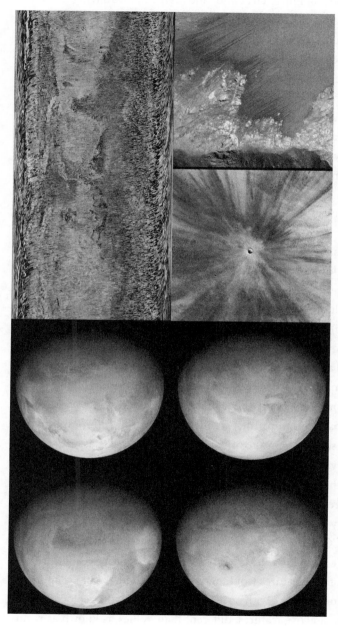

FIGURE 16.4 Cameras on NASA's Mars Reconnaissance Orbiter capture images at an impressive range of scales. *Left*: MARCI daily global color coverage at about one kilometer per pixel scale. *Top*: Global map with more than 99 percent coverage achieved as of 2017 by the CTX imager; this is the highest-resolution global imaging data set ever acquired of Mars, with every place here imaged at a resolution of six meters per pixel. *Bottom middle*: HiRISE false-color image of a fresh 30-meter-diameter impact crater, formed sometime between July 2010 and May 2012. *Bottom right*: HiRISE false-color image of recurring slope lineae in Coprates Chasma. The scene here is only about 500 meters across. NASA/JPL, MSSS, and University of Arizona.

with many of those that impacted at mid- and high latitudes having excavated shallowly buried ground ice, helping confirm the Odyssey-based GRS and NS inferences of ground ice from that previous mapping investigation. HiRISE resolution and the longevity of the MRO mission have also enabled tracking of the subtle movement of dark sand dunes across the surface for the first time (dunes were not observed to migrate in previous images at lower resolution, and so the observation that they move at all is a notable discovery). Curiously, dune migration rates observed in HiRISE time-lapse images are comparable to those of slowly moving dunes on Earth (around 1 to 10 meters per year), even though Mars has one-third the gravity and one-hundredth the atmospheric density.[37] While sand has clearly been a major agent of erosion and other surface changes, much is still not understood about the way it moves across the planet.

Another focus area of HiRISE observations has been the search for evidence of active flows in gullies or in other hill slope features, building on the initial discovery of these kinds of features in MGS MOC images, but at much better resolution. Specifically, HiRISE images (including many with stereo/3-D coverage) provide substantial additional detail about the color, depth, slopes, upslope source regions, and downslope terminal regions of thousands of gullies studied and monitored in detail over time. The enhanced resolution and extended time coverage have revealed a good correlation of gully activity with season on Mars, with gully activity observed in many places when seasonal CO_2 frost is also present.[38] While there are still significant puzzles remaining to be solved about the origin of gullies, and some aspects of their formation and evolution are still suggestive of the presence or involvement of liquid water, the broader emerging view based on HiRISE imaging is that these features are more likely to have been formed by (and are still being modified by) dry granular flows of dust and soil somehow initiated by the sublimation of CO_2 (dry ice) on or within fine-grained surface materials. If liquid water isn't required and dry ice turns out to play the leading role in gully formation, these features would represent yet another kind of uniquely Martian surface process, one with no known analog in Earth's geologic record.

Another class of active features, different than gullies, is known as recurring slope lineae or RSL. These features are characterized by dark material flowing down steep rocky slopes usually in the absence of an associated gullylike channel, and they have been observed to appear predominantly during seasonal times of peak solar heating (figure 16.4). RSL have been hypothesized to be small outbursts of shallow, salty groundwater that quickly sublimates as it flows downhill in the cold, thin Martian air—in which case Mars might have much more abundant groundwater in places than previously thought.[39] Alternatively, as has been the case for gullies, others have hypothesized that these features might actually also be dry granular flows: small dust avalanches triggered by wind, ice formation, or local shaking from occasional nearby small impacts.[40]

HiRISE imaging of the polar regions has captured active landslides, active geysers of trapped CO_2 gas jetting out from cracks in the ice, and the slow pattern of sublimation of the so-called permanent water-ice polar caps, especially in the south. According to some researchers, these observations indicate that Mars might be undergoing a period of slow global warming, perhaps caused by the same kinds of astronomical forcing (slow variations in the planet's tilt and orbital parameters) that have been responsible for many of the dramatic prehistoric changes in the climate on our own planet. (One example of this is the Milankovitch cycles responsible for Earth's ice ages.)[41]

Context Imaging and a Daily Weather Report

In most cases, HiRISE images are accompanied by lower-resolution (six meters per pixel) wider-field images from MRO's CTX camera, which—as the name implies—provides more of the regional geological context around the super high-resolution region imaged by HiRISE (figure 16.4). The CTX team at MSSS has performed its own heroics in this regard, by now having covered more than 99.7 percent of the planet's surface with images at CTX resolution.[42] Early in the mission it was simplest just to turn CTX on and off whenever possible to build up small spaghetti strips of surface imaging coverage as MRO orbited the Martian day side from south to north in its polar orbit. After a while, however, CTX started covering some of the same areas again. While this can be a good thing for building up stereo coverage of the terrain (which the CTX team has now done for about a third of Mars) or for searching for changes associated with dust storms, sand movements, new impact craters, and the like, it doesn't help build up overall global coverage. Also, seasonal dust storms and polar clouds started preventing some of the underlying surface from being in good view, further thwarting the attempt to map the entire planet. So the CTX team had to figure out exactly when during each orbit, as well as during each season, to turn CTX on and off to fill in holes (also called *gores*) in the global map without getting blocked by clouds. Ultimately, it took a decade of such gore-filling to get to about 90 percent surface coverage, and it's been a slow process since then creeping up on 100 percent. Some places, like the Hellas impact basin or the flanks of some of the tall Tharsis volcanoes, are persistently cloudy or dusty/hazy; it has sometimes taken years of attempts to clearly image such stubbornly uncooperative locales. Now that the global map is essentially complete, however, the team's persistence has paid off with a stunning data set that will be a legacy for Mars exploration for decades into the future.

Another remarkable data set acquired by MRO is the collection of daily global maps of Martian weather assembled by the MARCI camera team since orbital insertion in late 2006 (figure 16.4). MARCI's time-lapse global maps clearly show the day-to-day evolution of the planet's frequent dust storms and water-ice clouds (many of which follow specific "storm tracks" like some storm systems on Earth), as well as the season-to-season variations in the growth and decay of the polar caps. MARCI global mapping actually overlapped for a few months in late 2006 with similar global mapping being done from the MGS mission, including imaging by the MOC-WA (wide-angle) camera team, which involved many of the same team members as the MARCI team, since both were operated from Malin Space Science Systems Inc. in San Diego. Thus, the seven Mars years (so far) of still-active MARCI global weather maps can be combined with the nearly five Mars years of previous MOC weather maps (going back to late 1997), to characterize more than a Mars decade of daily, seasonal, and interannual variations of the active cycles of dust storms, water-ice clouds, and polar ice deposits on the Red Planet.

Occasionally during the MRO mission, as in 2007 and then more dramatically in 2018, nearly the entire planet becomes enshrouded in global dust storms that block the underlying surface from view. Indeed, the 2018 storm darkened the skies so much in places like Meridiani Planum that it led to the shutdown and eventual end of the mission of the solar-powered Opportunity rover exploring that area. While it is still not entirely clear what causes the planet's dust devils and smaller dust storms to occasionally scale up to vast regional events and even the truly monstrous planet-encircling and global-scale storms, the key to solving that mystery is no doubt captured within the MARCI images and other orbital data sets, for example the simultaneous near-daily MCS data on atmospheric pressures and temperatures. The ability to predict such extreme weather conditions on Mars will certainly be important to future explorers and settlers. While the planet-circling dust storms cannot yet be reliably predicted (though there are some promising hypotheses that purport to do just that),[43] the occurrence of smaller dust storm activity and other local weather effects from place to place is stunningly repeatable, with some of the same kinds of storms and clouds occurring in the same places from Mars year to Mars year, often within just a few sols of the earlier events. One can imagine a Martian calendar (of which many have already been proposed by enterprising calculators) for future residents being adapted to such events, with specific holidays, festivals, or planned shelter days eventually being synchronized to the planet's remarkably repeatable weather patterns.

Complementary mapping of the water and CO_2 ices in the polar regions that seasonally condense (go from vapor to solid) and sublime (go from solid to vapor)—as well as significant amounts that remain year-round—by MRO's subsurface-probing SHARAD radar instrument has also led to important discoveries. For example, radar probing of the seasonal and residual water-ice polar caps has enabled estimates of the total volume of water stored in the caps, which turns out to be a bit more than the volume of the Greenland ice cap on Earth.[44] Even more impressive, however, was the discovery from SHARAD data of 30 times as much volume of buried CO_2 ice down to more than a kilometer beneath the south polar near-surface H_2O and CO_2 ice deposits (plate 15). Those buried dry ice layers contain enough CO_2 to almost double the planet's atmospheric pressure if it were to sublime during times of more intense polar heating (for example, during times when the planet's variable axis tilt, or obliquity, was lower and thus solar heating at the poles less intense on average).[45] Subsequent detailed mapping and analysis has revealed that those buried layers of CO_2 ice may record more than a half-billion years of Martian climate variations, and that the periodic exchange of that buried CO_2 between the surface and the atmosphere as the planet's obliquity varied over time (mostly because of the gravitational influence of Jupiter) has likely resulted in significant periods of geological time when atmospheric pressure was higher and thus liquid water may have been more stable on the surface than during the present day.[46] In some ways, the MRO SHARAD data has provided the most dramatic demonstration to date of the critical role that the long-term cycling of CO_2 has had on the climate history of the Red Planet, first postulated by Robert B. Leighton and Bruce C. Murray back in 1966 based on Mariner 4 flyby data.

Another stunning data set being returned by MRO is a collection of moderate-resolution (for MRO) images of the surface being obtained in up to 544 different colors of the spectrum by the CRISM instrument. CRISM is an imaging spectrometer—both a camera and a spectrometer—that uses a diffraction grating to disperse sunlight reflected off Mars (or infrared heat radiation emitted from Mars) into its constituent components, much like a prism disperses white light into the many colors of the rainbow. Multiple detectors allow CRISM to image in both visible light and the longer (redder) wavelengths in the infrared, covering a range that is sensitive to the spectroscopic "fingerprints" of many different kinds of minerals known or suspected to occur on Mars.[47] Using the south-to-north motion of the spacecraft as well as a gimbal to compensate for downtrack spacecraft motion, CRISM builds up its coverage across the surface of the planet over time, with most of the coverage in lower resolution, fewer-color mode (100–200 meters per pixel in 72 to 262 colors), and a small fraction of the planet covered in all 544 wavelengths at the finest scale (about 18 meters per pixel).

CRISM has enabled numerous discoveries about the mineralogy of Mars,[48] building on the successful initial imaging spectroscopy experience of the Phobos 2 imaging spectrometer (ISM) and Mars Express OMEGA instruments, and complementing the results gleaned from the thermal infrared observations by the MGS Thermal Emission Spectrometer (TES) and Odyssey THEMIS instruments. In particular, CRISM provides maps not only of unaltered ("primary") volcanic rocks and minerals (the kinds of minerals most commonly mapped by TES and THEMIS), but also of a number of key minerals formed by the weathering, water-related alteration, and/or metamorphism of those precursor volcanic rocks and minerals, like hydrated iron oxides, sulfates, clays, various forms of silica, and carbonates.[49] As on Earth, the specific kinds of minerals detected, and even the specific ratios of chemical elements in those minerals, can sometimes provide unique information on the environmental conditions (temperature, pressure, salinity, or acidity of the water) prevalent at the time those minerals were formed, as well as compositional information useful for identifying future mission landing sites (plate 15).

For example, CRISM has expanded on the lower-resolution results from the TES investigation to map the distribution of primary volcanic minerals like olivine and pyroxene across a wide variety of nondusty volcanic terrains around the planet. Just like on Earth, variations in the detailed chemical composition of these minerals provide information on the evolution of surface volcanism, and thus of interior heating and activity, over time. In the case of Mars, changes in the chemistry of the primary volcanic minerals in regions of different relative ages observed by CRISM are consistent with the slow cooling, and thus decreased intensity of subsurface melting, of the planet's interior over time. On Earth, continuing intense internal heating and the recycling action of plate tectonics can result in extreme variations in the composition of volcanic minerals, including the creation of very high-silicon, quartz-rich rocks like granite. Data from CRISM, Mars Odyssey THEMIS, and earlier MGS TES mapping shows that such high-silicon minerals are extremely rare on Mars, however, implying that the crust and mantle probably have not undergone as much chemical and mineral processing as on our home world.

Still, CRISM has helped show that Mars is not completely dominated by volcanic rocks and minerals: important deposits of both sedimentary and hydrothermally altered igneous rocks have been identified and mapped. Sedimentary layers, first seen in detail from orbit in MGS MOC images and subsequently studied in even greater detail in HiRISE photos, contain minerals like hydrated iron-bearing sulfates; iron-, magnesium-, and aluminum-rich clay-like minerals (in the family called phyllosilicates); and even hydrated amorphous silica materials like the mineral opal.[50] When studied on the ground by rovers like Spirit, Opportunity, and Curiosity, these minerals have been observed to occur as layers of sandstones, siltstones, or mudstones, sometimes containing other associated minerals like a variety of iron oxides, and sometimes replete with sulfate-rich veins that point to the intimate interactions between these rocks and surface water and/or groundwater. The detection and mapping of clay minerals, in particular, continues to yield exciting discoveries about the history of specific past watery environments on Mars; some deposits were formed by water at the surface, and others were formed in hydrothermal systems of hot water underground. Some water on Mars today is still trapped inside these kinds of minerals. The fact that a number of regions of the surface (like deltas, ancient lakebeds, and hot springs environments) appear to have been persistently wet, for significant intervals of geological time, also excites astrobiologists thinking about the duration of past habitable environments and the implications for the origin and evolution of life on Mars.[51]

Most sedimentary rocks and water-formed minerals discovered on Mars by CRISM and other previous spectrometers have so far been found to occur only in the uppermost crust (upper few kilometers) of the planet. The absence of evidence to date for deeper sedimentary rocks buried beneath younger lava flows is consistent with the absence of plate tectonics. However, CRISM data has revealed significant deposits of hydrothermal and perhaps gently metamorphosed rocks and minerals, and HiRISE images reveal layered blocks of "megabreccia" (deeper subsurface materials exposed by impact cratering), both of which have apparently been excavated from great depths. The minerals that have been identified (like silica, chlorite, and zeolite) are products of low-grade (or relatively low temperature and low pressure) metamorphism. The absence of high-grade metamorphic minerals again suggests a crust and mantle with generally less heat and energy available for mineral alteration compared to Earth, and no high-pressure plate tectonics.

Legacy of a Decade of Mapping

The successes of the Odyssey, Mars Express, and MRO missions began in the first decade of the 21st century and have extended through the second, but the legacy of these first highly sophisticated orbital mapping missions and their panoply of modern instrumentation and high-bandwidth data transmissions back to Earth will surely extend for decades to come. These missions, and MGS immediately before them, have not only been able to successfully recover the science that was intended to have been achieved by the failed Mars Observer and Mars Climate Orbiter missions of the 1990s, they have been able to move Mars science and exploration beyond anything that could have been imagined by those earlier teams. By providing a truly global planetary perspective on the geology, composition, mineralogy, physical properties, meteorology, and volatile inventory of the planet, they have enabled scientists, engineers, and mission planners worldwide to understand Mars in a synergistic and holistic way that was almost completely unknown previously (and indeed, is only known for one other place in the solar system—Earth).

As a result, these missions have provided the key environmental knowledge, geographic and geotechnical information, and scientific puzzles that are driving the most important questions in Mars scientific studies and mission planning going forward. Indeed, these missions and the data analysis and interpretation that they have enabled have already had a profound effect on the field, with their mapping data sets playing critical roles in the selection of Gale crater as the landing site for the Mars Science Laboratory (MSL) Curiosity rover mission in 2012, and most recently in the selection of Jezero crater for the Mars 2020 Perseverance rover mission in 2021. And their legacy continues, as the aging, venerable, but still highly functional Odyssey, Mars Express, and MRO spacecraft are predicted to continue to collect new measurements from orbit, and to relay new data from the surface, well into the 2020s.

17

Living on Mars with Spirit, Opportunity, and Phoenix

Twin Rovers Are Born

The loss of both the Mars Climate Orbiter (MCO) and the Mars Polar Lander (MPL) in 1999 not only created an instant set of critical Mars science and infrastructure holes that needed to be filled, but perhaps more importantly created an overall existential crisis for NASA's still-young Mars Exploration Program. Building on the 1997 success of Mars Pathfinder and working on the assumption that the "better, faster, cheaper" mission model could be applied not only to NASA's Discovery program but also to the separate Mars program, NASA created the Mars Surveyor '01 orbiter and surface-landed mission opportunities, seeking out scientific payloads and investigations that could be conducted on lower-cost, higher-risk spacecraft. The 1999 failures put the brakes on Mars Surveyor '01 and forced a reassessment and reorganization of the Mars program. Only the orbital part of the 2001 missions survived, going on to become the highly successful Mars Odyssey mission. The lander part was mothballed, and NASA Headquarters and the Mars science community huddled for months in mid-2000 to figure out the most cost-effective *and* low-risk way to restructure the Mars program and get the needed science from the next mission to Mars (either an orbiter or a surface mission), perhaps as soon as the 2003 launch opportunity, when public interest in Mars would be at crescendo pitch already, as the planet came closer to Earth than at any time in over 60,000 years.

The person tasked with leading that restructuring effort was seasoned NASA Ames Research Center scientist and program manager G. Scott Hubbard, who was well known for helping promote the airbag strategy for landing on Mars (successfully deployed by Pathfinder) and who had moved from California to Washington, D.C., in early 2000 to become NASA's first Mars program director or, as he called himself, the "Mars Czar."[1] Hubbard quickly assembled a team of scientists, engineers, technologists, and project managers called the Mission Options Assessment Review or MOAR to assess the choices for both the 2003 and 2005 Mars mission opportunities, the first of which would have to launch only about three years after NASA Headquarters made a decision based on Hubbard's committee's recommendation. In the business of designing, building, and testing deep space interplanetary spacecraft, three years from the start of a project to its launch is an *extremely* short amount of time. Indeed, Hubbard and his team recognized that the development time for any possible 2003 mission was so short that whatever was selected to fly would already have to be pretty far along. They would essentially need off-the-shelf instruments, spacecraft, and components to meet such a demanding schedule. Luckily, they had a number of lower-risk options available that had already gone through a community-wide competition and assessment process, including the instruments from the failed MCO mission, as well as the payload from the canceled Mars Surveyor '01 lander.

Low cost and low risk were only two of the variables in NASA's decision process, however. Hubbard's group also gave equal weight to the potential science discoveries that the new mission would make, and especially made sure that they would be aligned with NASA's new Mars program mantra, "follow the water," in which priority was given to exploring locations on the planet associated with past or present sources of water. Other variables included making sure that the new mission would fit in with as well as feed-forward into NASA's longer (decades longer) program of planned Mars exploration, and that it would interest and excite the general public—which was footing the bill, ultimately.

MOAR settled on a number of options for more detailed study,[2] including a reattempt at a retrorocket-descent stationary lander using the MPL design, an orbital mission to recover lost MCO science using updated instruments on an orbiter like Mars Global Surveyor (MGS), and a mobile rover based on the Mars Pathfinder concept, carrying elements of the payload that had been selected for the Mars Surveyor '01 lander, and delivered to the surface using a similar airbag-assisted system. Inside Caltech's Jet Propulsion Laboratory (JPL) in Pasadena, rover designers initially called this the Mars Geologist Pathfinder mission.[3] Of course, another option, launch no mission to Mars at all in 2003 because the risks were still too high for the expected science return, also remained an unspoken contender. By the time of the MOAR meeting on May 3, 2000, the group had come to the conclusion that there just wasn't enough time to fully understand and verify the fixes to the landing problems that had doomed MPL, and so the MPL-like option was taken off the table. If there was to be a 2003 Mars launch, then, it would either be an orbiter or a rover.

Two engineers from JPL, Mark Adler and Rob Manning, played critical roles in MOAR's ultimate decision. Both had been instrumental in the design, development, and successful landing of Mars Pathfinder and the Sojourner microrover just three years earlier, and both were strong proponents of getting a larger, more capable rover to Mars as soon as possible, leveraging the Pathfinder experience. During that fateful May meeting, Adler and Manning pitched the idea to MOAR of a rover that would be based on the Sojourner design but would be larger (around the size of a small golf cart), with a sophisticated science payload based on the instruments widely known to have been proposed (but never selected) for a Mars rover Discovery-class mission called Athena, led by Steve Squyres.[4] Adler and Manning described to Hubbard and MOAR how they believed JPL could merge the Athena science payload with a scaled-up Pathfinder design. Intrigued, MOAR asked for an assessment of whether such a rover could be ready by 2003 or would need until 2005. Adler pointed out that 2005 would be a terrible year to land anything on Mars because the mission would have to land and operate during Mars's southern hemisphere summer, the dustiest time of the Martian year.[5]

Even though much of the committee, as well as the Mars science community in general, was torn over the choices, Hubbard ultimately recommended the rover option to NASA Headquarters in a critical meeting in July. The decision had really come down to MOAR since no one believed that the updated (and more complex) instruments that would be needed to make the biggest scientific advances from the orbiter would be ready in time to meet the tight schedule to the launchpad. Instead, it made more sense to aim to give those higher-resolution instruments more time to be ready for the 2005 launch opportunity.[6] Hubbard and others, including Adler, Manning, Squyres, and veteran Viking team member Ray Arvidson also believed that NASA needed much more experience with surface mobility, surface operations, and "ground truthing" of orbital data sets in the lead-up to the longer-term Mars program objective of robotic sample return. And the rover folks also had a bit of celestial mechanics luck on their side: because of the extraordinary proximity of Earth and Mars during the 2003 launch window (appendix D) and the specific Mars season of the 2004 landing opportunity, both the communications rate (and thus the number of transmitted bits) and the power available for a solar-panel-powered rover would be greater than they would otherwise be for another 18 years. The 2003 launch was the most favorable opportunity for a surface mission in decades, and so MOAR recommended that Adler and Manning's rover mission go forward in 2003, and an orbiter mission be put off until 2005. The road ahead would be challenging because so little time was available to design, build, and test a new rover and its science payload. Choosing the rover was sort of the least unsatisfactory option for the 2003 launch opportunity. Adler described winning the MOAR shoot-out as "winning with a pair of twos."[7]

The final decision was not Hubbard's, however. It was made in that July meeting by Dan Goldin, the architect of the "better, faster, cheaper" deep space mission paradigm that had succeeded so spectacularly with the Mars Pathfinder and the Near Earth Asteroid Rendezvous Discovery missions but that, ironically, was also partially responsible for the 1999 Mars mission failures that had led to the needed restructuring of the entire Mars program. Though in 2000 Goldin was coming to the end of his long career as NASA administrator, it was not safe to regard him as a lame duck, or assume that he would simply adopt MOAR's recommendation without any pushback. Hubbard and crew carefully prepared a presentation package with their reasoned arguments on cost, risk, science return, and the other variables that went into their rover recommendation. As Hubbard tells it though, they had barely presented their conclusion when Goldin, always an out-of-the-box thinker, had made up his mind to pursue a slightly different course. "Why didn't you propose TWO rovers?" he asked.[8]

At first it seemed like a crazy idea. With a tight federal budget and an ever-tighter schedule, doubling the needed work seemed like the least appropriate thing to do.[9] But then the merit of the idea starting sinking in to Hubbard and others. NASA had a long history of building and launching "twin" spacecraft (Mariners 1/2, 3/4, 6/7, and 8/9; Pioneers 10/11; Voyagers 1/2; and Vikings 1/2) as a risk-reduction measure, and it had paid off after the failures of Mariners 1, 3, and 8. It was also a chance to dramatically increase the science payoff, if both missions succeeded (as many had). And, perhaps paradoxically, building two copies of the same machine (and instruments) can sometimes *help* meet a super-tight schedule, because so many spare parts have to be built and parts can even be interchanged between the machines if needed to keep one on schedule until the other catches up. Indeed, pushed to its limits, the strategy could even make it possible to test two identical spacecraft in the same time it would take to test just one, by doing half the tests on one spacecraft and the other half on the other. Many such "twin logic" decisions would ultimately have to be made in the crazy sprint between the August 2000 press conference announcing the decision to launch what was now being called the Mars Exploration Rover (MER) mission and the actual launch of two sporty rovers from Cape Canaveral just under three years later.[10]

Athena and the Mars Rock

It seems axiomatic that almost all of the most successful NASA robotic missions of deep space exploration have, at one or more times during their conception or gestation, been canceled. Perhaps it is because they were destined for such success that they all rose from the ashes (some, multiple times) to eventually fly to space. MER would be no exception.

One of the proposed ideas for the next Mars rover mission after Path-finder was to send a slightly more capable Sojourner-class rover that would be deployed to the surface after traveling to Mars on the deck of a Mars Surveyor '98–class three-legged lander (which was to have its first demonstration on the MPL mission in 1999). This mission concept was pitched to NASA by Steve Squyres in a proposal responding to NASA's 1996 call for the next round of Discovery program missions. NASA's rules for Discovery proposals were specific in terms of variables like cost, launch mass, and development schedule, but they were agnostic as to destination and science goals. Those were up to the proposing team. Squyres and his deputy, Arvidson, saw Discovery as an innovative, lower-cost way to do great science at Mars, and the kind of science that they and other experienced field geologists were advocating was science enabled by *mobility*. The rover would be outfitted with long-range and microscopic cameras, compositional and mineralogical spectrometers, and a brushing/abrading tool. Squyres named their proposed mission Athena, after the Greek goddess of wisdom. The purpose of the mission was to explore the geology, geochemistry, and mineralogy of a landing site chosen on the premise of "follow the water," identified on the basis of MGS orbital data.[11]

Sojourner had been designed to be a technology demonstration component of Mars Pathfinder, but even with its relatively limited range, it successfully demonstrated the value of mobility, slowly trundling from rock to rock near the lander and measuring a diverse range of rock and soil compositions with its elemental chemistry instrument (chapter 15). Based on that experience, it was easy for Squyres, Arvidson, and others to imagine how much more science and exploration could be achieved if such capabilities could be scaled up; the Discovery program provided a way for such ideas to be pitched to NASA and to be competed against other ideas for exploration of other places.

Athena was indeed a wise mission to pitch, but sadly it was also a victim of some unfortunate bad timing that was associated with the biggest scientific story of 1996. As described earlier, that summer, meteorite researchers at NASA's Johnson Space Center held a press conference to claim that the Mars meteorite ALH84001 contained chemical and fossilized microscopic evidence of preserved ancient Martian life-forms. It was a sensational, extraordinary claim that created an instant international media frenzy. Even U.S. president Bill Clinton got involved, vowing after the announcement that "the American space program will put its full intellectual power and technological prowess behind the search for further evidence of life on Mars."[12] The administration (and Congress) followed through on the vow, and NASA received a big boost in funding for solar system research the following year. This was great news for planetary researchers in general but potentially bad news for the Athena team in particular, because now it meant that NASA could afford a much more capable and ambitious Mars surface mission than one constrained by the comparatively meager budgets of the Discovery program. Indeed, in fall 1997, NASA rejected the Athena proposal and instead selected a comet flyby and solar wind sample return mission as the next Discovery winners.

Frustrated but undaunted, Squyres and most of the Athena team re-proposed the rover's payload in 1997 as candidate for the Mars Surveyor '01 lander opportunity, in a project called the Athena Precursor Experiment (APEX). Still in the wake of the ALH84001 announcement, the timing was right for such a collection of instruments focused on the geology, chemistry, and mineralogy of the surface, and in late 1997 APEX was selected to fly on the Mars Surveyor '01 lander. The APEX team would spend several years designing, building, and testing prototypes and engineering models of the instruments. At this same time, the details of the mission itself evolved dramatically, morphing into a potential first attempt at Mars sample return using a backup of the Sojourner rover called Marie Curie and a lander-launched Mars ascent vehicle that would propel delivered samples into Mars orbit. The original dream of a much more capable rover using its extended mobility to bring high-tech instruments to a variety of outcrops across an exciting field site was quickly slipping away from Squyres and the Athena team: Marie Curie could only carry one instrument and some rudimentary cameras, and the rest of the high-tech payload would have to remain stuck in one spot on a stationary lander. Indeed, the house of cards came tumbling down in late 1999 after the crash of the MPL revealed systemic problems with that system's entry, descent, and landing design. The Mars Surveyor '01 lander and APEX were canceled, Squyres and his team were banished for the time being into the wilderness, and, as described above, MOAR was enlisted to restructure the whole Mars plan going forward.

Ultimately, Adler and Manning came to the rescue in spring 2000, and much of the original Athena payload was carried to Mars on the larger, more capable MER mission, albeit with some significant changes required to accommodate the mass, volume, power, and cost constraints of the new rover design (figure 17.1).[13] Each rover's payload included a pair of high-resolution cameras called Pancam designed to take color stereo panoramas to assess the geology and some aspects of the mineralogy; a Microscopic Imager, on the rover's arm, designed to take black-and-white photos of tiny parts of the surface at a scale comparable to a geologist's hand lens; a thermal infrared spectrometer for mineral mapping called Mini-TES that was based on the Thermal Emission Spectrometer (TES) instrument that had flown on MGS; an elemental chemistry spectrometer on the rover's arm called the Alpha Particle X-Ray Spectrometer (APXS) that was a direct (improved) descendant of the APXS instrument originally carried to Mars on the Sojourner rover; a Mössbauer spectrometer, also on the rover's arm, that could use the Mössbauer effect from nuclear physics (the recoil of gamma ray–emitting atomic nuclei) to detect specific iron-bearing minerals; a brushing and grinding Rock Abrasion Tool (affectionately called the RAT), also on the rover's arm, that could allow the instruments to get beneath weathering rinds and the ubiquitous dust that coats or covers everything on Mars; a series of magnets designed to characterize the properties of Martian dust; and several calibration targets designed to help guarantee the accuracy of the imaging and infrared spectroscopy data sets. Each rover was also outfitted with three more pairs of black-and-white stereo cameras to provide the engineering data needed to properly drive the vehicle and to accurately place the arm on rock and soil targets.

Where to Land?

The decision in fall 2000 to send two rovers to Mars carrying the previously vetted and partially developed Athena payload brought Squyres's team back to life (and back from a deep depression), and into the forefront of the international planetary exploration spotlight. Not only would the Mars Exploration Rover (MER) have to reestablish NASA's competence for landing on Mars, but it would also have to live up to the expectations of the media and a space-savvy public still whipped into a frenzy about the possibility of life on Mars by ALH84001. Even though subsequent analyses by others had not been able to confirm the interpretations of the original authors of the ALH84001 *Science* paper (and, indeed, those specific interpretations have still not been confirmed to this day, as of late 2020), the buzz created by the claimed discovery of ancient fossilized Mars microbes continued to reverberate through the scientific community.

Regardless, the combined team of JPL engineers, engineering and technology experts from other NASA centers and from aerospace companies around the world, and the Athena Mars science team had to race around the clock to

FIGURE 17.1 *Left:* The Spirit rover, one of the twin rovers designed for the MER mission, underwent drive tests at NASA's Jet Propulsion Laboratory in late 2002. The smaller Mars Pathfinder flight-spare Marie Curie rover was also on the test floor, for fun and for scale. *Right:* Cartoon view of Spirit (or Opportunity, its twin), with some major components indicated. NASA/JPL.

get the rovers to the launchpad. Called MER-A (or MER-2) and MER-B (or MER-1) for most of their first few years, the A and B rovers were eventually named Spirit and Opportunity by a student, in another worldwide naming contest sponsored by NASA. Spirit would be the first to launch, followed by Opportunity. With the prospect of operating two independent vehicles looming ahead, it turned out to be exceedingly wise that Squyres had chosen highly experienced planetary scientist Ray Arvidson as his deputy principal investigator for the team. While both scientists would end up working on both rovers throughout their missions, for a time Squyres would be regarded as the captain of the Opportunity rover, and Arvidson as the captain of Spirit.

For the first time, NASA and the global Mars science community had high-resolution imaging, elemental chemistry, and mineralogy data to help assess the choice of landing sites, thanks to MGS and Mars Odyssey orbiter observations. In addition, 20 years of significantly improved knowledge of the Martian weather, topography, orbital parameters, and gravity field, as well as advances in some entry, descent, and landing (EDL) systems at JPL (like special descent cameras designed to track the landscape automatically), helped shrink the sizes of the "error ellipses" where the MER landings could be targeted. Rather than all this data making the job easier, however, the choice of landing sites for the rovers became even more complex because it was now possible to land in so many more potential sites with interesting geology or mineralogy (as well as many more visible obstacles to avoid!). Shortly after the rover mission was announced, NASA began a multiyear landing site selection process that involved not only mission EDL engineers but also the global community of Mars researchers. Multiple workshops were convened, during which many potential sites were proposed, argued over, and winnowed to fewer and fewer numbers over time.[14]

The first rover, Spirit, would ultimately be targeted to a 165-kilometer-wide impact crater called Gusev, about 15° south of the equator (plate 16). The attractive thing about Gusev was that there is a long and meandering water-carved channel that flows into the crater from the south, but no obvious exit for the water that must have flowed in. Thus, the working hypothesis prior to landing was that Gusev might have hosted a crater lake early in Mars history. On Earth, such confined lakes are excellent places for geological research, where both physical and chemical/mineral evidence of life can be preserved in the lakebed sediments. The second rover, Opportunity, would be sent to the opposite side of the planet (helping simplify dual rover operations, if they both landed safely), to a broad dark region called Meridiani Planum, which had been visible for centuries in Earth-based telescopic observations (plate 16). Meridiani is dark because it is relatively dust free, thus potentially exposing rocky outcrops for the rover's study. But also, Meridiani was the place where the TES team on MGS had discovered the planet's highest concentration of coarse-grained hematite (Fe_2O_3), an iron oxide mineral that on Earth often forms in the presence of liquid water. In Gusev, there was a clear geological story telling the team to "follow the water" to this place; in Meridiani the planet was screaming the same message, but the story was mineralogical rather than morphological.

These two sites were judged to be not only scientifically interesting and aligned with NASA's "follow the water" strategic theme, but also safe for the rovers to land on (based on elevation, solar power availability, rock abundance, roughness/slopes, and other factors). However, the NASA Headquarters–level approval to land the rovers at these sites would only come after months of testing of the parachute and other landing system components by the JPL engineers. In fact, some of the most critical testing actually happened after launch, because up until then neither Gusev (too windy/rocky for the as-tested EDL system) nor Meridiani (too high in elevation for the as-tested parachute system) was deemed safe enough to land in.[15] Fortunately, and after heroic efforts by the rover engineering team, testing succeeded in validating the performance of the EDL system at both sites by the beginning of August, just in time for the postlaunch trajectory correction maneuvers needed to target each rover to its scientifically desired locale.

With Spirit in Gusev Crater

Spirit launched early in the 2003 Earth–Mars launch window, lifting off from Cape Canaveral on a Delta II launch vehicle on June 10. Opportunity launched about a month later on July 7. Typically (or, at least, hopefully) the six to nine months it takes spacecraft to cruise to Mars is an uneventful period, and this was generally true of the cruise phase of both Spirit and Opportunity. However, both spacecraft encountered unusually strong high-energy particle bombardments by solar flares during parts of their cruise phase, which occurred near the maximum of the Sun's 11-year cycle of activity. While mission operations engineers back on Earth could monitor for electronics problems created by such intense "solar storm" activity in many of the rovers' systems, certain parts of the rovers' computer memory could not be monitored. As a result, and at some nonzero risk to the mission, ground controllers completely reloaded both rovers' memory banks shortly before landing, to make sure that none of the software was corrupted.[16]

NASA had taken an important lesson away from the 1999 failure of MPL: monitor the EDL telemetry from a Mars landing vehicle *continuously* during the landing process, using whatever assets possible, so that if a failure occurred, engineers, managers, future mission designers, and all the relevant funding stakeholders (like Congress) would hopefully be able to tell exactly what had failed. To that end, NASA had arranged specific trajectory correction tweaks to the orbits of both the MGS and Odyssey orbiters to make sure that they were in direct radio contact above the rovers during their atmospheric entries, or could serve as rapid relays of initial data immediately after the rovers landed. The combination of direct-to-Earth signals from the rovers and relayed signals from the orbiters, all captured by the enormous radio telescopes that make up NASA's Deep Space Network and all displayed to a global audience in real time on television and the Internet, gave the world a front-row seat to the "six minutes of terror" needed to land safely on Mars.[17]

And what a show it was! The same heat shield then parachute then retrorockets then airbag landing system that JPL had devised for Mars Pathfinder seven years earlier was used again—with some modifications—to safely land both Spirit and Opportunity on Mars. Reconstructed telemetry after the landing, which was all choreographed by each rover's onboard computer to happen autonomously, showed that the airbag-encased Spirit lander and rover bounced about a dozen times on the surface before finally coming to rest on the rocky plains of Gusev crater. Within minutes the airbags were deflated and the lander's petals opened, flower-like, to expose the rover to Mars for the first time. Earth set below the horizon shortly after the landing, so the team held its collective breath as Mars Odyssey passed over Gusev, planning to relay the first pictures from the surface. This important experiment—relaying images and other data through a Mars orbiter and then on to Earth—worked spectacularly the very first time it was attempted, on Spirit sol 1, shortly after landing.[18] Television and computer screens around the world lit up with the first stunning views of the rover's new home (plate 17). The orbiter relay experiment would go on to work for both rovers many thousands of times more, using Mars Odyssey, MGS, and Mars Express to send back to Earth more than 100 times the data that could have been sent directly using the rover's own lower-data-rate radio system.

Soon after landing, it became clear, using the rover's many scientific instruments, that the plains of Gusev crater are dominated by relatively unaltered, bone-dry, volcanic rocks.[19] No evidence of the hoped-for lakebed sediments (or any evidence of there having ever been a lake at all) was found at or near the landing site. Even after driving the rover 600 meters away from the landing spot to visit a 200-meter-wide crater named Bonneville, on the hope that it would expose some ancient lakebed deposits beneath the ubiquitous lava flows, the team found no evidence of the hoped-for watery past. Gambling with a decision that would push the rover's endurance to its limits,[20] the team decided to head 2.5 kilometers farther on, with the legendary Indian American rover operator Vandana ("Vandi") Verma skillfully taking the rover up into a group of seven hills that had irresistibly beckoned from the landing site. Named the Columbia Hills after the seven astronauts killed in the 2003 Columbia Space Shuttle reentry tragedy, the hills appeared on the basis of orbital data to host older and different rocks from those on the plains.

Though Spirit's march toward the hills was determined, occasional (brief) interludes from its dogged exploration of its rocky surrounds were allowed. One of them, occurring two months after the landing at Gusev, was truly historic. In March 2004, Spirit captured the first image of Earth ever taken from Mars.[21] Mark Lemmon, an astronomer and planetary atmospheres researcher then at Texas A&M, and one of us (J. B.) did the planning for the Pancam observation, and also did a fair amount of advocacy with Squyres and mission management to get the rover to "wake up" early and use significant amounts of power to heat the cameras and other systems in the frigid predawn hours in Gusev crater. (Squyres, as a fellow disciple of Carl Sagan, didn't need much persuasion, and was actually a huge supporter of the idea. Once the managers and engineers were shown the historic value of the attempt—and its importance to the human spirit—they, too, got onboard.)

Imaging Earth from Mars was not without risks, however. The first was that it really is extremely cold at night on Mars, with predawn temperatures dipping down to −90°C to −105°C depending on season. So systems like the cameras had to use significant power to be heated up to the −30°C to −40°C minimum temperatures needed to operate them. For a solar-powered rover, this meant using the battery for nighttime observations, but of course the battery was primarily needed to run heaters inside the body of the rover to keep all of *those* systems warm *every single night*. Using battery power for science (or purely inspirational) measurements wasn't in the plan. Though the managers and engineers were supportive, they were well aware of the risks to the spacecraft; even so, they determined that the risks could be made acceptable by prepointing the rover's azimuth and elevation mast that carried the cameras to the right place in the sky in the afternoon, before the rover went to sleep, in order to get around having to heat the mast mechanisms during the following predawn hours. That saved some power, but it meant that a mosaic couldn't be taken—without heating the azimuth and elevation motors, only one camera pointing was possible. The team also agreed to use just one Pancam, not both. The left Pancam was selected because it had a broadband filter (to capture maximum possible light) that the right camera didn't. That helped save a little more power.

There was another risk: that the photo would be taken but would show nothing. Earth would certainly be there at the time of the observation; the team knew this, in part, because it used the astronomy software Starry Night to simulate a view from a vantage point in Gusev crater (the company has since used the rover's results in its marketing!). And the observation was timed so that Earth would be near greatest elongation from the Sun, and thus highest in the Martian sky. However, the dust in the Martian atmosphere causes the twilight sky to be really bright compared to Earth's twilight sky. "Astronomical twilight" as we know it here can occur as much as three hours before (or after) sunrise (sunset) on Mars, depending on how much dust is in the atmosphere! Since the dustiness cannot be perfectly predicted day by day, a chance just had to be taken, on the hope that the sky wouldn't be *too* bright and swamp out Earth. As insurance, three sets of images were taken about an hour before sunrise, from about 4:45 to 5:10 a.m. on sol 63 (March 7, 2004), using broadband, blue, green, and red filters. As expected, Earth was most visible in the first set, when the sky was darkest (but still bright! We all could have easily walked around in its glow . . .). A wide-angle "context" photo was also taken with the black-and-white left navigation camera, or Navcam. While Lemmon and team didn't expect to be able to see Earth in the Navcam photo (and they didn't; the camera is less sensitive than the Pancam), they were able to insert the Pancam photo of Earth into the right place in the Navcam view to get the landscape context of Earth just above the horizon. A little Photoshop cleanup work along the seams was needed to blend the images and create the product that went out to the world (figure 17.2). The result was sweet and emotional: everyone involved really enjoyed the historic "first" and, of course,

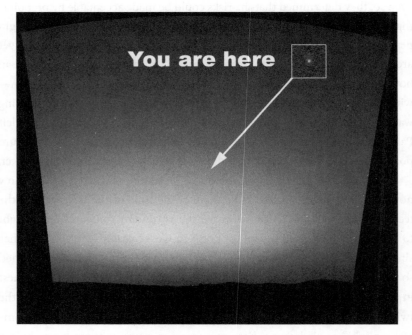

FIGURE 17.2 The "pale blue dot" of home. This photograph of Earth was taken by the Spirit rover one hour before sunrise on sol 63 (March 7, 2004). This image is a mosaic of shots taken by the rover's navigation camera showing a broad view of the sky, and a photo taken by the rover's panoramic camera of Earth. The contrast in the panoramic camera image was increased twofold to make Earth easier to see. It is the only thing visible in the bright predawn Martian sky. The resolution and sensitivity of the image is not sufficient to show the Moon. NASA/JPL/Cornell/Texas A&M.

truly appreciated the amazing and wonderful perspective that Spirit had been able to share. While Earth as a "pale blue dot" (in Sagan's phrase) had already been seen from beyond Neptune's distance by Voyager 1 back in 1990, Spirit's view of our home world would not be bested until July 2013, when the Cassini mission team produced its absolutely stunning view of Earth from Saturn. Spirit's shot of Earth was a giant, far-flung step outward from the famous *Earthrise* image taken by William A. Anders as he serendipitously captured our blue globe rising above the stark lunar landscape from Apollo 8 in December 1968, and it added a new chapter to humanity's understanding of just how tiny, insignificant, and fragile our world is in the immensity of the universe.

Spirit's cameras were pointed skyward once more, thanks to serendipitous circumstances, during late 2005.[22] By then it had scrambled up Husband Hill, a low peak about 100 meters (325 feet) above the plains of Gusev crater, and endured the rigors of the Martian winter. Though among the rover team's greatest concerns had been its solar-charged batteries being choked off by dust storms and the dim winter sunlight, by chance a dust devil had blown away many of the particles that had settled on the solar panels, and by the time the rover crested the summit, it was summer. Now there was a surplus of battery power, enough to run the rover's computer and cameras during the day but also at night, and Lemmon, Mike Wolff of the Space Science Institute, and co-author and Pancam team lead Bell were able to take advantage of the situation to turn Spirit into a kind of otherworldly backyard observatory. Over 30 nights, they obtained images of the Martian satellites, bright stars, and planets, and they searched for meteors, giving humanity its first-ever glimpse of what it would be like to be an astronaut viewing the night sky from the surface of the Red Planet.

Not surprisingly, the Martian satellites were the "star" performers in the sky over "Husband Hill Observatory." Especially Phobos. With a third of the apparent angular diameter of our full Moon and a period of revolution faster than that of the planet's rotation, it rises in the west and sets in the east. It shines brightly at about magnitude −9, but that's still only around 1/30th as bright as Earth's full Moon. Instead of being round in shape, Phobos is oblong, and the most obvious feature on its surface is the large impact crater Stickney, on its leading hemisphere. Like Earth's Moon, Phobos always holds the same face toward Mars. Deimos is also prominent, shining at magnitude −5.5, about 2½ times brighter than Venus appears in Earth's skies. Spirit was able to capture both moons' frequent "lunar eclipses" as they passed into and out of the shadow of Mars, and even detected a small amount of reddish "Mars shine" on Phobos during the few minutes of eclipse ingress and egress (once in shadow, they effectively disappeared). After the satellites, the most conspicuous objects in the Martian skies were Mars's neighboring planets—including Earth and Jupiter (at magnitudes −2.9 and −1.4, respectively), which rose together before dawn in the eastern sky in late 2005.

Back Down to Mars

Spirit's mobility system proved robust, though it more nearly resembled the proverbial tortoise than the hare, with a top speed of only about 4 centimeters per second (possible only during the few hours of peak sunlight) and a daily range of at most 30–40 meters. Through sheer persistence, the Spirit team managed over three months to guide the rover to its Columbia Hills destination. It then began to ascend the hills and to indeed allow the team to discover new things.[23] At least six different and new chemical and mineral categories of rocks were found in the hills. The rocks contained basaltic minerals common to many terrestrial volcanic rocks (minerals like olivine, pyroxene, and feldspar), but also iron-bearing sulfates, hydrated iron oxides, clays, and even carbonates formed by varying degrees of alteration of those basaltic rocks by watery fluids. The team really hit pay dirt, however, after guiding the rover down the other side of the hills to a small, circular, layered, flat-topped, and low mesa known as Home Plate. By this time, more than 750 sols into the mission, one of the rover's front wheels had stopped spinning. Forced to drive backward, dragging that stuck wheel through the soft soil, the rover accidentally exposed both buried hydrated sulfates and buried hydrated silica minerals in the vicinity of Home Plate. That evidence, plus outcrops of opaline silica, as well as the explosive volcanic deposits of Home Plate and the surrounding frothy volcanic rocks, led to the hypothesis that an ancient hydrothermal system had been present in the area.[24] The Spirit rover team would spend the rest of the mission exploring the details of this fascinating place.

While it took several Mars years and much more driving than had ever been anticipated, the MER team was eventually able to show that it was indeed "following the water" in its explorations of Gusev crater. The water story was not what had been anticipated from orbital reconnaissance, however (a recurring theme in modern Mars exploration).

Rather than lakebed deposits and ancient shorelines, the team uncovered subtle evidence of water-rock interactions in some places, and then dramatic evidence of hydrothermal interactions (hot water/steam and rock) at Home Plate. A reassessment of the opaline silica outcrops adjacent to Home Plate demonstrated their likely origin to be hot spring sedimentary deposits known as silica sinter, with provocative evidence for structures like those that form around hot springs on Earth through the interaction of geology and microbiology.[25] Thus, with evidence of liquid water and heat/energy sources, coupled with the inferred presence of organic molecules (from a long history of asteroid and comet impacts) and possible textural biosignatures, the Spirit rover team was able to postulate the exciting conclusion that much earlier in Mars history, Gusev crater could possibly have been a habitable environment for life.[26]

Spirit and Opportunity were designed to perform primary missions that lasted at least 90 sols on Mars, and to be able to drive at least 600 meters from the original landing location. Both rovers ultimately ended up far exceeding those goals. Indeed, Spirit's mobility allowed the team to move it from the photogenic but relatively less interesting environment in which it landed to a much more exciting and ultimately rewarding field site a significant distance away. Like all solar-powered systems, however, the rover was extremely sensitive to illumination conditions. Spirit remained operational (though with a series of partially disabling instrument and mobility problems over time) for more than 2,200 sols on the surface, and it drove more than 7.7 kilometers. Both rovers, but more so Spirit because of its dustier home, were occasionally granted life-extending solar power reprieves by swirling dust devils or by strong gusts of wind (often preferentially experienced on hills and ridges, just as for us on Earth) that almost magically "cleaned" some of the ubiquitous airfall dust off the solar panels, resulting in dramatic increases in available power. Sometimes in space exploration, you get lucky.

And sometimes you don't. Spirit's ultimate demise was initiated by a sand trap within a tiny crater near Home Plate that was concealed by a crusty surface layer and that gave way under the passing rover, trapping it in powdery Martian soil tens of centimeters deep. Months of effort with a mock-up in a

JPL testbed intended to try out extraction techniques, followed by weeks of wheel spinning on Mars, resulted in only centimeters of progress out of the trap. Dwindling power from the dimming autumn Sun left the team unable to free the rover. Spirit's last communication before the winter season finally sapped its power was received on Earth on March 22, 2010, more than six Earth years after it arrived in Gusev for its 90-sol mission.

Mars Time Again

From January through April 2004, the scientists, engineers, technicians, managers, and support staff working on both the Spirit and Opportunity missions lived on a disorienting daily schedule known as "Mars time." The length of a day on Mars (a "sol") is 24 hours, 39 minutes, 35 seconds, or about 40 minutes longer than a 24-hour solar day on Earth. Since the rovers were solar powered, they would automatically "wake up" in the midmorning when the Sun was high enough in the sky and then "go to sleep" starting in the late afternoon as it set. Just before the rover slept, that sol's stash of images and other data had to be uplinked to the Odyssey orbiter or the Mars Reconnaissance Orbiter (MRO) and from there downlinked to Earth. The team would work to process and interpret those measurements while the rover slept, coming up with the list of rover commands for the next sol that would have to be uplinked from Earth back to the rover after it woke up the next morning.

If the presleep downlink happened, say, around noon on Earth one day, it would happen around 12:40 p.m. the next Earth day, then around 1:20 p.m. the next Earth day, and so on.[27] After a few weeks, the rover was going to sleep in the middle of the night for team members keeping up with it back in Pasadena. After about a month on Earth, the cycle would be back to where it started. The result was a team of people living on Earth according to the solar cycles of another planet, slowly drifting their work schedules and circadian rhythms through the day to stay in sync with the rovers on Mars. For a subset of the team working on *both* rovers at the same time, there was little time for human sleep because the rovers were waking up and sleeping on opposite sides of the planet. It seemed like the Sun never set on the NASA Mars empire!

After three to four months of this distinctly discombobulating schedule, however, some team members began to get fatigued and short-tempered, and mistakes became more common. So project managers brought the team "back to Earth," setting work hours within sort of normal limits for a 40-hour work-week, and having the team plan multiple sols of rover uplink for times when the humans and the robots were too far out of sync. It was a strange and unusual early-mission experience (spawning a number of research projects on human sociology, sleep, and work cycles),[28] first demonstrated on Mars Pathfinder back in 1997, and since then, also used for the Curiosity rover (and planned for the Mars 2020 rover mission as well).

A Golden Opportunity in Meridiani

About three weeks after Spirit landed, that rover's twin, Opportunity, went through its own similar six minutes of terror en route to a safe landing in Meridiani Planum. Even though the rover carried an identical scientific instrument payload as Spirit, the Opportunity team would have a very different mission experience than their companions on the other side of the planet,[29] mostly due to the stark differences between Meridiani and Gusev (plate 18). Meridiani Planum is a dark and relatively dust-free (low-albedo, in telescopic parlance) equatorial region that has been monitored from Earth for centuries and was chosen as the prime meridian for Martian longitudes in the 19th century (hence the name Meridiani Planum for what had earlier been known as Dawes Forked Bay). It appears to be dust free because prevailing winds, at least in the modern Martian climate but perhaps for much longer, make this flat, expansive, relatively rock-free plain a place where *erosion* is the dominant agent of change over time, rather than *deposition*. Erosion on Mars, primarily by sandblasting under current climate conditions, removes the dust cover and grinds and planes down the surface over time. Hills, impact crater rims, and other positive topographic relief features are flattened out, and valleys, crater floors, and other low spots are filled in with some of that eroded debris.

Billions of years of erosion in Meridiani Planum have led to rocks that used to be in the subsurface now being exposed at the surface, as patches of flat, light-toned, layered outcrops. By chance, the airbag-enshrouded rover and lander happened to bounce many times across the flat plains and then roll into a small (auditorium-sized), shallow, and heavily eroded impact crater. While most of the surface showed up as a dark chocolate-brown color in the first color Pancam images sent back, brighter, buff-colored outcrop rocks happened to be exposed in the eroded walls of the crater, right in front of the rover. It was a fantastic stroke of luck for the Opportunity team—landing right next to the first true geological outcrop ever studied on Mars. In a celebratory nod to both golfers (what a hole in one!) and several past famous past ships of exploration, team members quickly dubbed their new home "Eagle crater."

The outcropping rocks in Eagle crater, with their clearly visible layers and light tone, *looked* different from any rocks previously seen on Mars. That impression turned out to be true in terms of their chemistry and mineralogy as well, once the rover was driven over to sample them directly with the arm-mounted spectrometers and other instruments.[30] Up close the outcrops were revealed to be layered sedimentary rocks, most likely sandstones, consisting of relatively soft (easy to grind into) altered minerals like iron- and magnesium-bearing sulfates and potentially other kinds of salts. The identification of the mineral jarosite (an iron-bearing hydrated sulfate) was particularly diagnostic of the presence of a watery past, because on Earth that mineral usually forms in watery but acidic (low pH) environments. The early mission excitement over the implications of surface or shallow subsurface water for the past habitability of Mars was soon tempered a bit, however, by the realization that the water might have been too acidic to support microbial life-forms like we see on Earth.

Another stunning discovery enabled by Opportunity's instrumentation was the existence of countless small spherical ball bearing–sized rocky grains littering the surface, both on the floor of Eagle crater and out on the plains beyond. Because of their shape, size, and bluer color than the more reddish sand and rock that they covered, the team began informally calling them "blueberries." Mysterious at first, the team quickly discovered after driving over to the outcrops that the native habitat of the blueberries is within the layers of these rocks (like blueberries in a muffin). Apparently, the outcrop rock is easier to erode than the blueberries, and so over time they pop out of the eroding outcrop (the fragments of which are carried away by the wind) and collect on the surface in enormous numbers. Mini-TES and Mössbauer spectrometer measurements on clusters of blueberries revealed their composition:

dominated by coarse-grained hematite. No one on the team had imagined that the surface manifestation of the hematite signature detected by the TES spectrometer on the MGS orbiter—a discovery that had played a major role in choosing to send Opportunity to Meridiani—would be millions and millions of tiny spherical iron-rich pebbles scattered across the ground.

But what *are* these blueberries? Based on their composition and other properties determined from Opportunity measurements, and by analogy to similar kinds of iron-rich, spherical mineral grains found in certain places on Earth, the team's favored hypothesis is that they are a kind of geological formation called concretions. On Earth, such iron-rich concretions (which can be found as large as ping pong balls in some places like southern Utah) form inside permeable, water-saturated, iron-rich sandstone rocks. As the water slowly recedes or evaporates, dissolved iron and other minerals begin to precipitate out of the solution, similar in some ways to how mineral pre-cipitation forms stalactites or stalagmites in a cave. But deep inside a porous rock, if everything is relatively symmetrical and slow, the minerals grow as spheres until the water is gone. If this hypothesis is true, then given the fact that blueberries have been found throughout Opportunity's traverse in Merid-iani Planum, the entire region might once have been saturated with (acidic) groundwater and perhaps even ephemeral ponds or lakes on the surface.[31]

But was this ancient aqueous environment limited to just the tiny region around Eagle crater where the rover had landed? How extensive in space and time had it been? To address those questions, the Opportunity team bet on the longevity and long-distance mobility potential of the rover and set out to explore other nearby impact craters. By trying to visit larger and larger craters, the team could be probing deeper into the Martian subsurface (because larger craters expose and excavate deeper materials) and potentially further back into Martian geological time. But it would be risky, and the team would face the same kind of uncertainties that the Spirit rover team had faced about the potential lifetime of the rover's driving mechanisms.

First, the rover was driven to the 800-meter-distant impact crater Endur-ance (130 meters diameter), where the team discovered spectacular layers of water-altered cross-bedded sandstones and maneuvered the rover to perform the first-ever robotic stratigraphic sampling investigation on another planet, using the RAT and other instruments to assess the chemistry and mineralogy as the rover slowly descended along the crater wall.[32] Next, the team decided to spend more than 1,000 sols attempting to drive nearly 6 kilometers south to Victoria crater (750 meters diameter), where they discovered more and ever-deeper evidence for similar kinds of water-saturated sandstones, concretions, and other outcrops, often exposed in some of the most dramatic cliffs ever photographed on another world.[33] Completing their reconnaissance of Victoria around mission sol 1685 (by now, approaching five Earth years into the mission), the team decided to set out for its biggest target yet, 22-kilometer-wide Endeavour crater, more than 12 kilometers to the southeast as the crow flies. For *three Earth years* the team guided the rover on a circuitous route around sand traps and other driving hazards, finally arriving at the outer rim of the crater in August 2011, after nearly 19 kilometers of driving!

Despite some significant wear and tear on the rover by then (including one unsteerable front wheel, a stuck shoulder joint on the arm, and no-longer-operational Mini-TES and Mössbauer spectrometers), Opportunity's arrival at Endeavour crater was like the beginning of a new scientific mission. Driving the rover along the heavily eroded rim, the team discovered bright gypsum-bearing veins crisscrossing the fractured rocks, as well as clay mineral coatings on a number of rocky outcrop surfaces. There was significant evidence for the extensive interaction between rocks and ancient liquid water at Endeavour (groundwater as well as potentially surface water), but the water there appeared to have been different from that at Eagle, Endurance, and Victoria craters. Specifically, the nature of the minerals found at Endeavour was consistent with *fresh* water, not salty or acidic such as had been seen before. With the large impact event that formed the crater also likely causing long-term heating of the region, the potential for other energy sources such as volcanic or geothermal heat, and the assumed supply of organic molecules delivered by the steady "rain" of asteroids and comets over time, Endeavour crater was found to be another site that might have been habitable long ago.[34]

Opportunity would continue to traverse and explore the rim of Endurance crater for the rest of its operational lifetime, continuing to enable the team to uncover additional evidence of the past watery history of this area,[35] and eventually completing the first Mars marathon and then some, with more than 45 kilometers of odometry, setting the solar system record for longest rover traverse on another world (including the Apollo lunar rovers and robotic Soviet

Lunokhods). An exciting twist during the last few years of the rover's mission was the team's decision to drive the rover down a shallow gully along the inner rim of Endeavour crater, to study whether the feature was formed by wind or water. The investigation had been only partly completed when, in the middle of 2018, an enormous planet-encircling dust storm darkened the skies for weeks over Meridiani, cutting off the rover's solar power supply and forcing the team to hunker down and put the rover to sleep to save energy. But without daily solar charging (during the peak of the storm, the daytime sky was so black that an astronaut there would have needed a flashlight to get around), the rover's batteries were likely drained so far that communications were no longer possible. Opportunity's last contact with Earth was in June 2018, at 5111 sols into its 90-sol mission. Sometime shortly after that, it bit the dust.

A Phoenix Rises from the Ashes

One more successful and historic Mars landing occurred during the eventful first decade of the 21st century. The Discovery program of small, competitively selected missions led by individual scientists was proving to be a success not only in terms of cost-effective science return, but also in terms of innovative ideas captured for new missions (the program typically attracted around 25–30 proposals from the planetary science community for each of the four open competitive opportunities that had been announced since 1994). However, the program presented an additional hurdle for Mars exploration: while Mars missions could be proposed to Discovery, they had to compete with outstanding mission proposals to the rest of the solar system and thus had low odds of success and couldn't be built into a more strategic component of NASA's Mars program. G. Scott Hubbard and others thus came up with the idea of creating a low-cost and high-innovation set of missions following the Discovery model, but specifically for NASA's long-term Mars Exploration Program. The resulting "Mars Scout" program announced its first mission proposal opportunity in 2002. Around 25 proposals were submitted, reinforcing the notion that the community had lots of great Mars-specific mission ideas to pitch to NASA.

The winner of that first Mars Scout competition, announced in 2003, was a mission called Phoenix, designed to study polar processes on the Red Planet using the mothballed copy of the failed MPL design that had been designated for the canceled Mars Surveyor '01 lander.[36] The mission was led by the University of Arizona (the first Mars mission ever led by a public university), in collaboration with JPL, Lockheed, NASA/Langley, and other partners. This time Peter Smith, the impish lead of the Mars Pathfinder camera team who had helped make that mission so successful and influential, would lead an entire mission of his own. Part of the mission's goal was to recover some of the science of MPL, but more importantly the mission was also designed to be a "ground truth" test of new results that had come from orbital imaging and spectroscopy since then. This time the mission would be targeted for a high northern latitude (68°N), partly because of celestial mechanics considerations, but also partly because Mars Odyssey Gamma Ray Spectrometer (GRS) and Neutron Spectrometer (NS) mapping data had supported the hypothesis that significant amounts of ground ice were buried just below the surface at those high latitudes (chapter 16). Could Phoenix verify the Odyssey results from the ground? If so, the Odyssey maps could be used with confidence to identify ice deposits elsewhere on the planet.

The Phoenix scientific instruments included a high-resolution color stereo camera system known as the Surface Stereo Imager (SSI) built by the University of Arizona, situated atop an extended mast at a height two meters above the ground in order to simulate the 3-D view that a tall person would have standing on the surface; another color camera on the lander's robotic arm for close-up imaging of the surface; a mass spectrometer called the Thermal and Evolved Gas Analyzer (TEGA) with eight small ovens that the arm could drop samples into for chemical analysis; a small "wet chemistry" lab called the Microscopy, Electrochemistry, and Conductive Analyzer (MECA), where chemistry and microscopy experiments could be performed on arm-delivered soil samples; special probes and sensors to measure the temperature, humidity, and other properties of the soil and atmosphere; and a special rotating digging/scraping tool called a rasp that was designed to enable some of the hoped-for ice to be scooped up and delivered to the lander's chemistry instruments.[37]

Phoenix was launched on August 4, 2007, and after a leisurely (compared to Spirit and Opportunity) and mostly uneventful nine-plus-month cruise it performed a picture-perfect landing in the high Martian arctic on May 25, 2008. The landing was monitored closely in real time by the MGS and Odyssey orbiters, as had become NASA's nervous tradition since the Mars mission failures in 1999. The successful landing of Phoenix was a testament to the significant additional time, money, and brain power that had gone into figuring out, between 2005 and 2007, why the original MPL system design had failed nine years earlier, as well as the skill of the JPL, Lockheed, and Langley team that was able to modify the system and then retest and validate the changes during that narrow window of time.

Within five sols of landing, Smith's team had deployed the arm and taken stunning photos with the arm camera of the "blast zone" made by the retro-rockets underneath the lander.[38] The platy white patches of ground looked just like the predicted buried water ice! Verification required more careful measurements, however, and so the arm was commanded to dig trenches around the lander to try to uncover and scoop up some ice directly. Several white patches were exposed in these trenches, and the observation that the patches faded into the background within a few days was consistent with exposed ice that sublimated (transitioned directly from solid to vapor) once it was exposed to the thin atmosphere (figure 17.3). Eventually, after some snags were resolved with the sample delivery system, fragments of dirty ice scraped up by the rasp were successfully dumped into TEGA to confirm—quantitatively—that the white material just beneath the dusty, pebbly surface was indeed water ice.[39] The Phoenix team had given the community the ground truth it was seeking.

Other important discoveries from the Phoenix chemical measurements included the detection of minor amounts of carbonates and perchlorate salts in the soils measured in the TEGA instrument ovens. Remote-sensing searches for evidence of carbonate minerals like calcite, dolomite, or ankerite on Mars have been conducted for decades, since the formation of such minerals in a CO_2-rich atmosphere could provide additional evidence for a putative warmer and wetter epoch earlier in the planet's history.[40] While some evidence for relatively small amounts of carbonate distributed widely in the bright global soils or locally in specific outcrops has been found from previous and ongoing infrared remote sensing measurements (as described earlier), the first direct detection of small amounts (3–6 weight percent) of carbonates came from measurements on the Phoenix landing site soils.[41] The origin of these minerals is uncertain, but it could be related to the formation of altered ejecta by the large impacts that created the northern lowlands, by the settling and mixing

FIGURE 17.3 Phoenix mission SSI mosaic of several of the approximately 10-centimeter-wide trenches dug by the lander's robotic arm, including an ice-bearing trench toward the left. NASA/JPL-Caltech/University of Arizona/Texas A&M.

of carbonate-bearing global dust in the soils at the site, or by the in-place weathering and alteration of precursor soil minerals with the aid of water or ice at the site.

Calcium and magnesium perchlorate minerals (chemical salts containing the ClO_4^- ion) were also discovered in small but significant abundances (up to 0.5 weight percent) in the soil.[42] Perchlorate is a strong oxidant, and is thought to form in the atmosphere and/or on the surface of Mars from the action of harsh ultraviolet (UV) radiation on chloride-rich dust or soils. Perchlorate has two important characteristics that make the discovery particularly relevant for Mars research. First, it can accelerate the breakdown of complex organic molecules, especially in environments with high UV radiation. This is why some researchers speculate that perhaps the presence of perchlorates could explain the lack of detection of organic molecules by the Viking Lander biology experiments. Second, perchlorate salts act as a kind of antifreeze when dissolved in water, significantly lowering the solution's freezing point and thus, potentially, allowing the water to remain liquid in certain places even in the normally well-below-freezing environment of the Martian surface. Some scientists thus speculate that perhaps the water-related activity possibly implicated in high-resolution images of gullies and other similar landforms was enabled by the mixture of perchlorates and groundwater.[43]

Phoenix was a solar-powered mission designed for just a three-month primary mission in the harsh low-Sun, low-temperature environment of the Martian polar regions. Ultimately, the mission lasted two months longer than planned before succumbing to the inevitable shortage of sunlight and solar heating as winter approached. The lander was likely crushed under the weight of a thick load of seasonal solid CO_2 (dry ice) that snowed out of the atmosphere as the polar night set in. Regardless, as the first demonstration of the new Mars Scout program of low-cost, focused-science Mars missions, Phoenix was judged to be a great success.

Raising the Bar for Mars Surface Operations

The first successful landed Mars missions of the early 21st century substantially advanced our understanding of and experience with the Martian environment because of a combination of their use of sophisticated instrumentation, their deployment to interesting and diagnostic landing sites that had been chosen based on similarly sophisticated orbital data sets, and, in the case of the MER rovers, their unprecedented mobility and longevity. During the early 2000s the practice of utilizing Mars infrastructure—communications relays built in to the MGS and Odyssey orbiters (and, eventually, also included on Mars Express, MRO, MAVEN, and others)—became commonplace, to the point where future landed missions would begin to bank on the continuing availability of that infrastructure to achieve mission success.

These missions also created a worldwide culture of virtual Martians, consisting not only of the many hundreds of professional scientists, engineers, technicians, mission managers, and others who were actively working on the operations of and rapid analysis/interpretation of data from these extraterrestrial vehicles, but also of the millions in the general public who watched and cheered during their successful landings and who then followed along for the latest raw and processed images and other results instantaneously released online by science teams willing and eager to share the adventure globally in near real time. The high level of public engagement and excitement created and nurtured by these missions is among their greatest legacies, helping to not only excite and train the next generation of Mars mission team members, but also demonstrate the scientific, educational, and inspirational value of space exploration to citizen stakeholders around the world.

18

Mountain Climbing with Curiosity

A Decadal Imperative

Once every 10 years, for several decades now, NASA asks the U.S. National Academy of Sciences to help determine the highest-priority scientific goals to achieve and robotic missions to fly in planetary sciences. The academy then convenes a blue-ribbon panel of subject matter experts and conducts a community-wide (global) survey of scientists, engineers, technologists, and managers to come up with a consensus ranking to present to NASA. These decadal surveys have also been conducted for NASA in the fields of astronomy, Earth science, and heliophysics (the study of the Sun), and they are highly regarded by leading members of Congress and presidential administrations as important guidance in their annual process of allocating funding for NASA.

The first planetary decadal survey of the 21st century was conducted in 2001–2 and published in 2003, and was intended to guide NASA's planetary science program for the decade up to about mid-2013. As part of the final report, the group of about 50 leading researchers who led and served on the various survey committees assembled a ranked program of recommended NASA robotic missions.[1] These recommendations represented what the survey authors believed to be a consensus for missions going forward, gleaned from dozens of white papers (informal group reports), town hall meetings, and a vast correspondence involving many hundreds of individual planetary scientists from around the world. One recommendation, based partly on the success of the 1997 Mars Pathfinder mission, was for NASA to continue to fly "better, faster, cheaper" Discovery-class missions, with science goals not dictated by the survey but instead bubbled up, grassroots style, from NASA's goals and the community's interests. The survey also recommended that NASA initiate a new class of similarly community-led and competitively selected missions called the New Frontiers series, which could potentially achieve even more science because they would be cost-capped at around $650 million (or around $1 billion today) not including the launch vehicle, twice the cap placed on Discovery missions. Mars missions were treated separately in the 2003 decadal survey, but the division of mission classes was similar, with the Mars Scout line (the first of which was the Phoenix lander) being comparable to Discovery-class missions. Medium-class Mars missions were deemed comparable to the New Frontiers–class mission line, and large-class Mars missions fell under the general category of flagship missions, multibillion-dollar projects chosen by NASA Headquarters (with the concurrence of Congress and the executive branch), rather than proposed by the community.

For Mars Scout missions, the survey recommended that NASA pursue an orbiter mission to explore Mars's upper atmosphere: this would be realized about a decade later with the success of the MAVEN mission, the second and, ultimately, last mission selected in the Mars Scout line (which was subsequently merged with the comparable Discovery mission program). For medium-class Mars missions, the survey's highest recommendation for missions beyond 2005 was for an enhanced roving vehicle called the Mars Science Laboratory (MSL), and its second recommendation was a mission to deploy a set of landers loaded with geophysical and meteorologic sensors across the surface, in a mission (never realized) called the Mars Long-Lived Lander Network. For the next large-class Mars mission, the survey recommended that NASA begin planning for a Mars Sample Return mission, so as to be able to implement it in the following decade (2013–23).[2]

Mars Science Laboratory Is Born

The idea of a larger, more capable Mars rover that would follow Spirit and Opportunity of the Mars Exploration Rover (MER) mission for a launch in 2007 or later had been bandied about within NASA for several years leading up to (and during) the decadal survey process, in part via a working group of scientists, engineers, and managers called the MSL Project Science Integration Group (PSIG). PSIG's final report, issued in May 2003, echoed and amplified the decadal survey's recommendations for the mission, but stressed that assessing the "habitability" of Mars should be the mission's primary goal.[3] That is, for the mission to be truly compelling, it would have to focus on proving that Mars had extended periods of time when (1) liquid water was stable on the surface, (2) atmospheric and surface conditions were favorable for the assembly of complex organic molecules, and (3) energy sources potentially suitable to sustain metabolism were available (the three main conditions required for habitability of life as we know it). PSIG was most enthusiastic about searching for evidence of habitability on ancient Mars; however, it also supported the idea of having lower-priority goals to assess the planet's current habitability as well.[4]

The scientific (and public) interest in habitability, plus the phenomenal scientific and public relations successes of Spirit and Opportunity, helped formally push NASA (and Congress) over the edge to follow the decadal survey's and PSIG's recommendations to design, build, launch, and operate the MSL rover. In spring 2004, just a few months after the successful landings of Spirit and Opportunity, the agency put out a call for scientists to propose instruments for MSL; 10 science payloads were selected by the end of the year.

The scientific goals of the MSL mission were originally defined at a high level as part of the 2003 planetary decadal survey report:

The Mars Science Laboratory (MSL) mission will conduct in situ investigations of a water-modified site that has been identified from orbit. It will provide ground truth for orbital interpretations and test hypotheses for the formation of geological features. The types of in situ measurements possible include atmospheric sampling, mineralogy and chemical composition, and tests for the presence of organics. The mission should either drill to get below the hostile surface environment or have substantial ranging capability. While carrying out its science mission, MSL should test and validate technology required for later sample return.[5]

A New Rover Takes Shape

The basic design of the MSL rover was generally based on the previous successful NASA designs of the Mars Pathfinder and MER, but it was scaled up to accommodate a more ambitious mission (figure 18.1). Specifically, while the Spirit and Opportunity rovers were designed for a minimum mission lifetime of 90 sols on Mars and a minimum roving distance of 600 meters, the MSL rover was being designed to operate for at least a full Mars year (668 sols, a little more than two Earth years), and to be able to traverse between 5 and 20 kilometers across its landing site region. The rover was also being designed to carry a much heavier and more voluminous payload, including a drill and miniaturized versions of modern laboratory instruments that could detect organic molecules and determine mineral compositions very accurately. Such instruments and expanded capabilities far exceeded the carrying capacity, cost limits, and lifetime expectations of the Spirit and Opportunity rovers, and thus the MSL rover had to be bigger—much bigger.

When the Jet Propulsion Laboratory (JPL) engineers and other mission designers sat down and scoped it out in the early 2000s, using the best available estimates for the masses and volumes and power needs of the kinds of instruments and systems that the PSIG and the decadal survey had envisioned, they quickly reached the conclusion that the vehicle would need to be the largest Mars rover yet built. Indeed, the MSL rover design ended up being about the size of a Mini Cooper rather than a golf cart (figure 18.1).[6] It would have to carry more scientific instruments, some of which would need to operate overnight, and thus would need more power (and steadier power) than the earlier rovers. Ultimately, the decision was made to make MSL nuclear powered, using the same kind of Radioisotope Thermoelectric Generator (RTG) as had been used on Voyager, Viking, and other NASA missions.[7] The freedom from reliance on solar power was predicted to be an important factor in allowing

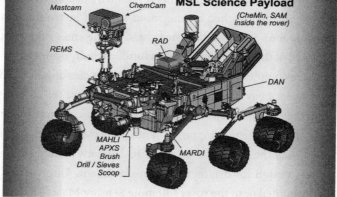

FIGURE 18.1 *Top*: Full-scale engineering models of all three generations of JPL Mars rovers, including the largest, the Mars Science Laboratory rover Curiosity, on display in the Mars Yard in Pasadena. *Bottom*: Cartoon showing the locations of the major science and engineering payloads carried to Mars by the Curiosity rover. NASA/JPL.

the rover to operate for much longer on Mars, and at a landing site potentially farther from the equator, than in the case of Spirit or Opportunity.[8]

The MSL rover seems like a car, but it is really a spacecraft (with a chassis instead of a bus), or more specifically only the final, deployed piece of a much more complex spacecraft system consisting of a cruise stage, an aeroshell, a descent stage, and the rover itself. The rover, embedded deep within a cocoon created by the aeroshell, is actually the brains of the whole operation, as its software controls all the functions of the other parts of the system. In addition to a computer and sophisticated software, the rover also has power, thermal, and mechanical/deployment systems comparable to those on many previous flyby, orbiter, and landed spacecraft, and indeed it relies on heritage from many of those earlier engineering systems.

Science Kit

Competition among the world's community of spacecraft science instrument providers is fierce. In response to NASA's 2004 Announcement of Opportunity: Mars Science Laboratory Investigations, which sought science payload elements for the mission, that community responded by submitting four dozen separate proposals that pitched an enormous range of instruments designed to meet the mission's advertised science goals. By the end of the year, NASA would choose 10 of those proposals (seven funded wholly or in part by NASA, plus three more completely funded by the space agencies of other nations) as the ultimate science payload to be carried by MSL (figure 18.1).[9]

The competed proposals resulted in five scientific-focused cameras or "eyes" being added to the rover, complementing the twelve other engineering-focused "eyes" that the mission team at JPL had already planned to build into the MSL rover design for driving and arm placement. Three of the science cameras are mounted onto the rover's Remote Sensing Mast (RSM) assembly, the one-time deployable "head" of the rover, which has a top height around two meters above the surface. Two of these cameras, called simply the Mast Cameras or Mastcams,[10] were proposed and built by now-veteran deep space camera maker Malin Space Science Systems Inc. (MSSS) as a complementary pair of zoomable color cameras, separated by about 24 centimeters to enable stereo imaging (these cameras, which would ultimately be scaled back from their original design, were together the first of the selected payload proposals). The third science camera on the mast is the Remote Micro-Imager (RMI), a black-and-white telephoto imaging system that uses the onboard Chemistry and Camera (ChemCam) laser spectroscopy instrument's telescope to identify the precise locations of laser shots.[11] The last two science cameras are the Mars

Hand Lens Imager (MAHLI), a color microscope that is capable of imaging features down to the scale of sand grains and is mounted on the rover's arm so that it can be placed right up against the surface;[12] and the Mars Descent Imager (MARDI), a color camera mounted to the bottom of the rover's body and designed to collect high-resolution images of the landing site during the final stages of the entry, descent, and landing (EDL) process.[13] Both MAHLI and MARDI were also designed, built, and operated by MSSS.

To complement MSL's record-setting 17 cameras in one Mars mission, an impressive number of spectrometers were selected as well. The ChemCam telescope at the top of the rover's mast, for example, feeds light from laser-zapped rock and soil targets down a series of fiber-optic cables and into a set of three small spectrometers mounted inside the rover body. ChemCam's spectrometers detect light from the ultraviolet to the near infrared, spanning wavelengths from about 240 nanometers to 900 nanometers (typical human vision is roughly 400–700 nanometers). The instrument uses a technique called Laser-Induced Breakdown Spectroscopy, or LIBS, to fire five-nanosecond pulses of a focused laser beam that ionizes a tiny portion of the nearby surface (spots only about one quarter to one third of a millimeter in size, at a maximum range of about seven meters from the rover). As the ionized atoms in the rocks and soils cool, they create narrow "emission lines" at specific wavelengths that are diagnostic of the chemical composition of those rocks and soils, similar to the way that stars and other astronomical objects emit lines at diagnostic wavelengths. ChemCam's spectrometers measure these lines, and then science team members compare their strengths and wavelengths to laboratory standards to determine the abundances of common geological elements like iron, silicon, sodium, magnesium, and aluminum.[14] The ChemCam design, fabrication, testing, and operations are a joint collaboration between the U.S. Los Alamos National Laboratory and the French national space agency, Centre Nationale d'Études Spatiale (CNES), in Toulouse.

Another spectrometer on MSL also measures the chemical composition of rocks and soils, while incorporating significant heritage and experience from previous Mars rovers. The Alpha Particle X-Ray Spectrometer, or APXS, is an enhanced version of a similar instrument carried to Mars on the Spirit and Opportunity rovers, which was itself an enhanced version of a similar instrument carried to Mars on the first rover, Sojourner.[15] The APXS on MSL is mounted on the rover's robotic arm and must be carefully positioned very close to or in contact with the surface region to be measured. APXS makes measurements of the "bulk" (average) chemical composition of circular regions of the surface around three centimeters in diameter and thus provides a high-fidelity set of bulk abundance measurements that complement the tiny point-to-point remote sensing measurements made by ChemCam. The APXS is led by a team from the University of Guelph and is funded by the Canadian Space Agency.

The MSL rover's third spectrometer package is an instrument suite called the Sample Analysis at Mars (SAM).[16] SAM is a high-tech analytic laboratory designed and built by NASA's Goddard Space Flight Center that combines a laser spectrometer and a mass spectrometer to deduce the chemical and isotopic composition of solid or atmospheric samples, and to search for both simple and complex organic molecules. SAM is one of the most complex and sophisticated instruments ever sent to Mars, consisting of a maze of tubes, wires, pumps, valves, heaters, 74 sample cups, numerous small motors, and an oven all packed into a microwave-sized box that takes up a significant fraction of the available science instrument volume within the Warm Electronics Box (WEB), or the body of the rover. Solid samples for SAM must be carefully delivered from the drill or scoop on the end of the rover's arm into a special funnel-like inlet on the rover deck that leads to the instrument itself. Because of the limited supply of sample cups in the instrument, many of which can only be used once, the team has to carefully consider and justify every potential sample measurement experiment, some of which can take up to three sols to prepare for and run on Mars. Atmospheric gases can also be directly fed into both of SAM's spectrometers to measure isotopes and to search for minor trace gases like methane.

The other analytic laboratory on MSL is called the Chemistry and Mineralogy instrument, or CheMin.[17] CheMin's unique attribute is that it can directly measure minerals, collections of atoms held together in specific structures that are often diagnostic of the environmental or chemical conditions in which they formed. CheMin uses a spectroscopic technique called X-ray diffraction (XRD) to measure the properties of powdered samples that, like the SAM samples, have to be delivered to the instrument, deep inside the WEB. A typical CheMin measurement takes about 10 hours to conduct, is conducted overnight, and is repeated many times over multiple nights in a row on the same sample. And like SAM, CheMin has only a limited supply of pristine sample holding cells (27), after which new measurements could end up "contaminated" by previous materials in the same cells. CheMin was designed, built, and tested at NASA's Ames Research Center.

One of the three completely non-NASA-funded instruments on the rover is called DAN, which stands for Dynamic Albedo of Neutrons.[18] "Albedo" is planetary science jargon for "how reflective is the surface?" (This was a contemporary iteration of the same question that had so intrigued telescopic observers over the centuries.) DAN measures how many neutrons are reflected out of the surface from either natural astronomical or rover RTG-created sources, or from a special neutron gun carried on the rover. In DAN's active mode, its neutron gun creates more than 100 million free neutrons every second using a tiny ion accelerator that smashes atoms of deuterium (an isotope of hydrogen with one proton and one neutron) into a target made of tritium (an isotope of hydrogen with one proton and two neutrons). These generated neutrons—or the more limited numbers of naturally occurring or rover-created ones that DAN detects in its passive mode—interact with atoms in the surface rocks and soils. The instrument can reveal the presence of hydrogen (for instance, in minerals like clays that contain H_2O or OH) because the neutrons lose energy when they collide with hydrogen atoms, which have very similar masses. DAN can distinguish between higher- and lower-energy neutrons, and it thus provides a sensitive way to search for evidence of hydration in the uppermost few tens of centimeters beneath the rover. DAN was designed, built, tested, and funded by Roscosmos, as the post-Soviet-era Russian space agency is known.

MSL also carries a meteorology package called the Rover Environmental Monitoring Station (REMS), which was designed, built, tested, and funded by a collaboration between the Centro de Astrobiología in Spain and the Finnish Meteorological Institute.[19] REMS is a weather station for Mars, monitoring temperature, pressure, wind speed, humidity, and UV index from sensors mounted onto the rover's mast and deck. Data from REMS complements a long history of similar meteorologic information about the current climate of Mars acquired from previous landed missions like Viking and Phoenix, and provides new inputs to models of both local and global weather phenomena.

Finally, MSL also carries an instrument designed not only to address the mission's science goals, but also to provide information needed for future human exploration of Mars: the Radiation Assessment Detector (RAD) instrument.[20] RAD was designed by the Southwest Research Institute in Boulder, Colorado, to collect the first measurements of potentially hazardous radiation at the surface of Mars, using sensitive detectors designed to register high-energy charged particles as well as neutrons and gamma rays, all of which could pose potential health threats to future astronauts and eventual settlers on Mars.

Overruns and Delays

Space missions are complex, expensive, technologically demanding projects that are often driven by grueling and strict schedule requirements. MSL was no exception, and as the design and initial testing of the rover's many systems and supporting components proceeded in 2005–8 for a planned launch in fall 2009, project engineers and managers encountered a series of vexing problems that required significantly more time and money to fix than originally budgeted.[21] And even though the problems were being fixed, implications of those fixes (and new problems that cropped up as a result) began stressing the schedule, literally to the breaking point.

For example, engineers working to design MSL knew, based on experience from Spirit and Opportunity, that the typical kinds of relatively small electrical motors (also called actuators) needed to move and steer the rover's wheels, or to point the mast and high-gain antenna, required significant heater power to keep their internal greasy lubricants from gumming up their steel gears in the frigid Martian environment. The team decided that actuators using lighter titanium gears and dry lubricants (which would work well in the cold on the rover's more than 50 motors) could significantly decrease the rover's power needs. It was a great idea in principle, motivated by reasonable engineering considerations. However, it proved to be a nightmare in practice, as the newly designed titanium gears and dry lubricant actuators weren't meeting some of the speed and other engineering requirements, and couldn't be proven to consistently work under Mars-like conditions over the expected lifetime of the mission.[22] Engineers would have to go back to the old, MER-like "wet lube" and steel-gear actuators, introducing additional cost and mass, and substantial lost time, into the rover's development.

As another example, in 2007 JPL's initial heat shield design for the rover's EDL system failed catastrophically during testing under simulated Mars-like conditions. Engineers had attempted to scale up the heat shield design used previously for the smaller and lower-mass MER missions, but those materials couldn't survive the higher speeds and higher temperatures needed for MSL. Fortunately, NASA was developing new heat shield technologies for its Orion entry capsule, designed for long-duration human deep space exploration after the Space Shuttle was retired, and that technology could be adapted for use by MSL. While it was a huge relief to find a solution that would work, there was a significant price to pay in both actual dollars and schedule delays.

Among the many straws that ultimately broke the schedule's back for MSL was yet another big development challenge, this one involving the drilling and sample processing system. Originally the system was designed as two mechanical arm assemblies, one that could drill intact core samples out of rocks, and a separate one that would crush/powder the samples for delivery to the SAM and CheMin instruments. However, early prototyping of these systems proved problematic, and the design didn't survive the preliminary design review process. Engineers had to go back to the drawing board and redesign the system, culling it down to a percussive drill that would create powdered drill tailings rather than a coring drill, and incorporating a powder handling, sieving, and delivery system directly into the rover's arm turret. The drill redesign, combined with the significant changes needed in the rover arm's capabilities, introduced even more schedule delays.

While MSL's many technical problems were getting solved, it was taking much more money than originally envisioned to solve them. Originally planned to be within the 2003 decadal survey's medium-cost mission class (with a budget not to exceed $650 million, or a little more than $900 million in 2020 terms), MSL had grown by 2007 to a $1.7 billion project (around $2.2 billion today), and by 2008 it was estimated to need up to $1.9 billion (around $2.3 billion today).[23] NASA Headquarters, saddled with a tight budget for the agency overall, became impatient with JPL's many requests for more money for MSL. Eventually, Alan Stern, a planetary scientist and the associate administrator of the

Science Mission Directorate at NASA, stepped in personally and directed JPL to cut back or "descope" a number of instruments and systems on the rover to lower overall costs and to try to keep the cost overruns to within NASA's Mars program budget.[24]

The descopes were viewed as draconian by some and absolutely necessary by others. Particularly painful were the proposed instrument descopes: removing the ChemCam and MARDI instruments entirely, and removing the zoom capability from the Mastcam cameras. ChemCam and MARDI were eventually restored with the help of significant foreign and other non-NASA funding, but the Mastcams were indeed pared down to ultimately be flown as one wide-angle camera (left eye) and one telephoto camera (right eye).[25] In retrospect, the descopes that were incorporated into MSL to save money arguably had little effect on the ultimate cost of the mission (which ended up around $2.5 billion by launch in 2011, or around $2.8 billion today; see appendix H), but they did negatively affect the operability of the mission and the morale of the team. For example, while the non-zoomable Mastcams have proven to work beautifully on Mars, it is nonetheless cumbersome to operate them as a stereo pair because of the 3× difference in resolution and field of view between the cameras.

As a result of these and other compounding smaller technical and cost challenges,[26] NASA Headquarters and MSL project management decided in late 2008 to pull the plug on the 2009 launch opportunity. There was no way that engineers and managers could fix all the problems and especially conduct all the needed tests of the fixes in time to get the rover to the launchpad with a high-enough assurance of mission success. Launch of MSL would have to be delayed until November 2011, giving the team 26 more months, until Earth and Mars lined up just right again, to get the rover and its systems ready to go. It was a crushing and expensive blow to the NASA Mars Exploration Program and to the thousands of people who were working in shifts around the clock to make the 2009 launch window. But building and testing complex machines and support systems for rocket launches, atmospheric entry and descent, complex landing maneuvers, and operations in a harsh environment like the surface of Mars is not easy. There is no established procedure or simple assembly manual for designing, building, and testing a Mars rover. The launch delay was a bitter pill to swallow, but it proved to be the right call for NASA and the MSL project.

If there was a slightly silver lining to the launch delay decision, it was perhaps only that everyone working on the project was finally able to call the rover by name. NASA had started a nationwide naming contest for the rover in late 2008, shortly before the launch delay decision was announced. The contest continued despite the delay, running through January 2009, and garnering suggestions from more than 9,000 K–12 students. In late May of that year, NASA announced that Kansas sixth-grader Clara Ma's entry was the winner. From that day forward, the MSL rover would be known as Curiosity because, in Clara's own words, "Curiosity is such a powerful force. Without it, we wouldn't be who we are today. Curiosity is the passion that drives us through our everyday lives. We have become explorers and scientists with our need to ask questions and to wonder."[27]

Where to Land—Again!

Well before the rover was actually built, where to send Curiosity for a now-planned August 2012 landing had been the subject of significant study and debate.[28] As it did for Mars Pathfinder and the MER missions, NASA conducted a set of public, community-wide landing site selection workshops that were designed to winnow down the choices using a combination of scientific and engineering "gates" that candidate sites had to pass through. The first workshop was held in 2006, and more than 60 potential landing sites that could address the mission's science goals were pitched to NASA by Mars researchers from around the world.[29] As is always the case, however, science potential is not the sole deciding factor in selecting a Mars landing site. Specifically, a number of critical engineering constraints were also imposed on the site selection process. Many of these constraints were similar to those imposed on previous NASA Mars landers and rovers (including elevation, surface slopes, rock abundance, and more), partly because of the basic similarity of the physics involved in *any* Mars landing attempt, and partly because the MSL landing system used some components and processes that were similar to those of previous NASA Mars landing systems. One constraint that was relaxed compared to previous landing site selection activities, however, was latitude: because of the use of the RTG power supply instead of solar panels, the selected site could be within a wider latitude zone, up to 30° north or south of the equator, and the power system would still keep internal and external rover temperatures within manageable limits. Finally, and most specific to the MSL landing system, the site had to be smooth enough (with potential topographic obstacles smaller than around 100 meters tall within regions around 1 kilometer wide) to stay within the allowed margin for altitude changes that could be tolerated during the final powered descent phase of the landing.

Just as in previous landing site selection efforts, there was a constant tension during the workshops between the need to land safely (and thus to choose relatively flat and boring landing sites) and the need to access exciting geological environments that are typically more challenging to land in but that might perhaps represent the best environments for past or present habitability. By the last workshop, in May 2011, the finalist sites had been narrowed down to just four that had been judged to best satisfy both the science and engineering requirements.[30] These included an ancient river and lake system in a crater named Holden, an ancient area of clay-rich sedimentary deposits in and around a valley called Mawrth, an ancient preserved delta deposit in the crater Eberswalde, and an enormous mound of sedimentary layered rocks within an ancient 154-kilometer-wide crater named Gale. Key common themes among these choices were the presence of ancient (three- to four-billion-year-old) exposed rocks, and sedimentary layers or other geological structures/patterns that indicated that surface water and/or shallow groundwater might have persisted at the site, perhaps for significant periods of geological time. All four sites satisfied the engineers, so the race to the finish line would be decided entirely by the science.

It was an agonizing choice to have to make: all four of the finalist sites (and indeed, many more of the original 60-plus potential sites) were compelling, and would be interesting places to explore. Ultimately, the finalist sites were voted on and ranked by the workshop participants by specific science criteria related to their geological diversity, their geological context, and their potential to preserve so-called biosignatures (certain textures, mineralogies, and compositions, including organic molecules, that could provide evidence for

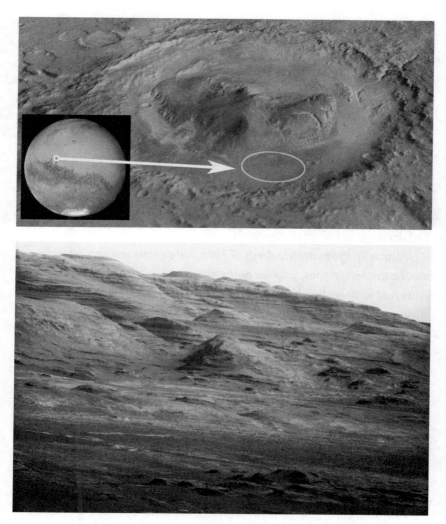

FIGURE 18.2 *Top*: Computer-generated orbital view of the Curiosity rover's landing site in 154-kilometer-wide Gale crater, created from Mars Odyssey THEMIS visible light images and Mars Global Surveyor MOLA topographic data. The rover's 20 by 25 kilometer landing ellipse is highlighted, as is the global context for Gale in the Hubble Space Telescope photo inset. *Bottom*: The telephoto lens view from the ground of Gale's mile-high central mound of layered sedimentary rocks, acquired by the Curiosity Mastcam imager on sol 17 of the mission. NASA/JPL/ASU/GSFC and NASA/JPL/MSSS.

past life) for billions of years. The effort to quantify these scientific criteria was admirable, but ultimately most members of the community knew that a significant amount of subjectivity and judgment goes into making such assessments, and so would be reflected in the final decision process. For many scientists, their choice of landing sites naturally had more to do with favoritism for their own specific processes or features prevalent at the sites (e.g., river and lake sediments, volcanic materials, key hydrated minerals). Or, the deciding factor might instead be their level of optimism versus pessimism about the lifetime of the mission, and thus the need either to have the rover's landing ellipse (20 by 25 kilometers) be right in the most interesting place to study, or to have the rover land in a safer location farther from the desired study region and drive out to the go-to sites from there.

Ultimately, the many hundreds of Mars community researchers were torn among the finalist sites, with the go-to lake-related sites Eberswalde and Gale emerging as only slightly more favorable (but essentially tied) in the voting. Around 50 members of the MSL Science Team also met in June 2011 to consider the finalist sites independently, from the perspective of the group of researchers who would have to actually carry out the mission at the chosen location. Led by project scientist and noted sedimentary geologist John Grotzinger from Caltech, the team was pretty evenly split between the spectacular delta at Eberswalde, and the spectacular layered mound of sediments (by now being informally called "Mount Sharp," after Caltech geologist and planetary scientist Bob Sharp) within Gale. In a close vote, the team favored Gale because of the greater evidence for both geological and mineralogical diversity of the site as seen from orbital data, the presence of abundant sedimentary rocks that might enable a more time-ordered story to be developed as the rover traversed, and a weak judgment that Gale's massive exposures of sedimentary rocks might preserve evidence for habitability better than Eberswalde's smaller delta deposits. Neither the research community nor the MSL team made the actual decision, however. Rather, those groups (and others) made recommendations to the NASA associate administrator for science, Ed Weiler, who was empowered by NASA as the "decisional authority" to decide where Curiosity would land. In July 2011, only about a year before landing, Weiler sided with both the community and the MSL team and made the official decision: Curiosity would go to Gale (figure 18.2).[31] Four months after this decision was made, on November 26, 2011, the spacecraft was lofted by an Atlas V rocket from Cape Canaveral. Playing hide-and-seek through partially cloudy skies, it set out on its nine-month journey to the Red Planet.

Safe on Mars!

NASA mission designers and EDL experts like JPL's Mars Program chief engineer Rob Manning and guidance and control engineer Miguel San Martín had long known that a vehicle the size of Curiosity would be too heavy to land on Mars using the kind of parachute, retrorocket, and airbag landing system that had worked for the smaller Spirit and Opportunity rovers. Fortunately, the engineers didn't have to start from scratch to design a new EDL system for Curiosity because, back in the early to mid-2000s, they had already studied a number of other relevant options for Mars landings, for rovers as well as for heavier potential future human landed missions.[32] Specifically, among the options that had been studied for a high-mass lander was a concept based on the way that heavy-lift helicopter pilots precisely control and "land" big payloads (like trucks and tanks) suspended from a deployed tether system. Among the most famous of such helicopters is the Sikorsky Skycrane, and thus the somewhat analogous EDL system that the NASA engineers ultimately converged on to land Curiosity came to be known as the "sky crane maneuver."

Much of Curiosity's sky crane landing system actually resembled earlier NASA EDL systems that had been used since Viking, because essentially all missions attempting to land on Mars need to solve the same basic problems. Thus, the Curiosity EDL team resurrected and scaled up a combined heat shield plus parachutes plus sky crane landing concept that had been seriously considered for the MER mission in the early 2000s. What was then known as the "rover on a rope" idea had been scuttled, however, in favor of the airbag landing system because of the inability to properly model and control the potential pendulum-like swinging of a rover as it dangled on an umbilical cord during a rocket-powered sky crane–style descent. By the mid-2000s, with more computing power (in the lab and on the new rover) and more mass available to enable the addition of a critical control system, the sky crane landing concept emerged as the favorite of the JPL EDL team. Specifically, the potential for chaotic swinging of the rover on a rope was mitigated substantially by sophisticated software that actively controlled eight large downward-pointed Mars Lander Engines by commanding eight-times-a-second thrusts. The thruster system was part of the rover's "descent stage" (which it wore like a backpack above the rover deck) and was designed to use real-time data from associated gyroscopes and active radar "pinging" of the distance to the surface to eventually zero out the descending vehicle's horizontal velocity. If that worked, the sky crane maneuver would then be able to lower the rover vertically, avoiding any problematic swinging.

The sequence of events needed to land Curiosity successfully on Mars—all of which had to occur perfectly, under prescribed and automated software control, 250 million kilometers from Earth—is mind-boggling.[33] Over the course of 431 seconds, from the first moment when its sensors could detect contact with the thin upper atmosphere, the EDL system had to decrease the rover's velocity from 5.8 kilometers per second (about 13,000 miles per hour) to zero—as gently as possible. During the initial entry phase, the rover remained cocooned inside a conical aeroshell (the "backshell") and behind a protective heat shield, which would reach a maximum temperature of 1000°C (1900°F) as atmospheric drag quickly slowed the rover down (peak deceleration reached 12.5 g's). While decelerating, the vehicle was also flying, however, using its eight small rocket thrusters to automatically maintain its angle of attack as well as to initiate several banking turns designed to ensure that the rover would arrive at the proper downrange location. This innovative guided-entry system worked perfectly and was the key to being able to land in Gale crater at all, in fact, as it helped reduce the size of the ultimate predicted uncertainty in the landing location from hundreds of kilometers across to just a few tens of kilometers.

At 259 seconds after the entry phase began, and now traveling at "only" about 1,600 kilometers per hour (1,000 miles per hour, which is Mach 1.7 in the Mars atmosphere), the rover deployed the largest parachute ever used in deep space to begin to slow the vehicle further. Measuring more than 15.6 meters across and trailing the rest of the descent stage by more than 50 meters, the nylon, polyester, and Kevlar parachute slowed Curiosity down to less than 320 kilometers per hour in vertical velocity over 116 seconds. During the parachute descent, the heat shield was separated and allowed to fall independently to the surface, exposing the bottom of the rover to the Martian environment for the first time.

From there, events deviated significantly from NASA's previous Mars landing processes. When the descent radar detected an altitude of only around one kilometer above the surface, the rover jettisoned the parachute and backshell, began free-falling to the surface along with the descent stage, and fired its powerful Mars Lander Engines to slow the fall. This powered descent phase of the mission lasted another 35 seconds, during which time the descent stage retrorocket thrusters successfully reduced the vehicle's horizontal velocity to nearly zero and vertical velocity to a very slow rate, ultimately getting the rover into a smooth 0.6 meter per second (1.3 miles per hour) vertical ride downward. The long-imagined sky crane maneuver then began with the rover only about 21 meters above the surface. It took five seconds for the rover to be gently lowered about 7.5 meters on nylon ropes, with the descent stage steadily thrusting above. The angles of the thrusters had been carefully chosen to be pointed away from the dangling rover below, so as to avoid damaging the sensitive hardware on its top deck. The rover's wheels—also its landing gear—were soon deployed, and after about 18 more seconds of slow descent, the rover's wheels contacted the surface. Detecting a change in the load on the thrusters, the rover's computer commanded the descent stage to stop descending, and it fired pyros to cut the ropes and the electrical connection to its longtime backpack companion.

Success! Curiosity had safely arrived on Mars, in Mars Year 31, $L_s = 150°$ (the terrestrial equivalent was August 6, 2012; see appendix F). Once freed, the descent stage used its own onboard computer to angle back and fly away from the rover in a safe manner; it traveled in a parabolic path for another 20 seconds and then crashed about 650 meters (0.4 miles) away.[34] Meanwhile, the rover executed a quick set of systems checks and took a few wide-angle snapshots of the landing site, relaying the data in real time to NASA's Mars Odyssey and Mars Reconnaissance Orbiter (MRO) spacecraft, which had been positioned to be in direct contact with the rover during most of the EDL process and for a few additional minutes afterward. Odyssey was able to relay the data back to Earth almost immediately (it had been configured for "bent pipe" radio signal relay mode), and after a travel time of almost 14 minutes— the time it took the radio signals to get to Earth at the speed of light—the first images and other data from Gale crater began to light up the engineering consoles at JPL.

The place went bonkers, with engineers and scientists jumping up and down, trading high fives, laughing joyously and openly weeping. All of it was carried live on the Internet and on several television networks, and viewers around the world, from living rooms to classrooms to science museums to Times Square shared the euphoria and celebrated along with the team.

The landing hardware and software had executed flawlessly, though only later, after all the telemetry was analyzed and a full survey was completed of all rover systems, was it realized that not everything had truly gone perfectly. The rover's top deck, for example, was littered with dirt and small pebbles, likely thrown up as part of a cloud of debris created by the retrorocket thrusters during the final sky crane descent to the surface. One result of that barrage was fatal damage to one of the wind sensors on the REMS weather station mounted to the stowed rover mast. A second wind sensor survived the landing, but without two working sensors it became impossible to determine the wind direction with the originally planned accuracy. Happily, clear protective and deployable covers that were designed to anticipate such potential damage had kept the most susceptible science and engineering cameras from getting dirty, enabling spectacular views of the rover's new home (figure 18.2).

Bradbury Landing

Shortly after landing, and following a tradition of memorialization established at five previous Mars landing sites, NASA decided to name Curiosity's new home Bradbury Landing, after science fiction icon Ray Bradbury, who had died just a few months earlier. The name was both fitting and widely endorsed, as many on the rover team had been fans of Bradbury's *The Martian Chronicles* since a young age. Bradbury himself would perhaps have recalled at this moment the epigram he had placed at the beginning of the book: "'It is good to renew one's wonder,' said the philosopher. 'Space travel has again made children of us all.'" It is safe to say that everyone, no matter how seasoned and experienced, felt for at least a moment the wonder of a child, as Clara Ma had doubtless anticipated when she had suggested the name Curiosity.

The EDL system had accurately guided the rover to a relatively flat and safe plains region inside Gale crater; Bradbury Landing turned out to be a few kilometers away from some large and potentially rover-trapping sand dunes, and only about five kilometers (as the proverbial crow would fly) from the base of the sedimentary deposits of Mount Sharp. The high-resolution Mastcam view from the surface of the nearby towering mountain with its series of sedimentary layers was both stunning and daunting: studying those layers would allow the team to decipher the past history of the climate of Mars, but getting into those layers meant a long drive (to go around or through the sand dunes) and then a long climb.

After the excitement of the landing itself, the engineering team at JPL began the relatively methodical process of checking out all of the rover's systems and configuring the rover for driving and science operations. The process took 15 sols to complete, and involved unstowing the high-gain antenna, the mast, and the arm, uploading updated flight software optimized for Mars surface operations instead of cruise and EDL, checking the health and proper operations of all the instruments, and verifying that the mobility system was fully functional after serving as landing gear. In the meantime, cameras on the rover began acquiring initial panoramic views of the surroundings so that the science team could start plotting a course toward Mount Sharp. On sol 16, Curiosity made its first drive on Mars, a tentative "bump" backward about seven meters.

While the spray of sand and dust created by the sky crane's powerful thrusters may have wreaked some havoc on the rover's deck and instruments, it turned out to create a relative scientific bonanza on the ground at the rover's precise touchdown spot. By blowing away much of the dust and fine sand at the site, the thrusters had helped reveal some of the underlying pebbly, cobbly, and bedrock exposures around the rover. The resulting high-resolution images enabled the first important discovery of the mission: much of the surface around the rover consisted of rocks made from pebbles and cobbles (a few millimeters to a few centimeters across) embedded within a matrix of fine-grained "cement" of some kind. The immediate interpretation that came to mind among the geologists on the team was that this was a conglomerate, a kind of sedimentary rock made out of coarse-grained clasts within a matrix of fine-grained material, and commonly found on Earth in streambeds, river channels, and shoreline deposits (plate 19). At Bradbury Landing, many of theses pebbles and cobbles are rounded, suggesting significant transport and erosion across the surface by the action of flowing water.[35] The hypothesis that the floor of Gale crater was once a streambed was exciting to the team, and was consistent with evidence for a number of delta-like sedimentary features that had been mapped from orbital images along the boundary between the floor and inner rim of Gale. The excitement was picked up by the media assembled to report on the mission's early science results, and the notion that Curiosity was driving around in an area that perhaps three to four billion years ago had been an ankle- to knee-deep river generated significant public interest and excitement.[36]

A Diversion to Yellowknife Bay

While the conglomerate discovery was invigorating, the team was eager to get on the road toward the primary exploration destination: the layered rocks of Mount Sharp (officially named Aeolis Mons by the International Astronomical Union, IAU, though the team and most of the world ignores that imposed moniker). While the base of Mount Sharp itself was not that far away, a traversable path (safe slopes and no insurmountable obstacles for the rover) up into those layered rocks was much farther away, to the southwest. This would mean embarking on a long drive, comparable to the long drives (10 kilometers or longer) that the Opportunity rover took between its detailed studies of large impact craters in Meridiani, and it would likely take many months or even years to reach the accessible base of the mountain. Even so, the team chose to initially drive away to the *east* instead of to the southwest. This decision was surprising to many.

However, if one knew the backstory, it made sense. One of the activities that project scientist John Grotzinger and mission long-term planning leads like geologist Dawn Sumner (of University of California, Davis) engaged the science team in, starting well before landing, was detailed mapping of the entire landing ellipse within which Curiosity might touch down (which had shrunk during the course of preparations from an initial area of 225 by 20 kilometers to one of 20 by 7 kilometers just before landing). The idea was that wherever the rover landed, that area and all its surroundings would be well known to whatever fraction of the team had chosen or been assigned to map it, and those folks could help lead the process of creating a consensus on what to do there and how to best proceed toward Mount Sharp. The landing ellipse had been divided into 151 so-called quadrangles of about 2.3 square kilometers (about 1 square mile) each. Bradbury Landing turned out to be in the quadrangle called Yellowknife, named after the city in the province of Northwest Territories, Canada, on the Canadian Shield, which has served as the starting point for many of the great geological expeditions that mapped some of the oldest rocks in North America (some over four billion years old).

Coincidentally, the Yellowknife quadrangle turned out to be extremely interesting geologically; it includes a confluence of many of the major kinds of terrains that occur across the floor of Gale crater near where the rover landed. Specifically, less than half a kilometer from Bradbury Landing, the team identified a topographically low point they named Yellowknife Bay, where three of the most common Gale materials all came together: "hummocky" plains, consisting of rounded cobbles and cross beds that suggested deposition by running water; rocks bedded and fractured along lines horizontal to the surface; and heavily cratered terrain (figure 18.3). In addition to simply being a single place nearby where a diversity of materials could be studied, Yellowknife Bay, as one of the deepest places on the entire floor of Gale crater, would also have been the place where standing water would have pooled for the longest amount of time in the ancient Martian past. Indeed, the bedded and fractured unit was hypothesized by the team to be potentially a lake deposit. Given the mission's focus on habitability and the ancient environment of Mars, it was a potential bonanza too hard to pass up. Here, then, is the reason the team decided, nonintuitively to those not following along closely, to head away from the Mount Sharp traverse path, eastward and downhill.

The diversion paid off, dramatically, with a second round of exciting scientific discoveries made by the rover team in Yellowknife Bay.[37] Panoramic and hand-lens-scale images revealed the widespread occurrence of sedimentary sandstones and mudstones, the latter being fine-grained rocks inferred to have been laid down in moderate to deep lake (what geologists call lacustrine) environments (plate 19). That these sediments were formed in liquid water seems unequivocal based on the detailed study conducted over several hundred sols by the rover team. Moreover, elemental and mineralogical data from the rover's quantitative chemistry instruments revealed that the water had likely been neutral in pH and low in salinity (and thus would have been drinkable, had we or anyone like us been there long ago), unlike the generally more acidic waters inferred from the Opportunity rover's data from Meridiani, and

FIGURE 18.3 Curiosity rover "selfie" mosaic from Yellowknife Bay, the topograph-ically lowest point visited by the rover within Gale crater, about 400 meters east of the Bradbury landing site. The outcrop rocks here consist of finely layered sandstones and mudstones that record a history of streams and shallow lakes on ancient Mars. Mosaic taken on MSL mission Sol 177 (February 3, 2013). NASA/JPL-Caltech/MSSS.

that the common chemical elements associated with biologic processes on Earth—carbon, hydrogen, nitrogen, oxygen, phosphorus, and sulfur—were all widely available in that environment.

The first absolute radiometric age dating of rocks by a spacecraft on another world revealed that the sediments on this part of the floor of Gale are around four billion years old, consistent with the age of formation of the ancient Martian crust, and thus recording a time when Mars is thought to have been much warmer and wetter than it is today.[38] While it was not possible using the rover's instruments at this location to directly determine the amount of time a more Earthlike environment persisted in Gale crater, estimates based on analogies with mudstone formation rates on Earth ranged from just a few hundred years up to a few tens of millions. The summary paper describing what the rover team found and interpreted at Yellowknife Bay was a tour de force led by

Grotzinger, titled "A Habitable Fluvio-Lacustrine Environment at Yellowknife Bay, Gale Crater, Mars." As the paper concluded:

> The surprising result is that the stratigraphy of Yellowknife Bay may not only preserve evidence of a habitable environment, but one that is relatively young by Martian standards. . . . This would indicate that times of sustained surface water, neutral pH, and authigenic [locally formed] clay formation extended later into Mars' history. . . . Curiosity's detection of a relatively young, and strikingly Earth-like habitable environment at Gale crater underscores the biologic potential of relatively young fluvio-lacustrine environments.[39]

Curiosity was designed and built to help humans search for specific and quantitative evidence of habitability on Mars, and lo and behold the team had found that evidence quickly in Gale crater, right near the landing site. But the mission's discoveries were far from over.

Mountaineering

With strong scientific winds at their back from the successful exploration of Yellowknife Bay, the Curiosity team finally set out around sol 324 (early July 2013) on the long-anticipated trek to the southwest, toward a promising gap in the field of dangerous sand dunes that prevented more direct access to the lower slopes of Mount Sharp. By around sol 753 (mid-September 2014) the rover would reach the Pahrump Hills, finally beginning to explore the informally named Bagnold Dunes (after early 20th-century desert explorer and aeolian geologist Ralph Bagnold, 1896–1990) and the Murray Buttes (after Bruce C. Murray), which mark the very lowest parts of Mount Sharp (plate 19). It would take the team more than 550 more sols (until around April 2017) to navigate the rover around and through a gap in the dunes and to start climbing into the steeper foothills. During that long traverse, occasional science campaign stops were planned to further characterize the geology of the Gale crater floor units, including numerous additional drill campaigns into sandstones and mudstones, discovery and analysis of outcrops rich with mineral-filled veins (evidence of the ubiquitous presence of groundwater within Gale crater's heavily fractured rocks; plate 20), and detailed though cautious study of the enormous dune field itself.

From 2017 to 2019, Curiosity continued to climb Mount Sharp, to an elevation of around 400 meters higher than at Bradbury Landing. During that time the team systematically explored higher and higher (younger and younger) stratigraphic features and layers (what geologists call "marker beds") in the sequence of sediments that make up Mount Sharp, including a resistant ridge of rocky outcrops named Vera Rubin ridge (after the noted U.S. astronomer, 1928–2016), which contains hematite, an iron oxide mineral often formed in aqueous environments on Earth, and an even higher set of layers that contain enhanced abundances of clay-bearing minerals, also usually formed in aqueous environments on Earth.

Curiosity continues going strong, and the rover has now (as of late 2020) traversed a little over a half-marathon, covering a total distance of more than 24.3 kilometers since 2012. The rover lost the capability to drill for a time because of a subsystem failure, but engineers at JPL reprogrammed the rover to be able to drill using a different combination of rotary and percussive motions, after which drilling and detailed chemical and mineral sampling of drill materials once again resumed. Curiosity has climbed about two-thirds of the way toward what has been envisioned as the highest, most important marker bed in Mount Sharp: layers of sulfur-bearing mineral deposits seen from orbit that could indicate the time in Martian history when the environment changed from one involving mineral formation by more neutral pH, low-salinity waters to one in which the waters were more acidic and mineral formation became dominated by sulfur-bearing phases (like those revealed by many of the Spirit and Opportunity measurements in Gusev and Meridiani). Why did the climate of Mars change from more Earthlike to its modern, extremely cold and extremely dry state? When and over what length of time did that transition occur? Did it occur continuously over time, or intermittently and in stages? Are clues as to the cause(s) of that transition preserved in the rock record?

It is possible that in continuing its exploration of Gale crater, Curiosity will discover answers to these critical questions in the exploration of our neighboring planet. But already Mars has been shown to be, in Alice's words from *Alice's Adventures in Wonderland*, "curiouser and curiouser!"

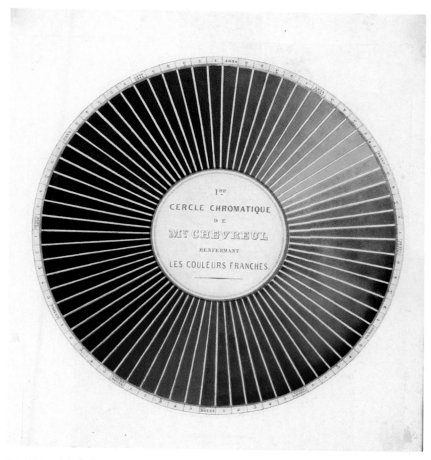

PLATE 1 Michel Eugène Chevreul's color wheel, as printed in *Expose d'un moyen de deinir et de nommer les coleurs* (1861). Courtesy of the Linda Hall Library, Kansas City, Missouri.

PLATE 2 Simultaneous contrast. Though the gray swatch is identical in all three images, it appears brighter against the black background than it does against the white background, and it appears slightly bluish against the Mars-colored background in the lower image. Courtesy of William Sheehan.

PLATE 3 Pastel drawings of Mars from the 1858 opposition by Father Angelo Secchi at the observatory of the Collegio Romano, Rome, showing the blue feature he referred to as the "Blue Scorpion" passing across the disc. Courtesy of Marco Faccini, Museo Astronomico e Copernicano dell'Istituto Nazionale di Astrofisica–Osservatorio Astronomico di Roma.

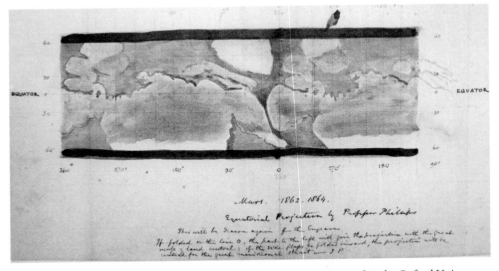

PLATE 4 Phillips's color map of Mars, 1864. South at top. Original in the Oxford University Museum of Natural History (OUMNH), Phillips Papers, Box 1, 1864. Courtesy of OUMNH.

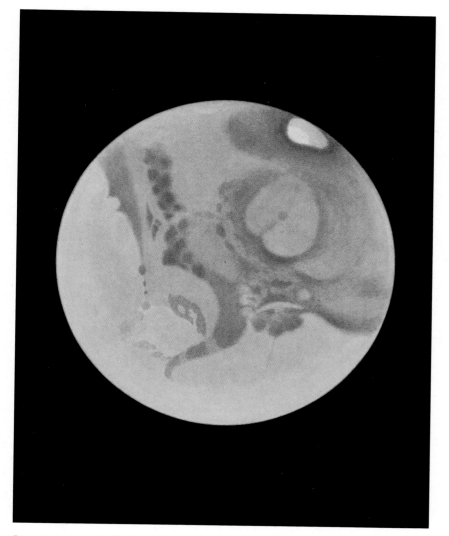

PLATE 5 Antoniadi's Mars, September 20, 1909, as it appeared on his first night of observations with Grand Lunette, the 83-centimeter Henry Brothers refractor at Meudon. Syrtis Major is just to the right and below center, and to the left is the leopard spotting he recorded in Mare Tyrrhenum. This is one of several versions produced by Antoniadi, and it is in the library collection of the Juvisy Observatory. Courtesy of William Sheehan.

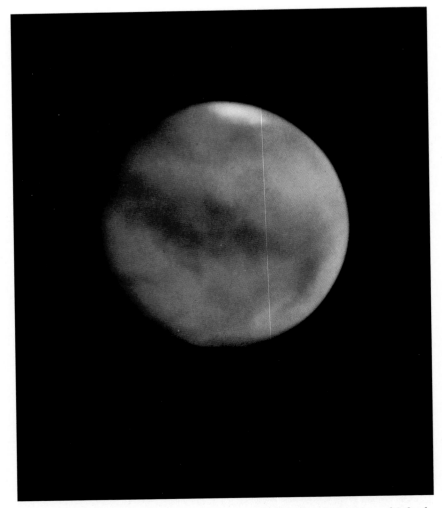

PLATE 6 Mars, photographed on August 24, 1956, by Robert B. Leighton of Caltech. JPL-Caltech, courtesy Dale P. Cruikshank.

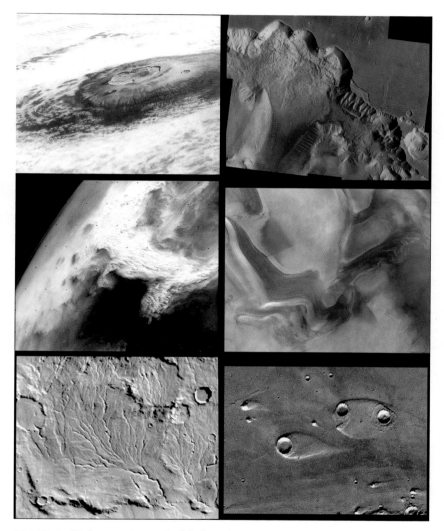

PLATE 7 Higher-resolution highlights of Mars imaging from the Viking Orbiters. *Top row, left to right:* Oblique view of Olympus Mons by artist Gordon Legg, based on a Viking Orbiter mosaic; and view of Ophir Chasma, one of the connected valleys of Valles Marineris. *Center row:* Image of a large dust storm over the Thaumasia region; and photo of layered terrains within the north polar cap. *Lower row:* 200-kilometer-wide mosaic of Warrego Valles, an extensive valley network in the ancient southern highlands; and streamlined islands carved around impact craters of roughly 10 kilometers in diameter near the mouth of Ares Vallis in Chryse Planitia. NASA/JPL.

PLATE 8 Examples of color imaging from the Viking Landers. *Top row:* First color mosaic from Viking Lander 1 (*left*); and Soil trenches dug by the lander's sampling arm. *Lower row:* First color mosaic from Viking Lander 2 (*left*); and morning frost photographed during the first winter at the landing site. NASA/JPL.

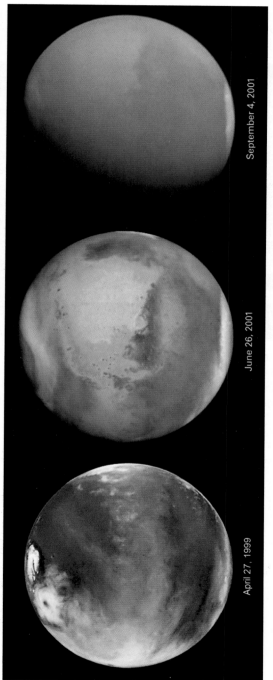

April 27, 1999   June 26, 2001   September 4, 2001

PLATE 9   *Left*: A blue light (410 nanometer) image taken with HST in 1999, showing a huge polar cyclone (to the left and just south of the bright residual north polar ice cap at the top of this view) more than 1,600 kilometers across at 65°N latitude and 85°W longitude. *Center and right*: HST imaging before and during the planet-encircling dust storm of 2001. Jim Bell, Steve Lee, Mike Wolff, and the NASA/ESA Hubble Heritage Team.

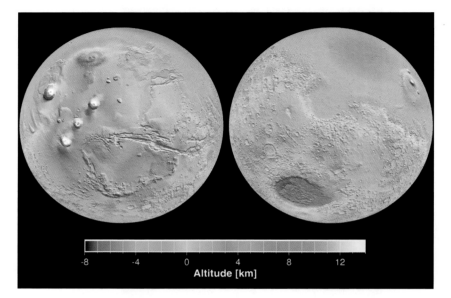

PLATE 10 Mars Global Surveyor's MOLA instrument was used to create this global map of the topography of the Tharsis (*left*) and Hellas hemispheres of Mars. Because there is no equivalent of "sea level" on Mars, elevations there are measured with "zero" defined as the elevation relative to the average radius of the planet. NASA/JPL/GSFC.

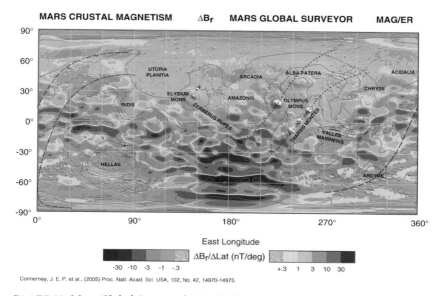

Connerney, J. E. P. et al., (2005) Proc. Natl. Acad. Sci. USA, 102, No. 42, 14970-14975.

PLATE 11 Mars Global Surveyor's MAG/ER instrument map of variations in the strength of the planet's surface magnetism as measured from the spacecraft's nominal 400-kilometer circular orbit. Bluer colors represent negative variations, redder colors positive. The overall range of variations spans more than a factor of 100 in field strength. NASA/JPL/GSFC.

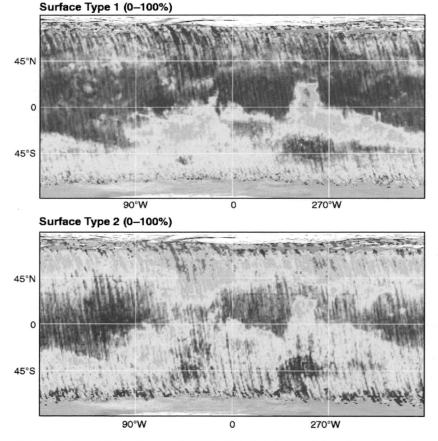

Surface Type 1 (0–100%)

Surface Type 2 (0–100%)

PLATE 12 Mars Global Surveyor's TES initial global map of the mineralogy of the low-albedo regions of Mars, including both the distribution of typical basalt compositions (Surface Type 1, *top*) and the distribution of what has been interpreted as either basaltic andesite or altered basaltic compositions. Redder colors are higher concentrations; bluer are lower. Many of the classical albedo features, such as Syrtis Major, Sinus Sabaeus, Meridiani Sinus, and Margaritifer Sinus, can be recognize as basaltic regions. NASA/JPL/ASU.

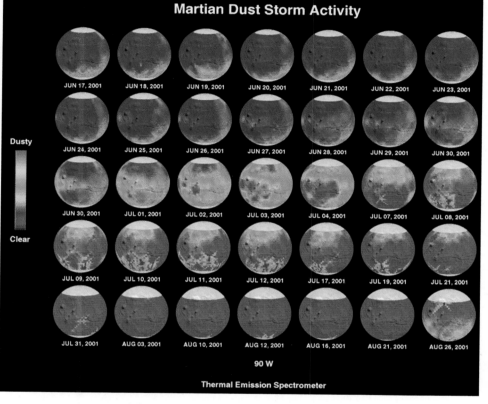

PLATE 13 Mars Global Surveyor's TES recorded the expansion of dust during the planet-encircling dust storm of June–July 2001. The warmer colors in this series of images of the planet, centered on 270°W longitude, show local atmospheric warming associated with the dust. NASA/JPL/ASU.

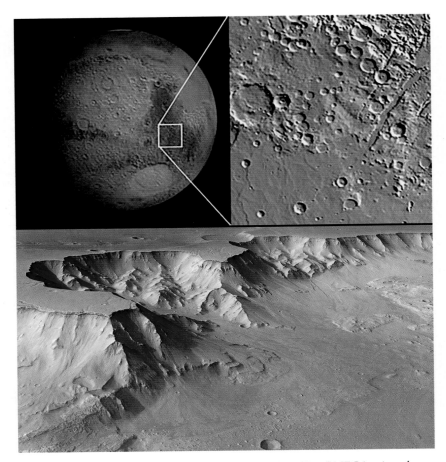

PLATE 14 Discoveries from the Mars Express mission. *Top:* OMEGA mineral map of small deposits of water-bearing clay minerals in the Syrtis Major region. *Bottom:* HRSC perspective 3-D view of landslides and debris flows along the walls of Aurorae Chaos, a highlands region a few hundred kilometers east of Valles Marineris. ESA/DLR/CNES/Brown University.

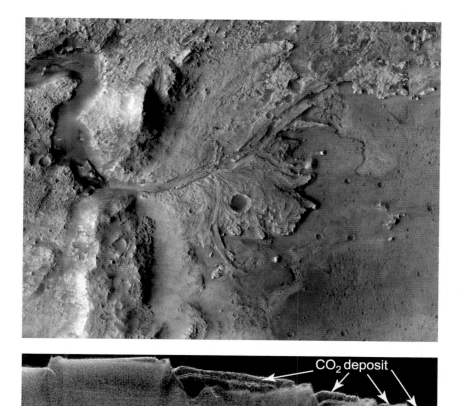

PLATE 15 *Top:* CRISM-derived mineral map of the delta in Jezero crater (the landing site of the Mars 2020 rover Perseverance) overlaid onto a CTX mosaic of the crater. Clay-bearing materials are shown here in green, olivine-bearing materials in yellow, and pyroxene-bearing materials in blue. Purplish colors are places with no distinct CRISM spectroscopic signatures. The small crater in the middle of the delta is about 1 kilometer across. *Bottom:* SHARAD "radargram" showing surface and buried layers of CO_2 ice along a 330-kilometer-long by 1-kilometer-deep transect across the south polar cap of Mars. NASA/JPL-Caltech/ASI/JHU-APL.

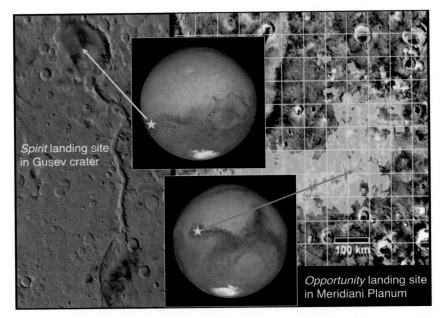

Spirit landing site in Gusev crater

Opportunity landing site in Meridiani Planum

100 km

PLATE 16 The landing sites for the Spirit rover's mission in 165-kilometer-wide Gusev crater, and for the Opportunity rover's mission in the dark plains of Meridiani Planum. Their global context is indicated in Hubble Space Telescope views (north is up). The Viking Orbiter mosaic (*left*) shows that Gusev is at the terminus of a long fluvial channel called Ma'adim Vallis. The MGS TES background map (*right*) shows the distribution of the mineral hematite (yellows and reds are higher abundances) near Opportunity's landing site. NASA/JPL/ASU/Space Telescope Science Institute.

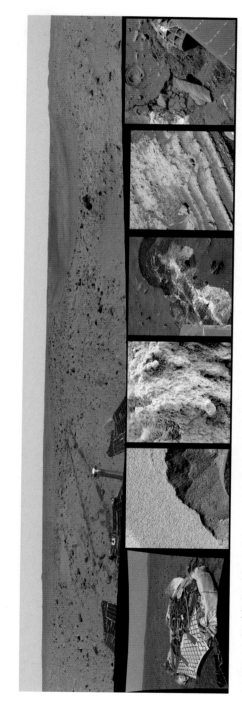

PLATE 17 Highlights from Spirit rover imaging in Gusev crater. *Top:* Pancam 360° panorama from sols 68–69 from the Gusev plains near the rim of Bonneville crater. *Bottom, left to right:* Sol 16 Pancam mosaic of the lander and (deflated) airbags that delivered the rover to the surface; sol 73 Pancam image of part of a trench dug by the rover wheels into a small sandy ripple, showing how only the uppermost surface is reddish and dusty; sol 163 Microscopic Imager closeup view of a three-centimeter-wide piece of nodular rock named Pot of Gold; sol 721 Pancam image of sulfate- and silica-rich soils exposed by the rover wheels near Home Plate; sol 751 Pancam view of deformed sediments and a potential "bomb sag" in layered deposits at Home Plate; final Spirit color Pancam photo, taken on sol 2191, showing colorful, layered rocks and soils near the rover wheel. NASA/JPL/Cornell/ASU/USGS.

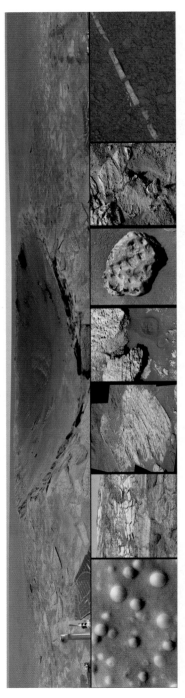

PLATE 18 Highlights from Opportunity rover imaging in Meridiani Planum. *Top*: Pancam 360° panorama from sols 117–123 taken from the rim of 130-meter-wide Endurance crater. *Bottom, left to right*: Sol 14 merged Microscopic Imager and Pancam false-color mosaic of 1- to 3-millimeter-wide "blueberries" on the floor of Eagle crater; sol 174 Pancam false-color photo of "RAT holes" ground into sedimentary layers during the drive into Endurance crater; sol 208 Pancam photo of polygonal desiccation features in outcrop rocks on the floor of Endurance crater; sol 237 Pancam image of dark sandy drifts next to light-toned sulfate outcrop rocks; sol 352 Pancam photo of an iron-nickel meteorite named Heat Shield Rock; sol 690 Pancam photo of finely laminated and cross-bedded layered sandstones in the Meridiani Plains; sol 2769 Pancam photo of a gypsum-bearing vein exposed on the rim of Endeavour crater. NASA/JPL/Cornell/ASU/USGS.

PLATE 19 Curiosity rover images from Gale crater. *Left:* Rover "selfie" taken on sol 1463 (September 17, 2016) using the arm-mounted MAHLI camera in the Murray Buttes at the base of Mount Sharp. *Inset, left:* High-resolution Mastcam view of finely layered sandstone and mudstone within the butte in the background. *Top right:* Rover tracks through a three-foot-tall sand dune in an area called Dingo Gap along the drive to Mount Sharp on sol 538 (February 9, 2014). *Center row, left to right:* Mastcam photo from sol 281 (May 21, 2013) of the 1.6-centimeter-wide drill hole called Cumberland, with a line of ChemCam laser pits visible in the drill tailings; sol 27 (September 2, 2012) Mastcam photo of rounded pebbles cemented into conglomerates at the Bradbury Landing site; Mastcam view of sunset on sol 956 (April 15, 2015). *Bottom right:* Mastcam photo of centimeter-scale bright mineral veins in outcrops at a region called Garden City, imaged on sol 929 (March 18, 2015). NASA/JPL/MSSS.

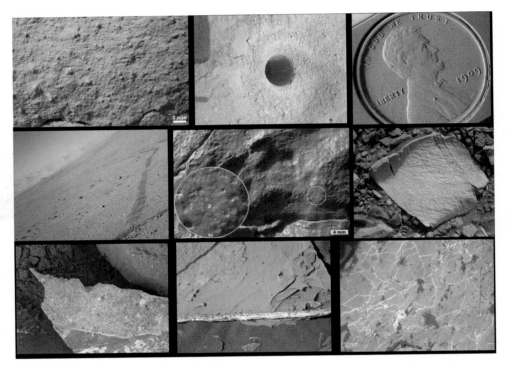

PLATE 20 High-resolution hand-lens-scale examples and highlights from MSL MAHLI imaging in Gale crater. *Top row, left to right:* Mafic sandstone texture and grains from Gillespie Lake, sol 132; bright veins and ChemCam laser pits in the John Klein drill hole, sol 270; a first-year U.S. "Lincoln penny" that was included on the MAHLI calibration target and sent to Mars (this sol 411 image shows the scale of MAHLI images as well as some individual clumps of dust grains). *Middle row:* "Landscape" view from sol 644 of rover tracks and the distant northeast rim of Gale crater; part of the surface of Kanosh boulder, a mafic sandstone apparently smoothed by aeolian sand grain abrasion, sol 942; photo from sol 1041 of the freshly broken surface of a small, finely laminated, high-silica stone named Lamoose, interpreted to be a mudstone rock. *Bottom row:* Part of a broken (by the rover wheels) and partially altered sandstone rock called Impalila, from sol 1345; a bright vein in the mudstone rock named Burnt Coat, from sol 1783 (the rock is partially covered by dark windblown sand at bottom); sol 2223 image of the gray, heavily fractured, vein-filled, and partly sand-covered mudstone rock called Highfield. NASA/JPL/MSSS/Ken Edgett.

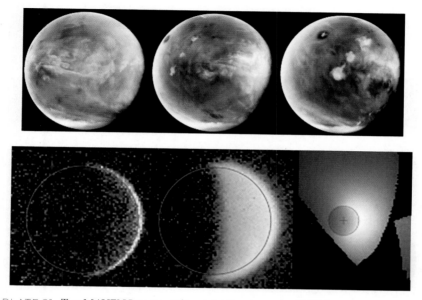

PLATE 21 *Top:* MAVEN Imaging Ultraviolet Spectrometer (IUVS) instrument views of cloud formation on Mars as the planet rotated below the spacecraft on July 9–10, 2016. *Bottom, left to right:* IUVS images of atomic carbon, atomic oxygen, and atomic hydrogen. All these elements are products of sunlight (coming from the right), which breaks down CO_2 and H_2O in the Martian upper atmosphere. The disc of Mars is outlined in red in each image. NASA/MAVEN/University of Colorado.

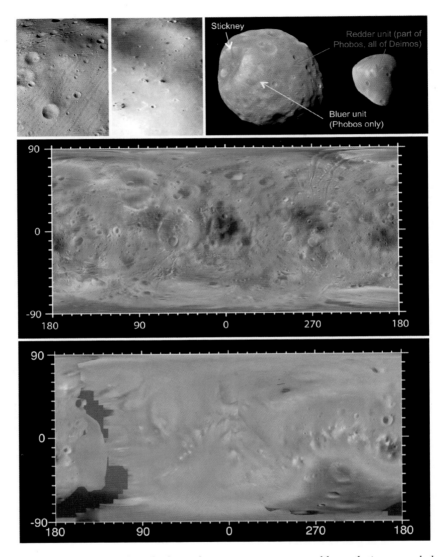

PLATE 22 Viking Orbiter high-resolution images at comparable resolutions revealed large differences in roughness between the rougher surface of Phobos with its well-preserved craters (*top left*) and the smoother surface of Deimos (*top center*) with craters infilled by regolith. Multispectral and hyperspectral imaging (*top right, see text*) revealed spatial variations in Phobos's spectral properties, especially associated with Stickney. Deimos's brightness variations are accompanied by less variation in color. The two maps show "dynamical height" or topography on Phobos (*middle*), which has a range of heights of 1.8 kilometers, and Deimos (*bottom*), which has a range of heights of 1.9 kilometers. In the maps, red is high, and blue is low. NASA/JPL.

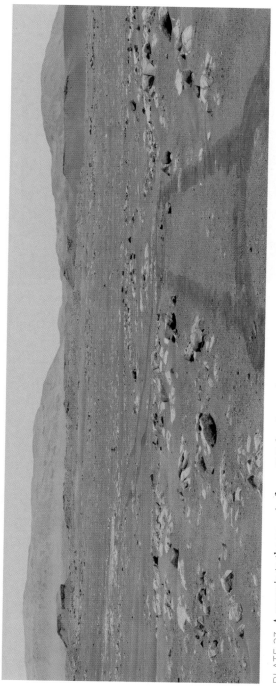

PLATE 23   A zoom in to the scene in figure 21.5 showing rocks strewn in the foreground, the front ridge of the Jezero crater delta beckoning (some two kilometers away), and the mountainous rim of the crater in the far distance (some four to five kilometers away). For scale, the isolated delta mesa near the upper left edge of the photo, which mission scientists have informally dubbed "Kodiak," is about 200 meters across and about 50 meters high. NASA/JPL-Caltech/ASU/MSSS.

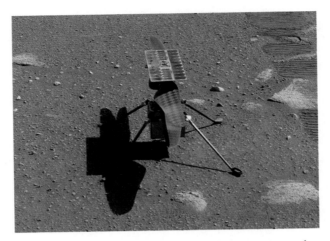

PLATE 24 Ingenuity on Mars. The innovative Mars helicopter is seen here right after drop off, in a close-up taken by Perseverance's zoomable Mastcam-Z on April 5, 2021, Sol 45 of the mission. NASA/JPL-Caltech/ASU/MSSS.

19

Atmospheric Explorers

Clearing the Air

While more detailed exploration of the surface of Mars was a critical imperative of the 2003 planetary decadal survey, the community also laid out important questions in that study about the planet's thin atmosphere that needed to be addressed by future missions.[1] These questions were focused on understanding the processes and history of the climate of Mars, from the ancient past to the present. The roles of water, other volatile components (like CO_2), and airborne dust in driving the weather and climate of Mars were judged to be particularly important for detailed, close-up study. Future missions would need to focus, for example, on examining the "sources, sinks, and reservoirs" of volatiles on Mars, on determining how the atmosphere evolves over long time periods, on determining whether there is an active water cycle on Mars, on directly measuring the dynamics of the middle and upper atmosphere of the planet, and on determining the rate of escape of different atmospheric gases (especially those associated with water and other volatiles). Many of these same topics and questions were echoed and amplified in the 2013 planetary decadal survey,[2] especially given the new knowledge about the planet that had been gained by missions like Odyssey, Mars Reconnaissance Orbiter

(MRO), Mars Express, Mars Exploration Rover (MER), and Phoenix during the first decade of the 21st century.

Among the major recommendations of the 2003 decadal survey was that NASA fly a dedicated Mars upper-atmospheric orbiter mission that would directly sample the abundance and composition of atoms, ions, and molecules high above the planet's surface and determine the ways that the atmosphere and ionosphere interact with the solar wind and its embedded magnetic field. Of particular importance would be measuring the rate that atoms and ions are slowly "leaking" out of the modern upper atmosphere (accelerated to escape velocities by interactions with the high-energy solar wind), and then using knowledge of the modern escape rates to extrapolate the atmospheric composition and thickness back to the early history of the planet.

MAVEN

While the 2003 decadal survey had advocated for a dedicated low-cost mission to study the upper atmosphere of Mars, it also reinforced the need to enable even lower-cost and competitively selected Mars missions via the Mars Scout program, which had been created back in 2002 by NASA's "Mars Czar" Scott Hubbard, and which had kicked off its first selected mission—the Phoenix lander—in 2003 (see chapter 17). The second (and, ultimately, last, as Mars Scout was absorbed into NASA's Discovery program in 2010) community-wide call for Mars Scout mission proposals went out in 2006, and again the global planetary science community responded enthusiastically, submitting more than 20 Mars mission proposals that included just about every imaginable mode of transport: orbiting, flyby sampling of the atmosphere, airplaning on gigantic wings above the surface, crash-landing hard into the surface in order to penetrate to the subsurface, or just plain landing. Ultimately, two proposals were selected as finalists for "Phase A" studies, with the expectation that one of them would eventually be selected for a launch in 2011. The two selected proposals were the Mars Atmosphere and Volatile EvolutioN (MAVEN) mission, and The Great Escape mission, both of which were orbiters focused on responding to the decadal survey's call for dedicated study of the upper atmosphere.

Unfortunately, the selection of the winner was delayed in late 2007 because of a serious conflict of interest involving one of the subcontractors that was helping both NASA in its evaluation of the proposals *and* one of the finalist teams in the preparation of its proposal.[3] Because the entire review process and review panel had to be completely reconstituted, the launch date for the second Mars Scout mission had to be delayed until 2013. Eventually, in September 2008 NASA selected MAVEN as the next mission to fly to Mars (figure 19.1).

Led by planetary scientist Bruce Jakosky from the University of Colorado Boulder, MAVEN was directly responsive to the decadal's call for a detailed, in situ study of Martian upper atmosphere composition and escape processes. More specifically, MAVEN was designed to test the hypothesis that Mars once had a thick, warming atmosphere (more like Earth's than today), but that almost all of that atmosphere was subsequently "lost" to space due either to the small size of Mars or to interactions with the solar wind in the absence of a shielding global magnetic field. How can a planet just lose its atmosphere? Many different ways have been identified, but they can be boiled down to two general categories: First, when ultraviolet (UV) light from the Sun interacts with a planetary atmosphere, it can provide enough energy to make particles near the top of the atmosphere escape, either via thermal escape (the gases are heated enough to increase their kinetic energy to the point where some of them reach escape velocity), or by photochemical escape (gas molecules are broken apart by sunlight, and some of the constituent ions are energized enough to reach escape velocity). Or second, ions in a planet's upper atmosphere can either be accelerated directly to escape velocities by electric and magnetic fields from a planet or from the solar wind, or be accelerated to high-enough speeds by those fields that when they crash into other atmospheric molecules they create a "splash" of other particles via a process called sputtering, whereupon some of those sputtered particles can then escape. Via these processes (or several other less efficient processes), atmospheric gases can be slowly "lost" from a planet. The key to testing whether atmospheric loss is actively occurring, or has been persistently occurring for long periods of time, is that these sorts of loss processes can be mass dependent, meaning that lighter-mass atoms and molecules are preferentially lost compared to heavier-mass ones. So instruments that can measure and detect, for example, anomalously high ratios of heavier versus lighter isotopes of the same atoms

FIGURE 19.1 Artistic renderings of the NASA MAVEN orbiter at Mars (*top*) and atmospheric ions being dramatically stripped from the upper atmosphere of Mars by a high-energy solar storm. MAVEN successfully measured the current rate of this process, and determined that it led to the loss of the planet's earlier, thicker atmosphere. NASA.

or molecules—like the ratio of heavy argon (^{38}Ar) to regular argon (^{36}Ar)—could find smoking-gun evidence of massive amounts of atmospheric loss on a planet like Mars.

MAVEN's scientific payload includes a suite of six sensitive "fields and particles" instruments to measure ions and electrons in the solar wind and the ionosphere of Mars as well as the properties of the interplanetary magnetic field and its interactions with the planet (Mars does not have a global magnetic field like Earth's, but the solar wind interacts directly with the atmosphere of Mars, leading to the formation of a weak Martian magnetosphere formed from the Sun's magnetic field draping itself around the planet and interacting with ionized upper-atmospheric gases). In addition, the mission includes an Imaging Ultraviolet Spectrometer (IUVS) that studies the chemistry and dynamics of the atmosphere, an Extreme Ultraviolet Monitor (EUV) that monitors the intensity of high-energy UV sunlight incident on the atmosphere, and a mass spectrometer that directly samples and measures the compositions (including isotopes) of Martian gases when the spacecraft makes occasional "deep dips" into the upper layers of the atmosphere. MAVEN is also the first NASA Mars mission without a dedicated camera, although scans from the ultraviolet spectrometer can be assembled into images that reveal atmospheric ions as well as features like clouds, hazes, and the surface polar caps (plate 21).

MAVEN was launched in November 2013 and was successfully inserted into a 4.5-hour, nearly polar elliptical orbit around Mars in fall 2014. During MAVEN's prime mission (one Earth year), and during extended missions since, the instruments have collected data on the energy input of the Sun at Mars, the planet's upper-atmospheric composition, and the rate of escaping atmospheric particles, using precession of the orbit to sample the full range of latitudes and longitudes at each orbit's periapse. Perhaps most exciting have been the deep dips performed on a small subset of the orbits that enabled the periapse to be lowered from the normal 150-kilometer altitude down to as low as 125 kilometers, thus sampling the deepest layers of the planet's upper atmosphere.[4]

Even after only one year of data collection, the MAVEN team had amassed enough information on atmospheric loss processes to confirm the hypothesis that the persistent stripping of atmospheric gases from the upper atmosphere's interactions with the solar wind and interplanetary magnetic field were likely responsible for the loss of most of the planet's early, thicker atmosphere.[5] Indeed, part of the excitement of the MAVEN results was that the slow loss of the Mars atmosphere can be actively measured *today*, as mission data reveals that, on average, about 100 grams of material per second is being stripped off the top of the planet's atmosphere by the electron pickup process alone.

While one of the leading hypotheses for why an early warmer, wetter Mars eventually transitioned into the much colder, drier world that it is today was confirmed, there were still (and continue to be) many surprises revealed by the MAVEN mission data.[6] For example, the stripping of the atmosphere by the solar wind does not occur in a constant way, but instead increases and decreases seasonally (roughly tracking the significantly changing distance of Mars from the Sun over the course of each year) and can change dramatically during periods of lower solar activity or more intense solar magnetic storms, decreasing by two times or increasing by up to 100 times, respectively. One important implication of this discovery is that the rate of atmospheric loss was likely much higher on average very early in the history of the solar system, when the Sun was a more violent younger star. Another surprise was the discovery that auroras on Mars are diffuse, occurring all over the planet as opposed to just in the polar regions like Earth's auroras. In retrospect, as was voiced by planetary scientist Nick Schneider from the University of Colorado Boulder, the lead scientist on MAVEN's UV spectrometer, MAVEN researchers experienced a "dope slap" moment when they realized that because Mars doesn't have a magnetic field like Earth's that guides auroral particles to the converging field lines at the poles, there's no special reason why auroras on Mars can't occur anywhere that the solar wind slams into and ionizes the upper atmosphere.[7] Such "dope slap" moments occur more frequently even among scientists than we like to admit, illustrating the truth of what George Orwell (in an entirely different context) once said: "To see what is in front of one's nose needs a constant struggle."[8]

In addition to the mission's science instruments, MAVEN also carries an Electra radio system that provides scientific information on the density of the upper atmosphere and enables the spacecraft to serve as a telecommunications relay for landers and rovers operating on the surface. In early 2019, NASA directed the MAVEN team to tighten the spacecraft's orbit, lowering both the periapse and apoapse using aerobraking in the upper atmosphere.[9]

The new orbit has a period of around 3.6 hours and a periapse altitude of around 132 kilometers, allowing the spacecraft to relay more frequent and higher-bandwidth communications passes from landers and rovers on the surface, while still being able to perform some of the extended-mission science originally intended for MAVEN.[10]

Mangalyaan

MAVEN was not the only Mars mission sent from Earth during the late 2013 launch window. India joined the small list of nations attempting to explore Mars with the launch that November of its Mangalyaan (Sanskrit for "Mars-craft") mission, also sometimes colloquially known as the Mars Orbiter Mission, or MOM (figure 19.2). This was the most ambitious deep space mission yet attempted by the Indian Space Research Organisation (ISRO), following its very successful 2008–9 Chandrayaan-1 lunar orbiter mission. ISRO was able to keep mission costs low (estimated at less than US$100 million) by basically recycling the Chandrayaan-1 design, by categorizing the mission as primarily focused on technology and capability demonstrations (developing the designs, planning, management, and operations of deep space interplanetary missions), and by keeping the scientific instrument payload much simpler than, for example, the instruments on the roughly $700 million MAVEN mission (appendix H).

That payload included fields and particles instruments and a mass spectrometer designed to study the upper atmosphere of Mars (with science goals consistent with those outlined in the 2003 decadal survey and those of the MAVEN mission), and a visible-wavelength color camera and thermal infrared spectrometer designed to study the surface. Mangalyaan was successfully launched into Earth orbit from India's Satish Dhawan Space Centre in Andhra Pradesh on November 5, 2013, and after a series of orbit-raising maneuvers was injected into a trans-Mars orbit on November 30. A series of three trajectory correction maneuvers during cruise readied the spacecraft for a Mars orbit insertion burn on September 22, 2014, just two days after MAVEN had arrived in orbit. The insertion was successful, and on September 24 Mangalyaan was captured into a highly elliptical orbit that takes it within about

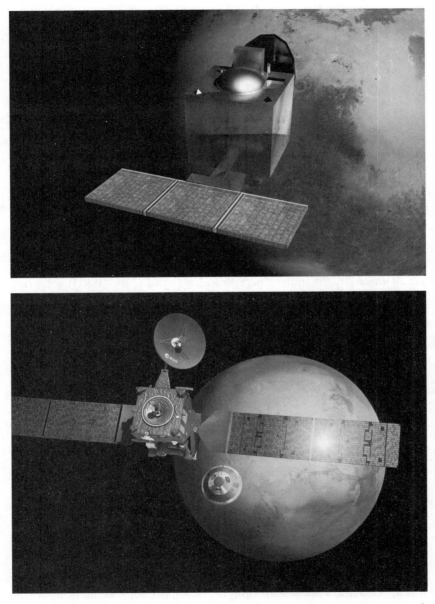

FIGURE 19.2 *Left*: Artistic depiction of the Indian Space Research Organisation's Mangalyaan Mars orbiter. Courtesy of Nesnad, CC BY-SA 4.0, posted to Wikimedia. *Right*: The European Space Agency's ExoMars Trace Gas Orbiter (TGO) and its deployment of the Schiaparelli lander spacecraft. ESA.

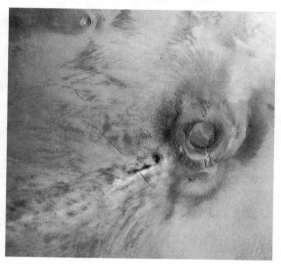

FIGURE 19.3 *Left*: Clouds and geological features around the Tharsis volcano Arsia Mons, from the Mangalyaan Mars orbiter. *Top*: Ice deposits in part of the 50-kilometer-wide, high northern latitude Korolev crater, from the ExoMars Trace Gas Orbiter (TGO). ISRO/ISSDC/Justin Cowart and ESA/Roscosmos/CASIS.

420 kilometers of the surface every 72 hours. India became the first nation to successfully get a spacecraft into Mars orbit on its first try.

Because Mangalyaan is a technology demonstration mission carrying instruments with modest capabilities compared to those of pricier, science-focused missions, expectations for its scientific discoveries have been tempered. Nonetheless, some good science has been done by ISRO researchers during the six-month primary mission as well as during the continuing extended mission. This includes measuring the vertical profiles of upper-atmospheric gases (CO_2, N_2, CO, and O) at unique local times of day compared to other missions (including MAVEN),[11] and mapping global-scale atmospheric phenomena and surface albedo and color changes at local times of day not routinely monitored by previous NASA or European Space Agency (ESA) missions (figure 19.3).[12] The real success of Mangalyaan, however, appears to be its extremely successful technological demonstration of ISRO's interplanetary launch, navigation, orbital operations, and communications capabilities, many of which are expected to be used to guide future robotic missions to Mars and other deep space planetary destinations.[13]

Mysterious Methane

The thin Martian atmosphere is dominated (95 percent) by carbon dioxide (CO_2), with much of the remainder (4.5 percent) being the inert gases nitrogen and argon. "Trace gases" account for the remaining 0.5 percent or so; these include molecules like oxygen, carbon monoxide, and water vapor. Many other important gases have been detected in the atmosphere of Mars at much lower levels, however. An example is methane (CH_4), which was initially determined by Earth-based telescopic and Mariner 9 orbital observations during the mid- to late 20th century to have an upper limit of no more than a few to 10 or so parts-per-billion (ppb) in the atmosphere.[14] The detection of methane would be important because it is easily broken down and destroyed in the Martian atmosphere by strong UV radiation from the Sun and reactions with other atmospheric gases, having a lifetime of only around 300 years. Thus, the detection of *any* methane on Mars implies that there must be one or more sources resupplying that gas to the atmosphere. Where it gets interesting is that while there are a number of geological processes that can release methane into the atmosphere, there are also (at least on Earth) biologic processes that produce methane as well. And on Earth the biologic ones are dominant.

The history of searching for methane on Mars has been anything but straightforward, and has led to a series of fascinating, perplexing, sometimes contradictory, and still inconclusive results. A landmark event in this history occurred in 2004, when researchers working with data from the ESA Mars Express mission's infrared spectrometer reported the detection of a globally averaged methane abundance around 10 ppb, and patchy occurrences of atmospheric methane with abundances upward of 30 ppb.[15] The discovery was met with significant skepticism in the planetary science community, because such (relatively) high abundances should have been detectable from Earth-based telescopic observations for decades. Then, in a second interesting twist in 2009, a team of planetary astronomers led by Michael Mumma from NASA's Goddard Space Flight Center announced that while measuring methane and water vapor simultaneously on Mars using high-dispersion infrared spectrometers with three ground-based telescopes in northern early and late summer 2003 and near vernal equinox 2006, they had in fact detected "strong release of methane" from specific regions of Mars.[16] The inferred abundances ranged upward of 50 ppb, compared to a normal background level of just a few ppb or less. Inferred "plumes" of methane were reported to have occurred in several regions, including parts of Sinus Sabaeus and Syrtis Major, not far from where the Mars Express detections were reported, and near where several orbital spacecraft had detected evidence of groundwater-formed hydrated minerals and potentially other volatiles. The reported ground-based observations remain the subject of significant controversy among Mars atmospheric researchers. Nonetheless, the fact that most of Earth's methane is generated by microscopic to macroscopic life-forms (live and fossilized) suggested the exciting possibility, widely picked up by the media, that there could be a connection between the putative methane detections and the presence of potentially habitable (or inhabited) subsurface regions on Mars.

The controversy followed the familiar lines of the disputed Viking biology experiments and the ALH84001 fossil meteorite announcements, as it was soon pointed out that there were several geological and atmospheric processes that could potentially form methane in the Martian environment without the involvement of biology.[17] For example, a number of potential water-rock weathering and oxidation reactions involving minerals known to occur in the Martian crust can release hydrogen, which can then go on to form methane gas via a variety of other chemical processes. Methane can also be created from the UV breakdown or other kinds of alteration of more complex atmospheric or surface hydrocarbons (such as those commonly delivered to planetary surfaces in meteorites and micrometeorites). Finally, some methane is released naturally from the interiors of terrestrial planets by volcanism. While the presence of significant amounts of methane in a planetary atmosphere could indeed be what astrobiologists call a potential biosignature, there are many viable nonbiological sources of methane as well.

The Sample Analysis at Mars (SAM) instrument on the Curiosity rover had been widely expected to resolve the confusion about methane on Mars using its Tunable Laser Spectrometer (TLS) subsystem, part of which was designed specifically to measure methane and other trace gases directly from the surface.[18] Indeed, initial measurements of atmospheric methane using TLS yielded essentially zero methane, with an upper limit only around 1 ppb.[19] However, instead of being resolved, the situation became even more confounding based on subsequent TLS measurements of atmospheric methane. Specifically, during three Mars years of observations, TLS measurements showed several "spikes" in methane, to levels up to nearly 10 times the typical background levels,[20] as well as apparent seasonal variations during the course of the planet's annual elliptical orbit.[21] While TLS has never measured methane abundances as high as the many 10s of ppb reported from some previous telescopic and spacecraft observations, the origin of both the ~10 ppb "spike" and the seasonal variations observed by TLS remain an enigma.

The reported discoveries of relatively abundant and time-variable methane, and especially its potential biological origin, have been met with substantial skepticism in the broader planetary atmospheric sciences community, with some researchers claiming that such spatial- and time-variable occurrences of methane at the levels reported are physically implausible,[22] or simply indicative of measurement errors due to terrestrial atmosphere or rover systems contamination, or to noisy data.[23]

While the Indian Mangalyaan orbiter carried an instrument that had been designed to search for methane in the Martian atmosphere, that instrument unfortunately failed to work properly from Mars orbit. Significant attention turned, then, to the next "definitive" attempt at resolving the methane mystery, using instruments on the ExoMars Trace Gas Orbiter mission.

TGO and Schiaparelli

The ExoMars Trace Gas Orbiter (TGO) mission is a joint collaboration between the Russian space agency Roscosmos, which provided the launch vehicle and several science instruments, and the ESA, which built the spacecraft and the rest of the science instruments. TGO was originally envisioned back in 2009 as a collaboration between ESA and NASA, which was also planning to initiate a new mission called Mars Science Orbiter (MSO). The idea was that NASA would launch both TGO and MSO together, allowing for complementary science observations of the planet.[24] However, TGO was also supposed to carry

ESA's planned ExoMars rover and a small landing demonstration module, causing costs to eventually soar because the missions had to be split into two NASA launches. Because of those projected overruns and continuing general NASA cost overruns associated with delays and technical problems with Mars Science Laboratory (MSL) and the James Webb Space Telescope, NASA pulled out of the joint TGO agreement with ESA in 2012 as part of a general budget-slashing exercise.[25] NASA's unfortunate termination of the earlier agreement forced ESA to restructure its efforts, and spawned its eventual collaboration with the Russians.

ESA and Roscosmos pressed ahead together with the development of TGO and successfully launched the mission on board a Proton rocket in March 2016. TGO carries four science investigations: two ultraviolet and infrared spectrometers designed to accurately measure the abundances of atmospheric methane and other trace gases; a color imaging system designed to study the surface geology and topography (figure 19.3); and a neutron detector designed to expand the search for subsurface water or ice. Perhaps the most anticipated investigation would be the efforts by the spectrometers to resolve the decades-long methane mystery, including via near-simultaneous measurements between Curiosity on the ground and TGO from orbit. It was with great anticipation, then, that TGO arrived safely in a high elliptical Mars orbit in October 2016. During the next 11 months the spacecraft team used thruster firings and gentle dips into the planet's upper atmosphere to aerobrake the spacecraft into its final 400-kilometer circular near-polar orbit, from where science activities began in April 2018.

Initial scientific results from TGO are only recently beginning to stream in, but the early results regarding methane are particularly striking. Over the course of four months of collection of exquisite spectroscopic data sets, the TGO team reported that it "did not detect any methane over a range of latitudes in both hemispheres, obtaining an upper limit for methane of about 0.05 parts per billion by volume."[26] This result is stunning because that level of methane is some 10 to 100 times lower than previously reported values from Earth-based and Curiosity measurements. Thus, the community remains perplexed about the Mars methane mystery, with even the TGO authors themselves grasping at straws, claiming that reconciliation of the different observations and results "would require an unknown process that can rapidly remove or sequester methane from the lower atmosphere before it spreads globally."[27] While TGO continues to search for evidence of the elusive methane (and to obtain data on other trace gases and surface properties), it's not clear that the mystery will be resolved any time soon. At the time of writing (late 2020), a big question mark hangs over the whole matter.

About three days before TGO's capture into Mars in October 2016, the orbiter released a small technology demonstration lander named Schiaparelli (figure 19.2), which was designed to be ESA's next attempt after the failed Beagle 2 lander of 2003 to test entry, descent, and landing (EDL) technologies that would be used for future Mars surface missions, like the ExoMars rover, now scheduled for a 2022 launch. Schiaparelli was a small, 1.65-meter-diameter circular palette lander designed to use a heat shield, parachute, and retrorockets to achieve a soft landing in Meridiani Planum (not far from the NASA Opportunity rover landing site).[28] The lander was designed to last at least a few sols on Mars, and it carried a meteorology package to characterize the current Martian environment, as well as a descent camera and laser range reflector. While separation from TGO and atmospheric entry and parachute deployment appear to have all been successful, contact with the lander was lost around a minute before the planned landing, pointing potentially to a failure in the parachute/backshell ejection and/or retrorocket deceleration process (verified by a subsequent ESA investigation). During the week after the attempted landing, photographs of the region taken by the Mars Reconnaissance Orbiter's High Resolution Imaging Science Experiment (HiRISE) camera revealed the crash site and scattered debris from the lander, heat shield, and parachute. While the Schiaparelli landing was not successful (and invoked painful memories of Beagle 2), ESA declared much of the demonstration a success, as many key elements of the EDL process to be used on future missions worked as planned. The failure did, however, serve as a reminder that, as Scott Hubbard says, "Mars really is hard, and landers are extremely hard."[29]

20

Shooting the Moon(s)

Spacecraft Exploration of Phobos and Deimos

Orbits and Magnitudes

For almost a century after U.S. Naval Observatory astronomer Asaph Hall discovered the satellites of Mars in 1877 (see chapter 6), few details were known about these bodies, which Edgar Rice Burroughs would so evocatively describe as "hurtling moons." Hall was able to determine the periods of their almost perfectly circular, near-equatorial orbits, and in the process discovered the remarkable fact (which had first confused him into thinking there might be several inner moons) that Phobos zips around Mars around 3.25 times every sol, even as the outer moon (Deimos) orbits much more slowly, around once every 1.25 sols. Commenting on their orbital properties, the editors of the *Observatory* noted in 1877 that "the phenomena presented to an inhabitant of Mars must be very remarkable, for the outer satellite will remain above the horizon for two and a half days and nights, and the inner will rise in the west and set in the east twice in the course of the night."[1]

In fact, trying to picture the phenomena of the moons of Mars proved to be very stimulating to the imaginations of astronomers in the immediate aftermath of Hall's discovery. Edward Emerson Barnard, who went on to become one of the most distinguished observational astronomers of all time but was then a young and struggling amateur, produced a manuscript (never published) called "Mars: His Moons and His Heavens" in 1880, in which he worked out in detail the eclipses, transits, and mutual occultations of the moons. As earlier noted, Mars is blessed with many more of these phenomena than Earth; for instance, there are 1,330 Phobos eclipses every year. Such eclipses, transits, and even mutual occultations have now been imaged by the rovers from the surface of Mars itself—something Barnard could hardly even dream of. Barnard also tried to imagine as best he could the magnificent bird's-eye view of Mars that would be seen from someone on Phobos:

> In 7h 39m 15s [the period of revolution of Phobos around Mars] the planet would pass through all the phases our moon goes through—from new moon to new moon. What a magnificent spectacle! A huge crescent over five thousand, five hundred times as large as our crescent moon—stretching from the horizon nearly half way to the zenith, behind which the stars are swiftly passing to reappear again at the west; whole constellations of these bright fires melting away before its rapid approach! What an awe-inspiring sight to witness from this satellite the rapid sweep of sunlight over this stupendous globe! Its waxing and waning! See the sun sink behind it and darkness reign for nearly one hour at the time of the Martian equinoxes; then the swift emergence of the brilliant orb from the planet's edge quickly followed by a thin rim of light glinting along the edges of the great globe.[2]

The orbital period of Phobos relative to the rotational period of Mars was (and remains) unprecedented among the natural satellites of our solar system, and the similarity of the orbital period of Deimos and the spin rate of Mars make that moon comparable to what modern astronomers would call a geostationary (or perhaps areostationary) satellite. In addition, Hall's discovery observations of Phobos and Deimos immediately led—via the Newtonian form of Kepler's third law—to a much better estimate of Mars's mass than had been available during the "moonless" era. Previous estimates, based on Mars's gravitational pull on the other planets, had ranged from 0.09 to 0.13 times the mass of Earth, with the best estimate having been given by the mathematical co-discoverer of Neptune, Urbain Jean Joseph Le Verrier. Le Verrier's determination that the mass of Mars was about 1/118th that of Earth, "was the product of a century of observations and several years of laborious calculation by a corps of computers; whereas from the measures of the satellite on four nights only, ten minutes' computation gave a value of the planet's mass . . . more trustworthy than Le Verrier's."[3] Indeed, the value published by Hall is within 0.2 percent of the modern accepted value of 0.107 Earth masses (appendix C). Subsequent observations of the moons through the middle of the 20th century provided the best pre-spacecraft-era knowledge of their orbits and orbital precession rates, the mass of Mars, and the planet's axis tilt (obliquity).[4]

Based on the moons' faint magnitudes and lack of a resolved disc, Hall and other 19th-century astronomers knew that the moons of Mars must be quite small, indeed "much smaller than the minor planets hitherto discovered."[5] An accurate assessment of just how small the moons are remained elusive, however, because a larger, lower-albedo (darker) unresolved body could have the same observed overall brightness (magnitude) as a smaller, higher-albedo (brighter) unresolved body. Still, assuming reasonable average albedo values typical of Mars or of terrestrial volcanic materials, astronomers thought that the dim magnitudes of Phobos and Deimos meant that they were probably no more than about 25 kilometers and 15 kilometers in diameter, respectively. Similar assumptions went into new, early to mid-20th-century telescopic searches for additional moons of Mars, all of which came up empty down to predicted size limits of only around 1.4 kilometers in diameter.[6]

Shooting the Moons

It would take the advent of the Space Age, and the early flyby and orbital missions to Mars, to reveal the actual sizes and shapes of tiny Phobos and Deimos. The first resolved image of Phobos was somewhat accidentally acquired during the Mariner 7 flyby of the planet in 1969. In just one of the images (frame 7F91), a dark, elongated speck less than 10 pixels across appears near the edge of the field of view in silhouette against the bright plains of Arabia Terra, just to the west of Syrtis Major (figure 20.1). That speck turned out to be the only Mariner 7 image of Phobos where the moon spanned more than a few pixels in size. Mariner 7 imaging team member (and future Voyager camera lead) Brad Smith was able to use the image and knowledge of the spacecraft's position to estimate the moon's shape as roughly ellipsoidal,

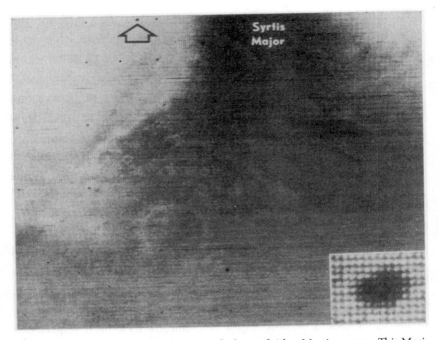

FIGURE 20.1 The first resolved spacecraft photo of either Martian moon. This Mariner 7 photo shows Phobos (*arrowed*) as an elongated speck against the plains of Arabia Terra (the inset is a 15× enlargement). The other round specks in the photo are camera calibration markers or other artifacts. NASA/JPL/Brad Smith.

with dimensions larger than the earlier magnitude-based telescopic estimates, around 18 by 22 kilometers.[7] An additional discovery reported by Smith from the Mariner 7 image, and the explanation for why Phobos turned out to be larger than expected, was that Phobos was very dark, with an albedo of only around 6.5 percent (for comparison, Earth's Moon averages 12 percent). The reason for such a relatively low albedo (compared to average lunar and Mars values) was not understood, though speculation focused on the possibility that the surfaces of such tiny bodies might be relatively dust free because of the extremely low gravity (only around a few thousandths of Earth's gravity).

Perhaps the most interesting speculations from the few dozens of pixels of Phobos captured in the Mariner 7 frame was Smith's musings about what the elongated shape of Phobos, rather than a spherical shape, might mean in terms of the origin of that moon: "It does suggest, however, that Phobos did not form by accretion as it orbited within the planetesimal cloud around primordial Mars, but may have been captured in its present form at some later time. That Phobos may have had its origin within the asteroid belt is not an unreasonable supposition."[8]

More detailed information (and more informed speculation) would come from better imaging of both Phobos and Deimos from Mars orbiters. The first truly revealing images came in 1971 and 1972 from Mariner 9, when mission navigators took images of both Phobos and Deimos whenever the spacecraft's highly elliptical orbit took it within about 7,000 kilometers of either object (figure 20.2). They ultimately acquired more than 200 images of the moons (including the first images to resolve Deimos) over the course of the mission, with scales down to as small as 30 meters per pixel in the images taken from the closest ranges using the narrow-angle camera from the Mariner 9 television system.[9]

The more systematic imaging of both moons from Mariner 9 enabled the first true "shape models" of extraterrestrial moons to be created. The basic

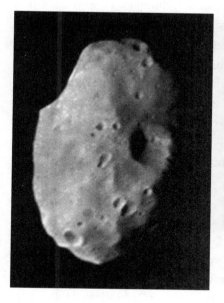

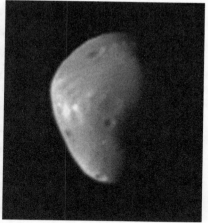

FIGURE 20.2 Some of the best Mariner 9 views of Phobos (*left*, from December 1971) and Deimos (from January 1972), shown scaled to their approximate relative sizes. The crater to the right of center in this view of Phobos is called Hall, and Stickney is the "bite" taken out of the upper left. The two nearly overlapping craters near the center of this view of Deimos are Voltaire (larger, muted) and Swift (smaller, sharper), which are still—surprisingly—the only two named surface features on Deimos. NASA/JPL/Emily Lakdawalla.

shape of these models was a 3-D ("triaxial") ellipsoid, with the longest axis always pointing toward Mars because the moons (like our Moon and many others across the solar system) are locked in synchronous rotation with Mars, always keeping the same "face" pointed at the planet. The dimensions of the best-fit Mariner 9–based triaxial ellipsoid for Phobos was 27.0 by 21.6 by 18.6 kilometers, and for Deimos was 15.0 by 12.2 by 11.0 kilometers (Deimos was thus indeed shown to be smaller than Phobos, consistent with its comparable albedo but dimmer telescopically determined magnitude).[10] For scale, Phobos has around the same surface area as the Hawaiian island of Oahu; Deimos has around a third of that.

In addition to determining the sizes of the moons to high precision, Mariner 9 imaging of Phobos and Deimos also helped mission planners perfect a new image-based spacecraft guidance approach called "optical navigation," using cameras to measure the relative positions of stars, planets, and moons to navigate with high precision.[11] Most deep space planetary missions since then have relied on such optical navigation techniques. However, not only do spacecraft orbiting Mars require accurate navigation to allow the scientific instruments to record desired data, the navigators need to make sure the spacecraft do not go anywhere that is not safe. For example, in May 1977, navigators calculating the predicted future orbital path of the Viking Orbiter 1 could not guarantee that it would miss Phobos a few days hence, so before the close encounter the spacecraft was commanded to perform a small rocket burn that did guarantee a miss. In 2017 the MAVEN spacecraft also had to do a similar small maneuver to make sure it did not collide with Phobos.[12]

Mapping Tiny Worlds

Mariner 9 imaging of Phobos and Deimos enabled the creation of the first cartographic maps of the moons of another planet. More extensive imaging of both moons by the Viking Orbiters beginning in 1976 allowed those maps to be updated and refined using higher-resolution data. These small worlds are not perfect ellipsoids but are instead quite irregular. Indeed, their surfaces preserve numerous impact craters formed during violent past collisions with small asteroidal and cometary fragments. While Phobos is more nearly ellipsoidal than Deimos, in places the deviations of both moons approach two to three kilometers from a perfect ellipsoid. The most significant deviations range from a hemispheric-sized impact crater on Phobos to the simple bowl-shaped impact craters that litter the surfaces of both moons down to the resolutions of the available images.

As detailed surface features began to be identified on Phobos and Deimos from Mariner 9 images, it fell to the International Astronomical Union (IAU) and specifically its nomenclature committee (chaired, at the time, by Carl Sagan) to devise a naming scheme for those features.[13] For Phobos, Sagan's committee chose to name surface features after individuals involved with the discovery, dynamics, or properties of the Martian satellites. The largest impact crater on Phobos is named Stickney (the maiden name of Asaph Hall's heroic wife, Angeline, who, as noted in chapter 6, urged him to the telescope for one last try, which proved successful). Other large craters are named Hall, D'Arrest, Roche, Todd, Shklovsky, and Sharpless. In addition, several large craters—Gulliver, Grildrig, and Clustril—are taken from Jonathan Swift's *Gulliver's Travels* (1726), since Swift attributed to the Laputan astronomers two satellites of Mars, saying: "They have likewise discovered two lesser stars, or satellites, which revolve about Mars, whereof the innermost is distant from the centre of the primary planet exactly three of his diameters and the outermost five; the former revolves in the space of ten hours and the latter in 21½." It was an astonishingly good guess, especially for Phobos, but that is all it was; no telescope on Earth could have shown the tiny moons in the eighteenth century. For Deimos, the IAU committee decided on names for authors or artists who had alluded to the Martian satellites, but the available list seems to have been rather small. To date, only two craters on Deimos have formally been named by the IAU, one Swift and the other Voltaire (who in his *Micromégas, Histoire Philosophique*, described the visit of a giant from Sirius with a Saturnian friend to the neighborhood of Mars, and also mentioned the planet's "deux lunes").[14]

As noted above, the general physiography of both Phobos and Deimos is dominated by impact craters. As of the mid-2010s, tens of thousands of craters of more than 10 meters in diameter had been identified on Phobos, and more than 500 craters of greater than 20 meters in diameter had been found on Deimos, from Mariner 9, Viking Orbiter, Mars Express, and Mars

Reconnaissance Orbiter (MRO) spacecraft imaging campaigns.[15] While there is no clear consensus on the inferred absolute ages of Phobos and Deimos based on their crater statistics, both surfaces are likely to be at least several billion years old, and potentially nearly as old as the solar system itself. Their ancient nature comes from the observation that they have high crater densities (numbers of craters per square kilometer), around half the crater density of the ancient highlands of the Moon. Interestingly, Phobos and Deimos have approximately equal surface crater densities, suggesting that whatever their absolute ages are, they are comparable.

The large numbers of impact craters on both moons imply that significant amounts of surface material have been pulverized over time. That implication, combined with infrared observational data that shows that both moons are covered by fine-grained, relatively porous materials with very low thermal conductivity, suggests that both their surfaces are covered by regolith.[16] On Phobos, the regolith is estimated to be from a few tens to about 100 meters thick, depending on how much crater ejecta is retained after impact events and how much escapes entirely from the low-gravity surface environment. With lower gravity than Phobos, Deimos is expected to have a regolith with a thickness about five times less,[17] although for the same reason it appears that this regolith is more porous, as is indeed indicated by radar observations.[18]

A major surprise from the higher resolution and more extensive surface coverage of the Viking Orbiter images of Phobos was the discovery of a network of long linear depressions or "grooves" that span much of the surface of the satellite. Some of the grooves are tens of kilometers long, and they are typically 100–200 meters wide and 10–20 meters deep. Many appear to consist of aligned adjacent, partially merged pits. The grooves are deepest and widest adjacent to the large crater Stickney and in places even crosscut it, and so are hypothesized by some to be associated with the formation of that crater. While no definitive explanation has yet emerged to explain the origin of the grooves, leading hypotheses include the idea that the Stickney-forming impact caused enormous amounts of fracturing of Phobos,[19] or that large boulders ejected from Stickney during the impact event carved the grooves in the regolith as they slid, bounced, and rolled away from the crater and across the surface.[20] Other hypotheses include the grooves being "secondary crater chains" formed when material ejected from impacts on Mars slammed into Phobos,[21] similar to crater chains seen on the Moon, or else being cracks caused by tidal stresses from Mars starting to slowly disrupt the asteroid.[22] Lots of ideas have been floated, but there is still no consensus as to the true origin of these enigmatic Phobosian features.

While Phobos is relatively uniform in terms of its brightness and crater density, Deimos has a number of patches, typically elongated, of surface material that are up to 30 percent brighter than their surroundings. Stereo images have been used to reveal the topography of Deimos, enabling the determination of surface slopes and thus the sometimes-nonintuitive definition of which way is "downhill" on such an irregularly shaped microgravity world. That kind of analysis has shown that the bright patches on Deimos are places where the regolith is creeping or flowing downhill, perhaps instigated by seismic shaking produced by impact cratering events.[23] These areas could be brighter because finer-grained materials are brighter than and more easily mobilized than coarser-grained materials, as with the sands and dust on Mars itself. Indeed, the significant mobility of "sediments" on Deimos is thought to be responsible for the widespread partial infilling and generally more muted appearance of impact craters on Deimos compared to Phobos. One hypothesis is that the lower gravity of Deimos means that fine-grained materials ejected during impact cratering events travel more extensively across the surface (perhaps even circumnavigating the surface), helping create a widespread and relatively uniform global layer of dusty "sediments" that settle out last and cap the regolith.[24] Still, the details of impact ejecta transport and granular flow processes on Deimos are not well understood. However, as plausibly suggested by Cornell small-body geology expert Peter Thomas, the dramatically different appearances of the surfaces of Phobos and Deimos may simply reflect the dominant role of the youngest large impact crater and its ejecta (Stickney on Phobos; a 10-kilometer-wide, unnamed depression in the southern hemisphere of Deimos) in shaping the currently observed surface.[25] Specifically, the large southern hemisphere crater on Deimos is much larger relative to Deimos's radius than Stickney is to Phobos's radius, and so is likely to have had a much more dramatic effect. As Thomas notes:

The amount of covering possible from Stickney on Phobos is nominally only about one-third that possible from the large crater on Deimos (31 m vs 81 m if evenly spread). Even though the ejecta are concentrated near the source craters, nearly half the area of Deimos outside the large crater is predicted to have an initial cover of over 50 m, while only about one-sixth the area of Phobos outside of Stickney probably had as thick a cover. Additionally, any seismic effects of the impact on the morphology of preexisting craters would be much more severe on Deimos, as essentially the whole surface is within 1.5 crater radii of the rim.[26]

Failures and Successes in Spacecraft Exploration: 1998–2011

Even more detailed imaging and other studies of Phobos, including the first attempted landing, were planned as part of the Soviet Phobos 1 and Phobos 2 missions, both launched in summer 1988. Phobos 1 failed due to a commanding error during the cruise to Mars, but Phobos 2 successfully went into Mars orbit in late January 1989 (see chapter 14). The mission acquired some unique and useful Mars science data over a period of several months before communications were lost only about 100 kilometers short of an attempted landing on Phobos. Still, during the approach, the mission was able to acquire about three dozen images of Phobos that revealed parts of the surface at higher resolution than any previous Mariner 9 or Viking Orbiter imaging. Those photos were used to improve knowledge of the shape of Phobos, as well as of its orbital dynamics, color, composition, and texture.[27]

Phobos 2 marked the end of an era; it was the last Soviet space probe to Mars. After the fall of the Soviet Union in December 1991, the Soviet space agency became Roscosmos, whose first Mars mission was the Mars-96 probe, launched with high hopes in November 1996. Mars-96 was a large and highly ambitious mission designed primarily to study Mars though also intended to perform close flybys of Phobos and Deimos. Unfortunately, Mars-96 failed shortly after launch.

The other late 1996 Mars mission, NASA's Mars Global Surveyor (MGS) orbiter, successfully reached Mars in 1997. Its imaging of the moons was infrequent but still scientifically important, especially images of Phobos, which achieved resolutions down to a stunning four meters per pixel.[28] MGS optical navigation imaging of the moons as well as laser altimeter observations of the shadow of Phobos cast onto the Martian surface also provided new ways to constrain the orbital properties and magnitude of the Mars tidal effects on the satellites (and, by inference, help provide some new constraints on the interior properties of Mars itself).[29]

Starting in 2004 and continuing to the present, the European Space Agency (ESA) Mars Express Orbiter and NASA MAVEN orbiter have also conducted occasional focused studies of the moons, especially when those orbiters' elliptical polar orbits cross the orbit of Phobos. Mars Express data includes high-resolution color and stereo images that now cover major fractions of the surfaces of both moons (figure 20.3); ultraviolet, near-infrared, and thermal infrared spectra; radar sounding data; and gravity and orbital dynamics measurements.[30] MAVEN data captured Phobos in the ultraviolet, with the goal of potentially linking its composition to specific classes of asteroids and

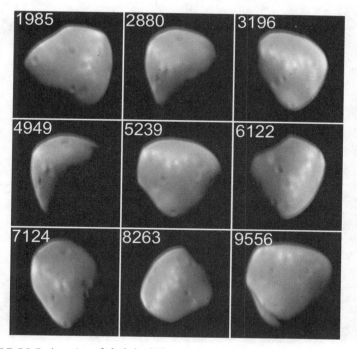

FIGURE 20.3 A series of slightly different views of the sub-Mars hemisphere of Deimos, acquired by the European Space Agency's Mars Express Orbiter between 2005 and 2011 (orbit number indicated). ESA.

meteorites.[31] Additional images of Phobos at a resolution of around 6 meters per pixel, and of Deimos at around 20 meters per pixel, were also acquired by the High Resolution Imaging Science Experiment (HiRISE) camera on NASA's MRO mission in 2008 and 2009, providing additional viewing angles and perspectives on the geology of these small worlds (figure 20.4).[32]

After the failure of Mars-96, Roscosmos regrouped and designed its next mission, called Phobos-Grunt ("Phobos-Ground"), to be specifically focused on analyzing and returning a sample to Earth from the innermost Martian moon. Phobos-Grunt was an extremely ambitious mission, designed to be launched in 2009 and to not only collect and return several hundred grams of soil from Phobos, but also to collect data about Mars, to conduct experiments

FIGURE 20.4 High-resolution enhanced false-color view of Phobos from NASA's Mars Reconnaissance Orbiter HiRISE imaging system in March 2008. The image has a scale of 6.8 meters per pixel. The large crater Stickney and its famous series of linear grooves are visible at lower right. NASA/JPL/University of Arizona.

on and near Phobos, and to deploy a Mars orbiter named Yinghuo-1, which was to be the first deep space interplanetary mission conducted by the China National Space Administration. Technical difficulties delayed the launch of Phobos-Grunt to November 2011. The spacecraft was successfully launched into an initial elliptical low Earth orbit, but unfortunately the upper-stage engine failed to burn and inject the spacecraft into its planned interplanetary trajectory. Without adequate power or the ability to control the propulsion system, Phobos-Grunt fell back to Earth in January 2012. Mars continued to be as elusive and frustrating for the Russian space program as it had been for the Soviet one.

What Are Phobos and Deimos Made Of?

While imaging observations from multiple space missions have been successful in revealing many details of the geology and topography of Phobos and Deimos (plate 22), it has been much more challenging to measure and tease out the details of the composition, mineralogy, and other bulk physical properties (appendix C) of those small worlds.[33]

Estimates of the masses of Phobos and Deimos far more accurate than any attainable from Earth were realized from the tiny deflections they produced in the paths of the Viking Orbiters during their close flybys of the moons in the late 1970s. Estimates of the masses of these moons could then be combined with estimates of the volume (from shape models created from images) to derive initial estimates of their densities.[34] Phobos and Deimos were shown to have relatively low densities well below the values for typical silicate rock compositions (the modern accepted values are around 1.88 and 1.47 grams per cubic centimeter, respectively). Given their overall rocky appearance and their locations in the inner solar system, where it is too warm for icy bodies to remain stable over geological time, these relatively low densities imply that both moons have significant porosity in their interiors.

Parenthetically, the satellites' porous interiors recall a long-standing enigma and show how easy it is to draw provocative, if incorrect, conclusions by pushing data further than its accuracy allows. In 1945, after analyzing measurements of the positions of the satellites, mostly based on visual micrometric data going back to their discovery in 1877, U.S. Naval Observatory astronomer

Bevan Percival Sharpless (1904–50) announced that Phobos appeared to be rapidly spiraling inward toward Mars. It seemed that such acceleration could only be produced by drag of some sort, and in 1959 a Soviet astronomer, Iosif Shklovsky (1916–85), concluded that the drag was with the outer atmosphere of Mars.[35] This was reasonable enough, as at the time Mars's atmosphere was thought to be much denser than is now known; however, in order to explain the rapid rate of acceleration, Shklovsky went a step further and made the rather ridiculous suggestion that Phobos was not merely porous but was actually hollow inside, and might even be an artificial space station! A reanalysis of the observations later failed to confirm the acceleration reported by Sharpless, and attributed his result partly to errors of the observations and partly to the parameters for Phobos's orbit he adopted.[36]

Many clues to the composition of the Martian moons have come from whole-disc telescopic observations of their visible-wavelength colors as well as spectra of sunlight reflected from their surfaces in the visible and near-infrared parts of the spectrum. Both satellites have a generally gray to slightly reddish visible color as measured by ground-based telescopes and the Mariner 9, Viking Orbiters, and Viking Lander 1 (from the surface) imaging systems. The initial disc-resolved spacecraft-based observations noted only very small color variations observed across either surface (the bright regions on Deimos are brighter but essentially the same color as the average background materials).[37] Additionally, ground-based infrared data was not able to detect evidence for significant hydration of the surfaces of either moon.[38] Many of the whole-disc and early space-based color and spectroscopic data sets available for both Phobos and Deimos supported the idea that both moons could be similar to so-called C-type asteroids, which are thought to be similar in composition to carbonaceous chondrite meteorites, or perhaps the D-type asteroids, which are interpreted as primitive bodies like the C-types, though with redder spectra. However, alternative interpretations for their composition were certainly not ruled out, especially in view of their apparently anhydrous nature (that is, lacking absorption features indicative of water or hydroxyl on their surfaces; this trait is potentially inconsistent with carbonaceous chondrite materials, which often include small amounts of water).

Higher-resolution visible-color imaging and infrared spectroscopic observations of Phobos were acquired in 1989 by instruments on the Phobos 2 mission, and that data enabled the measurement of subtle color and compositional variations.[39] Specifically, most of the surface was observed to have properties consistent with previous lower-resolution color and whole-disc spectroscopy measurements, but some parts of the interior and ejecta of the large crater Stickney were observed to be significantly lower albedo and "bluer" than the average ("redder") materials. One interpretation of the bluer places was that they represented material excavated from deeper inside the satellite, while the redder material represented surficial materials that are likely more "space weathered" (exposed to the constant bombardment of the solar wind and other high-energy particles, which leads to localized melting, physical breakdown, and formation of submicroscopic metallic iron at grain boundaries).[40] Other researchers speculated that the color variations simply reflect the overall heterogeneity of the surface.[41] Detailed assessment of the Phobos 2 infrared spectra supported the idea that the best asteroid analogy to Phobos might not be the C-types, but instead the poorly understood T-type asteroids,[42] which exhibit low albedo and somewhat reddish and featureless spectra, and also appear anhydrous.[43]

Even higher-resolution visible to near-infrared spectroscopic measurements (with more extensive spatial coverage) were acquired for both Phobos and Deimos using the Observatoire pour la Minéralogie, l'Eau, les Glaces et l'Activité (OMEGA) instrument on Mars Express and the Compact Reconnaissance Imaging Spectrometer for Mars (CRISM) instrument on MRO.[44] The results were generally consistent with previous telescopic and spacecraft observations, in that the surfaces of both moons are dominated by materials with a reddish but relatively featureless visible to near-infrared signature (the "redder" unit previously identified on Phobos), and "bluer" materials are uniquely associated with Stickney crater and its ejecta. The average color properties of the moons as derived from those data sets appeared inconsistent with highly space-weathered ordinary chondrites or Martian crustal basaltic materials. Rather, and consistent with many previous results, the spectra of Phobos and Deimos from OMEGA and CRISM have been interpreted to indicate that both moons have primitive compositions, consistent with C-type or D-type asteroids and carbonaceous chondrite meteorites.[45]

Additional analyses have also revealed weak visible to near-infrared absorption features on both Phobos and Deimos that have been interpreted as evidence either for the formation of extremely fine-grained iron oxides as part of the space weathering process, or for the presence of a class of clay silicate minerals known as phyllosilicates.[46] Phyllosilicates are formed by the aqueous alteration of precursor silicate minerals and have been detected on low-albedo, reddish-colored asteroids that have been interpreted as having primitive bulk compositions (including C-types and D-types). Interestingly, phyllosilicates have also been detected on Mars itself from a variety of missions and instruments.

The Viking Orbiters, MGS, Mars Odyssey, and Mars Express have also observed the Martian moons in the thermal infrared wavelength region, where the measured signal is not from reflected sunlight but from energy directly emitted from an object proportionally to its temperature. Thermal infrared measurements have been used to determine the ways that the surfaces of Phobos and Deimos warm up and cool down during the day, confirming the fine-grained nature of their surfaces.[47] Multifilter and spectroscopic observations have also provided additional constraints on the composition and mineralogy of the moons. For example, MGS Thermal Emission Spectrometer (TES) instrument data of Phobos primarily covering the Stickney crater region has been modeled as being dominated by a fine-grained silicate composition (with spectral features similar to those seen in fine-grained basalts), but also with a phyllosilicate component, consistent with detection in some near-infrared spectra.[48] Mars Odyssey Thermal Infrared Imaging System (THEMIS) measurements of Phobos—the highest spatial resolution infrared images yet obtained of either Martian moon—are also dominated by the thermal signature of fine-grained silicates.[49] However, the TES and THEMIS data sets of Phobos do not match each other, perhaps because TES primarily measured near Stickney crater (the "bluer" unit from previous visible to near-infrared observations), while THEMIS imaged other regions of the moon (the average, "redder" unit). The lack of a match, then, could suggest significant spatial variability among the thermally derived composition or physical properties of Phobos.

Origin and Destiny

There is still significant debate and uncertainty about the origin of Phobos and Deimos, even after more than 40 years of study by a variety of space missions observing the moons up close from within the Martian system. The two leading and directly contradictory hypotheses are that they are both captured asteroids (perhaps from the nearby main asteroid belt, or potentially even from the outer solar system), or that they formed in place, in Mars orbit, either at the same time that Mars was forming or perhaps later from the merging of debris launched into Mars orbit by a giant impact on the planet.[50]

Evidence for the "capture hypothesis" comes from the interpretation of Phobos and Deimos as primitive objects, with compositions thought to be similar to several classes of primitive asteroids and meteorites based on visible to near-infrared spectra.[51] A number of researchers have pointed out the ad hoc nature of such a hypothesis, however, and especially the seemingly unlikely odds of two captured asteroids that were coming in on elliptical trajectories to have both settled into extremely circular, equatorial orbits. Evidence for the "Mars co-forming" or "Mars impact ejecta-forming" hypothesis is partially based on the interpretation of thermal emission spectra that the moons have a primarily basaltic nature, similar to the primarily basaltic nature of Mars itself.[52]

Uncertainties about the origins of Phobos and Deimos reflect those about other bodies in the solar system. Gigantic impacts certainly were commonplace, and Earth's Moon, the retrograde rotation of Venus, and the catawampus tilt of the axis of Uranus have all been explained as due to giant impacts early in the solar system's history. The moons of Mars could have been formed initially with Mars (or by an impact on early Mars) but then had primitive materials added to their surfaces by subsequent impacts, disguising their actual origins.[53] Or perhaps they really are captured primitive bodies that have somehow themselves captured impact-ejected Martian crustal materials on their surfaces.[54] One hypothesis even postulates that the Phobos and Deimos we observe today are the remnants of multiple cycles of one or more previously larger moons that spiraled toward the planet and were broken apart by tides to form a ring, part of which re-accreted into new, smaller moons.[55] It will require additional data analysis from more close flybys or dedicated orbital/lander missions, and probably even future sample return missions, to definitively test all of these competing hypotheses.

While the past histories of Phobos and Deimos remain uncertain, we have much better predictions of the future history of these small, enigmatic moons based on both theoretical calculations and the continued monitoring of their changing orbits. Even though they are small, the Martian moons still do produce a small tidal bulge on Mars itself (just as our own Moon creates a small tidal bulge in the solid Earth, not just in the oceans). Because Phobos orbits the planet faster than Mars spins on its axis, the tidal bulge raised by Phobos lags behind the moon's orbital position, exerting a net force in the direction opposite the moon's orbital motion. That force pulls Phobos ever so slightly toward the surface (by an average of around 10 centimeters per year), causing it to accelerate in its orbit as it gets closer to Mars. Deimos undergoes the opposite effect; because it orbits slower than Mars spins, the tidal bulge is in front of that moon and pulls Deimos forward, and thus outward, causing it to slowly decelerate as it recedes.[56]

Both the acceleration of Phobos and the deceleration of Deimos have been measured by spacecraft in Mars orbit and on the surface, beginning with the first sets of resolved spacecraft images of the moons by Mariner 9.[57] Monitoring continued during the Viking Orbiter missions, but the lack of systematic spacecraft observations of the moons for nearly 20 years after Viking ended meant that uncertainties were growing on the predicted positions of both moons (but especially Phobos) based on extrapolations of orbital parameters last formally updated in the 1980s. In part for this reason, both Spirit and Opportunity rovers began in 2004 to conduct timing observations of transits of Phobos and Deimos across the disc of the Sun as seen from the surface of Mars.[58] Indeed, while Deimos had not strayed significantly compared to its predicted position, Phobos was observed in the rover images to be ahead of its predicted position by about 11 kilometers (one Phobos radius) and below (out of plane) by about 0.5 kilometers. Similar deviations were noted by the Mars Express optical navigation team, which had to compensate for these prediction errors in its efforts to image Phobos at high resolution.[59] Continued monitoring of the positions of the moons by orbiters and rovers (including Curiosity beginning in 2012) has helped keep the predicted positions of the satellites accurate for additional higher-resolution imaging and other observations.

Thus, despite the earlier errors in Sharpless's calculations, Phobos has, in fact, subsequently been shown to exhibit a small but significant acceleration of its motion, produced by tidal friction rather than drag with the atmosphere of the planet. Indeed, the rate of acceleration of the orbit of Phobos is remarkable, and implies that it will eventually, in less than 100 million years, spiral into Mars and be destroyed.[60] This is perhaps secondary evidence against the "Mars co-forming" or "Mars impact ejecta-forming" hypothesis for the origin of the moons, as it seems uncanny (but certainly possible) that the orbit of Phobos could have been evolving inward for more than four billion years and we just happen to be catching it in its last few tens of millions of years of existence. Conversely, the rate of deceleration of Deimos is remarkably slow, to the point that it is thought to have moved outward only by a very minor fraction of its current orbital radius since its formation (or capture), and it is not predicted to move outward significantly more over the remaining lifetime of the solar system.[61]

The two tiny moons of Mars continue to puzzle and amaze almost a century and a half after their unexpected discovery from Foggy Bottom. It may well be that we will have definitive answers to many of the outstanding questions about them only when we obtain samples of material and can determine whether the stuff they are made of more resembles ancient rock from Mars or some of the asteroids. Meanwhile, at least for a few more tens of millions of years, they will continue to hurtle on.

21

Ongoing and Upcoming Missions

The 2020s

The Beat Goes On

The history of Mars exploration in the Space Age continues to evolve and expand. Mars has been and will continue to be—well into the future—the most visited destination for robotic space mission exploration beyond the Earth-Moon system (figure 21.1). At the time of writing (March 2021), Mars has become a very crowded place indeed. Eight orbiters—Mars Odyssey, Mars Express, Mars Reconnaissance Orbiter (MRO), the Mars Orbiter Mission (Mangalyaan), MAVEN, the Trace Gas Orbiter, the Tianwen-1 orbiter, and the Hope orbiter—provide a constant stream of information about the Martian environment, as well as, in the cases of Odyssey, MRO, and MAVEN, orbital communications relay support to active missions on the surface. Without active relay satellites like those, rovers and landers on the surface would only be able to send back a small fraction of their normal daily data volume because of power and thus bandwidth limitations on their onboard communications systems. The orbiters high above them, however, with their large solar panels and comparatively giant power supplies (much like Earth's own communications satellites, which are even more power hungry), can communicate at relatively high bandwidth with surface landers and rovers because they are

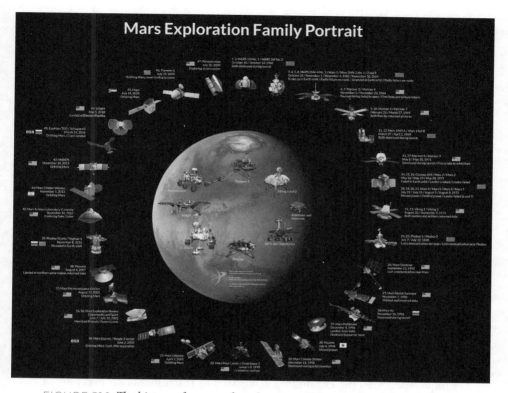

FIGURE 21.1 The history of spacecraft exploration of Mars as of mid-2021. This history includes 47 missions (more than 40 percent of which have failed) from eight different nations/organizations, and ten successful landers or rovers, including three missions that successfully arrived in orbit and/or on the surface in winter/spring 2021. NASA, JPL, Roscosmos, JAXA, ESA, ISRO, Jason Davis, and The Planetary Society.

close, and then can use their higher-power transmitters to send that data back to Earth at a comparably higher data rate.

Two NASA rovers, Curiosity and Perseverance, are alive and well and exploring Gale crater and Jezero crater, respectively. The stationary lander InSight, with its heat flow probe and seismometer, continues to gather information about the deep interior structure of Mars. Remarkably, from the early days when only the United States and the Soviet Union sponsored interplanetary missions, five major space agencies are now represented by these missions: NASA, the European Space Agency (ESA), the Indian Space Research

Organisation (ISRO), the United Arab Emirates Space Agency, and the China National Space Administration. The latest missions are Perseverance, Tianwen-1, and Hope, which arrived on Mars in February 2021. Meanwhile, there are grizzly veterans, which continue to function years after their expected lifetimes: these include Mars Odyssey, which arrived in Martian orbit in 2001, and Mars Express, which arrived in 2003. And Curiosity, which landed at Gale crater in August 2012, is continuing its slow trek from the older to younger strata of its sedimentary landscape, and seems likely to remain functional for another five years. (It fortunately survived the monster global dust storm of 2018, to which its cousin Opportunity succumbed.)

Still other missions are deep in the planning stages and await the next available launch windows. Another rover (a joint mission of the ESA and Russia) is scheduled to launch in 2022, and another orbiter and Phobos lander (from the Japanese space agency) is being planned for a 2024 launch. At the same time, new Mars science continues to be done with ground-based radio telescopes, which are busy making radio-wavelength observations of atmospheric gases. Eagerly anticipated is the infrared James Webb Space Telescope (JWST), long plagued by delays and cost overruns but currently scheduled for launch on October 31, 2021, which promises new viewing geometries and unique spectroscopic capabilities compared to instruments on previous or current missions.

New Insight

The first mission dedicated exclusively to exploring the interior, as opposed to the atmosphere or surface of Mars, was NASA's InSight lander, which arrived safely on Mars in November 2018, and is now well into its prime mission (figure 21.2). InSight, which is a backronym that represents "Interior Exploration using Seismic Investigations, Geodesy and Heat Transport," was competitively proposed to NASA's lower-cost Discovery mission program in 2010, and was selected for flight in 2012. Originally scheduled for a 2016 launch (which was delayed to early 2018 because of science instrument problems), InSight is a mission designed to primarily study the interior of the Red Planet, searching for evidence of marsquakes and attempting to measure, from a single

FIGURE 21.2 *Top*: The InSight Mars lander with its solar panels deployed during assembly and testing at Lockheed Martin Space Systems, Denver. *Right*: The InSight lander's seismometer and wind cover (*center*) and heat flow "mole" (*center left*), deployed on the surface in Elysium Planitia. One of the lander's footpads can be seen at lower right. NASA/JPL/Lockheed Martin.

landed station, the amount of heat flowing from the interior to the surface.[1] To reduce cost and risk, the lander is essentially the same Lockheed Martin lander that successfully carried the Phoenix mission's experiments to the surface in 2008 (chapter 17). And because the mission's main goal is geophysical—studying the subsurface of Mars rather than the geology of the surface—the InSight team intentionally aimed for a relatively safe, flat, sandy surface with few rocks. Team members settled on a region in the vast lava plains of Elysium Planitia, along the boundary between the heavily cratered southern highlands and the smoother northern lowlands. While the site is historically relatively dusty, the most important aspect of site selection was that the shallow subsurface be loose and sandy enough ("unconsolidated" to a geologist) for the lander's heat flow probe or "mole" to be able to burrow into the surface, but that the deeper subsurface contain more rocky or consolidated materials that could efficiently transmit any seismic waves generated in the deeper crust or mantle of the planet.[2]

After a successful heat shield–, parachute-, and retrorocket-assisted landing on November 26, 2018, imaging assessment of the landing site revealed that it is indeed a relatively flat, sandy, impact-fragmented plain with few rocks (figure 21.2)—frankly, somewhat geologically boring, but just what the mission engineers and geophysical scientists were hoping for.[3] After about 100 sols (by February 2019) of careful robotic-arm work to unpack instruments from the deck of the lander and deploy them down onto the surface, InSight's prime mission of searching for marsquakes was finally ready to begin. Planetary geophysicists were giddy about the potential discoveries that could be forthcoming, as it had been more than 40 years since seismometers had been operated on another world (with five successful lunar seismic stations deployed by the Apollo astronauts that operated from 1969 to 1977).[4]

Indeed, the InSight team detected the first (weak) marsquake very soon after the seismometer became fully operational, and 173 more were detected in the first half of the lander's prime mission (this mission was set for one Mars year).[5] The quakes detected are relatively small compared to earthquakes (with magnitudes ranging from around 2 to 4 on the terrestrial earthquake scale), but they are not insignificant, nor are they consistent with the kinds of quakes expected from small meteoroid impacts, which are known to occur frequently. Indeed, the surprising and exciting early mission result, reported in the initial peer-reviewed research papers by InSight principal investigator and JPL planetary geophysicist Bruce Banerdt and colleagues, is that InSight's detections indicate that Mars is "a seismically active planet," with tectonic activity much higher than the Moon's and only slightly below that experienced in the central regions of Earth's major tectonic plates, far from the much more active plate boundaries where larger earthquakes originate.

More problematic has been the planned measurement of the heat flowing out of the interior of Mars by the lander's self-digging mole, which became stuck only about 35 centimeters deep into its planned five-meter descent. Slow-going efforts to dislodge and reset the mole using the robotic arm showed promise, but digging ultimately ceased in late 2020 with no significant mole descent.

Two other delightful surprise results from the InSight mission so far deserve special note. The first is the amazing success of the "weather station" experiments designed to monitor winds, temperatures, and pressures at the landing site. A lesson from terrestrial and lunar seismometer experiments has been that super-sensitive instruments like those deployed by InSight can be influenced by wind or other atmospheric pressure changes, and so monitoring the weather very accurately is critical to telling those signals apart from signals generated by subsurface quakes. In the process of meeting this primary need, however, InSight's weather instruments have characterized the meteorology of Mars in unprecedented detail, revealing for example a new class of "dustless dust devils" that generate low-frequency infrasound vibrations picked up by the seismometer when these vortices pass by. Planetary meteorologists on the team are struggling (in a delighted way) to keep up with new hypotheses and new models to understand and explain the data. In addition, the lander's magnetometer has measured a much stronger field at the surface than had been anticipated, a result that is also still not understood but that could indicate that ancient Mars had a stronger, and/or longer-lived core-generated magnetic field than once thought.[6] Similarly puzzling but exciting insights about the history of the magnetic field of Mars are coming in from recent MAVEN mission results as well.[7]

The second delightful surprise from InSight actually happened back on the day of landing. Two tiny spacecraft, cereal-box-sized "CubeSats" created for the independent mission called Mars Cube One or MarCO, were launched along with InSight but sent on slightly different trajectories to Mars.[8] MarCO A and MarCO B, as the spacecraft are known, were designed not to land on but to fly past Mars—like tiny versions of the earlier flyby Mariners—with the flybys timed so that they could hopefully relay radio signals in real time directly back to Earth during InSight's landing (in a change from NASA practice that had been adopted since the failures of Mars Climate Orbiter [MCO] and Mars Polar Lander [MPL] back in 1999, none of the other Mars orbiters were able to be positioned to relay InSight's signals during and immediately after the entry, descent, and landing events). It was an ultra low-cost, high-risk experiment, as no CubeSats had ever operated in deep space before (hundreds have been launched into Earth orbit). The experiment worked beautifully, and the entire world was treated to telemetry and the mission's first postlanding images in near real time (except for the eight-minute one-way travel time for radio signals coming from Mars then). MarCO demonstrated that tiny spacecraft can play important roles in deep space exploration, a lesson that NASA intends to test even more extensively when it launches 13 science and technology mission CubeSats to the Moon as part of the first Space Launch System demonstration flight (now known as Artemis I) in 2021 or 2022.[9]

2020 Vision: A Perfect Mars Trifecta

The 2020s have opened on a new series of Earth–Mars launch windows, occurring a few months before each of this decade's Earth-Mars oppositions in October 2020, December 2022, January 2025, February 2027, and March 2029. Five new missions—from five different space agencies!—are already underway or in advanced stages of planning for the first three of these launch windows.

The first one to set off was an orbiter mission called the Emirates Mars Mission from the United Arab Emirates' Mohammed bin Rashid Space Centre (at Al Khawaneej, Dubai). Primarily focused on atmospheric science, the orbiter was launched on July 19, 2020, from the Tanegashima Space Center near Minamitane, Japan. It carries three science instruments (a camera and two spectrometers) to study Martian weather and composition. The orbiter and the instruments were designed and built in collaboration with the University of Colorado Boulder, the University of California, Berkeley, and Arizona State University. The orbiter itself was named Hope (al-Amal in Arabic) because "it sends a message of optimism to millions of young Arabs," according to Sheikh Mohammed bin Rashid Al Maktoum, the ruler of the Emirate of Dubai.[10] The orbiter successfully entered orbit around Mars on February 9, 2021, to begin a mission that is expected to last two Earth years (one Mars year).

Following close on the heels of Hope was Tianwen-1, whose name means "quest for heavenly truth" and is taken from a long poem by the ancient Chinese poet Qu Yuan (c. 340–278 BCE). Its quest is in part to recover at least some of the science lost when China's earlier piggybacking Yinghuo-1 orbiter was lost during the launch failure of the Russian Phobos-Grunt mission in 2012. Tianwen-1 set out on a Long March 5 heavy-lift launch vehicle from the Wenchang Spacecraft Launch Site in Wenchang, Hainan, China, on July 23, 2020. It successfully entered orbit around Mars on February 10, 2021. Though its scientific objectives are broad, and include an effort to characterize the Martian environment and demonstrate technologies for future Chinese Mars missions, its most ambitious component is a rover, Zhurong (named for the Chinese god of fire), which successfully landed in Utopia Planitia at 109.7°E (250.3°W) longitude and 25.1° latitude on May 14, 2021. The location was southwest of where Viking 2 landed in 1976 and included an area believed to contain vast amounts of water ice beneath the surface. The Chinese space program became the fourth, after the USSR, the United States, and the European Space Agency, to attempt a soft landing on Mars, but it was only the second (after the United States) to succeed.[11]

And then, on July 30, 2020, NASA began its Mars 2020 mission by launching a new rover named Perseverance toward Mars. While it is a new mission, the rover is mostly built from old parts; in order to fit the new rover into a tight budget, NASA built Perseverance out of more than 90 percent spare parts left over from the assembly of Curiosity in the mid-2000s. For this reason, Perseverance outwardly looks a lot like Curiosity. But the resemblance is only skin deep, as it is actually outfitted with a quite different complement of instruments and systems, subserving a mission much different from Curiosity's.[12] Specifically, Perseverance is intended to be the first step in an international robotic Mars Sample Return campaign, whose aim is to collect and cache several dozen carefully chosen cigar-tube-sized rock cores and sediment samples for future missions to eventually return to Earth (see below). This much more ambitious mission has driven a different set of tools compared to previous rovers, including a coring capability in its drill, enhanced stereo/zoom camera capabilities for more efficient driving and sample selection, and new kinds of spectrometers to help identify the most compelling places to collect samples. Another innovation was a small drone helicopter, Ingenuity, whose development was led by the Burmese American engineer MiMi Aung. The helicopter was designed to make the first powered, controlled flight beyond Earth (and succeeded in doing so on April 19, 2021, showing the feasibility of such technology for future missions) (figure 21.6).

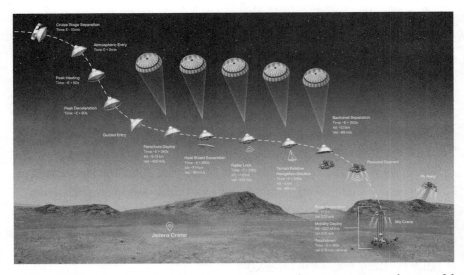

FIGURE 21.3 Entry, descent, and landing operations resulting in Perseverance's successful soft landing at Jezero crater on Mars, February 18, 2021. NASA/JPL-Caltech.

On February 18, 2021, after a seven-month journey from Earth to Mars, and with millions watching with bated breath on NASA Television and YouTube, Perseverance executed a complicated sequence of entry, descent, and landing maneuvers from the time it reached the top of Mars's thin atmosphere until it successfully landed on the surface at its target site, Jezero crater (figure 21.3). (Since Mars was eleven light minutes away, all of the intricate navigation procedures had to be carried out automatically and in real time by instruments on board the spacecraft, without any input from Earth.) At the point of encounter with the top of Mars's atmosphere, the spacecraft was traveling 20,000 kilometers per hour. Atmospheric drag slowed it down significantly while heating the external surface of the heat shield to a peak temperature of about 1,300°C (2,370°F). As soon as the spacecraft had been slowed by atmospheric friction with the heat shield to under 1,600 kilometers per hour, its supersonic parachute was deployed, using a new technology called Range Trigger to calculate the distance to the landing site and open the parachute at exactly the right time. The now-redundant heat shield was then discarded. With the rover now exposed to the Martian air for the first time, and using another new technology, Terrain-Relative Navigation, it began to use a special camera to correlate features on the surface with those on previously prepared onboard

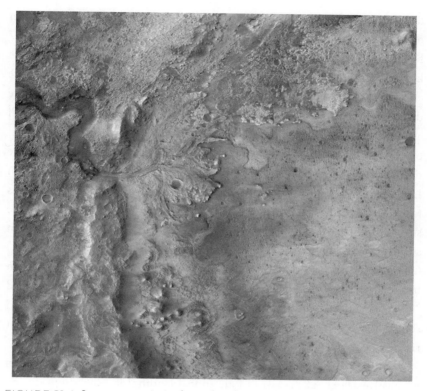

FIGURE 21.4 Jezero crater sits in the midst of an ancient delta on Mars, making it a promising site in which to look for signs of microbial life that might have flourished during the planet's warmer, wetter early history. This image was taken with the High Resolution Stereo Camera aboard the ESA Mars Express Orbiter. ESA/DLRFreie Universität Berlin.

maps to exactly locate the targeted landing site. Since Mars's atmosphere is too thin to slow the rover to a safe landing speed, the parachute was jettisoned and the rover lowered to the surface by means of eight retrorockets, until—on reaching its final descent speed of 2.7 kilometers per hour—the rover was brought safely to the surface by means of the same sky crane maneuver used to lower the Curiosity rover to the surface eight and a half years earlier. Some 20 meters above the surface, and 12 seconds before touchdown, the descent stage lowered the rover on a set of six-meter-long cables, as the rover deployed its hitherto-stowed wheels into the landing position. On sensing contact with the surface, the rover cut the cables connecting it to the descent stage, and the latter was released to make an uncontrolled (explosive) landing at a safe distance from the rover. Everything was carried out flawlessly; the rover set down at roughly 1 kilometer southeast of the center of its landing ellipse (7.7 by 6.6 kilometers), at 12:55 Pacific standard time on February 18, 2021. (The landing was more accurate than any previous Mars landing, a testament to the effectiveness of the new technology used, and, in another first, all stages of the descent were dramatically captured by cameras on board the spacecraft as well as by the Mars Reconnaissance Orbiter's HiRISE camera, looking down from overhead.) As soon as it was clear that Perseverance was safely on the surface of Mars, a giant cheer went up at mission control at NASA's Jet Propulsion Laboratory. The first image from one of the so-called Hazard Cameras was taken at 15:53:58 on the mission clock (local mean solar time) and showed the surface immediately around the rover. It was late afternoon in early Martian spring. (Mars had just passed through its northern vernal equinox, L_s = 5.2°.) On Sol 4, a 360° panorama stitched together from 79 high-resolution zoom images from Mastcam-Z showed that the rover—by now inevitably nicknamed "Percy," a name with deep resonances in the history of Mars exploration—had settled in a "sweet spot," able to "see" the front ridge of Jezero crater's ancient river delta only two kilometers away (figure 21.5 and plate 23).

Why Jezero crater? The location had been chosen in October 2018 after an exhaustive search, at the last of four annual landing site workshops during which candidate sites were gradually and painstakingly winnowed from 30 to eight, then to three, and finally to just this one. Jezero is an impact crater 49 kilometers across, located in one of the most geologically complex and interesting areas on Mars (figure 21.4). It lies near the western rim of the 1,200-kilometer-wide Isidis impact basin, close to the latter's interface with the dark, much younger low-relief shield volcano Syrtis Major Planum. Isidis itself is a gigantic, now largely dust-filled, basin, and with Hellas and Argyre among the most prominent impact feature on the planet; of Noachian age, it was probably the last giant impact basin to form, perhaps 3.9 billion years ago. Sometime later another impact formed Jezero crater, and then, probably about 3.5 billion years ago, during Mars's warmer and wetter times, water from river channels spilled over the crater wall to form a lake (as we see also at Gale crater, where Curiosity has been exploring). We see both the river outflow channel and the delta of its inflow channel, where the presence of clays and carbonates attest that sediments were carried by water. On Earth,

FIGURE 21.5 Perseverance on Mars. The highest resolution of 360° panorama yet acquired on Mars, showing the view from "Van Zyl Overlook" near the Perseverance landing site in Jezero crater. This panorama was stitched together with 992 Mastcam-Z right-eye and 110-millimeter zoom images taken from Sol 53 to 64 (April 15–26, 2021). NASA/JPL-Caltech/ASU/MSSS.

when running water reaches an open body of water, it slows down and drops a significant amount of the sediments it is carrying; but the fines settle farther out, beyond the mouth of the river. Again on Earth, as for instance in the Gulf of California, the fines form sediments that contain the remains of microorganisms. If—as some scientists believe—microbial life originated long ago on Mars, we may actually be able to discover quantitative evidence of its existence in samples of those lakebed materials. Indeed, at Jezero, sediments from three environments potentially converge: those from the streams flowing into the lake, those from the lake itself, and any that groundwater may have surfaced following evaporation of both lake and streams.

In addition to obvious analogies to river deltas on Earth, like the Mississippi delta, Jezero crater shows a tantalizing resemblance to more exotic features, such as the hypersaline paleolake Pilot Valley Basin in the Great Salt Desert of northwestern Utah. This was a fairly deep ancient freshwater lake long ago, but it is now highly salinated as it lost its water to evaporation as the climate changed. Nevertheless, sediments remain, and as discovered by astrobiologist Kennda Lynch, the hypersaline paleolake still teems with a diversity of perchlorate-reducing extremophilic microbes able to thrive by consuming perchlorates in the sediments. (As noted earlier, instruments aboard both the Phoenix lander and Curiosity rover detected perchlorates in the Martian soil. It would seem that nutrients for these kinds of seemingly bizarre organisms might just exist on the Red Planet.)

Lynch represents a new generation of scientists involved in Mars research. She was named one of the 2020 winners of NASA's Astrobiology Program Early Career Collaboration Award, and is currently (as of March 2021) working in collaboration with other young scientists like the University of Florida's Amy Williams on sample analysis and Georgetown University's Sarah Stewart Johnson on subsurface life detection.[13]

Further Plans

Two more ambitious Mars missions are also in the works for the following two launch opportunities. In 2022, ESA and the Russian Space Agency plan to launch a lander and rover that would expand the Exobiology on Mars

(ExoMars) program, which deployed Trace Gas Orbiter (TGO) to Mars back in 2016.[14] The plan is for a Russian lander named Kazachok ("little Cossack") to deliver ESA's rover, named Rosalind Franklin after the English chemist and DNA pioneer. The solar-powered Russian lander will include a package of scientific instruments designed to study the surface, subsurface, and atmosphere at the landing site in Oxia Planum, a location chosen because it shows evidence for ancient sedimentary rocks based on orbital remote sensing. The solar-powered ESA rover will deploy off the lander and carry a package of exobiology instruments to search for any chemical or mineral signatures of past life, and a 2-meter drill to extract samples for its onboard instruments. And in 2024, the Japanese Aerospace Exploration Agency (JAXA) is planning to launch its second dedicated mission to Mars (after the failure of its 1998 Nozomi orbiter mission), in an ambitious project called Martian Moons Exploration, or MMX. The MMX mission will initially orbit Mars and use its international payload of scientific instruments to study the planet's surface and atmosphere before then switching its focus to detailed studies of the Martian moons Phobos and Deimos.[15] MMX will deploy a small European rover onto Phobos as part of one or more "touch and go" exercises designed to try to collect samples from that enigmatic moon (building on the technology and experience from the JAXA Hayabusa and Hayabusa-2 near-Earth asteroid sampling missions). If all goes well, MMX will send its sample return capsule, with its 10 grams or more of precious Phobos cargo, back to Earth in 2029.

In addition to these five new Mars missions approved for development and launch in the 2020 through 2024 launch windows, two more science-focused missions are being studied and could be approved soon by their respective space agencies. Specifically, the Indian Space Research Organisation is studying a follow-on mission to its successful 2013 Mars Orbiter Mission (MOM) and Mangalyaan spacecraft.[16] Called MOM-2, the mission could launch another orbiter, and potentially a lander and small rover, to Mars as soon as 2024. And NASA is contemplating a new science orbiter tentatively called Mars Ice Mapper that could launch as soon as 2026 and would use a sensitive radar mapper to scout for surface and shallow subsurface water ice at high resolution.[17] In addition to its science mission, the new NASA orbiter would also serve as a much-needed replacement and upgrade for the surface communications relay infrastructure based on the current (aging) orbiters at Mars.

Mars Sample Return

Beyond the in-the-works near-future missions, mission studies, and continuing Earth-based observations, plans are also being devised for an international, collaborative robotic Mars Sample Return mission that would potentially bring cached samples from NASA's 2020-launched Perseverance rover back to Earth in the early 2030s.[18] While only the first step has been approved (Perseverance will collect and cache the samples, starting in 2021), plans are actively being developed and negotiated between NASA, ESA, and others for the next two required steps in the process (figure 21.6). First, a dedicated Sample Return Lander (SRL, potentially built by NASA) and Sample Return Rover (SRR, potentially built by ESA) would be launched as early as 2026, to land in Jezero crater and use the rover to collect the several dozen or more rock core, soil, and atmospheric sample tubes cached by the Perseverance team. The SRL would also serve as a launchpad for a NASA Mars Ascent Vehicle (MAV) that it would carry. The MAV would be a small rocket capable of reaching Mars

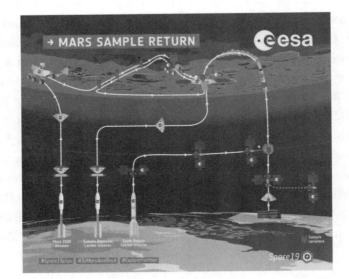

FIGURE 21.6 Cartoon of the currently envisioned plan for a robotic Mars Sample Return mission, starting with the launch of the Mars 2020 rover Perseverance, then with the launch of a Sample Return Lander, Sample Return Rover, and Mars Ascent Vehicle as well as an Earth Return Orbiter in 2026, and culminating in the return of a sample capsule to Earth in 2031. K. Oldenburg/ESA.

orbit. After the SRR collects the samples and delivers them back to the SRL, they would be transferred to a sample canister in the MAV and launched into Mars orbit. Meanwhile, also in 2026, an Earth Return Orbiter (ERO, potentially built by ESA) would be launched to Mars so that it would be prepared to locate and rendezvous with the MAV's sample canister, potentially in 2028 or 2029. The ERO would transfer the sample canister to a special Earth reentry capsule and then head back home. According the currently evolving plan, the ERO would return to Earth and deploy the reentry capsule for a hard landing in the Utah desert in 2031.

What could possibly go wrong? Despite the complexity and technical challenges of the concept, and the significant costs associated with the three launches of four additional spacecraft that would be needed to realize such a plan, many space scientists, engineers, managers, and even space-savvy politicians recognize the importance of getting samples back from Mars relatively soon. The drivers are both scientific and explorative. From a science perspective, any potential biosignatures or other evidence for past life identified in the samples or at the sampling locations using the instruments on Perseverance would likely be equivocal at best, because of current pragmatic limits on the fidelity and capabilities of the relevant instruments that can be miniaturized and ruggedized for deployment on Mars. Analysis of those samples would thus need to occur in much more advanced and sophisticated Earth laboratories to support a more detailed assessment of their implications for the habitability of Mars, including any potential biosignatures that they might contain. And from an exploration perspective, engineers want to be able to fully characterize the physical, compositional, and geotechnical properties of Martian surface and atmospheric samples in detail before humanity takes the next giant leap in Mars exploration in the decades ahead: sending the first people to the Red Planet.

22

Our Future Mars

One voyage there is I fain would take
While yet a man in mortal make;
Voyage beyond the compassed bound
Of our own earth's returning round . . .
Yet still I sit in my silent dome,
Wharf of this my island home,
Whence only thought may take passage to
That other island across the blue,
Against hope hoping that mankind may
In time invent some possible way
To that far bourne that while I gaze
Through the heaven's heaving haze
Seems in its shimmer to nod me nay.

—Percival Lowell, "Mars," unpublished poem, early 1894

Here we reach the end of our book.[1] We have not, obviously, come close to reaching the end of our exploration of the Red Planet, though we may hope that we have at least reached the beginning of the end—perhaps, if nothing else, "to know the place, for the first time" as T. S. Eliot (1888–1965) intoned.

As we might have remarked at the beginning, when we evoked the millennia when Mars was only a fiery red star in the sky and everything stretched before us in long swells, the rise (as with other things in life) would become more perceptible on looking back.

We look back on the long swells we have climbed so far, then forward to what still lies ahead. Much of what lies ahead depends on what we decide to do here on Earth. Mars has been a force to conjure with in the history of our attempt to explore other worlds, and in so doing to better understand our own. What will its role be in the future? What part of that future will play out on Mars?

The Russian pioneer of rocketry Konstantin Tsiolkovsky once said (in Carl Sagan's paraphrase), "The Earth is the cradle of mankind, but one cannot stay in the cradle forever."[2] Even so, for the foreseeable future—except as we are able to escape by means of our automated probes—at least most of us will have to do, for better or worse, with the cradle. Just like Percival Lowell, who was a literary talent of the first order whatever his shortcomings as an observer, we must sit in our silent domes, looking out from the wharves of our island home.

But the urge to leave the cradle—quite apart from the practical means of doing so—is of a piece with one of the most fundamental things that make us human: our deep-seated curiosity and drive to find out what lies beyond the next hill. It was this that explains the nomadic thrust of our species from its origin in Africa across the vast Eurasian landmass (and from Southeast Asia via primitive boats to Australia, and, much later, from Siberia via the Beringia bridge into the Americas) as well as the later surge, at first driven partly by the thirst for gold, from the static, confining, and oppressive medieval European world (ravaged by the Black Death) to the clean, expansive, and seemingly unsullied New World that from 1500 to about 1900 beckoned as a Great Frontier. Those who emigrated there were stirred, as Walter Prescott Webb (1888–1963) said, "to mighty deeds, achievements, and sacrifices,"[3] as well as, tragically, to much selfishness, waste, and bloodshed, not least against Indigenous peoples or those who were forcibly relocated in servitude and slavery.

Curiosity has always been blended in unison with a significant alloy of fear—fear of facing the unknown, of going beyond the bounds of the familiar, of leaving the apron strings for danger. In some, it has been so much more overpowering that instead of urging an advance to adventure it has summoned a reflexive desire to retreat: to pull up the drawbridge, to close off all avenues of the outside world within the security of the walled city (or the walled nation), to ward off the threat of the unknown by limiting knowledge to the dogmas of the past. That alloy of curiosity pulling us forward and fear holding us back has been ever in evidence in the history of our complicated relationship with Mars. Given humanity's long history of cosmic self-centeredness, the first, more instinctive reaction, is perhaps fear of the unknown, fear of others like or unlike ourselves.

The pointed use that H. G. Wells made of humanity's shyness of (or rather stark terror toward) anything not resembling itself in *The War of the Worlds* was nothing short of genius, as he conjured the possibility that on Mars there might be "minds that are to our minds as ours are to those of the beasts that perish, intellects vast and cool and unsympathetic, regarded this earth with envious eyes, and slowly and surely drew their plans against us."[4] But perhaps, with their being more advanced than ourselves, their acquaintance might be worth the making. Perhaps our species could benefit from a bit of cosmic humility. As Lowell put it: "If astronomy teaches anything, it teaches that man is but a detail in the evolution of the universe, and that resemblant though diverse details are inevitably to be expected in the host of orbs around him. He learns that, though he will probably never find his double anywhere, he is destined to discover any number of cousins scattered through space."[5] This still seems, perhaps, as close to a truism as can be: If astronomy teaches us anything, it is that *Earth* is but a detail in the evolution of the universe. Astronomy has taught us—and Mars in particular has taught us profoundly—that though we may never find Earth's double anywhere, we may discover any number of our cousins scattered through space. And one such cousin is Mars.

God of War

Despite Mars's warlike associations—it is, after all, the color of blood, and was worshiped as the god of war—the attempts to work out the motions of the planets, which culminated in Johannes Kepler's "war on Mars," and the early efforts to map its lands and oceans using telescopes, were peaceful enough. To the extent that these efforts involved a quest to expand the empires of terrestrial powers, such as those of Spain and Portugal, the Netherlands, France, and England, they did so in only figurative terms. Christiaan Huygens in the Netherlands and Giovanni Domenico Cassini in France were the first to sketch markings on its tiny disc. And at the height of an empire so vast that it could boast the Sun never set over it, the great 19th-century Mars observer Giovanni Virginio Schiaparelli pointedly rejected the imperial nomenclature of his contemporaries and instead invested his map of the planet with the gauzy golden names of the ancient half-mythical Mediterranean world, thus mythologizing Mars forever for the human imagination.

Mentioning Schiaparelli reminds us that Mars has always been a master of illusions. The dark areas on its disc, of which the most prominent, Syrtis Major, Huygens first made out on November 28, 1659—305 years to the day before Mariner 4 was launched—appeared to later telescopic observers as intensely blue-green. Though the colors were illusory, they led astronomers to believe that Mars, like our world, consisted of lands and seas. Later, when belief in water-filled basins could no longer be sustained, they became the ghosts of seas, dry seabeds. Yet the white polar caps still glistened as repositories of water, to be harnessed by the Martians and pumped to the equatorial regions through the vast engineering works of the canals. That the dark areas changed their outlines seemed explicable only in terms of the growth and decay of vegetation. It was only later still that it became apparent that the argument for vegetation was also based on illusion: in reality, Mars was a world of windblown dust, and the changes were nothing more than the far-off shiftings of powder on a planetary Etch A Sketch board, in which bright fine dust was blown onto and then off exposed outcrops of basalt. The dreams of finding in Mars a living world also seemed to turn to dust.

There was another illusion. The dust, suspended in the thin atmosphere of Mars and brightening its sky, deceived astronomers into thinking that Mars's atmosphere was much denser than it really was, and thus that its climate was more hospitable than we now know it to be. The historic flyby of Mariner 4 in July 1965 showed that instead of the 85 millibars or more that previous astronomers had estimated, the average atmospheric pressure on Mars was only 6 millibars, less than 1 percent of Earth's surface pressure. Moreover, the craters scattered over its surface in Mariner 4's rather stark gray television images—with nary a canal!—made it seem a lunar desolation, and not even, in those rather poor-quality images, a magnificent one. For a moment, even the water in the polar caps was given up; it seemed instead that they might be nothing more than vast deposits of frozen carbon dioxide.

But strangely, this Mars, too—seemingly based on unsentimental hard-headed and irrefutable fact—proved to be an illusion. As the view of a world of seas and lands, and later of deserts crisscrossed by canals and burgeoning fields of vegetation, had been conjured up from views of Mars at too-low resolution, so the desolate moonlike Mars was an illusion based on too little coverage. Mariner 4 had imaged only a narrow strip of the surface and had happened, by sheer bad luck, to image only some of the least interesting part, the ancient cratered plains. If it seemed that Mariner 4 had been the incarnation of the curiosity that killed the cat, then Mariners 6 and 7 (the flyby Mariners of summer 1969) and more so Mariner 9 (the first spacecraft to successfully orbit Mars and study it at close range for an extended period of time, in 1971–72) were the satisfaction that gave it another life. Thanks to Mariner 9, we learned that in addition to the ancient cratered plains, Mars shared many of the geological features found on Earth, including volcanic landscapes, desert canyons, dry riverbeds, towering dunes, and poles covered in caps of seasonal dry ice and more permanent water ice. All that was revealed from orbit.

The flyby Mariners reached Mars just after Neil Armstrong and Buzz Aldrin set foot on the Moon with Apollo 11, while Mariner 9 ended its orbital reconnaissance of Mars in October 1972, less than two months before Apollo 17, so far the last human mission to the Moon, left the lunar surface. These overlaps of timelines invite a thoughtful interlude, suggested by the fact that the objective of so much of humanity's yearning for other worlds has so long been centered on a planet associated with war. For unfortunately, and perhaps as an unforgivable indictment of our species, much of humanity's ingenuity has been expended in the interest of our warfaring instincts. Rockets to Mars were no exception, having initially been perfected as missiles meant for destruction and dominance, not science and salvation. But swords *can* be beaten into plowshares. We can at least hope for (if not yet fully work toward) the realization of a world where the exploration of Mars has advanced beyond its origins in the service of the unquenchable god of war.

World War II saw the development of radar and infrared detectors, used in Earth-based astronomical research, as well as the first rockets, originally (as described in chapter 11) designed as vengeance weapons, plausibly capable of reaching the Moon and planets. Their mastermind was Wernher von Braun, a member of the Allgemeine SS whose rockets (built with slave labor at Mittelwerk) were launched at Allied targets in a futile attempt to stave off the inevitable during the last days of the war. In the *Marsprojekt* of 1948, von Braun envisaged a mission to Mars that was extravagant only when not compared to the utterly wasteful extravagance of the wars that had just ended: a fleet of 46 reusable space shuttles of 39-ton lift capacity (compared to the modest 20 tons of NASA's later Space Shuttle) would, with a turnaround time of 10 days, make 950 flights to orbit in eight months, and assemble 10 fully fueled spaceships, reaching a mass of 3,720 metric tons, ready to ferry 70 persons on a 260-day voyage to Mars.[6] It seems staggeringly optimistic. And yet . . .

For von Braun and everyone else of that era, the dream of Mars was more alluring than any other. He did not live to see it fulfilled, but he did make a great step in that direction through his greatest achievement, the Saturn V rocket, which carried the Apollo astronauts to the Moon (the relatively easy way station on the difficult path to Mars).

Apollo, which in retrospect seems perhaps humanity's greatest and most hopeful achievement of the 1960s (or, arguably, ever), took place against the background of the Cold War, the smoldering conflict between the two nuclear-powered countries at the time, the United States and the Soviet Union. For John F. Kennedy—who in May 1961 threw down the gauntlet to send men to the Moon and return them safely before the decade was out—getting to the Moon was all about beating the Soviets. For that matter, though we tend to forget it now, the vast majority of the American public wasn't all that interested in space at the time either. Apollo was widely scorned as a costly boondoggle. At the time of Kennedy's speech to Congress, only about 40 percent endorsed the estimated $40 billion cost of such a project. There was certainly a great deal of interest in beating the Soviets. But once Armstrong and Aldrin set foot on the lunar surface, public interest rapidly dwindled. It was an American triumph; it proved American exceptionalism. Once done, there was no need to do it again. Only a few saw it, as von Braun did, as an evolutionary step forward "equal in importance to the moment when aquatic life came crawling up on the land."[7] Indeed, many Americans feel similarly disenchanted today, echoing, for example, even moonwalker Aldrin's 2003 opinion that sending astronauts back to the Moon was "more like reaching for past glory than striving for new triumphs,"[8] or President Barack Obama's 2010 blunt remark that "we've been there before."[9]

No one could deny there was an aspect to it of beating swords into plowshares—or rather intercontinental ballistic missiles into spacefaring vehicles. But it also tapped into one of our strongest (and oldest) species traits, tribalism, and one of our most refined skills, warfare. Indeed, critics might argue that except for the details this was old hat, the familiar performance of conquest that humans had practiced and cherished for millennia: a little like Hannibal's "Moonshot" of 218 BCE of launching 40,000 soldiers, thousands of horses and pack animals, and 37 elephants over the Alps in early winter to attack and slaughter Romans.

In Akron, Ohio, a newspaper survey of public opinion done just after the Moon landing found that 50 percent felt that the Moon voyage wasn't very important, scientifically or otherwise, though at least it showed we were ahead of the Soviets; 25 percent thought the whole thing somehow blasphemous; and a full 9 percent didn't think it had even happened. And this in a state that bragged about being Neil Armstrong's birthplace.

When, shortly before Apollo 11, President Richard Nixon appointed a Space Task Group to explore future manned space objectives, NASA administrator Thomas O. Paine sought von Braun's involvement. The task group explored a number of future projects that harked back to the visionary *Marsprojekt* of 40 years earlier: reusable shuttles, a large orbiting space station, continuing lunar exploration, and a crewed mission to Mars (to be accomplished by 1982). What happened? According to Paine:

> Von Braun contributed to all these plans but none were pursued; the "Moon Race" was won, and national attention had turned elsewhere. The divisive Vietnam conflict made high-tech programs suspect, and science education came to seem elitist. With no future US manned mission in prospect, Saturn V production was terminated and the space program slumped back to a third of its 1960s peak. At the same time American universities experienced a steady decline in young people pursuing graduate work in science and technology.[10]

It didn't help when, in April 1970, an oxygen tank on Apollo 13 blew up and the astronauts' lives hung precariously in the balance. As the life-or-death stakes of the astronauts' attempt to return to Earth played out tens of thousands of kilometers from Earth, television networks naturally interrupted their regular programming to issue reports, only to be beset by angry callers impatient to get back to soap operas (of which the top-rated ones at the time were, ironically, *As the World Turns* and *Another World*). By September of that same year, Paine had left NASA, taking with him any thoughts of crewed missions to Mars for the far-foreseeable future.

To Mars! But How?

More than 50 years on from Apollo 13 (this is being written in late 2020), there have been plans, seemingly more serious than the perennial ones that are offered by each presidential task group but that always seem to go nowhere, to send humans back to the Moon. Then, too, the question is always asked: where do you go after you've been to the moon?[11] The answer is always: to Mars.

A great deal of thought has been given to the logistics of crewed missions to Mars in the decades since von Braun's *Marsprojekt*. There has indeed been no shortage of "official" concepts or plans studied for getting humans to the Red Planet.[12] For example, in the early 1960s the NASA Future Projects Office conducted a study called the Early Manned Planetary-Interplanetary Roundtrip Expeditions (EMPIRE) project, investigating ways to use a potential more powerful follow-on rocket to the Saturn V called "Nova" to enable piloted flyby/orbiter missions to Mars and Venus. Around the same time, NASA's Manned Spacecraft Center (renamed the Johnson Space Center or JSC in 1973) initiated its own study of a piloted Mars *landing* that would use the Saturn V itself as the initial launch vehicle (with von Braun's support), along with a new Mars Excursion Module lander. NASA's Ames Research Center was also working on an independent piloted Mars lander study around the same time. The EMPIRE, JSC, and Ames studies were mostly completed by the time of the first Mariner 4 robotic flyby of Mars in 1965. The successful flyby dealt a blow not only to the hopes for life on Mars, but also to the momentum that had been building for piloted flyby or lander missions to Mars. Mariner 4 (as well as the Mariner 2 flyby of Venus and the Ranger probes to the Moon) had demonstrated that robotic missions could do useful basic reconnaissance of deep space targets. Those successes, combined with late 1960s NASA budget decreases, the tragic Apollo 1 fire, and the decreasing level of public interest in the space program in the United States (partly because of the unrest associated with the Vietnam War, race-related riots, and the assassinations of Martin Luther King Jr. and Robert F. Kennedy) combined to truly take the wind out of NASA's future human exploration sails—not just for Mars, but overall.

NASA's future human spaceflight plans would subsequently focus, beginning in the 1970s, on the Space Shuttle and later the space station. While successful in deploying and maintaining the Hubble Space Telescope, deploying the Magellan, Galileo, and Ulysses deep space missions (and a number of Earth-orbiting satellites), and enabling the completion of the International Space Station, the Shuttle program would ultimately prove to be a 50-year (at least) diversion of NASA's goals for human exploration of deep space.

Still, in the past few decades, additional concepts ("architectures," in the modern space exploration lingo) and plans ("roadmaps") have continued to be pitched for getting humans to the Red Planet. For example, in the 1980s Buzz Aldrin (who had been dubbed "Dr. Rendezvous" by his fellow astronauts because of his MIT doctoral degree in orbital dynamics) developed and advocated a "Mars Cycler" architecture for future human missions to the Red Planet. Aldrin's concept is to cycle twin crew-carrying spacecraft continuously between Earth and Mars, each one taking about five and a half months to get from one planet to the other, with smaller interceptor spacecraft utilized to ferry crew to and from Earth and Mars (orbit and/or surface).[13] Buzz most recently presented a roadmap for Mars settlement, including the Cycler concept, to NASA in 2015.

Another concept, called "Mars Direct," has been advocated for more than 30 years by aerospace engineers Robert Zubrin (the outspoken and passionate founder and president of the nonprofit Mars Society) and David Baker from Martin Marietta, an aerospace company that merged with Lockheed in the mid-1990s and is now the industry giant Lockheed Martin. As originally proposed in the Mars Direct architecture,[14] a Saturn V–class heavy-lift rocket would send an uncrewed Earth Return Vehicle (ERV) to Mars. The ERV would carry a large supply of hydrogen and, after landing on Mars, would convert that hydrogen to methane and oxygen via chemical reactions with Mars atmospheric CO_2. Once the ERV was known to have produced enough methane to fuel the return trip, a second launch—presumably 26 months after the first—would send a crew in a Mars Habitat Unit (MHU) to land on the surface for an 18-month stay, using the oxygen created by the ERV for life support during their stay, and the methane to fuel the return trip. Additional uncrewed ERVs and crewed MHUs could continue to be sent in future favorable Earth–Mars launch opportunities, approximately biannually. The concept could potentially have had the first crews on Mars by 1999.

Another nonprofit group, The Planetary Society, has also pitched humans-to-Mars architecture ideas, based on a 2015 workshop involving key stakeholders.

Its orbit-first concept would take a page from the old EMPIRE study in the 1960s, first sending intermediate missions to the Moon and near-Earth interplanetary space in the 2020s, and then sending missions to Mars orbit (including Phobos encounters) in the early 2030s to demonstrate and validate the technologies, hardware, and techniques needed to eventually get people that far into deep space and eventually (by the late 2030s) down onto the Martian surface.[15] Inherent in that concept is also a set of Principles for Human Spaceflight advocated by the society to guide NASA's roadmap.[16] The first of these principles is that NASA should get on with once again sending humans *beyond low Earth orbit*, with the ultimate goal of landing humans on Mars.

At the same time, the for-profit commercial space world also continues pitching architectures and roadmaps for getting humans to Mars. For example, Lockheed Martin's 2017 "Mars Base Camp" concept would also focus on incremental crewed missions: first to near-lunar space, then to high and low Mars orbits (including associated encounters with Deimos and Phobos) in the 2020s, and then to the surface of Mars in the 2030s.[17] One possible advantage of the Lockheed architecture is that it would leverage much of the existing NASA Space Launch System rocket development (the new heavy-lift, Saturn V replacement), as well as NASA's Orion deep space crew capsule development.

The most recent nongovernmental humans-to-Mars architecture has been pitched by Elon Musk, the visionary billionaire engineer and technology entrepreneur who founded and is the CEO of SpaceX. SpaceX's mission statement is "to revolutionize space technology, with the ultimate goal of enabling people to live on other planets." Musk's plan is to focus on infrastructure and technology development—like reusable launch vehicles, on-orbit propellant refueling, advanced human-rated spacecraft development, and in situ resource utilization—to enable the first settlements on Mars in the next 10–20 years.[18] So far SpaceX has made incredible and innovative advances in reusability and crew/cargo spacecraft for Earth orbital and limited deep space applications (including the stunning 2018 launch of its Falcon Heavy rocket propelling a red Tesla Roadster convertible into a Mars-crossing heliocentric orbit, and the 2020 initial testing of a proposed new "Starship" heavy-lift rocket), but its Mars achievements have remained aspirational and somewhat close to the vest. Still, SpaceX's efforts are fueled by Musk's deep pockets, passion for innovation, and demonstrated ability to attract the best and the brightest into his sphere of influence. It's not impossible that the first people on Mars could be SpaceX employees (or customers), rather than NASA astronauts. On the other hand, Musk's most recent vision, which involves "sending 1 million people to Mars by 2050, using no less than three Starship launches per day (with a stash of 1,000 of these massive spacecraft on call)," seems farfetched to say the least.[19]

At the official government level, NASA itself has embarked on many studies of potential human missions to Mars beyond the early ones in the 1960s, which were associated with the potential use of the Saturn V or a related (imagined) follow-on.[20] Among the most noteworthy in recent decades have been efforts initiated or publicly advocated by several presidential administrations. For example, on July 20, 1989—the 20th anniversary of the Apollo 11 landing—President George H. W. Bush announced a Space Exploration Initiative (SEI) that would complete the space station initiated by his predecessor Ronald Reagan in the 1990s, as well as send American astronauts back to the Moon in the first decade of the new century and then send on to Mars after 2010. It was a bold, forward-looking roadmap—with an estimated total price tag of $400–$500 billion—but ultimately only part of the proposed space station development and construction work had the momentum to continue making some forward progress. The Moon and Mars parts of the SEI (and the space station work that would make that facility an integral part of future Moon/Mars missions) died on the vine by 1993 because Congress refused to fund it and NASA refused to back off on its baseline programs. Ultimately, the political support needed to mount such an expensive and expansive effort never materialized.

The 1996 announcement of the putative discovery of evidence of ancient life preserved in the famous Martian meteorite ALH84001 energized the public and the Mars science and exploration communities, and it initiated another round of NASA studies of potential human missions. However, while the excitement ultimately led to an expansion of the Mars robotic exploration program (and thus much of the spectacular new knowledge about the Red Planet's current environment and ancient past history described in this book), the buzz wasn't followed up with enough funding (or political support) to advance the development of human missions to the Red Planet.

A more recent presidential attempt to kick-start eventual human missions to Mars was called the Vision for Space Exploration and was pitched in 2004 by President George W. Bush and his administration.[21] The vision called for NASA to complete what had by then become the International Space Station and to retire the Space Shuttle fleet by 2010, to develop a new human-rated transport spacecraft (later named Orion) by 2008 and conduct its first human spaceflight mission by 2014, and to explore the Moon with robotic missions by 2008 and with human missions by 2020. In that architecture, human missions to the Moon would help develop and test new approaches and technologies for eventual exploration of Mars, though no specific timetable was proposed for the first trips to the Red Planet. Like presidential efforts before, however, the vision was not adequately funded. While the space station was essentially completed and the Space Shuttle retired by 2011, the development of Orion has been substantially delayed (it has still not flown in deep space). Human lunar exploration plans were then scuttled by the Obama administration in 2009, in favor of potential human missions to near-Earth asteroids leading to a goal of piloted missions to Mars orbit in the 2030s, followed by a Mars landing. And while the pendulum for near-term human exploration of deep space swung back to the Moon during the Trump administration, the long-term U.S. goal still remains getting humans to Mars.

While NASA hasn't yet adopted its own or anyone else's specific plan for getting astronauts to Mars, the idea of "living off the land" (formally known in NASA parlance as "in situ resource utilization" or ISRU) advocated in many of the previously proposed architectures has merit, and inevitably will be required for long-term stays and eventual human settlement of Mars. Indeed, one of the experiments on the Perseverance rover is designed to finally demonstrate ISRU on Mars by extracting oxygen from the CO_2-rich atmosphere via simple chemical reactions.

Robots Versus Humans

Thus, in the absence of a specific commitment and plan, NASA continues to make only incremental advances toward sending humans to Mars. The obstacles are formidable even apart from the challenges in mustering the political will and gaining the funding. The first and most obvious is that humans aren't very well adapted to long-range space travel. When one of us (W. S.) was in medical school, he heard a saying: "the brain is a computer made of meat." Why, if one were designing to go to Mars, would one ever choose meat, a perishable good, as the material out of which to make one's computer? Getting to Mars would be extraordinarily hazardous for computers made of meat, and the bodies that provide the brain's life-support system. One serious concern is radiation from galactic cosmic rays and solar flares, the latter especially intense during periods of peak solar activity, as demonstrated by numerous robotic deep space missions. The predicted total radiation dose for a typical human voyage to Mars, a 500-day stay on the surface, and a return trip to Earth, based on actual radiation measurements made by the Curiosity rover, is more than 10 times the average total dose received by astronauts in six months on the International Space Station, and more than 125 times the dosage of an average CT scan.[22] Significant shielding will be needed, but heavy shielding such as lead is problematic; though such shielding is weightless in orbit (where presumably a Mars spacecraft would be assembled), as soon as one attempts to accelerate a heavy-shielded spacecraft one has to deal with inertial mass, and this means greatly increasing the energy requirements. The acceleration could of course be done very slowly and incrementally, but this would increase the time to Mars—and the duration of a mission is already formidable (one-way cruises to Mars by the most efficient route, a Hohmann transfer ellipse, take 200–300 days for ordinary spacecraft). Alternative shielding schemes using liquid water—which is a good radiation mitigator that the crew would have to carry anyway for life support—offer potential lower-mass solutions that are being explored.

Though very long Earth-orbital missions have been carried out (at the time of writing, the record for longest single human spaceflight is held by the Russian cosmonaut Valeri Polyakov, who spent nearly 438 consecutive days in the Mir space station from January 1994 to March 1995), being in Earth orbit is still being tethered to Mother Earth's apron strings, and enjoying the protection of its magnetic field and radiation belts. The ancient Greek navigators seem to have carefully hugged the shores of the Mediterranean and were afraid of traveling across the open sea—for good reason. Traveling across the open seas of space is even more daunting, and should disaster strike, such is Mars's distance that it would be a minimum of three minutes (usually much more) to communicate a distress signal, and another three minutes for mission controllers to respond (and many months to years for any help to arrive). These are some of the difficulties just in getting to Mars, to say nothing of all of the hardships to be encountered once one actually gets there.

Because humans are involved, moreover, risks have to be kept to a minimum. No one would accept the kind of odds that robotic missions have experienced, not far from 50/50, the flip of a coin. This means building in a lot more redundancy and safety, thus greatly increasing the cost on top of the additional life-support needs. "Better, faster, cheaper" has limited application when it comes to human spaceflight. As Donna Shirley, the Mars Pathfinder rover project manager, wrote in 1998: "The Apollo missions to the Moon cost the equivalent of more than a hundred billion in today's dollars. Until politicians decide to commit that kind of money to space exploration by humans again, spacecraft and robots will have to do our exploring for us."[23] Of course, it's still people who are doing the exploring, even if this exploration is virtual, using robots; even so, Shirley's point about the dramatically lower cost of robotic exploration is valid.

In fact, some wit has amended the Dan Goldin motto to read "faster, better, cheaper—pick two," on the grounds that, as engineers know, you can't have it all (some have even gone further and suggested, "faster, better, cheaper—pick one"). Adding humans to the mix certainly wouldn't necessarily allow either faster or cheaper missions. And "better" is the source of significant argument and debate. Perhaps in the 1970s it would have seemed self-evident that humans are much better at exploring than any robotic probe could ever be, but in 2020, it is harder to argue that—simply because the spacecraft and robots have done so much, so well.

On the one hand, some argue that there's nothing our brains could do in person on Mars that they couldn't just as well do in the comfort of an office in front of a computer terminal receiving inputs from orbiters, landers, and rovers that are sending back information feeds across what Percival Lowell liked to call "the intervening void." On the other hand, the Apollo missions, plus centuries of field experience by expert geologists on Earth, have demonstrated that human intuition, specific expertise, high-level tactical awareness, and fine motor skills are often required to explore new terrains, or even just to incrementally advance the exploration of already discovered places. Robots have been exceedingly good (and cost efficient) at doing basic reconnaissance in deep space: How large is that crater? Where do deposits of this particular mineral occur at the outcrop scale? How many rocks are on that hill? But no one has yet figured out how to program a robot to tell us *why* the landscape has evolved to the state that we presently observe, or to react tactically to choose a specific new measurement or sample that can provide a key clue about *how* the current climate works or the past climate changed. And a robot cannot factor in the question of *ethics* in rapidly needed decisions about where and how to explore, sample, drill, or otherwise change the pristine environment of a new world. (Admittedly, our track record on Earth isn't very impressive there either.) Indeed, it's not clear that robots will *ever* be able to achieve the level of interpretative or adaptive sophistication of the human brain (though that doesn't stop science fiction authors from speculating that they inevitably will, of course). As well, and while this is actively changing on our own planet, some large-scale kinds of activities, like mining or construction, involve levels of complexity and coordination that are just not possible for robotic automation today, or perhaps ever. Even much of the basic reconnaissance work that robots do so well could be done much more quickly and efficiently by trained professionals in the field, although that is not adequate justification on its own to support the risk of human lives and the expenditure of scarce resources.

An Earthlike World?

Once we have gone to all the trouble and expense (and justification) of getting to Mars, what kind of world would the astronauts find? This, too, we know—and again, it's thanks to robotic spacecraft.

At first sight, they would find themselves amid scenery that looks strikingly like Iceland, or the Dry Valleys of Antarctica, or Arizona—even with the occasional dust devil sweeping across the desert landscape. Another Martian illusion! Mars's air is stratospherically thin, the planet has no ozone layer or magnetic field to blunt radiation (hard ultraviolet and cosmic rays), it is bone dry, and it is bitterly cold. There's nothing there to eat, breathe, or drink (unless you happen to be at high-enough latitude to scoop up or dig for accessible water ice, with the right equipment to extract and melt it). It is far more hostile than Antarctica, the summit of Everest, or any other place on Earth—by orders of magnitude. Indeed, at a time when we are busily visiting environmental disaster on our home planet and some ponder whether Mars might provide an escape (at present at least for a few extremely wealthy people able to afford the cost of getting there), the reality is that Mars offers no refuge. No matter how badly we wreck Earth, and no matter how hard we try to "terraform" Mars, for hundreds or more likely thousands of years—and perhaps forever—Earth will still be far more habitable than Mars will ever be.[24]

What we have learned from the robotic spacecraft exploration of Mars is vast, including provisional answers to some of the questions that have been among the oldest and most important that humans have asked themselves. Above all, the question: is there life elsewhere in the universe? Mars has long been the most accessible place in the solar system to furnish us with answers to that question.

It has been clear for quite some time that life is tied to the availability of liquid water, and that one of the essential elements for the emergence and survival of life on a planet is the availability of surface and/or groundwater over sufficiently long geological time periods. We know that many features on Mars, including degraded craters and dry rivers and lakes and possibly (in the northern hemisphere) even the basin of a Martian ocean, attest to the presence of liquid water at one time (early) in Mars's geological history. There is also evidence of groundwater below the surface, as we have discussed in many places in this book.

On Earth, water seems essential to the origin of life, but it has also become clear that life may have originally formed in very harsh environments, such as hydrothermal vents near volcanic regions, where water is found at temperatures up to about 400°C (750°F). Life has continued to survive and thrive in the extremophiles such as the acidophilic archaea or eukaryoa like *Cyanidium*, which seem to have gotten started in such environments—and still survive in them—or the highly radiation-tolerant bacterium *Deinococcus radiodurans*. Life has been found in extreme environments kilometers beneath the surface of Earth, and in frozen rocks in Antarctica. On Earth, the history of life can be read almost all the way back to the beginning in sedimentary rocks. Of course, as described in this book, some of the sites we have explored on Mars have also been found to contain sedimentary rocks, including fine sandstones and mudstones that were laid down in shallow water environments. Indeed, numerous places on Mars have been shown from orbital data as well to have had an extensive history of sedimentary deposition.

Though any ancient Martian ocean would have been lost early on, lakes and streams may have persisted—or perhaps returned intermittently, depending on the long astronomical cycles. As well, Mars has experienced significant episodes of both explosive and effusive volcanism, most extensively early on, but evidence exists for volcanic eruptions having occurred even relatively recently in Mars history (within the past few tens of millions of years or less). And of course, Mars and all the planets have been continually bombarded by asteroids and comets, delivering significant quantities of the carbon- and hydrogen-rich precursors to complex organic molecules to the surfaces of these worlds. Thus, the idea that life-forms similar to the extremophiles now found on Earth could have gotten a foothold on Mars is hardly far-fetched—especially given the fact that life seems to have gotten started on Earth very early, within only a few hundred million years of when it became possible.

If life did get started on Mars, it may well have adapted to the changing surface conditions (as has life on Earth), retreated underground as surface conditions became too harsh, and perhaps even continued to survive there up to the present day. However, after Mars lost most of its atmosphere and surface water, "any primitive life-forms would have found evolution toward greater complexity challenging if not impossible."[25]

Mars is now so forbidding that even if (or when) humans someday reach the planet, establishing short-term bases and especially permanent settlements will be extremely challenging—or, in a more pessimistic view, perhaps even impossible. Schemes for terraforming the planet seem far-fetched given the available technology that could be applied to such a planetary-scale enterprise today or in the near future. It would be necessary to greatly thicken the atmosphere, presumably by releasing CO_2 and H_2O frozen in the caps in order to produce greenhouse warming to the point where water can once more exist in liquid form on the surface. (However, there are probably not enough of those volatiles on the whole planet to make that happen.)[26] Even if such a formidable engineering achievement were possible, can humans change an entire planet's atmosphere? We know we can, since we have already done so (though in a mostly harmful sense). Since the start of the Industrial Revolution, we've increased the relative amount of CO_2 in our own atmosphere by more than

40 percent. We'd likely need to somehow increase Mars's atmospheric CO_2 by 10 to 100 times more than that (hopefully without the associated smog and other pollution), and thus it would likely take centuries, perhaps millennia, for the climate to change to the required degree. Turning Mars back into a more Earthlike world would be a climate engineering project unequaled by any previous species since the blue-green algae began pumping huge amounts of oxygen into Earth's Proterozoic atmosphere 2.5 billion years ago, poisoning much of the existing life on the planet, but (fortunately for us) forever changing its destiny.

No Place Like Home

And so we come again to the point that T. S. Eliot captured so well in his 1942 poem "Little Gidding":

We shall not cease from exploration
And the end of all our exploring
Will be to arrive where we started
And know the place for the first time.[27]

It may seem like a cliché, but the most important thing we have gained from the exploration of Mars from the time when the Babylonians called it the Star of Death to the latest results from our orbiter, lander, and rover avatars, is the view Mars has given us of Earth, both literally and figuratively. We were long in quest of a world that appeared as a beautiful opal in the telescope—an Earthlike world of lands and seas—but *that* Mars was always in large part a mirage. We now see Mars as it really is. Not better or worse than we imagined; but different, and distinctly Martian.

But from Mars we also see Earth as it really is: a Morning and Evening Star, shining like a sapphire. From Mars, it is small and round and peaceful and serene, "the Globe we groan in, fairest of their evening stars" as Alfred, Lord Tennyson, wrote, imagining seeing Earth from afar. Or, in the words of John Milton, Earth from Mars would appear as "this pendant world, in bigness as a star of smallest magnitude close by the Moon." And indeed, the Moon would

FIGURE 22.1 Earth and the Moon, as imaged from a distance of 142 million kilometers from orbit around Mars. The image was taken with the HiRISE camera of the Mars Reconnaissance Orbiter on October 3, 2007. NASA/JPL-Caltech/University of Arizona.

be quite visible from Mars in binoculars or small telescopes next to its far more splendid companion (figure 22.1).[28]

Writing during the time of the Covid-19 pandemic (mid-2021), with much of the world on lockdown and, as a result, on an at least temporary pause in the roaring and the raving, we—who have loved Mars so well—nevertheless admit, on grounds quite unscientific, that for all its faults there still is no place like home.

Acknowledgments

The authors are immensely grateful to the numerous friends, colleagues, historians, scientists, and observers (both amateur and professional) who have contributed thoughtful conversations, shared memories, scholarly works, informal reviews, pointers to key (and new) resources, and images/photos relevant to the long and rich history of telescopic and spacecraft studies of Mars. We also heartily acknowledge the colleagues who helped us pull together some of the information collected in the appendixes, particularly Tenielle Gaither (appendix E), Emily Lakdawalla (appendixes F and G), and Casey Dreier (appendix H), whose contributions are all outstanding.

J. B. acknowledges the incredibly helpful reviews of early drafts of the spacecraft-era chapters by Ray Arvidson, Dave Brain, Ken Edgett, Bethany Ehlmann, Jim Erickson, Abby Fraeman, Matt Golombek, Bill Hartmann, Ken Herkenhoff, John Logsdon, Rob Manning, Scott Murchie, Steve Ruff, Larry Soderblom, Peter Thomas, Ashwin Vasavada, and Mike Wolff. Their corrections, additions, and recollections of the relevant histories and results from these historic voyages of robotic (and, via robotics, human) exploration have been absolutely critical to this book. I also thank Gunter Krebs, Phil Stooke, and Ted Stryk for their selfless help tracking down or creating some of the early Mars mission photos and images featured here. In addition, I would like to extend a more general sense of deep gratitude and appreciation for the tens of thousands of men and women—engineers, scientists, managers, technicians, administrators, students, and their supporting staff at universities, government labs, aerospace companies, and the major engineering and research centers run by NASA, ESA, and other space agencies—who, since the first attempted robotic missions to Mars back in 1960, right up to the present moment and the current preparation of new spacecraft for Mars operations, have worked so diligently, professionally, and passionately to extend the reach of our vision and understanding all the way out to the elusive and evocative Red Planet. I also thank my family and friends for their support of my occasional flights like this to Mars, and especially the ever-patient and loving Jordana Blacksberg, who has been my own shining beacon in the night sky.

W. S. acknowledges the many historians, archivists, astronomers, and psychologists, a number now deceased, who have so generously given their time and insights during the writing of this book (and earlier Mars books as well), including Leo Aerts, Lauren Amundson, Antoinette Beiser, Victor Baker, Richard Baum, Judith Bausch, Klaus Brasch, Henri Camichel, Clark Chapman, Mike Conley, Dale Cruikshank, Audouin Dollfus, Richard Dreiser, Joel Hagen, Bill Hartmann, Laurie Hatch, Art Hoag, Bill Hoyt, Roger Hutchins, Sarah Stewart Johnson, Reiichi Konnai, Leonard Martin, Richard J. McKim, Masatsugu Minami, Tony Misch, Patrick Moore, Donald E. Osterbrock, David Portree, William Lowell Putnam III, Kevin Schindler, Matthew J. Sharps, Brad Smith, Rem Stone, David Strauss, Paolo Tanga, Clyde Tombaugh, and Andy Young. The staff and archivists of the Brera Observatory, the British Astronomical Association, the Mary Lea Shane archives of the Lick Observatory, Lowell Observatory, Meudon Observatory, Nice Observatory, Paris Observatory, Rome Observatory, Royal Astronomical Society, Société Astronomique de France, and Yerkes Observatory have regarded no request as too small (or large), and have provided much advice and assistance over many years. Finally, my family—Debb, Brendan, and Ryan—have provided me invaluable support.

Both of the authors would like to express their appreciation to our extraordinary colleagues at the University of Arizona Press, led by our editor Allyson Carter, for their support, not only in this book but in many others over the last several decades of literary activity. We would also like to express our thanks for the above-and-beyond effort of Jessica Hinds-Bond, who has gone far beyond in copyediting the book and made many improvements, and has also provided the index.

Appendix A

Chronology of Mars Exploration Mission Launches

1960[1]

    October 10: Marsnik 1 (Mars 1960A), attempted Mars flyby (launch failure), USSR (chapter 11)

    October 14: Marsnik 2 (Mars 1960B), attempted Mars flyby (launch failure), USSR (chapter 11)

1961

1962

    October 24: Sputnik 22, attempted Mars flyby, USSR

    November 1: Mars 1, Mars flyby (contact lost), USSR (chapter 11)

    November 4: Sputnik 24, attempted Mars lander, USSR

1963

1964

    November 5: Mariner 3, attempted Mars flyby, United States (chapter 11)

    November 28: Mariner 4, Mars flyby, United States (chapter 11)

    November 30: Zond 2, Mars flyby (contact lost), USSR (chapter 11)

1965

    July 18: Zond 3, lunar flyby, Mars test vehicle, USSR

1966

1967

1968

1969

    February 25: Mariner 6, Mars flyby, United States (chapter 11)

    March 27: Mariner 7, Mars flyby, United States (chapter 11)

    March 27: Mars 1969A, attempted Mars orbiter (launch failure), USSR

    April 2: Mars 1969B, attempted Mars orbiter (launch failure), USSR

1970

1971

    May 9: Mariner 8, attempted Mars flyby (launch failure), United States (chapter 12)

    May 10: Cosmos 419, attempted Mars orbiter/lander, USSR

    May 19: Mars 2, Mars orbiter and attempted lander, USSR (chapter 12)

    May 28: Mars 3, Mars orbiter/lander, USSR (chapter 12)

    May 30: Mariner 9, Mars orbiter, United States (chapter 12)

1972

1973

    July 21: Mars 4, Mars flyby (attempted Mars orbiter), USSR

    July 25: Mars 5, Mars orbiter, USSR

    August 5: Mars 6, Mars lander (contact lost), USSR

    August 9: Mars 7, Mars flyby (attempted Mars lander), USSR

1974

1975

    August 20: Viking 1, Mars orbiter and lander, United States (chapter 13)

    September 9: Viking 2, Mars orbiter and lander, United States (chapter 13)

1976

1977

1978

1979

1980

1981

1982

1983

1984

1985

1986

1987

1988

> July 7: Phobos 1, attempted Mars orbiter / Phobos landers, USSR (chapters 14 and 20)
>
> July 12: Phobos 2, Mars orbiter and attempted Phobos landers, USSR (chapters 14 and 20)

1989

1990

1991

1992

> September 25: Mars Observer, attempted Mars orbiter (contact lost), United States (chapter 14)

1993

1994

1995

1996

> November 7: Mars Global Surveyor, Mars orbiter, United States (chapter 14)
>
> November 16: Mars-96, attempted Mars orbiter/landers, Russia (chapter 20)
>
> December 4: Mars Pathfinder, Mars lander and rover (Sojourner), United States (chapter 15)

1997

1998

> July 3: Nozomi (Planet-B), Mars orbiter, Japan (chapter 16)
>
> December 11: Mars Climate Orbiter, attempted Mars orbiter, United States (chapter 16)

1999

> January 3: Mars Polar Lander, attempted Mars lander, United States (chapter 16)
>
> January 3: Deep Space 2 (DS2), attempted Mars penetrators, United States (chapter 16)

2000

2001

> April 7: Mars Odyssey, Mars orbiter, United States (chapter 16)

2002

2003

    June 2: Mars Express, Mars orbiter and lander (Beagle 2), ESA and UK (chapter 16)

    June 10: Spirit (MER-A), Mars rover, United States (chapter 17)

    July 8: Opportunity (MER-B), Mars rover, United States (chapter 17)

2004

2005

    August 12: Mars Reconnaissance Orbiter, Mars orbiter, United States (chapter 16)

2006

2007

    August 4: Phoenix, Mars Scout lander, United States (chapter 17)

2008

2009

2010

2011

    November 8: Phobos-Grunt, attempted Martian moon Phobos lander, Russia (chapter 20)

    November 8: Yinghuo-1, attempted Mars orbiter, China (chapter 20)

    November 26: Mars Science Laboratory, Mars rover (Curiosity), United States (chapter 18)

2012

2013

    November 5: Mangalyaan, Mars orbiter, India (chapter 19)

    November 18: MAVEN, Mars Scout mission orbiter, United States (chapter 19)

2014

2015

2016

    March 14: ExoMars 2016, Mars orbiter (Mars Express) and lander (Schiaparelli), ESA (chapter 19)

2017

2018

    May 5: InSight, Mars lander, United States (chapter 21)

2019

2020

    July 19: Hope, Mars orbiter, United Arab Emirates (chapter 21)

    July 23: Tianwen-1, Mars orbiter and rover, China (chapter 21)

    July 30: Mars 2020, Mars rover (Perseverance) and rotorcraft (Ingenuity),
        United States (chapter 21)

2021

2022

    August–October: ExoMars 2022, Mars rover, Russia/ESA (chapter 21)

    August–October: lander, Russia (chapter 21)

    TBD: EscaPADE, dual Mars orbiting spacecraft, United States

2023

2024

    September: Martian Moons Exploration (MMX), Japan (chapter 21)

    TBD: Phobos Sample Return mission, Japan

Appendix B

Mission and Instrument Acronyms

AAP	Apollo Applications Program
AEP	Apollo Extension Program
AI	artificial intelligence
ALH84001	Allan Hills (Antarctica) meteorite sample #1, identified in 1984
APEX	Athena Precursor Experiment
APL	Applied Physics Laboratory of Johns Hopkins University
APXS	Alpha Proton (or Particle) X-Ray Spectrometer
ASI/MET	Atmospheric Structure Instrument/Meteorology package
ASPERA	Analyzer of Space Plasmas and Energetic Atoms instrument
ASU	Arizona State University
BAA	British Astronomical Association
BAAS	British Association for the Advancement of Science
BCE	Before the Common Era (before the year 0)
CASIS	Center for the Advancement of Science in Space
CCD	charge-coupled device
CE	Common Era (after the year 0)
ChemCam	Chemistry and Camera (instrument on MSL)
CheMin	Chemistry and Mineralogy (instrument on MSL)
CNES	Centre Nationale d'Études Spatiale (French space agency)
COSTAR	Corrective Optics Space Telescope Axial Replacement

CRISM	Compact Reconnaissance Imaging Spectrometer for Mars
CTX	Context Camera (instrument on MRO)
DAN	Dynamic Albedo of Neutrons (instrument on MSL)
DLR	German Aerospace Center
EDL	entry, descent, and landing
EMPIRE	Early Manned Planetary-Interplanetary Roundtrip Expeditions (NASA project)
ERO	Earth Return Orbiter (NASA/ESA concept study)
ERV	Earth Return Vehicle (NASA/ESA concept study)
ESA	European Space Agency
EUV	Extreme Ultraviolet Monitor (instrument on MAVEN)
EVA	extravehicular activity (typically a "space walk")
GRS	Gamma Ray Spectrometer (instrument on Mars Odyssey)
GSFC	Goddard Space Flight Center
HiRISE	High Resolution Imaging Science Experiment
HRSC	High Resolution Stereo Camera
HST	Hubble Space Telescope
IAU	International Astronomical Union
ICBM	Intercontinental Ballistic Missile
IFOV	instantaneous field of view (pixel scale)
IMP	Imager for Mars Pathfinder
InSight	Interior Exploration using Seismic Investigations, Geodesy and Heat Transport
IPP	International Planetary Patrol
IR	infrared
ISAS	Institute of Space and Aeronautical Science (Japan)
ISSDC	International Space Settlement Design Competition
ISM	Imaging Spectrometer for Mars (instrument on Phobos 2)
ISRO	Indian Space Research Organisation
ISRU	in situ space resources
IUVS	Imaging Ultraviolet Spectrometer (instrument on MAVEN)
JAXA	Japanese Aerospace Exploration Agency
JSC	Johnson Space Center
JPL	Jet Propulsion Laboratory of Caltech
JWST	James Webb Space Telescope

LANL	Los Alamos National Laboratory
LIBS	Laser-Induced Breakdown Spectroscopy
LOX	liquid oxygen
LPL	Lunar and Planetary Laboratory of the University of Arizona
MAG/ER	Magnetometer/Electron Reflectometer (instrument on MGS)
MAHLI	Mars Hand Lens Imager (instrument on MSL)
MARCI	Mars Color Imager
MARDI	Mars Descent Imager (instrument on MSL)
MARIE	Martian Radiation Experiment (instrument on Mars Odyssey)
MARSIS	Mars Advanced Radar for Subsurface and Ionosphere Sounding
MAV	Mars Ascent Vehicle (NASA concept study)
MAVEN	Mars Atmosphere and Volatile EvolutioN mission
MAWD	Mars Atmospheric Water Detector
MCO	Mars Climate Orbiter mission
MCS	Mars Climate Sounder instrument
MECA	Microscopy, Electrochemistry, and Conductive Analyzer
MER	Mars Exploration Rover mission
MESUR	Mars Environmental Survey
MGS	Mars Global Surveyor mission
MHU	Mars Habitat Unit
Mini-TES	Miniature Thermal Emission Spectrometer instrument
MIT	Massachusetts Institute of Technology
MMX	Martian Moons Exploration mission
MOAR	Mars Options Assessment Review committee
MOC	Mars Observer Camera, or Mars Orbiter Camera
MOLA	Mars Orbiter Laser Altimeter
MOM	Mars Orbiter Mission
MPL	Mars Polar Lander mission
MRO	Mars Reconnaissance Orbiter mission
MSL	Mars Science Laboratory (Curiosity rover) mission
MSO	Mars Science Orbiter (NASA mission concept)
MSSS	Malin Space Science Systems Inc.
MSWG	Mars Science Working Group
MVACS	Mars Volatile and Climate Surveyor
NACA	National Advisory Committee for Aeronautics

NASA	National Aeronautics and Space Administration
NEAR	Near Earth Asteroid Rendezvous mission
NSSDCA	NASA Space Science Data Coordinated Archive
NOAA	National Oceanic and Atmospheric Administration
NS	Neutron Spectrometer (instrument on Mars Odyssey)
OMB	Office of Management and Budget
OMEGA	Observatoire pour la Minéralogie, l'Eau, les Glaces et l'Activité
PI	principal investigator
PFS	Planetary Fourier Spectrometer
PMIRR	Pressure Modulated Infrared Radiometer instrument
PSIG	Project Science Integration Group (MSL mission committee)
RAD	Radiation Assessment Detector (instrument on MSL)
RAT	Rock Abrasion Tool
REMS	Rover Environmental Monitoring Station (instrument on MSL)
RMI	Remote Micro-Imager (part of the MSL ChemCam instrument)
RSL	recurring slope lineae
RSM	Remote Sensing Mast
RTG	Radioisotope Thermoelectric Generator
SAF	Société Astronomique de France
SAM	Sample Analysis at Mars (instrument on MSL)
SEI	Space Exploration Initiative (NASA concept study)
SHARAD	Shallow Radar instrument
SLS	Space Launch System
SNC	shergottites, nakhlites, and Chassigny (meteorite types)
SPICAM	Spectroscopy for the Investigation of the Characteristics of the Atmosphere of Mars
SRL	Sample Return Lander (NASA concept study)
SRR	Sample Return Rover (NASA concept study)
SSB	Space Science Board (of the National Academy of Sciences)
SSI	Surface Stereo Imager
TEGA	Thermal and Evolved Gas Analyzer instrument
TES	Thermal Emission Spectrometer (instrument on MGS)
TGO	Trace Gas Orbiter mission
THEMIS	Thermal Infrared Imaging System (instrument on Mars Odyssey)

TIROS	Television InfraRed Observation Satellite
TLS	Tunable Laser Spectrometer subsystem on SAM instrument
UA	University of Arizona
UCLA	University of California, Los Angeles
UHF	ultra-high frequency
UK	United Kingdom
U.S.	United States
USGS	United States Geological Survey
USSR	Union of Soviet Socialist Republics
UV	ultraviolet
WEB	Warm Electronics Box
XRD	X-ray diffraction

Appendix C

Physical and Orbital Characteristics of Mars, Phobos, and Deimos

TABLE A3.1 Physical and Orbital Characteristics of Mars, Phobos, and Deimos

	Mars	
Property[a]	Value	Notes
Mass	6.42×10^{23} kg	10.7% of Earth's mass
Average radius	3,389.5 km	53.2% of Earth's radius
Core radius	1,700–1,800 km	$\approx$ 50% of Earth's core radius
Surface area	144,800,000 km²	$\approx$ the same as Earth's continental land area
Average density	3.93 g/cm³	Earth's = 5.51 g/cm³
Surface gravity	3.72 m/sec²	38% of Earth's surface gravity
Escape velocity	5.0 km/sec	Earth's = 11.2 km/sec
Length of day (sol)	24 h 39 m 34.9 s	1 sol = 1.027 Earth days (39.6 minutes longer)
Obliquity	25.2°	Earth's = 23.5°
Planetary albedo	$\approx$ 0.16	Earth's $\approx$ 0.30
Average surface temp.	210K = −63°C = −82°F	Earth's = 287K = +14°C = +57°F
Daily temp. range	184K to 242K (−89°C to −31°C)	At the Viking Lander 1 site
Surface pressure	4.0 to 8.7 mbar	Average = 6.4 mbar (Earth's = 1014 mbar)
Wind speeds	2–7 m/s to 17–30 m/s	Average to dust storms (Earth's = 0–100 m/s)
Atmospheric gases	CO_2 (95.3%), N_2 (2.7%), Ar (1.6%)	Earth: N_2 (78.1%), O_2 (21.0%)
Atmos. water vapor	$\approx$ 210 ppm (0.02%)	Earth: $\approx$ 10,000 ppm (1%)
		(*continued*)

TABLE A3.1 (*continued*)

Property[a]	Value	Notes
	Mars	
Semimajor axis	1.52 AU (227.9 million km)	Earth's = 1 AU = 149.6 million km
Eccentricity	0.093	Earth's = 0.017
Orbital inclination	1.85°	Earth's = 0.00° to the ecliptic
Perihelion (P)	1.38 AU (206.6 million km)	Earth's = 147.1 million km
Aphelion (A)	1.67 AU (249.2 million km)	Earth's = 152.1 million km
Solar irradiance	Average = 589 W/m²	Earth's = 1,367 W/m²
Change in irrad., P to A	45% less solar energy at aphelion	Earth's = 7% less at aphelion
Orbital period	687.0 Earth days (668.6 sols)	Earth = 365.25 days
Synodic period	779.9 days	25.6 Earth months between oppositions
Distance from Earth	0.37 to 2.68 AU	55.7 to 401.3 million km
Apparent magnitude	−2.9 to +1.9	Venus: −4.9 to −3.0; Jupiter: −2.9 to −1.7
Angular diameter	3.5" to 25.1"	Venus: 9.7" to 66.0"; Jupiter: 29.8" to 50.1"

Property	Phobos	Deimos
	Satellites	
Semimajor axis	9,378 km (2.8 Mars radii)	23,459 km (6.9 Mars radii)
Eccentricity	0.02	0.00
Orbital inclination	1.08°	1.79°
Orbital period	7.65 h (3.22 orbits/sol)	30.31 h (0.81 orbits/sol)
Spin period	7.65 h (synchronous)	30.31 h (synchronous)
Diameter (ellipsoid)	26.8 × 22.4 × 18.4 km	15.0 × 12.2 × 10.4 km
Density	≈ 1.9 g/cm³	≈ 1.5 g/cm³
Albedo	≈ 7%	≈ 8%
Surface gravity	0.6 cm/sec² (1/1700th Earth's)	0.3 cm/sec² (1/3300th Earth's)

Sources: NSSDCA, "Mars Fact Sheet," last updated November 25, 2020, http://nssdc.gsfc.nasa.gov/planetary/factsheet/marsfact.html; NSSDCA, "Earth Fact Sheet," last updated November 25, 2020, http://nssdc.gsfc.nasa.gov/planetary/factsheet/earthfact.html.
[a] Where appropriate, all orbital and spin parameters are relative to reference epoch J2000.

Appendix D

Oppositions of Mars, 1901–2099

Table A4.1 gives the opposition date, the planet's distance in AU, and the declination north and south of the celestial equator. This last is especially important in planning observations, since for northern observatories, a far southerly declination means that the planet must be viewed through a longer path in Earth's atmosphere, which obviously will have a deleterious effect on the "seeing."

Because Earth passes the perihelion of Mars's orbit on August 28 each year, the most favorable oppositions are those that occur within a month or two on either side of this date. The opposition of August 28, 2003, was thus exceptional, in occurring within a few hours of Mars's perihelion passage, and so the planet approached closer to Earth than it had in 60,000 years. The cycle of oppositions repeats, within a few days, after 79 years, and even more precisely after 284 years. Thus, the opposition of 2003 was a near counterpart to that of 1924, that of 2020 was close to that of 1941, and so on. The perihelic oppositions occur at intervals of 15 to 17 years: thus in 1909, 1924, 1939, 1956, 1971, 1988, 2003, 2018, and so forth.

The period several months before opposition also defines the favorable "launch windows" for spacecraft missions to Mars. Spacecraft launched during these intervals are able to follow the shortest, most fuel-efficient trajectories to Mars.

Note that owing to the slight inclination of Mars's orbit to that of Earth, the minimum separation between the two planets can actually occur a few days before or after the opposition date.

TABLE A4.1 **Oppositions of Mars, 1901–2099**

Opposition date	Distance to Earth (AU)	Declination
1901 Feb 22	0.678	+02° 20'
1903 May 29	0.640	−00° 05'
1905 May 08	0.543	−16° 57'
1907 Jul 06	0.411	−27° 59'
1909 Sep 24	**0.392**	−04° 13'
1911 Nov 25	0.517	+21° 43'
1914 Jan 05	0.625	+26° 33'
1916 Feb 10	0.675	+19° 08'
1918 Mar 15	0.662	+05° 55'
1920 Apr 21	0.588	−10° 21'
1922 Jun 10	0.462	−25° 55'
1924 Aug 23	**0.373**	−17° 40'
1926 Nov 04	0.465	+14° 26'
1928 Dec 21	0.589	+26° 39'
1931 Jan 27	0.663	+22° 24'
1933 Mar 01	0.675	+11° 26'
1935 Apr 06	0.624	−03° 52'
1937 May 19	0.515	−20° 39'
1939 Jul 23	**0.389**	−26° 42'
1941 Oct 10	0.414	+03° 09'
1943 Dec 05	0.545	+24° 24'
1946 Jan 14	0.641	+23° 35'
1948 Feb 17	0.678	+16° 25'
1950 Mar 23	0.652	+02° 20'
1952 May 01	0.564	−14° 17'
1954 Jun 24	0.432	−27° 40'
1956 Sep 10	**0.379**	−10° 08'
1958 Nov 01	0.494	+19° 07'
1960 Dec 03	0.610	+26° 49'
1963 Feb 04	0.671	+20° 42'
1965 Mar 09	0.669	+08° 08'

(continued)

TABLE A4.1 (*continued*)

Opposition date	Distance to Earth (AU)	Declination
1967 Apr 01	0.605	−07° 42'
1969 May 31	0.486	−23° 57'
1971 Aug 10	**0.376**	−22° 14'
1973 Oct 25	0.441	+10° 17'
1975 Dec 15	0.570	+26° 02'
1978 Jan 22	0.654	+24° 05'
1980 Feb 25	0.677	+13° 27'
1982 Mar 31	0.638	−01° 21'
1984 May 11	0.537	−18° 05'
1986 Jul 10	0.406	−27° 44'
1988 Sep 28	**0.396**	−02° 06'
1990 Nov 27	0.523	+22° 37'
1993 Jan 07	0.628	+26° 16'
1995 Feb 12	0.676	+18° 10'
1997 Mar 17	0.661	+04° 40'
1999 Apr 24	0.583	−11° 37'
2001 Jun 13	0.456	−26° 30'
2003 Aug 28	**0.373**	−15° 49'
2005 Nov 07	0.470	+15° 54'
2007 Dec 24	0.593	+26° 46'
2010 Jan 29	0.664	+22° 09'
2012 Mar 03	0.674	+10° 16'
2014 Apr 08	0.621	−05° 08'
2016 May 22	0.510	−21° 39'
2018 Jul 27	**0.386**	−25° 29'
2020 Oct 13	0.420	+05° 26'
2022 Dec 08	0.550	+24° 59'
2025 Jan 16	0.644	+25° 06'
2027 Feb 19	0.678	+15° 22'
2029 Mar 25	0.649	+01° 04'
2031 May 04	0.559	−15° 28'
2033 Jun 28	0.427	−27° 49'
2035 Sep 15	**0.383**	−08° 02'
2037 Nov 19	0.500	+20° 16'
2040 Jan 02	0.615	+26° 41'
2042 Feb 06	0.672	+19° 49'
2044 Mar 11	0.668	+06° 55'
2046 Apr 17	0.601	−08° 59'
2048 Jun 03	0.480	−24° 44'

(*continued*)

TABLE A4.1 (*continued*)

Opposition date	Distance to Earth (AU)	Declination
2050 Aug 14	**0.374**	−20° 43'
2052 Oct 28	0.446	+11° 58'
2054 Dec 17	0.574	+26° 20'
2057 Jan 24	0.656	+23° 26'
2059 Feb 27	0.677	+12° 20'
2061 Apr 02	0.635	−02° 37'
2063 May 14	0.532	−19° 11'
2065 Jul 13	0.402	−27° 18'
2067 Oct 02	**0.400**	−00° 01'
2069 Nov 30	0.529	+23° 26'
2072 Jan 11	0.631	+25° 54'
2074 Feb 14	0.677	+17° 11'
2076 Mar 19	0.659	+03° 25'
2078 Apr 27	0.579	−12° 50'
2080 Jun 16	0.450	−26° 57'
2082 Sep 01	**0.375**	−13° 52'
2084 Nov 10	0.476	+17° 16'
2086 Dec 27	0.597	+26° 47'
2089 Jan 31	0.666	+21° 20'
2091 Mar 06	0.673	+09° 06'
2093 Apr 11	0.617	−06° 25'
2095 May 26	0.504	−22° 36'
2097 Jul 31	**0.383**	−24° 26'
2099 Oct 18	0.424	+07° 19'

Note: Perihelic oppositions are in bold.

Appendix E

Mars Nomenclature

CONTRIBUTED BY TENIELLE GAITHER, U.S. GEOLOGICAL
SURVEY, FLAGSTAFF, ARIZONA

Humans have been naming the stars and planets for thousands of years, and many of these ancient names are still in use. For example, the Romans named the planet Mars for their god of war; the satellites Phobos and Deimos, discovered in 1877, were named for the twin sons of Ares, the Greek god of war. In this age of orbiters, rovers, and high-resolution images, modern planetary nomenclature is used to uniquely identify a topographic, morphological, or albedo feature on the surface of a planet or satellite so that the feature can be easily located, described, and discussed by scientists and laypeople alike.

The International Astronomical Union's (IAU) Working Group for Planetary System Nomenclature (WGPSN) is responsible for the task of approving new names for planetary surface features, rings, and natural satellites. The members of the WGPSN and the associated task groups (one each for Mercury, Venus, Mars, the outer solar system, and small bodies) work to provide a clear, unambiguous system of nomenclature that represents cultures and countries from all regions of Earth. The six task groups and the WGPSN comprise 38 volunteer members, representing 13 countries and one city-state: China, Finland, France, Germany, Japan, New Zealand, Norway, Russia, Spain, Switzerland, Vatican City State, Ukraine, the United Kingdom, and the United States. This international composition supports the equitable distribution of names from different ethnic groups, countries, and gender on each planetary body.

The Gazetteer of Planetary Nomenclature (https://planetarynames.wr.usgs .gov) displays the database of all names of topographic and albedo features on planets and satellites (and some planetary ring and ring-gap systems) that the IAU has named and approved from its founding in 1919 through the present time. The Gazetteer of Planetary Nomenclature database and website are maintained by the U.S. Geological Survey's (USGS) Astrogeology Science Center in Flagstaff, Arizona. NASA has provided funding for the USGS Astrogeology's administration of the planetary nomenclature project since 1982.

A Very Brief History of Planetary Nomenclature

With the invention of the telescope in 1608, astronomers from many countries began studying the Moon and other planetary bodies. Some of these astronomers began applying names to the features they observed, creating several different systems of nomenclature. Most of these early naming systems were applied to the Moon because it is close enough for its surface features to be seen clearly. In particular, Giovanni Battista Riccioli, Johann Hieronymus Schroeter, and Johann Heinrich von Mädler created and distributed lunar maps showing three different sets of nomenclature. By the early 1900s, the lack of correlation of the names used by different selenographers led to considerable confusion, making clear the need for a single system of nomenclature.

When the IAU was founded in 1919, one of its first actions was to form a committee to standardize lunar nomenclature. The IAU nomenclature committee presented its recommendations in 1932 in the form of the classic publication "Named Lunar Formations" by the largely self-taught English astronomer Mary Adela Blagg (1858–1944) and the Viennese amateur Karl Müller (1866–1942). The IAU adopted the names included in this publication in 1935. Because this was the first systematic listing of lunar nomenclature, it set the stage for future systems that would be adopted by the IAU for names on other planets and satellites in our solar system.

Martian nomenclature was similarly standardized in 1958, when an ad hoc committee of the IAU recommended for approval the names of 128 albedo features (bright, dark, or colored) observed through ground-based telescopes. Though as discussed in the main text, there were several different systems of nomenclature in use for Mars, including that introduced on an 1867 map by Richard Anthony Proctor based on the 1864/65 drawings of the Reverend William Rutter Dawes, and modifications of Proctor's scheme by Camille Flammarion (in 1875) and Nathaniel Green (in 1879), by the end of the 19th century most astronomers had accepted the scheme of nomenclature introduced by Giovanni Virginio Schiaparelli in his "Osservazioni astronomiche e fisiche sull'asse di rotazione e sulla topografia del pianta Marte" (in *Atti della R. Accademia del Lincei, Memoria della cl. di scienze fisiche*, Memoria 1, ser. 3, vol. 2, 1877–78, pp. 308–439). This was later expanded and revised by Schiaparelli in his subsequent memoirs published between 1880 and 1910, as well as by a number of later observers, including Percival Lowell and Eugène Michel Antoniadi. Again, as with the earlier history of lunar nomenclature, considerable confusion was introduced over the years, partly because many of the features ("canals") were of uncertain identification and permanence.

Antoniadi introduced the first map without canals in 1903, and only focused on the more definite albedo markings. An important landmark in the history of Martian nomenclature during this era was the publication of Antoniadi's great book *La Planète Mars* (Paris, 1930) and its spectacular albedo feature maps (figure A5.1). The authoritative reference on the history of Martian nomenclature is Jürgen Blunck, *Mars and Its Satellites: A Detailed Commentary on the Nomenclature*, 2nd ed. (Smithtown, N.Y.: Exposition Press, 1982).

As the flyby Mariners began to obtain images of the geologically diverse Martian surface in the mid-1960s, and especially after Mariner 9's highly successful orbital mission in 1972, it became apparent that the existing nomenclature for Mars, based on albedo features, was inadequate for the vast number of new topographic features identified (craters, mountains, and valleys). The IAU was now asked to sort through the confusion and approve one nomenclature system for Mars. In response to this request, a Working Group for Martian Nomenclature was established in 1970, and it presented its recommendations to the IAU at its 1973 meeting in Sydney, Australia, thus setting the stage for the single recognized system of names for Mars that exists today.

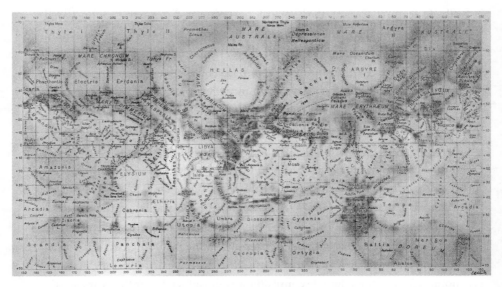

FIGURE A5.1 Albedo map of Mars by Eugène Michel Antoniadi, which provides the basis of classical albedo names now used, in conjunction with a descriptor term, to provide new names for morphological features identified by modern planetary scientists in spacecraft images. For example, the albedo name Argyre is used in the IAU-approved feature names Argyre Cavi, Argyre Mons, Argyre Planitia, and Argyre Rupes. Reworked map by Joel Hagen; originally published in *La Planète Mars* (Paris: Hermann and Cie, 1930), plates 2, 3, and 4.

Descriptor Terms and Themes

The Gazetteer of Planetary Nomenclature lists descriptor terms, or feature types, that are used to describe named morphological, topographic, or albedo features. Determining the most appropriate descriptor for a feature is one of the critical initial steps in the naming process, and is usually, but not always, straightforward. For example, the descriptors *vallis* (valley; canyonlike, often sinuous trough) and fossa (long, narrow depression; ditch; narrow, linear trench) seem distinct and allow many morphological features to fit one description or the other very neatly, but some individual features may have characteristics of both descriptors. Nature does not always produce features that fit tidily into our little boxes!

Once the feature has been assigned a descriptor, a name is chosen that fits the naming theme assigned to that descriptor. Naming themes allow for a large number of potential names to be in reserve for future name requests. For example, small (50 kilometers or smaller) craters on Mars are named for small towns and villages of the world. No commemoration of specific modern towns or villages is intended; this theme is simply a large source of international names.

Table A5.1 lists the descriptor terms for all the currently named features on Mars, Deimos, and Phobos, along with the number of features of that type. Tables A5.2 and A5.3 list the nomenclature themes for features in each descriptor category.

Representative Features and Maps

Tables A5.4 through A5.7 list representative features of a variety of features on Mars: albedo features, craters, montes, and valles.

Digital versions and links to modern maps of numerous planets, moons, and asteroids can be found online at the USGS Astrogeology Science Center's maps page (https://www.usgs.gov/centers/astrogeology-science-center/maps), Arizona State University's Mars Space Flight Facility global maps page (https://www.mars.asu.edu/data), The Planetary Society's website (http://planetary.org), and many other places. Some specific examples for active Mars researchers as well as historians of Mars research include:

- Emily Lakdawalla's article "Mapping Mars, Now and in History," The Planetary Society website, February 26, 2009, https://www.planetary.org/articles/1858
- Viking colorized global mosaic at 232 meters per pixel at the equator (figure A5.2), USGS Astrogeology website, https://astrogeology.usgs.gov/search/map/Mars/Viking/MDIM21/Mars_Viking_MDIM21_ClrMosaic_global_232m
- MGS MOLA global shaded relief colorized topography map, USGS Astrogeology website, https://astrogeology.usgs.gov/search/map/Docs/Globes/i2782_sh1

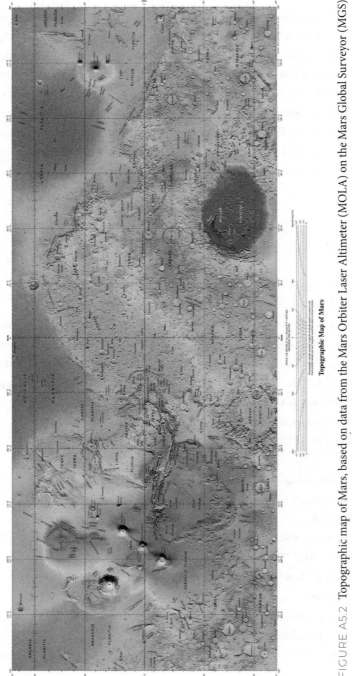

Topographic Map of Mars

FIGURE A5.2 Topographic map of Mars, based on data from the Mars Orbiter Laser Altimeter (MOLA) on the Mars Global Surveyor (MGS) spacecraft. The image used for the base of this map represents more than 600 million measurements gathered between 1999 and 2001, with the average accuracy of each point about 100 meters in horizontal position and about 1 meter in elevation relative to an areoid determined from the Martian gravity field. USGS.

- MGS Thermal Emission Spectrometer global albedo map, ASU Mars Space Flight Facility website, http://www.mars.asu.edu/data/tes_albedo
- Mars Odyssey THEMIS daytime infrared global mosaic at 100 meters per pixel at the equator, ASU Mars Space Flight Facility website, http://www.mars.asu.edu/data/thm_dir_100m
- MRO CTX global mosaic at 5 meters per pixel, Murray Lab website, http://murray-lab.caltech.edu/CTX
- Map all Mars landing sites, failed and successful, The Planetary Society website, https://www.planetary.org/space-images/mars_landing_site_map_lakdawalla

For further information on planetary nomenclature, please visit the Gazetteer of Planetary Nomenclature website (https://planetarynames.wr.usgs.gov), or email questions to gs-astro_nomenclature@usgs.gov.

TABLE A5.1 Descriptor Terms for Morphological Features on Mars, Deimos, and Phobos

Descriptor / Plural	Description	Number
Albedo feature	Geographic area distinguished by amount of reflected light	126
Catena, catenae	Chain of craters	16
Cavus, cavi	Hollows, irregular steep-sided depressions usually in arrays or clusters	28
Chaos, chaoses	Distinctive area of broken terrain	30
Chasma, chasmata	A deep, elongated, steep-sided depression	25
Collis, colles	Small hills or knobs	22
Crater, craters	A circular depression	1,147
Dorsum, dorsa	Ridge	35
Fluctus, fluctūs	Flow terrain	3
Fossa, fossae	Long, narrow depression	59
Labes, labēs	Landslide	7
Labyrinthus, labyrinthi	Complex of intersecting valleys or ridges	6
Lingula, lingulae	Extension of plateau having rounded lobate or tongue-like boundaries	5

(*continued*)

TABLE A5.1 (*continued*)

Descriptor / Plural	Description	Number
Macula, maculae	Dark spot, may be irregular	1
Mensa, mensae	A flat-topped prominence with cliff-like edges	36
Mons, montes	Mountain	52
Palus, paludes	"Swamp"; small plain	5
Patera, paterae	An irregular crater, or a complex one with scalloped edges	28
Planitia, planitiae	Low plain	11
Planum, plana	Plateau or high plain	34
Regio, regiones	A large area marked by reflectivity or color distinctions from adjacent areas, or a broad geographic region	1
Rupes, rupēs	Scarp	26
Scopulus, scopuli	Lobate or irregular scarp	13
Serpens, serpentes	Sinuous feature with segments of positive and negative relief along its length	3
Sulcus, sulci	Subparallel furrows and ridges	14
Terra, terrae	Extensive landmass	11
Tholus, tholi	Small domical mountain or hill	22
Unda, undae	Dunes	6
Vallis, valles	Valley	153
Vastitas, vastitates	Extensive plain	1

TABLE A5.2 Nomenclature Themes for Features on Mars

Feature type	Theme
Albedo features	Names from classical mythology assigned by Schiaparelli and Antoniadi
Large craters (approximately 50 km and larger)	Scientists, especially those who have contributed significantly to the study of Mars; writers and others who have contributed to the lore of Mars
Small craters (approximately 50 km and smaller)	Small towns and villages of the world with populations of approximately 100,000 or less. This category is simply a large source of crater names. No commemoration of specific towns or villages is intended

(*continued*)

TABLE A5.2 (*continued*)

Feature type	Theme
Smaller features within a larger named feature	The normal naming convention shall be suspended and instead their names shall be chosen so that they bear a mnemonic relationship to the given name of the larger feature, followed by the appropriate descriptor term (crater, cavus, patera, etc.)
Large valles	Name for Mars/star in various languages
Small valles	Classical or modern names of rivers
Other features	From a nearby named albedo feature on Schiaparelli or Antoniadi maps. If no nearby albedo feature name is available, then from a nearby named crater

TABLE A5.3 Nomenclature Themes for Features on Deimos and Phobos

Planetary body	Theme
Deimos	Authors who wrote about Martian satellites
Phobos	Scientists involved with the discovery, dynamics, or properties of the Martian satellites, and people and places from Jonathan Swift's *Gulliver's Travels*

TABLE A5.4 Representative Albedo Feature Names on Mars

Feature name	Center latitude, longitude	Name origin
Arabia	19.78°N, 30°E	Country bordering on Aeria (Egypt)
Arcadia	44.66°N, 260°E	Mountainous region in southern Greece
Elysium	24.74°N, 150°E	Home of the blessed on western edge of world
Hellas	39.67°S, 70°N	Greece
Mare Australe	59.71°S, 350°E	"South Sea"
Mare Boreum	59.71°N, 180°E	"North Sea"
Utopia	49.67°N, 110°E	Greek, meaning "nowhere"; ideal state

TABLE A5.5 **Representative Crater Names on Mars**

Feature name	Center latitude, longitude	Name origin
Gale	5.37°S, 137.81°E	Walter F.; Australian astronomer (1865–1945)
Gusev	14.53°S, 175.52°E	Matvei M.; Russian astronomer (1826–66)
Huygens	13.88°S, 55.58°E	Christiaan; Dutch physicist-astronomer (1629–95)
Jezero	18.41°N, 77.69°E	Town in Bosnia and Herzegovina
Schiaparelli	2.71°S, 16.77°E	Giovanni V.; Italian astronomer (1835–1910)

TABLE A5.6 **Representative Montes Names on Mars**

Feature name	Center latitude, longitude	Name origin
Aeolis Mons	5.08°S, 137.85°E	Classical albedo feature name
Elysium Mons	25.02°N, 147.21°E	Classical albedo feature name
Hadriacus Mons	31.29°S, 91.86°E	Classical albedo feature name
Olympus Mons	18.65°N, 226.2°E	Classical albedo feature name
Tharsis Montes	1.57°N, 247.42°E	Classical albedo feature name

TABLE A5.7 **Representative Valles Names on Mars**

Feature name	Center latitude, longitude	Name origin
Ares Vallis	10.29°N, 334.39°E	Word for "Mars" in Greek
Gediz Vallis	4.85°S, 137.44°E	River in Turkey
Mawrth Vallis	22.43°N, 343.03°E	Welsh word for "Mars"
Neretva Vallis	18.55°N, 77.2°E	River in Bosnia and Herzegovina and Croatia
Valles Marineris	14.01°S, 301.41°E	General name of the system of canyons honoring the scientific team of the Mariner 9 mission

Appendix F

A Seasonal and Historical Almanac for Mars

CONTRIBUTED BY EMILY LAKDAWALLA, THE LAKDAWALLA GROUP LLC

As Mars spins and orbits the Sun, it experiences patterns of days and nights and shifting seasons that are remarkably similar to Earth's. Mars's solar days are only 3 percent longer than Earth days (appendix B), so Mars explorers have translated Earth timekeeping to Mars, using a 24-hour clock and a definition of local solar time (LST) for each lander location. Mars's axial tilt is also similar to Earth's (25.19° and 23.44°, respectively), so Mars experiences equinoxes and solstices that bound spring, summer, fall, and winter, opposite in each hemisphere, with season-long nights and days at north and south poles.

However, while Earth's orbit is nearly circular, Mars's orbit has a relatively large eccentricity (0.09). The eccentricity causes variations in Mars's solar distance that strongly affect both the length and severity of its seasons, driving dramatic seasonal weather patterns. The seasonal changes also affect the availability of solar energy and heating for landed missions. The weather is more extreme in the south than the north, because Mars's perihelion occurs very close to its northern winter solstice. Parameters describing Mars's orbit and spin as of January 1, 2000, at 12:00:00 (a date that is referred to as the J2000 epoch), are listed in appendix B.

Julian Dates

Astronomers often employ the Julian Date convention for tracking the passage of time. Named for Julius Scaliger (*not* Caesar), Julian Date (JD) is the number of days since noon on January 1, −4712 (that is, January 1, 4713 BCE).[1] The J2000 epoch is equivalent to JD 2451545.0.

There are a few different flavors of Julian Date. The Modified Julian Date (MJD) reduces the number of digits needed to bookkeep dates, and makes the date turn over at midnight rather than noon, by subtracting 2400000.5 from the Julian Date. (In other words, it records time in days since midnight on November 17, 1858.) Either MJD or JD may refer to either Terrestrial Time (TT, which was known until 1984 as Ephemeris Time or ET) or Coordinated Universal Time (UTC). TT differs from UTC in that UTC occasionally has leap seconds inserted to correct for changes in Earth's rotational speed. The difference as of 2020 is 65 seconds (that is, UTC runs 65 seconds behind TT). If minute- or second-level precision is required in timekeeping applications, it's important to know whether the TT or UT convention is being used. Allison and McEwen provide an analytic approximation to the difference between TT and UTC in their article "A Post-Pathfinder Evaluation of Areocentric Solar Coordinates."[2]

Mars Days

Mars's days have been colloquially referred to as "sols" since the Viking Lander missions in order to differentiate them from Earth days.[3] Martian hours, minutes, and seconds do not have special Mars names, at least none that are in widespread use. Instead, Mars simply has its own hours, minutes, and seconds, all of which are 1.02749125 times longer than their Earth counterparts.[4] Because the rhythms of daytime operations are crucial in landed mission planning, mission timelines are kept according to sol number, counting up sols since landing. Some missions have considered landing day to be sol 0, others sol 1.

Mars Years and Seasons

Most of the Mars community has adopted a Mars Year convention first articulated by Clancy et al. in 2000 such that Mars Year 1 began at the equinox of April 11, 1955.[5] The date is convenient because it predates the Space Age and also because during Mars Year 1 (abbreviated MY1), astronomers around the world observed a global dust storm. More recently, in 2015, Piqueux et al. extended the convention to MY0 (which began on May 24, 1953) and to negative-numbered Mars years before that.[6]

There is no consensus definition of months or any equivalent for Mars. Instead, the community communicates time within a Martian year by stating planetocentric solar longitude, abbreviated L_s and pronounced "ell sub ess." Distinct from heliocentric solar longitude, L_s is defined with respect to a planet's equinoxes: it is 0° at northern vernal equinox, 90° at northern summer solstice, and so on. Table A6.1 provides an almanac relating Mars seasonal events, Earth dates, and selected significant mission and weather events.

Seasonal Weather

Several aspects of Mars's orbit act to produce very different seasonal weather patterns in northern and southern hemispheres. Mars's northern winter solstice ($L_s = 270°$) is almost coincident with perihelion ($L_s = 251°$), when Mars is closest to the Sun (1.38 AU). Northern summer solstice nearly coincides with aphelion (1.66 AU). At perihelion, Mars experiences sunlight that is 45 percent more intense than at aphelion. Hence, northern seasons are moderated by Mars's dramatic swings in solar distance. Southern seasons are made more extreme.

In addition, the planet moves more slowly at aphelion, adding to the length of the moderate northern summer and intense southern winter. The southern hemisphere is also mostly at higher topographic elevations than the north, which makes average temperatures cooler. As a result of these contrasts, southern winters are especially harsh.

TABLE A6.1 Mars Year/L_s and Earth Date Almanac of Solstices, Equinoxes, and Other Historical Events in Mars's History

MY	L_s	Earth Date	MJD	Other Notable Events in Same Mars Year (MY)[a]
0	0°	1953 May 24	34521.504	187.7° / 1954 Jun 24: Opposition; C/A 0.43 AU on Jul 02
	90°	1953 Dec 09	34720.174	
	180°	1954 Jun 10	34903.471	
	270°	1954 Nov 04	35050.060	
1	0°	1955 Apr 11	35208.456	249.0° / 1956 Aug 18: GLOBAL DUST STORM ONSET[b]
	90°	1955 Oct 27	35407.144	263.0° / 1956 Sep 10: Opposition; C/A 0.38 AU on Sep 07
	180°	1956 Apr 27	35590.500	
	270°	1956 Sep 21	35737.080	
2	0°	1957 Feb 26	35895.414	328.8° / 1958 Nov 16: Opposition; C/A 0.49 AU on Nov 08
	90°	1957 Sep 13	36094.088	
	180°	1958 Mar 15	36277.459	
	270°	1958 Aug 09	36424.091	
3	0°	1959 Jan 14	36582.415	332.7° / 1960 Oct 10: Launch of Marsnik 1 (launch failure)
	90°	1959 Aug 01	36781.048	334.9° / 1960 Oct 14: Launch of Marsnik 2 (launch failure)
	180°	1960 Jan 31	36964.400	
	270°	1960 Jun 26	37111.039	
4	0°	1960 Dec 01	37269.398	014.2° / 1960 Dec 30: Opposition; C/A 0.61 AU on Dec 25
	90°	1961 Jun 18	37468.021	
	180°	1961 Dec 18	37651.382	
	270°	1962 May 13	37797.994	
5	0°	1962 Oct 19	37956.351	002.3° / 1962 Oct 24: Launch of Sputnik 22 (launch failure)
	90°	1963 May 05	38154.988	006.2° / 1962 Nov 01: Launch of Sputnik 23 (Mars 1)
	180°	1963 Nov 05	38338.343	(contact lost shortly before Mars flyby)
	270°	1964 Mar 30	38484.989	007.7° / 1962 Nov 04: Launch of Sputnik 24 (launch failure)
				050.3° / 1963 Feb 04: Opposition; C/A 0.67 AU on Feb 03
6	0°	1964 Sep 05	38643.323	029.0° / 1964 Nov 05: Launch of Mariner 3 (launch failure)
	90°	1965 Mar 22	38841.971	039.7° / 1964 Nov 28: Launch of Mariner 4
	180°	1965 Sep 22	39025.306	040.3° / 1964 Nov 30: Launch of Zond 2 (contact lost shortly after launch)
	270°	1966 Feb 15	39171.928	084.1° / 1965 Mar 09: Opposition; C/A 0.67 AU on Mar 12
				143.1° / 1965 Jul 15: Mariner 4 closest approach (flyby at 9,846 km)
				144.6° / 1965 Jul 18: Zond 3 launch (lunar flyby, Mars test vehicle)
7	0°	1966 Jul 24	39330.293	120.1° / 1967 Apr 15: Opposition; C/A 0.60 AU on Apr 21
	90°	1967 Feb 07	39528.942	265.0° / 1967 Dec 27: Mariner 4 contact lost
	180°	1967 Aug 10	39712.319	
	270°	1968 Jan 03	39858.909	

(continued)

MY	L_s	Earth Date	MJD	Other Notable Events in Same Mars Year (MY)[a]
8	0°	1968 Jun 10	40017.237	117.6° / 1969 Feb 25: Mariner 6 launch
	90°	1968 Dec 25	40215.896	132.3° / 1969 Mar 27: Mariner 7 launch
	180°	1969 Jun 27	40399.290	134.7° / 1969 Apr 02: Launch of Mars 1969B (launch
	270°	1969 Nov 20	40545.917	failure)
				165.3° / 1969 May 31: Opposition; C/A 0.48 AU on Jun 09
				199.7° / 1969 Jul 31: Mariner 6 closest approach (flyby at 3,431 km)
				202.7° / 1969 Aug 05: Mariner 7 flyby
9	0°	1970 Apr 28	40704.215	176.5° / 1971 May 09: Mariner 8 launch (launch failure)
	90°	1970 Nov 12	40902.864	177.1° / 1971 May 10: Cosmos 419 launch (launch failure)
	180°	1971 May 15	41086.248	182.5° / 1971 May 19: Mars 2 launch
	270°	1971 Oct 08	41232.895	187.6° / 1971 May 28: Mars 3 launch
				189.0° / 1971 May 30: Mariner 9 launch
				232.3° / 1971 Aug 10: Opposition; C/A 0.38 AU on Aug 12
				260.0° / 1971 Sep 23: GLOBAL DUST STORM ONSET[c]
				292.4° / 1971 Nov 14: Mariner 9 orbit insertion during dust storm
				300.2° / 1971 Nov 27: Mars 2 orbit insertion; lander crashed
				303.5° / 1971 Dec 02: Mars 3 lander landing/EOM
				320.6° / 1972 Jan 01: Mars orbit clears sufficiently for Mariner 9 mapping to begin
10	0°	1972 Mar 15	41391.209	073.0° / 1972 Aug 22: Mars 2 and 3 missions declared over
	90°	1972 Sep 29	41589.836	102.1° / 1972 Oct 27: Mariner 9 contact lost
	180°	1973 Apr 01	41773.242	247.3° / 1973 Jul 21: Mars 4 orbiter launch (followed by
	270°	1973 Aug 25	41919.863	Mars 5 orbiter and 6 and 7 landers on Jul 25, Aug 05, and Aug 09)
				249.8° / 1973 Jul 25: Mars 5 orbiter launch
				256.8° / 1973 Aug 05: Mars 6 lander launch
				259.3° / 1973 Aug 09: Mars 7 lander launch
				300.0° / 1973 Oct 13: GLOBAL DUST STORM ONSET[d]
				306.8° / 1973 Oct 25: Opposition; C/A 0.44 AU on Oct 17
11	0°	1974 Jan 31	42078.191	004.9° / 1974 Feb 10: Mars 4 flyby (failed orbit insertion)
	90°	1974 Aug 17	42276.808	005.8° / 1974 Feb 12: Mars 5 orbit insertion; contact lost 2
	180°	1975 Feb 17	42460.216	days later
	270°	1975 Jul 13	42606.860	017.9° / 1974 Mar 09: Mars 7 begins descent prematurely, flies past Mars
				019.5° / 1974 Mar 12: Mars 6 impacts surface; successfully transmitted data from atmosphere
				293.5° / 1975 Aug 20: Viking 1 launch
				305.4° / 1975 Sep 09: Viking 2 launch
				358.2° / 1975 Dec 15: Opposition; C/A 0.57 AU on Dec 09

(*continued*)

MY	L_s	Earth Date	MJD	Other Notable Events in Same Mars Year (MY)[a]
12	0°	1975 Dec 19	42765.159	083.1° / 1976 Jun 19: Viking 1 orbit insertion
	90°	1976 Jul 04	42963.788	097.0° / 1976 Jul 20: Viking 1 lander arrival
	180°	1977 Jan 04	43147.160	104.8° / 1976 Aug 07: Viking 1 orbit insertion
	270°	1977 May 30	43293.810	117.6° / 1976 Sep 03: Viking 2 lander arrival
				204.0° / 1977 Feb 14: GLOBAL DUST STORM ONSET
				268.0° / 1977 May 27: GLOBAL DUST STORM ONSET[e]
13	0°	1977 Nov 05	43452.135	036.8° / 1978 Jan 22: Opposition; C/A 0.65 AU on Jan 19
	90°	1978 May 22	43650.770	118.6° / 1978 Jul 25: Viking 2 orbiter last contact
	180°	1978 Nov 22	43834.170	255.0° / 1979 Mar 25: Unconfirmed global dust storm
	270°	1979 Apr 17	43980.768	onset[f]
14	0°	1979 Sep 23	44139.076	071.0° / 1980 Feb 25: Opposition; C/A 0.68 AU on Feb 26
	90°	1980 Apr 08	44337.719	091.0° / 1980 Apr 11: Viking 2 lander last contact
	180°	1980 Oct 09	44521.159	151.4° / 1980 Aug 17: Viking 1 orbiter last contact
	270°	1981 Mar 04	44667.783	
15	0°	1981 Aug 10	44826.043	105.5° / 1982 Mar 31: Opposition; C/A 0.64 AU on Apr 05
	90°	1982 Feb 24	45024.668	208.0° / 1982 Oct 13: Unconfirmed global dust storm
	180°	1982 Aug 27	45208.097	onset[g]
	270°	1983 Jan 20	45354.763	226.5° / 1982 Nov 13: Viking 1 lander last contact
16	0°	1983 Jun 28	45513.039	146.0° / 1984 May 11: Opposition; C/A 0.53 AU on May 19
	90°	1984 Jan 12	45711.634	
	180°	1984 Jul 14	45895.058	
	270°	1984 Dec 07	46041.703	
17	0°	1985 May 15	46200.011	202.8° / 1986 Jul 10: Opposition; C/A 0.40 AU on Jul 16
	90°	1985 Nov 29	46398.614	
	180°	1986 Jun 01	46582.050	
	270°	1986 Oct 25	46728.688	
18	0°	1987 Apr 01	46886.965	227.8° / 1988 Jul 07: Phobos 1 orbiter/landers launch
	90°	1987 Oct 17	47085.591	(contact lost shortly after launch)
	180°	1988 Apr 18	47269.018	231.4° / 1988 Jul 12: Phobos 2 launch
	270°	1988 Sep 11	47415.685	280.3° / 1988 Sep 28: Opposition; C/A 0.39 AU on Sep 22
				350.4° / 1989 Jan 29: Phobos 2 orbit insertion
19	0°	1989 Feb 16	47573.959	018.5° / 1989 Mar 27: Phobos 2 last contact
	90°	1989 Sep 03	47772.575	209.3° / 1990 Apr 25: Hubble Space Telescope launches;
	180°	1990 Mar 06	47956.009	often photographs Mars near opposition
	270°	1990 Jul 30	48102.632	340.4° / 1990 Nov 27: Opposition; C/A 0.52 AU on
				Nov 20
20	0°	1991 Jan 04	48260.920	329.7° / 1992 Sep 25: Mars Observer orbiter launch (con-
	90°	1991 Jul 22	48459.533	tact lost just prior to Mars orbit insertion)
	180°	1992 Jan 22	48643.008	
	270°	1992 Jun 16	48789.632	

(*continued*)

MY	L_s	Earth Date	MJD	Other Notable Events in Same Mars Year (MY)[a]
21	0°	1992 Nov 21	48947.868	022.7° / 1993 Jan 07: Opposition; C/A 0.63 AU on Jan 03
	90°	1993 Jun 08	49146.482	
	180°	1993 Dec 08	49329.956	
	270°	1994 May 04	49476.625	
22	0°	1994 Oct 09	49634.854	057.9° / 1995 Feb 12: Opposition; C/A 0.68 AU on Feb 11
	90°	1995 Apr 26	49833.442	
	180°	1995 Oct 26	50016.907	
	270°	1996 Mar 21	50163.575	
23	0°	1996 Aug 26	50321.836	034.6° / 1996 Nov 07: Mars Global Surveyor launch
	90°	1997 Mar 13	50520.413	038.3° / 1996 Nov 16: Russian Mars 96 orbiter/lander
	180°	1997 Sep 12	50703.893	launch
	270°	1998 Feb 06	50850.540	046.5° / 1996 Dec 04: Mars Pathfinder launch
				091.7° / 1997 Mar 17: Opposition; C/A 0.66 AU on Mar 20
				142.7° / 1997 Jul 04: Mars Pathfinder landing
				179.5° / 1997 Sep 12: Mars Global Surveyor orbit insertion
				followed by long aerobraking period
				188.0° / 1997 Sep 27: Mars Pathfinder contact lost
				225.3° / 1998 Jul 03: Japanese Nozomi orbiter launch
				(distant flybys; failed to reach Mars orbit)
24	0°	1998 Jul 14	51008.795	068.7° / 1998 Dec 11: Mars Climate Orbiter launch
	90°	1999 Jan 29	51207.382	078.8° / 1999 Jan 03: Mars Polar Lander/Deep Space 2
	180°	1999 Jul 31	51390.848	launch
	270°	1999 Dec 25	51537.531	107.3° / 1999 Mar 09: Mars Global Surveyor routine
				mapping begins
				129.0° / 1999 Apr 24: Opposition; C/A 0.58 AU on May 01
				211.5° / 1999 Sep 23: Mars Climate Orbiter impacts Mars
				256.3° / 1999 Dec 03: Mars Polar Lander impacts Mars
25	0°	2000 May 31	51695.773	142.2° / 2001 Apr 07: 2001 Mars Odyssey launch
	90°	2000 Dec 16	51894.376	177.7° / 2001 Jun 13: Opposition; C/A 0.45 AU on Jun 21
	180°	2001 Jun 17	52077.819	184.7° / 2001 Jun 26: GLOBAL DUST STORM ONSET[h]
	270°	2001 Nov 11	52224.466	258.4° / 2001 Oct 24: 2001 Mars Odyssey orbit insertion
26	0°	2002 Apr 18	52382.737	196.7° / 2003 Jun 03: Mars Express/Beagle 2 launch
	90°	2002 Nov 03	52581.358	200.9° / 2003 Jun 10: MER Spirit launch
	180°	2003 May 05	52764.856	217.5° / 2003 Jul 08: MER Opportunity launch
	270°	2003 Sep 29	52911.471	249.9° / 2003 Aug 28: Opposition; C/A 0.37 AU on
				Aug 27
				322.0° / 2003 Dec 25: Mars Express orbit insertion into
				250 × 11,560 km orbit with 7.566 hr period; Beagle 2
				impacts Mars
				327.1° / 2004 Jan 03: MER Spirit landing
				339.1° / 2004 Jan 25: Opportunity landing

(*continued*)

MY	L_s	Earth Date	MJD	Other Notable Events in Same Mars Year (MY)[a]
27	0°	2004 Mar 05	53069.686	266.2° / 2005 Aug 10: Mars Reconnaissance Orbiter
	90°	2004 Sep 20	53268.306	launch
	180°	2005 Mar 22	53451.834	319.9° / 2005 Nov 07: Opposition; C/A 0.46 AU on Oct 30
	270°	2005 Aug 16	53598.497	
28	0°	2006 Jan 21	53756.684	023.2° / 2006 Mar 10: Mars Reconnaissance Orbiter orbit
	90°	2006 Aug 08	53955.264	insertion
	180°	2007 Feb 07	54138.774	129.2° / 2006 Nov 02: Mars Global Surveyor contact lost
	270°	2007 Jul 04	54285.458	131.6° / 2006 Nov 07: Mars Reconnaissance Orbiter science operations begin (crossing equator at 15:00 LT)
				189.9° / 2007 Feb 25: Rosetta Mars flyby (250 km)
				261.5° / 2007 Jun 20: GLOBAL DUST STORM ONSET[1]
				289.2° / 2007 Aug 04: Phoenix lander launch
29	0°	2007 Dec 09	54443.676	007.1° / 2007 Dec 24: Opposition; C/A 0.59 AU on Dec 24
	90°	2008 Jun 25	54642.236	076.7° / 2008 May 25: Phoenix lands
	180°	2008 Dec 25	54825.750	151.1° / 2008 Nov 02: Phoenix contact lost
	270°	2009 May 21	54972.406	211.9° / 2009 Feb 18: Dawn Mars flyby at 542 km
30	0°	2009 Oct 26	55130.635	044.3° / 2010 Jan 29: Opposition; C/A 0.66 AU on Jan 27
	90°	2010 May 13	55329.204	067.1° / 2010 Mar 22: Spirit last contact (sol 2211)
	180°	2010 Nov 12	55512.721	
	270°	2011 Apr 08	55659.403	
31	0°	2011 Sep 13	55817.601	026.6° / 2011 Nov 08: Russian/Chinese Phobos-Grunt orbiter/lander and Yinghuo-1 orbiter launch (launch failure)
	90°	2012 Mar 30	56016.182	
	180°	2012 Sep 29	56199.670	
	270°	2013 Feb 23	56346.353	034.8° / 2011 Nov 26: Mars Science Laboratory Curiosity rover launch
				078.1° / 2012 Mar 03: Opposition; C/A 0.67 AU on Mar 05
				150.7° / 2012 Aug 06: Curiosity lands
32	0°	2013 Jul 31	56504.580	045.3° / 2013 Nov 05: Mars Orbiter Mission launches
	90°	2014 Feb 15	56703.159	051.3° / 2013 Nov 18: MAVEN launches
	180°	2014 Aug 17	56886.679	055.7° / 2013 Nov 28: Comet ISON passes within 1.8 million km of Mars (successfully observed by MRO)
	270°	2015 Jan 11	57033.314	113.3° / 2014 Apr 08: Opposition; C/A 0.62 AU on Apr 14
				200.6° / 2014 Sep 22: MAVEN arrives in orbit
				201.7° / 2014 Sep 24: Mars Orbiter Mission arrives
				217.4° / 2014 Oct 19: Comet C/2013 A1 (Siding Spring) passes within 138,000 km of Mars (successfully observed by MRO, MAVEN, and Mars Express)
33	0°	2015 Jun 18	57191.524	122.3° / 2016 Mar 14: ExoMars Trace Gas Orbiter and Schiaparelli lander launch
	90°	2016 Jan 03	57390.111	
	180°	2016 Jul 04	57573.661	156.3° / 2016 May 22: Opposition; C/A 0.50 AU on May 30
	270°	2016 Nov 28	57720.323	244.9° / 2016 Oct 19: ExoMars Trace Gas Orbiter arrives; Schiaparelli lander impacts surface

(*continued*)

MY	L_s	Earth Date	MJD	Other Notable Events in Same Mars Year (MY)[a]
34	0°	2017 May 05	57878.490	125.9° / 2018 Feb 06: Elon Musk's Tesla Roadster launched
	90°	2017 Nov 20	58077.073	onto trans-Mars trajectory on a Falcon-9 Heavy
	180°	2018 May 22	58260.613	170.5° / 2018 May 05: InSight launches along with MarCO
	270°	2018 Oct 16	58407.306	CubeSats
				190.5° / 2018 Jun 10: Opportunity contact lost amid dust
				storm (sol 5111)
				218.7° / 2018 Jul 27: Opposition; C/A 0.38 AU on Jul 31
35	0°	2019 Mar 23	58565.481	112.4° / 2019 Nov 26: InSight lands; MarCO CubeSats
	90°	2019 Oct 08	58764.042	fly past Mars, successfully relaying InSight landing
	180°	2020 Apr 08	58947.600	telemetry
	270°	2020 Sep 02	59094.272	241.9° / 2020 Jul 19: Emirates Mars Mission "Hope"
				launch
				244.0° / 2020 Jul 23: Tianwen-1 orbiter and lander/rover
				launch
				248.6° / 2020 Jul 30: Mars 2020 Perseverance rover and
				Ingenuity helicopter launch
				295.2° / 2020 Oct 13: Opposition; C/A 0.41 AU on Oct 06
36	0°	2021 Feb 07	59252.467	000.8° / 2021 Feb 09: Emirates Mars Mission "Hope"
	90°	2021 Aug 25	59451.011	orbiter Mars orbit insertion
	180°	2022 Feb 24	59634.581	001.3° / 2021 Feb 10: Tianwen-1 orbiter and lander/rover
	270°	2022 Jul 21	59781.261	enter Mars orbit
				005.2° / 2021 Feb 18: Landing of Mars 2020 Perseverance
				rover and Ingenuity helicopter
				~039°–066° / 2021 May–June: Estimated landing attempt
				of Tianwen-1 lander and rover
				276.7° / 2022 Aug 01: Approximate date of launch period
				opening for ExoMars 2022 Kazachok lander and Rosa-
				lind Franklin rover
				350.7° / 2022 Dec 08: Opposition; C/A 0.54 AU on Dec 01
37	0°	2022 Dec 26	59939.430	044.8° / 2023 Apr 01: Approximate date of planned
	90°	2023 Jul 12	60137.983	ExoMars 2022 rover arrival
	180°	2024 Jan 12	60321.519	064.2° / 2023 May 15: Approximate date of Psyche mis-
	270°	2024 Jun 07	60468.226	sion Mars flyby (500 km)
38	0°	2024 Nov 12	60626.407	030.8° / 2025 Jan 16: Opposition; C/A 0.64 AU on Jan 12
	90°	2025 May 29	60824.969	
	180°	2025 Nov 29	61008.513	
	270°	2026 Apr 25	61155.167	
39	0°	2026 Sep 30	61313.354	065.1° / 2027 Feb 19: Opposition; C/A 0.68 AU on Feb 20
	90°	2027 Apr 16	61511.924	
	180°	2027 Oct 17	61695.515	
	270°	2028 Mar 12	61842.176	

(*continued*)

MY	L_s	Earth Date	MJD	Other Notable Events in Same Mars Year (MY)[a]
40	0°	2028 Aug 17	62000.309	099.4° / 2029 Mar 25: Opposition; C/A 0.65 AU on Mar 29
	90°	2029 Mar 03	62198.873	
	180°	2029 Sep 03	62382.461	
	270°	2030 Jan 28	62529.169	
41	0°	2030 Jul 05	62687.304	138.2° / 2031 May 04: Opposition; C/A 0.55 AU on May 12
	90°	2031 Jan 19	62885.838	
	180°	2031 Jul 22	63069.414	
	270°	2031 Dec 16	63216.112	
42	0°	2032 May 22	63374.285	191.2° / 2033 Jun 28: Opposition; C/A 0.42 AU on Jul 05
	90°	2032 Dec 06	63572.820	
	180°	2033 Jun 08	63756.409	
	270°	2033 Nov 02	63903.083	
43	0°	2034 Apr 09	64061.239	266.8° / 2035 Sep 15: Opposition; C/A 0.38 AU on Sep 11
	90°	2034 Oct 24	64259.801	
	180°	2035 Apr 26	64443.388	
	270°	2035 Sep 20	64590.096	
44	0°	2036 Feb 25	64748.227	331.7° / 2037 Nov 19: Opposition; C/A 0.49 AU on Nov 11
	90°	2036 Sep 10	64946.785	
	180°	2037 Mar 13	65130.369	
	270°	2037 Aug 07	65277.045	
45	0°	2038 Jan 12	65435.203	
	90°	2038 Jul 29	65633.752	
	180°	2039 Jan 29	65817.377	
	270°	2039 Jun 25	65964.037	
46	0°	2039 Nov 30	66122.148	016.0° / 2040 Jan 02: Opposition; C/A 0.61 AU on Dec 28
	90°	2040 Jun 15	66320.702	
	180°	2040 Dec 16	66504.335	
	270°	2041 May 12	66651.040	
47	0°	2041 Oct 17	66809.131	052.0° / 2042 Feb 06: Opposition; C/A 0.67 AU on Feb 05
	90°	2042 May 03	67007.659	
	180°	2042 Nov 03	67191.279	
	270°	2043 Mar 29	67337.998	
48	0°	2043 Sep 04	67496.121	085.8° / 2044 Mar 11: Opposition; C/A 0.67 AU on Mar 14
	90°	2044 Mar 20	67694.629	
	180°	2044 Sep 20	67878.259	
	270°	2045 Feb 13	68024.952	
49	0°	2045 Jul 22	68183.082	121.9° / 2046 Apr 17: Opposition; C/A 0.60 AU on Apr 24
	90°	2046 Feb 05	68381.596	
	180°	2046 Aug 08	68565.224	
	270°	2047 Jan 01	68711.945	

(continued)

TABLE A6.1 (*continued*)

MY	L_s	Earth Date	MJD	Other Notable Events in Same Mars Year (MY)[a]
50	0°	2047 Jun 09	68870.052	167.7° / 2048 Jun 03: Opposition; C/A 0.47 AU on Jun 12
	90°	2047 Dec 24	69068.581	
	180°	2048 Jun 25	69252.178	
	270°	2048 Nov 18	69398.886	

Sources: Spacecraft event dates are from D. R. Williams, "Chronology of Mars Exploration," NSSDCA, last updated December 15, 2020, https://nssdc.gsfc.nasa.gov/planetary/chronology_mars.html; dust storm timing information is from M. Kahre et al., "The Mars Dust Cycle," in *The Atmosphere and Climate of Mars*, ed. R. M. Haberle (Cambridge: Cambridge University Press, 2017), table 10.1, https://doi.org/10.1017/9781139060172.010.

[a] In the last column, C/A refers to closest approach.

[b] S. Miyamoto, "The Great Yellow Cloud and the Atmosphere of Mars: Report of Visual Observations During the 1956 Opposition," *Contributions from the Institute of Astrophysics and Kwasan Observatory, University of Kyoto* 71 (1957): 239–281; E. C. Slipher, *Mars: The Photographic Story* (Cambridge, Mass.: Sky, 1962); L. J. Martin and R. W. Zurek, "An Analysis of the History of Dust Activity on Mars," *JGR* 98 (1993): 3221–3246; R. W. Zurek and L. J. Martin, "Interannual Variability of Planet-Encircling Dust Storms on Mars," *JGR* 98 (1993): 3247–3259.

[c] L. J. Martin, "The Major Martian Yellow Storm of 1971," *Icarus* 22, no. 2 (June 1974): 175–188; L. J. Martin, "The Major Martian Dust Storms of 1971 and 1973," *Icarus* 23, no. 1 (September 1974): 108–115.

[d] Martin, "The Major Martian Dust Storms of 1971 and 1973."

[e] G. A. Briggs, W. A. Baum, and J. Barnes, "Viking Orbiter Imaging Observations of Dust in the Martian Atmosphere," *JGR* 84 (1979): 2795–2820; T. E. Thorpe, "The Mars Opposition Effect at 20°N. Latitude and 20°W. Longitude," *Icarus* 37 (1979): 389–398; J. A. Ryan and R. D. Sharman, "Two Major Dust Storms, One Mars Year Apart: Comparison from Viking Data," *JGR* 86 (1981): 3247–3254; Zurek and Martin, "Interannual Variability of Planet-Encircling Dust Storms on Mars."

[f] C. B. Leovy, "Observations of Martian Tides over Two Annual Cycles," *Journal of the Atmospheric Sciences* 38 (1981): 30–39; Ryan and Sharman, "Two Major Dust Storms."

[g] J. E. Tillman, "Mars Global Atmospheric Oscillations: Annually Synchronized, Transient Normal-Mode Oscillations and the Triggering of Global Dust Storms," *JGR* 93 (1988): 9433–9451; Zurek and Martin, "Interannual Variability of Planet-Encircling Dust Storms on Mars."

[h] B. A. Cantor, M. C. Malin, and K. S. Edgett, "Multiyear Mars Orbiter Camera (MOC) Observations of Repeated Martian Weather Phenomena During the Northern Summer Season," *JGR* 107, no. E3 (2002), https://doi.org/10.1029/2001JE001588; M. D. Smith et al., "Thermal Emission Spectrometer Observations of Martian Planet-Encircling Dust Storm 2001A," *Icarus* 157 (2002): 259–263; M. J. Strausberg, "Observations of the Initiation and Evolution of the 2001 Mars Global Dust Storm," *JGR* 110, no. E2 (2005), https://doi.org/10.1029/2004JE002361; B. A. Cantor, "MOC Observations of the 2001 Mars Planet-Encircling Dust Storm," *Icarus* 186 (2007): 60–96.

[i] B. A. Cantor et al., "Observations of the Martian Atmosphere by MRO-MARCI, an Overview of 1 Mars Year" (Third International Workshop on the Mars Atmosphere: Modeling and Observations, Williamsburg, Va., 2008), abstract #9075; M. D. Smith, "THEMIS Observations of Mars Aerosol Optical Depth from 2002–2008," *Icarus* 202, no. 2 (2009): 444–452, https://doi.org/10.1016/j.icarus.2009.03.027; H. Wang and M. I. Richardson, "The Origin, Evolution, and Trajectory of Large Dust Storms on Mars During Mars Years 24–30 (1999–2011)," *Icarus* 251 (2015): 112–127, https://doi.org/10.1016/j.icarus.2013.10.033.

Finally, there are patterns of annual variations in atmospheric opacity that arise from the Mars dust cycle.[7] Mars typically enjoys a relatively clear season, spanning from $L_s \approx 0°$ to $135°$, when less dust is lifted into the atmosphere by surface winds. Dust loading usually increases between $L_s \approx 135°$ and $160°$, and along with it, the atmospheric opacity. Data that has been acquired continuously since Mars Global Surveyor's arrival in mid-MY24 suggests the presence of three recurring waves of dust storm activity. Most of the observed 8 to 35 regional storms that have occurred each Mars year happen within these windows. Early-season activity ($L_s \approx 135°$ to $180°$) is primarily in the southern hemisphere. A pre-solstice wave ($L_s \approx 180°$ to $236°$) and a post-solstice wave ($L_s \approx 308°$ to $336°$) can produce storms that cross between hemispheres.

Once in a while—perhaps nine times since MY1 (1955)—global dust storms (or at least what many in the research community call "planet-encircling dust storms") have shrouded the planet in dust so thick that the surface can't be seen from telescopic or orbital observations. All of these global storms began between $L_s = 185°$ and $300°$, bracketing the northern winter solstice. Global storms are listed in table A6.1.

Appendix G

Timekeeping on Mars

CONTRIBUTED BY EMILY LAKDAWALLA, THE LAKDAWALLA GROUP LLC

Time on Mars is easily divided into days based on its rotation rate. Sols, or Martian solar days, are only 39 minutes and 35 seconds longer than Earth days. For convenience, sols are divided into a 24-hour clock. Each landed Mars mission keeps track of local solar time, or LST, at its landing site, because local solar time relates directly to the position of the Sun in the sky and thus the angle from which camera views are illuminated. The time of day, local solar time, depends on the lander's longitude on Mars.

Mean and True Solar Time

The following discussion relies heavily on Allison and McEwen's "A Post-Pathfinder Evaluation of Areocentric Solar Coordinates" (2000, hereinafter referred to as AM2000), which has guided timekeeping on Mars since the end of the Pathfinder mission. The equations in it underlie Robert Schmunk's Mars24 Sunclock application, which is in popular use for mission timekeeping.[1]

Mean Solar Time

For convenience (not to mention sanity), the sol calendar and 24-hour clock used by Mars mission operations usually employ sols, hours, minutes, and seconds of fixed length, 1.02749125 times longer than their Earth equivalents. This is "mean solar time," or MST. With a single reference date for midnight MST on any Earth date, it's simple arithmetic to calculate the MST (though not solar longitude) for any other Earth date. Conveniently, on JD 245149.5 (January 6, 2000, at 00:00), it was midnight MST at Mars's prime meridian. To find the MST for any Earth date and time, multiply the number of Earth days elapsed since the epoch by 1.02749125 and convert the fractional part to a 24-hour clock.

True Solar Time

Because Mars's rotation rate is constant but its velocity around the Sun is not, the timings of sunrises, noons, and sunsets—that is, "true solar time," or TST—diverge from mean solar time. When the planet moves faster in its orbit, close to perihelion, Mars's solar days lengthen. The differences add up to significant deviations between the actual position of the Sun in the sky and the fictional position that is bookkept in MST. The deviations are described in a mathematical expression called the "equation of time," which states the number of minutes that true solar time is ahead of or behind mean solar time.

Converting Between Earth and Mars Dates and Times

For a given Julian date, AM2000 provides an analytic recipe for determining Mars's solar longitude (AM2000 equations 16–19) and equation of time (AM2000 equations 20 and 21). This recipe is accurate to within 0.03 degrees of solar longitude and 8 seconds of solar time.

True and mean solar time coincide only twice each Martian year, near aphelion and perihelion, at solar longitudes of 57.7° and 258.0°. True and mean solar time diverge most at a solar longitude of 187.9°, when mean solar time runs behind true solar time by 39.9 minutes, and at 329.1°, when mean solar time is ahead of true solar time by 51.1 minutes. These differences matter for operating landed spacecraft at safe temperatures and/or sufficient power levels, for avoiding solar damage to delicate optical instruments, and for timing surface-based observations of the Sun, Phobos, Deimos, and stars.

AM2000 does not use the Mars Year convention (which was, after all, only defined the same year).[2] Therefore, it is necessary to make a small modification to AM2000 equation 14 to convert from Mars Year n and solar longitude L_s to Julian Date:

$$JD(n, L_s) = 2451508 + 1.90826 \times (L_s - 251°) - 20.42 \times \sin(L_s - 251°) + 0.72$$
$$\times \sin[2 \times (L_s - 251°)] + \{686.9726 + 0.0043 \times \cos(L_s - 251°)$$
$$- 0.0003 \times \cos[2 \times (L_s - 251°)]\} \times [\text{integer part of } (n - 24)]$$

AM2000 also predates the Mars longitude convention change from positive-west to positive-east that happened in 2001.[3] Before employing AM2000 equations 23 and 24 to determine local true solar time (LTST) from MST, take care to convert modern positive-east longitudes to positive-west longitudes by subtracting west longitude from 360°.

Tracking Time on Mars Missions

MST is defined for the Mars prime meridian. Time of day at a location away from the prime meridian depends on longitude. Most Mars landed missions have employed an offset local mean solar time (LMST) convention. (The exception is Mars Pathfinder, which employed an offset LTST convention for mission operations.) LMST is functionally a mission-specific time zone, MST with a longitude adjustment. Like most time zones on Earth, the difference between LMST and LTST is not more than an hour, and the steady and predictable march of the LMST convention is more convenient and close enough for most applications (especially mission operations planning). Ideally, the LMST reference longitude would be the same as the longitude of the landing location, but in practice, missions have to define the LMST convention before landing, so the reference longitude is never identical to the landing location.[4]

Earth and Mars epoch dates for successful Mars landers are listed in table A7.1, along with other landing date, time, and location information. To convert between Earth and mission dates and times, calculate the number

TABLE A7.1 Epoch Times and Other Date and Location Information for Mars Landed Missions

Lander	Launch Date[a]	Landing Date/Time[a]	Last Contact[a]	Landing Site Latitude / East longitude (deg)	Landing Site Elevation (m)	First Sol	LMST Ref., E. Longitude	Earth Date/Time on First Sol, 00:00 LMST
Mars 3	28 May 1971	2 Dec 1971 13:50:35[b]	2 Dec 1971	−45.044 / 202.019	+1626	—	—	—
Viking 1	20 Aug 1975	20 Jul 1976 11:53:06	13 Nov 1982	22.269 / 312.048[c]	−3637	0	316.845	19 Jul 1976 19:39:56
Viking 2	9 Sep 1975	3 Sep 1976 22:58:20	11 Apr 1980	47.643 / 134.288[d]	−4495	0	139.9515	3 Sep 1976 12:48:48
Mars Pathfinder and Sojourner rover	4 Dec 1996	4 Jul 1997 16:56:55 / 6 Jul 1997 05:40	27 Sep 1997	19.33 / 326.47[e]	−3681[f]	1	N/A	4 Jul 1997 14:24:40 (= LTST midnight)
MER Spirit	10 Jun 2003	3 Jan 2004 04:35	22 Mar 2010	−14.5692 / 175.4729[g] to −14.6004 / 175.5254[h]	−1936[f] to −1926[h]	1	175.655	3 Jan 2004 12:53:14
MER Opportunity	8 Jul 2003	24 Jan 2004 04:54:22	10 Jun 2018	−1.9462 / 354.4734[i] to −2.3309 / 354.6510[j]	−1387[f] to −1483[j]	1	354.65	24 Jan 2004 14:28:55
Phoenix	4 Aug 2007	25 May 2008 23:53:44	2 Nov 2008	68.2188 / 234.2508[k]	−4130	0	233.35[l]	25 May 2008 06:51:20[m]
MSL Curiosity	26 Nov 2011	6 Aug 2012 05:32	—	−4.5895 / 137.4417[n] to −4.7293 / 137.3911[a]	−4501 to −4128[l]	0	137.6188	5 Aug 2012 13:49:59[o]
InSight	5 May 2018	26 Nov 2018 19:52:59	—	4.502 / 135.623[p]	−2613	0	135.97[l]	26 Nov 2018 05:10:50[q]
Tianwen-1	23 July 2020	April or May 2021	—	19 to 30 / 90 to 134[r]	unknown	—	—	—
Mars 2020 Perseverance	30 Jul 2020	18 Feb 2021 20:00	—	18.44 / 77.50[r]	−2640	0	77.43[l]	18 Feb 2021 ~09:12
ExoMars	2022	2023	—	18.14 / 335.7[r]	−3000	—	—	—

Note: Information in italics indicates predictions for events that had not yet taken place as of the date of this writing. All Earth times and dates are Earth Received Time (TT until 1984, UTC after) except for the first sol date, which is Spacecraft Event Time. The LMST reference east longitude is akin to a time zone longitude for each mission. Unless otherwise indicated, these longitudes were derived by using the equations in AM2000 to compute the equation of time from imaging start times treated as TT for each mission's navigational or primary camera data set and then choosing a longitude for LMST that minimized the difference between the computed LTST and the LTST value included in the image metadata. These longitudes recover the metadata

LTST to within 22 Mars seconds for the Viking landers and within 7 Mars seconds for Spirit, Opportunity, and Curiosity. Pathfinder did not use LMST but rather a LTST convention based on dividing the time between local true solar midnights into 24-hour sols (AM2000). The MER mission defined a "Hybrid Local Solar Time" (HLST) convention intended to make LMST match LTST 45 days into the mission (B. Semenov, "MER Time Tag Issues," July 29, 2004, updated July 27, 2005, https://pds-imaging.jpl.nasa.gov/data/mer/opportunity/merln0_0xxx/document/mer_time_issues.txt). MER HLST can be treated as LMST with the stated reference longitude.

[a] Launch, landing, and last contact dates are from the NSSDCA Master Catalog, https://nssdc.gsfc.nasa.gov/nmc/SpacecraftQuery.jsp.

[b] Mars 3 apparently landed successfully in the middle of the global dust storm of MY9, and only transmitted data for about 20 seconds before contact was lost forever. It was long enough to partially transmit an image that indicated available light levels of 50 lux, barely brighter than twilight.

[c] NSSDCA, "Viking 1 Lander," accessed May 11, 2020, https://nssdc.gsfc.nasa.gov/nmc/spacecraft/displayTrajectory.action?id=1975-075C.

[d] NSSDCA, "Viking 2 Lander," accessed May 11, 2020, https://nssdc.gsfc.nasa.gov/nmc/spacecraft/displayTrajectory.action?id=1975-083C.

[e] P. Stooke, The International Atlas of Mars Exploration: The First Five Decades (Cambridge: Cambridge University Press, 2012).

[f] Tom Stein and Feng Zhou, pers. comm.

[g] R. E. Arvidson et al., "Localization and Physical Properties Experiments Conducted by Spirit at Gusev Crater," Science 305 (2004), https://doi.org/10.1126/science.1099922.

[h] F. Calef, pers. comm., August 31, 2020.

[i] R. E. Arvidson et al., "Localization and Physical Properties Experiments Conducted by Opportunity at Meridiani Planum," Science 306 (2004), https://doi.org/10.1126/science.1104211.

[j] Tim Parker, pers. comm., September 1, 2020.

[k] T. L. Heet et al. "Geomorphic and Geologic Settings of the Phoenix Lander Mission Landing Site," JGR 114 (2009), https://doi.org/10.1029/2009JE003416.

[l] M. Allison and R. Schmunk, "Technical Notes on Mars Solar Time as Adopted by the Mars24 Sunclock," NASA, updated March 8, 2020, https://www.giss.nasa.gov/tools/mars24/help/notes.html.

[m] Mark Lemmon, pers. comm., September 2, 2020.

[n] E. Lakdawalla, The Design and Engineering of Curiosity: How the Mars Rover Performs Its Job (Cham: Springer-Praxis Books, 2018).

[o] PDS NAIF, "MSL SCLK File Implementing LMST at GC Landing Site," NASA, August 8, 2012, https://naif.jpl.nasa.gov/pub/naif/pds/data/msl-m-spice-6-v1.0/mslsp_1000/data/sclk/msl_lmst_opsl20808_v1.tsc.

[p] M. Golombek et al., "Geology of the InSight Landing Site on Mars," Nature Geosci. 11 (2020).

[q] PDS NAIF, "INSIGHT SCLK File Implementing LMST at opsl81206 Landing Site," NASA, December 17, 2018, https://naif.jpl.nasa.gov/pub/naif/pds/data/insight_spice_kernels/sclk/insight_lmst_opsl81206_v1.tsc.

[r] P. W. Yao, C. Li, and B. Li, "The Spatial and Temporal Probability of Dust Storm Activity in Isidis, One of the Tentative Landing Areas of China's First Mars Probe" (51st Lunar and Planetary Science Conference, Woodlands, Tex., 2000), abstract #1905. The actual landing site will be at one of three possible sites within the defined box.

[s] J. A. Grant et al., "The Science Process for Selecting the Landing Site for the 2020 Mars Rover," Planet. Space Sci. 164 (2018): 106–126, https://doi.org/10.1016/j.pss.2018.07.001.

[t] M. A. Ivanov et al., "Geomorphological Analysis of ExoMars Candidate Landing Site Oxia Planum," Solar System Research 54 (2020), https://doi.org/10.1134/S0038094620010050.

of Earth days since the epoch date and multiply by 1.02749125. The result is the number of Mars sols that have elapsed in the same period; the fractional part can be converted to the 24-hour clock. Adding the value of the equation of time (AM2000 equations 20 and 21) plus a correction for the difference between the lander's actual and reference longitude to LMST gives LTST for the same date. (Again, take care to convert to west longitude before performing any calculations with AM2000 equations.)

Of course, Mars rovers move, which tends to take them away from their reference longitudes. Conveniently, one degree of longitude on Mars is almost exactly 60 kilometers. Moving one degree westward would therefore delay true solar time events by four minutes, so at the equator, 250 meters' westward driving delays true solar time events by about one second. These differences are usually negligible for mission planning. As of June 2020, the difference for Curiosity was about 10 seconds.

Appendix H

NASA's Historical Investment in Mars Exploration

CONTRIBUTED BY CASEY DREIER, CHIEF ADVOCATE AND SENIOR SPACE
POLICY ADVISOR, THE PLANETARY SOCIETY

Uniquely among national space agencies, NASA publicly reports its spending on major projects in annual reports to Congress. This allows for refined inflation adjustments that enable direct comparisons in mission costs over the space agency's entire history.

Table A8.1 reports the total mission costs (prime and extended missions) for every Mars project by NASA, broken out by the costs to design and build the spacecraft (Spacecraft Development), launch the spacecraft (Launch Services), and operate the spacecraft through its entire duration (Mission Operations). Also included is the reported life-cycle cost of the mission in original dollars as predicted at launch time. All values in table A8.1 are in millions of dollars. They are inflation-adjusted to 2020 dollars via NASA's New Start Index.[1] The reported life-cycle costs are in original dollar amounts, not adjusted for inflation, and are normalized to include launch costs and prime mission operations. The mission operations column includes both prime and extended operations through fiscal year 2019. Note that after fiscal year 2004, NASA implemented "full-cost" accounting, which assigned agency overhead proportionally to its projects, increasing reported costs.

Table A8.2 and figure A8.1 present NASA's cumulative annual spending on all Mars-related research, missions, and management in both real dollars

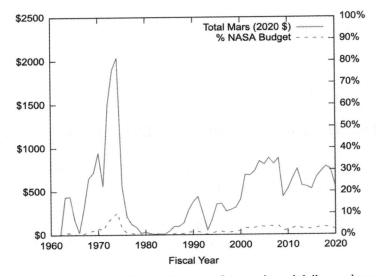

FIGURE A8.1 NASA's spending on Mars in inflation-adjusted dollars and as a percentage of agency appropriations. Courtesy of Casey Dreier, The Planetary Society.

and inflation-adjusted dollars, and as a percentage of NASA's total budget. All values in table A8.2 are in millions of dollars. Totals include reported spending on missions, research and analysis, and Mars-related project management. Fiscal year 2020 values are estimates from final congressional appropriations.

TABLE A8.1 **NASA's Historical Mars Mission Costs**

Mission	Launch Year	Reported Life-Cycle Cost (original $)	Spacecraft Development (2020 $)	Launch Services (2020 $)	Mission Operations (2020 $)	Mission Ops/Year (2020 $)	Total Mission Cost (2020 $)
Mariner 3 & 4	1964	$83.2	$875	$176[a]	$34	$34	$1,086
Mariner 6 & 7	1969	$148	$1,115	$193[a]	$32	$32	$1,341
Mariner 8 & 9	1971	$149	$895	$174[a]	$195	$50	$1,265
Viking 1 & 2	1975	$1,016	$6,038	$539[a]	$517	$91	$7,094
Observer	1993	$904	$1,149	$500	$41	$41	$1,690
MGS	1996	$249.6	$236	$94	$161	$16	$491
Pathfinder	1996	$265.6	$364	$88	$24	$24	$476
MPL/MCO	1999	$326.6	$329	$148[a]	$24	$23	$501
Odyssey	2001	$297	$275	$89	$260	$14	$623
MER	2003	$820	$973	$155[a]	$438	$26	$1,566
MRO	2005	$720	$784	$132	$528	$38	$1,444
Phoenix	2007	$457	$548	$119	$16	$16	$683
MSL Curiosity	2011	$2,476	$2,801	$261	$450	$64	$3,511
MAVEN	2014	$671	$431	$222	$120	$20	$773
InSight	2018	$813.8	$660	$190	$24	$16	$874
Perseverance	2020	$2,725.8	$2,330	$243	$292	$73	$2,865

Source: "Planetary Exploration Budget Dataset," compiled by Casey Dreier for The Planetary Society, accessed February 10, 2021, https://www.planetary.org/space-policy/planetary-exploration-budget-dataset.

Note: All values are in millions of dollars.

[a] Total cost for the two launches.

TABLE A8.2 NASA's Total Investment in Mars

Fiscal Year	Total Mars (original $)	Total Mars (2020 $)	% of NASA Budget
1960	$0.0	$0.0	0.0%
1961	$0.0	$0.0	0.0%
1962	$0.0	$0.0	0.0%
1963	$36.6	$433.4	1.0%
1964	$38.9	$440.8	0.8%
1965	$16.2	$177.5	0.3%
1966	$2.3	$24.1	0.0%
1967	$31.1	$306.7	0.6%
1968	$69.7	$651.8	1.5%
1969	$81.2	$718.3	2.0%
1970	$114.5	$946.9	3.1%
1971	$72.6	$564.9	2.2%
1972	$203.3	$1,496.9	6.1%
1973	$272.4	$1,897.7	8.0%
1974	$313.4	$2,036.7	10.3%
1975	$95.2	$558.4	2.9%
1976	$39.5	$212.5	1.1%
1976TQ[a]	$11.6	$61.1	1.2%
1977	$25.4	$123.4	0.7%
1978	$20.0	$90.1	0.5%
1979	$5.0	$20.6	0.1%
1980	$8.5	$31.4	0.2%
1981	$4.4	$14.9	0.08%
1982	$2.9	$9.1	0.05%
1983	$4.3	$12.7	0.06%
1984	$2.9	$8.1	0.04%
1985	$16.1	$43.5	0.2%
1986	$36.9	$96.8	0.5%
1987	$38.7	$97.4	0.4%
1988	$57.6	$137.7	0.6%
1989	$124.7	$284.6	1.2%
1990	$174.0	$380.1	1.4%
1991	$208.0	$438.5	1.5%
1992	$125.6	$251.7	0.9%
1993	$27.8	$53.5	0.2%
1994	$95.3	$177.7	0.7%
1995	$194.8	$354.1	1.4%
1996	$203.4	$360.8	1.5%

(*continued*)

TABLE A8.2 (*continued*)

Fiscal Year	Total Mars (original $)	Total Mars (2020 $)	% of NASA Budget
1997	$154.8	$270.7	1.1%
1998	$169.3	$289.0	1.2%
1999	$188.5	$314.0	1.4%
2000	$254.3	$407.1	1.9%
2001	$447.6	$691.5	3.1%
2002	$456.9	$686.7	3.1%
2003	$500.4	$736.1	3.2%
2004	$596.3	$847.9	3.9%
2005	$588.5	$812.1	3.6%
2006	$662.2	$886.0	4.0%
2007	$634.9	$817.8	3.9%
2008	$709.3	$882.4	4.1%
2009	$361.7	$441.3	1.9%
2010	$438.2	$527.6	2.3%
2011	$550.4	$652.2	3.0%
2012	$649.1	$760.7	3.7%
2013	$492.2	$568.5	2.9%
2014	$491.3	$556.6	2.8%
2015	$475.0	$527.3	2.6%
2016	$605.1	$662.6	3.1%
2017	$679.3	$730.2	3.5%
2018	$752.3	$789.9	3.6%
2019	$735.8	$754.9	3.4%
2020	$558.3	$558.3	2.5%
Total	$14,927	$28,686	2.3% average

Source: "Planetary Exploration Budget Dataset," compiled by
Casey Dreier for The Planetary Society, accessed February 10, 2021,
https://www.planetary.org/space-policy/planetary-exploration-budget-dataset.

Note: All values are in millions of dollars.

[a] TQ means Transition Quarter and refers to the three-month period (July 1–September 30, 1976) between fiscal year 1976 and fiscal year 1977. At that time, the fiscal year definition shifted from July 1–June 30 to October 1–September 30.

Notes

Abbreviations

AA	*Antiquarian Astronomer*
AJ	*Astronomical Journal*
AN	*Astronomsiche Nachrichten*
ApJ	*Astrophysical Journal*
Bull. Soc. Astron. France	*Bulletin Société Astronomique de France*
GRL	*Geophysical Research Letters*
JBAA	*Journal of the British Astronomical Association*
JGR	*Journal of Geophysical Research*
JHA	*Journal for the History of Astronomy*
Mem. BAA	*Memoirs of the British Astronomical Association*
Mem. S.A. It.	*Memorie della Società Astronomica Italiana*
MNRAS	*Monthly Notices of the Royal Astronomical Society*
Nature Geosci.	*Nature Geoscience*
PASP	*Publications of the Astronomical Society of the Pacific*
Phil. Trans. R. Soc. Lond.	*Philosophical Transactions Royal Society London*
Planet. Space Sci.	*Planetary and Space Science*
Pop. Ast.	*Popular Astronomy*
Proc. R. Soc. Lond.	*Proceedings Royal Society London*
Sov Astron. AJ	*Astronomical Journal of the USSR*
Space Sci. Rev.	*Space Science Reviews*
Transactions IAU	*Transactions of the International Astronomical Union*

Preface

1. C. Sagan, *The Cosmic Connection* (New York: Doubleday, 1973), 69.

Chapter 1

1. I. Semple, "Neanderthals—Not Modern Humans—Were First Artists on Earth, Experts Claim," *Guardian*, February 22, 2018.

2. See R. J. Bliwise, "Seeing Red," *American Scholar*, Spring 2011.

3. J. D. Barrow, *The Artful Universe* (Oxford: Clarendon Press, 1995), 184.

4. It is worth adding that the degree of ruddiness is likely to be more vivid to children and younger individuals, since the lens of the eye has not yet yellowed with age as it inevitably does.

5. B. Novakovic, "Senemut: An Ancient Egyptian Astronomer." *Publications of the Astronomical Observatory of Belgrade* 85 (2008): 19–23.

6. C. Flammarion, *Astronomy for the Amateur* (originally *Astronomy for Women*), trans. F. A. Welby (London: T. Nelson, 1904), 137.

7. O. Neugebauer, *The Exact Sciences in Antiquity* (1957; repr., New York: Dover, 1969), 81.

8. N. M. Swerdlow, *The Babylonian Theory of the Planets* (Princeton, N.J.: Princeton University Press, 1998), 3.

9. H. J. Nissen and P. Heine, *From Mesopotamia to Iraq: A Concise History* (Chicago: University of Chicago Press, 2009), 22.

10. The name means "when Anu and Enlil . . . ," from the opening words of the first tablet. Anu and Enlil were Sumerian gods.

11. Swerdlow, *The Babylonian Theory of the Planets*, 6.

12. S. Parpola, *Letters from Assyrian and Babylonian Scholars* (Helsinki: Helsinki University Press, 1993), Letter 381.

13. Swerdlow, *The Babylonian Theory of the Planets*, 3.

14. J.-P. Changeux and P. Ricoeur, *What Makes Us Think?*, trans. M. B. DeBevoise (Princeton, N.J.: Princeton University Press, 2000), 41.

15. We would need to note the overthrow of the Old Babylonian Empire by the Hittites from Asia Minor; the overthrow of the Hittites by the Kassites from the Zagros Mountains in what is now Iran; the arrival of the Akkadians of South Mesopotamia, at a time when Babylonia was weakly ruled and subject to Assyrian domination; and—after 911 BCE—the assertion of direct Assyrian control over Babylonia, which continued for three centuries; after which came more upheavals, revolts, the deportation or enslavement of whole populations (including the Jews), the destruction of cities, the rebuilding of cities, the collapse of empires, and the establishment of new empires. The New Babylonian Empire, established after the destruction of the Assyrian Empire in 612 BCE, was in turn overthrown by the Persian ruler Cyrus the Great in 539 BCE, who set the Jews

free and made a revitalized Babylon his administrative capital. Under Cyrus, the old arts of recording and interpreting omens continued to flourish; the omens from 652 BCE are found in the "Astronomical Diaries," which continued to be kept up for several more centuries (until 61 BCE), covering the entire period in which the Babylonians would achieve their heights in mathematics and astronomy and in turn pass their legacy on to the Hellenistic Greeks of the next great empire, that of Alexander the Great, who died at—where else?—Babylon in 323 BCE.

16. Parpola, *Letters from Assyrian and Babylonian Scholars*, Letter 8.

17. Parpola, Letter 104.

18. A. Pannekoek, *A History of Astronomy* (1961; repr., New York: Dover, 1989), 48.

19. Swerdlow, *The Babylonian Theory of the Planets*, 28–29.

20. Swerdlow, 56.

21. The Seleucid Empire was founded in 312 BCE by Seleucus I Nicator. On Alexander the Great's death in Babylon in 323 BCE, his vastly expanded Macedonian Empire was divided among his three generals, with Seleucus receiving Babylonia. The Seleucid Empire, which lasted until 63 BCE, was a major center of Hellenistic culture that maintained the preeminence of Greek customs and was dominated by a Greek political elite, and so during this period Babylonian astronomical data and Greek geometry were brought into close—and fruitful—contact.

   At some point earlier than this, probably the early fifth century BCE, the Babylonian astronomers had already discovered the so-called saros from observations of lunar eclipses. See Clemency Montelle, *Chasing Shadows: Mathematics, Astronomy, and the Early History of Eclipse Reckoning* (Baltimore, Md.: Johns Hopkins University Press, 2011).

22. An even more exact period for Mars's oppositions is 284 years + 2/5 days. Thus, the August 28, 2003, opposition was nearly identical to that of August 27, 1719.

23. Swerdlow, *The Babylonian Theory of the Planets*, 27.

24. M. Tsikritsis, X. Moussas, and D. Tsikritsis, "Astronomical and Mathematical Knowledge and Calendars During the Early Helladic Era in Aegean 'Frying Pan' Vessels," *Mediterranean Archaeology and Archaeometry* 15, no. 1 (2015): 135–149.

25. W. M. Ivins Jr., *Art and Geometry: A Study in Space Intuitions* (1946; repr., New York: Dover, 1964), 42.

26. M. Livio, *Is God a Mathematician?* (New York: Simon & Schuster, 2009), 12.

27. A. Einstein, foreword to Galileo, *Dialogue Concerning the Two Chief World Systems*, trans. S. Drake, 2nd rev. ed. (Berkeley: University of California Press, 1967), xv.

28. As a young man, Eudoxus immigrated to Athens. He is said to have been so impoverished that he could only afford accommodation in the Port of Piraeus, and had to walk the 11 km distance each way to Plato's lectures. After a few months, he left because of a disagreement and traveled—with the help of funds

raised by his friends—to Heliopolis, Egypt, to pursue his study of astronomy and mathematics. After another sixteen months, he traveled to Cyzicus, on the south shore of Propontis (now known as the Sea of Marmara), and later visited the court of Mausolus. In time he acquired many students of his own, and he and they returned to Athens and, finally, to his native Cnidus, where he built his observatory, whose instruments presumably included a gnomon, a klepsydra, and perhaps even wooden sighting tubes.

29. The first person of modern times to work out Eudoxus's system in detail from the accounts of it available in ancient documents was G. V. Schiaparelli. See G. V. Schiaparelli, "Le sfere omocentriche di Eudosso, di Callippo e di Aristotele," *Pubblicazioni del R. Osservatorio di Brera in Milano*, no. 9 (Milano, 1875).

30. E. A. Burtt, *The Metaphysical Foundations of Modern Physical Science: A Historical and Critical Essay* (London: Kegan Paul, Trench, Trubner, 1925), 24.

31. M. T. Wright, "Epicyclic Gearing and the Antikythera Mechanism," 2 parts, *Antiquarian Horology* 27 (2003): 270–279; 29 (2005): 52–63.

32. A. Van Helden, *Measuring the Universe: Cosmic Dimensions from Aristarchus to Halley* (Chicago: University of Chicago Press, 1985), 10.

33. C. Ptolemaeus, *Ptolemy's Almagest*, trans. G. J. Toomer (Princeton, N.J.: Princeton University Press, 1998), 36–37.

34. This line was reported by T. Carlyle, *History of Frederick the Second, Called Frederick the Great* (New York: Lovell, Coryell, 1887), book 2, chapter 7, Project Gutenberg.

35. O. Gingerich, *The Eye of Heaven* (New York: American Institute of Physics, 1993), 55.

36. A taste of this curious work can be had from the following extract from *Tetrabiblos*, trans. J. M. Ashland (London: Davis and Dixon, 1822), 114. Ptolemy is describing the "diseases of the mind," and says: "Epilepsy generally attaches to all persons born when Mercury and the Moon may be unconnected either with each other, or with the oriental horizon, while Saturn and Mars may be in angles and superintend the scheme; that is to say, provided Saturn be so posited by day, and Mars by night: otherwise, when the converse may happen in these schemes, viz. when Saturn may have dominion by night, but Mars by day (especially if in Cancer, Virgo or Pisces), the persons born will become insane. And they will become demoniac, and afflicted with moisture, in the brain, if the Moon, being in face to the Sun, should be governed by Saturn when operating her conjunction, but by Mars when effecting her opposition; and particularly when it may happen in Sagittarius and in Pisces." Clearly Ptolemy's ideas about epilepsy were less advanced than those of Hippocrates five centuries earlier.

37. E. Gibbon, *The History of the Decline and Fall of the Roman Empire* (New York: Harper, 1836), chapter 1, Project Gutenberg.

38. Pannekoek, *A History of Astronomy*, 161.

39. W. Durant and A. Durant, *The Story of Civilization*, vol. 3, *Caesar and Christ* (New York: Simon and Schuster, 1944), 390.

40. Plutarch, "On Listening to Lectures," in *Moralia*, vol. 1, trans. F. C. Babbitt, Loeb Classical Library 197 (Cambridge, Mass.: Harvard University Press), 205.

41. Plutarch, "On Tranquillity of Mind," in *Moralia*, vol 6, trans. W. C. Helmbold, Loeb Classical Library 337, (Cambridge, Mass.: Harvard University Press), 241.

42. J. Milton, *Paradise Lost*, book 8, lines 83–84.

43. Quoted in Gingerich, *The Eye of Heaven*, 4.

Chapter 2

1. See T. Carlyle, *History of Frederick the Second, Called Frederick the Great* (New York: Lovell, Coryell, 1887), book 2, chapter 7, Project Gutenberg.

2. O. Gingerich, *The Eye of Heaven* (New York: American Institute of Physics, 1993), 197.

3. B. Russell, *History of Western Philosophy*, 2nd ed. (London: George Allen and Unwin, 1961), 483.

4. D. J. Boorstin, *The Discoverers: A History of Man's Search to Know His World and Himself* (New York: Random House, 1983), 299.

5. "The Timid Canon" is the title of part 3 of Arthur Koestler's *The Sleepwalkers* (New York: Grosset and Dunlap, 1959).

6. In the cathedral records for April 1513 appears the note: "Doctor Nicolaus paid into the treasury of the chapter for 800 bricks and a barrel of chlorinated lime from the Cathedral work-yard." Dava Sobel to William Sheehan, pers. comm., March 3, 2009.

7. In *De Revolutionibus*, he says he kept his work secret not only for nine years (alluding to the maxim of Horace) but for four times nine. The statement cannot refer, of course, to *De Revolutionibus*, which must have been completed no later than 1531, since he uses observations made in 1529 but not 1532. Since *De Revolutionibus* was published in 1543, four times nine years would take us back to 1506, the end of his Italy period.

8. G. J. Rheticus, *Narratio Prima*, in E. Rosen, *Three Copernican Treatises* (1939; repr., New York: Dover, 1959), 136–137.

9. N. Copernicus, *Revolutions of the Heavenly Spheres*, book 1, chapter 10. The French historian Alexandre Koyré points out that "it is not always, or perhaps not sufficiently, appreciated that by placing the Sun at the center of the Universe in virtue of its dignity, Copernicus . . . completely overthrew the hierarchy of positions in the ancient and medieval Cosmos, in which the central position was not the most honorable, but, on the contrary, the most unworthy. It was, in effect, the lowest, and consequently appropriate to Earth's imperfection. Perfection was located above in the celestial vault, above which were 'the heavens' (Paradise),

whilst Hell was deservedly placed beneath the surface of the earth." See A. Koyré, *The Astronomical Revolution: Copernicus-Kepler-Borelli*, trans. R. W. Maddison (1961; repr., New York: Dover, 1992), 115.

10. Quoted in W. M. Ivins Jr., *Art and Geometry: A Study in Space Intuitions* (1946; repr., New York: Dover, 1964), 70.

11. N. Copernicus, *Commentariolus* (Little Commentary), in Rosen, *Three Copernican Treatises*, 77. The date of 1514 comes from the fact that a copy—of which only a few were circulated in Copernicus's lifetime—is listed in the catalog of one Mathias de Miechow of Cracow in that year. It is likely, however, that the manuscript had been written at least a year or two before.

12. Copernicus, 57. The strict adherence to uniform circular motions was merely a preference. Though Copernicus regarded it as one of the essential advantages of his system, it was not, as Koyré points out, "bound up with heliocentrism, as Copernicus believed—if indeed he did believe it. In fact, as was brilliantly shown by Kepler, it is always possible to replace the 'equant circle' by an epicycle; similarly, it is always permissible to substitute a concentric circle with an extra epicycle for an eccentric circle, as Copernicus shows on several occasions, following the example of Ptolemy." See Koyré, *Astronomical Revolution*, 49–50.

13. O. Neugebauer, *Exact Sciences in Antiquity* (1957; repr., New York: Dover, 1969), 202.

14. Quoted in J. L. E. Dreyer, *Tycho Brahe: A Picture of Scientific Life and Work in the Sixteenth Century* (Edinburgh: Adam and Charles Black, 1890), 14.

15. V. E. Thoren, *The Lord of Uraniborg: A Biography of Tycho Brahe* (Cambridge: Cambridge University Press, 1990), 16.

16. Dreyer, *Tycho Brahe*, 27.

17. H. Dingle, "Tycho Brahe," in *Astronomy*, ed. S. Rapport and H. Wright (New York: Washington Square Press, 1964), 42.

18. R. A. Rosenfeld, "From Uraniborg to Yerkes: A Fragile Monumentality," *Griffith Observer*, February 2009, 5.

19. Kepler, *New Astronomy*, trans. W. H. Donahue (Cambridge: Cambridge University Press), 232.

20. Kepler, 184.

21. M. Caspar, *Kepler*, trans. C. Doris Hellman (New York: Dover, 1993), 106.

22. "Ne frustra vixisse videar" is the actual phrase he used. See Dreyer, *Tycho Brahe*, 309.

23. W. Shakespeare, *Hamlet*, in *The Riverside Shakespeare*, ed. G. B. Evans et al. (Boston: Houghton Mifflin, 1974), act 1, scene 5, lines 95–104.

24. Kepler, *New Astronomy*, 256.

25. Kepler, 495.

26. Caspar, *Kepler*, 369.

27. Caspar, 371–372.

28. "Witches" made convenient scapegoats. Remarkably, the voluminous proceedings against Katharina Kepler, from her accusation in 1615 to her acquittal in 1621, including the stages of Kepler's intervention on her behalf, have been preserved in the state archives in Stuttgart. They were first unearthed by J. L. C. von Breitschwert, and published in their entirety in 1870 by Ch. Frisch in the eighth volume of his complete edition of Kepler's works. For a recent popular account, see U. Rublack, *The Astronomer and the Witch: Johannes Kepler's Fight for His Mother* (Oxford: Oxford University Press, 2015).

29. Caspar, *Kepler*, gives the Latin: "Mensus eram coelos, nunc terrae metior umbras. / Mens coelestis erat, corporis umbra jacet."

Chapter 3

1. M. Caspar, *Kepler*, trans. C. Doris Hellman (New York: Dover, 1993), 201.

2. Caspar, 195.

3. Longomontanus (C. Severin), *Introductio in Theatrum Astronomicum* (Copenhagen, 1639); quoted in C. J. Cunningham, *Studies of Pallas in the Early Nineteenth Century: Historical Studies in Asteroid Research*, 2nd ed. (Cham, Switzerland: Springer, 2017), 7.

4. Galileo Galilei, *Sidereus Nuncius, or The Sidereal Messenger*, trans. A. Van Helden (Chicago: University of Chicago Press, 1989), 3.

5. Galileo, 3.

6. W. Shakespeare, *Merchant of Venice*, in *The Riverside Shakespeare*, ed. G. B. Evans et al. (Boston: Houghton Mifflin, 1974), act 1, scene 1, line 9.

7. J. Kepler, *Conversation with Galileo's Sidereal Messenger*, trans. E. Rosen (New York: Johnson Reprint, 1965), 14. Later, when Galileo announced another discovery in an anagram (to protect his priority),

smaIsmrmIlmepoetaleumIbunenugttauIras

Kepler, assuming the discovery had to do with Mars, transposed the letters to read,

Salue umbistineum geminatum Martia proles, or
"Hail, twin companionship, children of Mars."

Kepler, however, had misconstrued the message. In fact, Galileo's anagram concerned Saturn, and the correct rearrangement was,

Altissimum planetam tergeminum observavi, or
"I have observed the more distant planet to have a triple form."

The basis of the triple form—Galileo's imperfect view of what was shown by Christiaan Huygens to be the ring system of Saturn—would take another half century to reveal. Meanwhile, Kepler's two satellites would live on in memory, and may have helped inspire the fantasy in Jonathan Swift's *Gulliver's Travels*.

8.   Galileo Galilei, *Le Opere di Galileo Galilei*, Edizione Nationale, ed. Antonio Favaro, vol. 10 (Florence: G. Barbera, 1900), 503.

9.   C. Flammarion, *Camille Flammarion's The Planet Mars*, trans. P. Moore, ed. W. Sheehan (Cham, Switzerland: Springer, 2015), 4–5.

10.   Spherical aberration is due to the fact that on passing through a lens of spherical curvature, rays near the periphery tend to focus at a point closer to the lens than those near the center. Chromatic aberration occurs because, on passing through the lens, the light is refracted into all the colors of the spectrum, and different colors come to focus at different points, which produces prismatic splendors around bright objects such as the Moon or Venus. In either case, it makes it impossible to bring the image to a sharp focus.

11.   C. Flammarion, *Camille Flammarion's The Planet Mars*, 13.

12.   P. Lowell, *Mars* (Boston: Houghton Mifflin, 1895), 21.

13.   H. Kamen, *Empire* (New York: HarperCollins, 2003), 505.

14.   Voltaire, *The Age of Louis XIV and Other Selected Writings*, trans. J. H. Brumfitt (New York: Washington Press, 1963), 133.

15.   Flammarion, *Camille Flammarion's The Planet Mars*, 15.

16.   R. Hooke, "The Particulars of Those Observations of the Planet Mars, Formerly Intimated to Have Been Made at London in the Months of February and March, anni 1666," *Phil. Trans. R. Soc. Lond.* 1, no. 14 (London, 1665–66): 239.

17.   Flammarion, *Camille Flammarion's The Planet Mars*, 18.

18.   Galileo had already said this in 1632. See Galileo Galilei, *Dialogue Concerning the Two World Systems*, trans. Stillman Drake, 2nd ed. (Berkeley: University of California Press, 1957), 63.

19.   B. le Bovier de Fontenelle, *Conversations on the Plurality of Worlds*, trans. H. A. Hargreaves (Berkeley: University of California Press, 1990), 52.

20.   C. Huygens, *The Celestial Worlds Discover'd; or, Conjectures concerning the Inhabitants, Plants and Productions of the Worlds in the Planets* (London: Timothy Childe, 1698), 21.

21.   Quoted in J. Ashbrook, *The Astronomical Scrapbook: Skywatchers, Pioneers and Seekers in Astronomy* (Cambridge, Mass.: Sky, 1984), 127.

22.   Fontenelle, *Conversations on the Plurality of Worlds*, xvii–xviii.

23.   See R. McKim, "Telescopic Martian Dust Storms: A Narrative and Catalogue," *Mem. BAA* 44 (1999): 14–15.

Chapter 4

1.   The best mirror-making material available at the time was a form of bronze known as speculum metal, in which the ratio of copper to tin is about two-to-one. This metal is heavy and, being rather hard and brittle, difficult to figure. It

also suffers from a significant loss of light at each reflective surface, and readily tarnishes, especially in damp weather such as is commonly encountered in the British Isles, so that mirrors made of this material need to be regularly repolished. Herschel found that his mirrors were unusable after three or four months, and so he eventually produced two or three speculums for each of his telescopes, which he deployed relay-style while reworking the tarnished members. Knowing all this can only increase our admiration for what he managed to accomplish— especially since, during the first part of his career, he was kept busy with his musical career as well as astronomy.

2.   Herschel's Mars observations from 1777 and 1779 are published in W. Herschel, "Astronomical Observations on the Rotation of the Planets Round their Axes, Made with a View to Determine Whether the Earth's Diurnal Motion is Perfectly Equable," *Phil. Trans. R. Soc. Lond.* 71 (1781): 115–138; and also in *Scientific Papers of Sir William Herschel*, ed. J. L. E. Dreyer, vol. 1 (London: Royal Society and Royal Astronomical Society, 1912), 16–29. A fuller account is found in W. Herschel, "On the Remarkable Appearances of the Polar Regions of the Planet Mars, the Inclination of Its Axis, the Position of Its Poles, and Its Spheroidical Figure; With a Few Hints Relating to Its Real Diameter and Atmosphere," *Phil Trans. R. Soc. Lond.* 74 (1784): 233–273; and in *Scientific Papers*, 1:131–136. The comment about the analogy between Mars and Earth is in *Scientific Papers*, 1:148; and the comment about the inhabitants enjoying a situation similar to our own is in *Scientific Papers*, 1:156.

3.   Schroeter's astronomical interests must have cost him a king's ransom. Not only did he suffer from a never-satisfied hunger for larger and larger telescopes, he also published his observations in a series of large and handsome editions, all at his own expense. As a writer, he was long winded and expressed himself in a prose that is turgid and almost unreadable. Also, the copperplate engravings of his drawings, by Tischbein of Bremen, are clumsy and inelegant, though Tischbein does not deserve the blame, as the engravings are faithful enough to Schroeter's originals. For all that, Schroeter did succeed in establishing lunar and planetary astronomy on a comprehensive basis, and achieved many significant results.

4.   J. H. Schroeter, *Areographische Beiträge zur genauern Kenntnis und Beurtheilung des Planeten Mars*, ed. H. G. van de Sande Bakhuyzen (Leiden: Leidener Sternwarte, 1881), 1.

5.   Schroeter, 2.

6.   Schroeter's two volumes on the Moon are J. H. Schroeter, *Selenotopographische Fragmente über den Mond*, vol. 1 (Helmstedt, 1791) and vol. 2 (Göttingen, 1802). His principal work on Venus is J. H. Schroeter, *Aphroditographische Fragmente zur genauern Kenntniss des Planeten Venus* (Helmstedt, 1796). The debate with Herschel is documented in a series of papers (in English) in *Phil. Trans. R. Soc. Lond.* 82 (1792) and 83 (1793). A complete bibliography and much detail about

Schroeter and his observatory are found in D. Gerdes, *Die Lilienthaler Stern-warte 1781 bis 1818: Machinae Coelestes Lilienthalienses die Instrumente; eine zeit-geschichtliche Dokumentation* (Lilienthal: Heimatverein Lilienthal, 1991). For a recent appreciation of Schröter, see W. Sheehan and R. Baum, "Observation and Inference: Johann Schroeter, 1745–1816," *JBAA* 105 (1995): 171–175.

7. The 47-centimeter telescope was the largest in Germany at the time, and the result of a collaboration between Schroeter and his gardener, Harm Gefken. In order to improve the reflectivity of the mirror, a thin coating of vaporized arsenic was added to the speculum metal surface. Unfortunately, this was a hazardous practice, and poor Gefken died in 1811, at the age of 55, owing to complications of chronic arsenic poisoning.

8. For a description of the destruction of Lilienthal, see R. Baum, "The Lilienthal Tragedy," *JBAA* 101 (1991): 369–371.

9. Schroeter was in the process of turning over his drawings and text about Mars to his engraver, Tischbein of Bremen, for publication when he died. The work was forgotten for several decades, but finally François Joseph Charles Terby, a tutor in physics at the University of Louvain and keen amateur astronomer, who was collecting all the drawings of Mars he could get his hands on for his monograph, *Aréographie, ou étude comparative des observations faites sur aspect physique de la planète Mars depuis Fontana (1636) jusqu'à nos jours* (Bruxelles: F. Hayez, 1875), managed to track them down through one of Schroeter's neph-ews. Terby published extracts in his *Aréographie*, then deposited the manuscripts in the Leiden University Library, where they were taken up by the director of the Leiden Observatory, Hendricus Gerardus van de Sande Bakhuyzen, who finally edited and published them in 1881.

The curved, hooklike marking was referred to by Bakhuyzen as "Spitze B." Joseph Ashbrook has called its briefly coming into prominence before disap-pearing altogether "the most striking change yet recorded on the surface of the Red Planet." See J. Ashbrook, *The Astronomical Scrapbook: Skywatchers, Pioneers and Seekers in Astronomy* (Cambridge, Mass.: Sky, 1984), 290. The fact that the marking's history is known at all is largely owing to the indefatigable astronomer of Lilienthal and his candid records.

10. H. de Flaugergues, "Les taches de la planète Mars," *Journal de Physique* 69 (Paris, 1809): 126. Flaugergues, the son of the magistrate at Viviers, became interested in astronomy at an early age, and though he never studied formally or attended a university, he developed a correspondence with such famous astronomers as Jérôme Lalande and Franz Xaver, baron von Zach. He was offered the director-ship of the Toulon Observatory in 1797, but refused. He was offered the direc-torship of the Marseille Observatory in 1810, but again refused. When later the French government awarded him the Légion d'honneur for his discovery of the

Great Comet of 1811, he was invited to Paris; he again refused, on the grounds that Paris would "never give him the beautiful sky of Viviers." See *Dictionnaire de la conversation et de la lecture: Répertoire des connaissances usuelles* 61 (1847): 167.

11. G. K. F. Kunowksy, "Einige physiche Beobachtungen des Mondes, des Saturns, und Mars," *Astronomische Jahrbuch für 1825* (Berlin, 1822), 225.

12. On Beer, see J. Blünck, *Wilhelm Beer: Genius der Astronomie und Ökonomie, 1797–1850* (Berlin: Ausstellungskataloge/Staatsbibliothek zu Berlin-PK, 1997). On Mädler, see D. B. Hermann, *Johann Heinrich Mädler, 1794–1874* (Berlin: Akademie Verlag, 1985).

13. Beer and Mädler's lunar studies are discussed in W. P. Sheehan and T. A. Dobbins, *Epic Moon* (Richmond, Va.: Willmann-Bell, 2001), 95–117.

14. C. Flammarion, *Camille Flammarion's The Planet Mars*, trans. P. Moore, ed. W. Sheehan (Cham, Switzerland: Springer, 2015), 89.

15. L. Hatch to W. Sheehan, pers. comm., September 18, 2009.

16. W. Beer and J. H. Mädler, "Physische Beobachtungen des Mars bei seiner Opposition im September 1830," *AN* 191 (1831): 448.

17. Beer and Mädler, 450.

18. Noting that it was two minutes less than the period William Herschel had published, they accounted for the discrepancy after reviewing Herschel's records from 1777 and 1779, finding that Mars had completed one more rotation than he had realized. As soon as this was factored in, the agreement with their own results was excellent.

19. Flammarion, *Camille Flammarion's The Planet Mars*, 99.

20. Flammarion, 102.

21. J. F. W. Herschel, *Outlines of Astronomy* (Philadelphia: Richard and Lea, 1849), 300–301.

22. See M. J. S. Rudwick, *The Great Devonian Controversy: The Shaping of Scientific Knowledge Among Gentlemanly Specialists* (Chicago: University of Chicago Press, 1985).

23. Goethe regarded his *Theory of Colours* as his greatest work, and went so far as to say to Johann Eckermann, in 1824, "As to what I have done as a poet, . . . I take no pride in it. . . . But that in my century I am the only person who knows the truth in the difficult science of colors—of that, I say, I am not a little proud, and here I have a consciousness of a superiority to many." He was vehemently opposed to the analytic treatment of color given by Isaac Newton in the *Optics*, and instead characterized color as arising from the dynamic interplay of light and darkness through the mediation of a turbid medium. Although his theory enjoyed little favor from scientists, Goethe's ideas did influence artists, notably the English painter J. M. W. Turner. See J. W. Goethe, *J. W. Goethe's Conversations with Eckermann (1823–1832)*, trans. John Oxenford (San Francisco: North Point Press, 1984), 28.

24. Perhaps the best biographical account of Mitchel is still that written by his son: F. A. Mitchel, *Ormsby Macknight Mitchel: Astronomer and General* (Cambridge, Mass: Riverside Press, 1887).

25. Mitchel, 52.

26. O. M. Mitchel, "Observations of Mars," *Sidereal Messenger* 2 (1846): 101.

27. Images from the Mars Global Surveyor show this region to be somewhat elevated, with rough, heavily cratered southern highlands. There are no mountains, and other areas nearby at a similar elevation do not retain frost well into the southern spring. However, the Mountains of Mitchel do feature a prominent, south-facing scarp that, being somewhat protected from sunlight, tends to retain frost. Though the regression of the Mountains of Mitchel is affected by dust, the regression was noted—again in Mars Global Surveyor images—to be more rapid in 2001, when a large dust storm occurred, compared to 1999, when the Martian atmosphere was relatively dust free. See B. P. Bonev, P. B. James, J. E. Bjorkman, and M. J. Wolff, "Regression of the Mountains of Mitchel Polar Ice after the Onset of a Global Dust Storm on Mars," *GRL* 29 (2002): article 13.

28. D. Le Conte, "Warren De la Rue—Pioneer Astronomical Photographer," *AA* 5 (February 2011): 14.

29. The church was originally built in the 17th century, but funds to complete the dome were never found. To cover up this fault, the painter Andrea Pozzo (1642–1709) was hired to create the bold tromp l'oeil paintings for which the church has long been celebrated.

30. P. A. Secchi, *Descrizione del Nuovo osservatorio del collegio romano* (Rome: Tipografia delle Belle Arti, 1856), 158.

31. M. J. Crowe, *The Extraterrestial Life Debate, 1750–1910* (Cambridge: Cambridge University Press, 1986), 423.

32. P. A. Secchi, "Extract of a Letter to George Rennie, Esq., F.R.S., Containing Explanatory Remarks on a Drawing of the Lunar Spot 'Copernicus,' Presented by Him to the Royal Society," March 13, 1856, *Proc. R. Soc. Lond.* 8 (1856–57): 72.

33. J. Phillips, "Notes on the Drawing of 'Copernicus,' Presented to the Royal Society by P. A. Secchi," *Proc. R. Soc. Lond.* 8 (1856–57): 73.

34. Quoted material from Flammarion, *Camille Flammarion's The Planet Mars*, 118. It may not be wide of the mark to mention that this was a great era of canal-building enterprises on Earth; the Ohio and Erie Canal, connecting Akron, Ohio, with the Cuyahoga River, had been finished in 1832, and was still busy with freight traffic despite the fact that railroads would increasingly take over this role. Meanwhile, Ferdinand de Lesseps's Suez Canal Company was organized at the end of 1858; in ten years it would complete, at great expense and human cost, the 164 kilometer canal connecting the Mediterranean to the Red Sea. So canals were certainly of topical interest at the time Secchi introduced the term to describe some features on Mars.

35. Flammarion, 119.
36. P. A. Secchi, "Osservazioni di Marte, fatte durante l'opposizione del 1858," *Memorie dell'osservatorio del collegio romano* (Rome: Tipografia delle Belle Arti, 1859), quoted in Flammarion, 117.
37. Flammarion, *Camille Flammarion's The Planet Mars*, 118.
38. On June 18, Secchi first saw a shadowy vertical mark over the region now known as Cyclopia. On June 20, a dark marking that had been clearly visible on June 18 now appeared to have been covered by a light cloud, while the dark following tip of Mare Cimmerium that had been visible on the 18th was apparently obscured. On the 24th, the Cyclopia extension was recognizable, but Mare Cimmerium was extremely obscured. According to Richard McKim, this is the first clear evidence of Martian dust activity since 1847, when Heinrich Schwabe, at Dessau, recorded an apparent regional obscuration in the southern hemisphere. See R. McKim, "Telescopic Martian Dust Storms: A Narrative and Catalogue," *Mem. BAA* 44 (1999): 19–20.
39. P. A. Secchi, quoted in Flammarion, *Camille Flammarion's The Planet Mars*, 117.

Chapter 5

1. F. A. Mitchel, *Ormsby Macknight Mitchel: Astronomer and General* (Cambridge, Mass: Riverside Press, 1887), 204. Mitchel would be in his grave by then. He had been trained at West Point, and had taught at the military academy for several years before resigning his position on his marriage and move to Cincinnati. In 1861, with the start of the Civil War, he reentered active military service as a brigadier general in charge of volunteers from Ohio. His wife suffered a final, fatal stroke the next day. Mitchel would serve in several campaigns in the war, and he achieved some notoriety for his role in the "great locomotive chase" in Alabama, which involved taking over a train to cut telegraph wires and destroy bridges in order to disrupt the Confederate Army's supply lines. He was honored with the rank of major general of volunteers, and placed in command of the Union's Department of the South at Hilton Head, South Carolina. There, he contracted yellow fever, from which he died on October 30, 1862.
2. E. Lias, *L'Espace cèleste et la Nature Tropicale* (Paris: Garnier Frères, Libraires-Éditeurs, 1865).
3. P. A. Secchi, *Osservazioni del pianeta Marte, Memorie dell'Osservatorio del collegio Romano*, n.s., vol. 2 (Rome: Tipografia delle Belle Arti, 1863), quoted in C. Flammarion, *Camille Flammarion's The Planet Mars*, trans. P. Moore, ed. W. Sheehan (Cham, Switzerland: Springer, 2015), 125, 127.
4. M. Faccini to W. Sheehan, pers. comm., October 26, 2018. Secchi managed to accomplish an extraordinary amount of valuable work, especially given that he had a shoestring budget (compared, for instance, to his British and French rivals), and against a backdrop of political and social unrest. These years saw the

founding of the Kingdom of Italy, which was strongly anticlerical. In 1870, when the occupation troops of Savoy arrived in Rome, the Collegio Romano was first used as a barracks for sharpshooters, then closed, with the symbol of the Society of Jesus chiseled from the doors. Three years later, with the extension of Savoyan rule to Rome itself (except the Vatican), the libraries, astronomical observatory, scientific laboratory, and Kircher Museum were appropriated by the new government. The Jesuits were finally removed. However, such was Secchi's great prestige that—though he shared his friend Pius IX's exile in the Vatican—he was allowed to continue his work undisturbed, and even had access to the observatory until his death in 1878. Only then was his telescope taken over by the government, which set it up at the Rome Observatory atop Monte Mario. See I. Chinnici, *Decoding the Stars: A Biography of Angelo Secchi, Jesuit and Scientist* (Leiden: Brill, 2018). The antipathy toward the papacy (and the Jesuits) was shared by the hero of Italian unification, Giuseppe Garibaldi, who at the 1867 Congress of Peace in Geneva is said to have referred to "that pestilential institution which is called the Papacy," and proposed giving the "final blow to the monster." J. Ridley, *Garibaldi* (New York: Viking Press, 1976), 576–577. On the political situation in Italy during Secchi's time, see D. I. Kertzer, *Prisoner of the Vatican* (Boston: Houghton Mifflin Harcourt, 1996); and D. M. Smith, *Italy and Its Monarchy* (New Haven, Conn.: Yale University Press, 1992).

5.   A. L. Cortie, "Sir Norman Lockyer, 1836–1920," *ApJ* 53 (1921): 234.

6.   Cortie, 241.

7.   L. Huxley, *The Life and Letters of Thomas Henry Huxley*, vol. 1 (London: Macmillan, 1900), 202.

8.   The authoritative biography of Phillips, though containing only a few pages on his astronomical interests and activities, is J. Morrell, *John Phillips and the Business of Victorian Science* (Aldershot: Ashgate, 2005). On Phillips's astronomy, see: R. Hutchins, "John Phillips, Geologist-Astronomer, and the Origins of the Oxford University Observatory, 1853–1875," *History of Universities* 13 (1994): 193–249.

9.   Phillips's lunar studies are described in W. P. Sheehan and T. A. Dobbins, *Epic Moon* (Richmond, Va.: Willmann-Bell, 2001), chapter 9.

10.   J. Phillips, "Ashmolean Society," report of meeting, March 2, 1863, Ashmolean Natural History Society, Department of Western Manuscripts, Bodleian Library, University of Oxford.

11.   The next globe after Phillips's was constructed by the instrument maker John Browning and presented to the Royal Astronomical Society in 1868. It adopted Proctor's nomenclature and was described by Proctor in a 15-page pamphlet, "Remarks on Browning's Stereograms of Mars," commissioned from Proctor as a work for hire and privately published by Browning in 1869. Proctor noted in a footnote that "Mr. Browning's globe was not actually the first ever constructed

to represent the ruddy planet, Professor Phillips of Oxford having, some years before, exhibited a Martial globe. But so many more details were shewn in Mr. Browning's globe, that it may be looked upon as the first really satisfactory attempt to represent the features in this manner." As far as author W. S. has been able to determine, the oldest Mars globe still extant is that constructed in 1882 by Camille Flammarion, which uses his system of nomenclature. Another, by L. Niesten, which follows the nomenclature of the English observer Nathaniel Green, appeared in 1892. A series of Mars globes constructed by Percival Lowell, which utilize Schiaparelli's nomenclature as modified and augmented by Lowell, commence with that constructed after the opposition of 1894, and are preserved at Lowell Observatory.

12. J. Phillips, "On the Telescopic Appearance of the Planet Mars," *Proc. R. Soc. Lond.* 12 (1863): 434.

13. J. Herschel, *Treatise on Astronomy* (1833; repr., Cambridge: Cambridge University Press, 2009), 279.

14. Phillips, "On the Telescopic Appearance of the Planet Mars," 435–436.

15. R. Hutchins, "John Phillips's Astronomy 1852–67, a Pioneering Contribution to Comparative Planetology," *AA*, no. 6 (January 2012): 44–58.

16. Phillips, "On the Telescopic Appearance of the Planet Mars," 435.

17. Hutchins, "John Phillips's Astronomy 1852–67," 53.

18. J. Phillips, "The Planet Mars," *Quarterly Journal of Science* 2 (London, 1865): 369–381.

19. Phillips, 377.

20. See R. McKim and R. A. Mariott, "Dawes' Observations of Mars, 1864–65," *JBAA* 98 (1988): 294–300.

21. R. A. Proctor, *Half-Hours with the Telescope* (New York: G. P. Putnam's Sons, 1873), 83.

22. K. M. D. Lane, *Geographies of Mars: Seeing and Knowing the Red Planet* (Chicago: University of Chicago Press, 2011), 26–27.

23. G. V. Schiaparelli to F. Terby, May 11, 1886. Most of Schiaparelli's correspondence about Mars has been published in the two-volume *Corrispondenza su Marte*; see vol. 1, *1877–1889* (Pisa: Domus Galilaeana, 1963); and vol. 2, *1890–1900* (Pisa: Domus Galilaeana, 1976).

Chapter 6

1. C. Flammarion, *Camille Flammarion's The Planet Mars*, trans. P. Moore, ed. W. Sheehan (Cham, Switzerland: Springer, 2015), 216.

2. A. Hall, *An Astronomer's Wife: The Biography of Angeline Hall* (Baltimore, Md.: Nunn, 1908), 38.

3. Hall, 38.

4. Her abolitionist views were already strong at age 14, when she wrote the following for a class essay: "Slavery or holding men in bondage is one of the most unjust practices. But unjust as it is even in this boasted land of liberty many of our greatest men are dealers in buying and selling slaves. Were you to go to the Southern states you would see about every dwelling surrounded by plantations on which you would see the half clothed and half starved slave and his master with whip in hand ready to inflict the blow should the innocent child forgetful of the smart produced by the whip pause one moment to hear the musick of the birds inhale the odor of the flowers or through fatigue should let go his hold from the hoe. . . . I hope these cruelties will soon cease as many are now advocating the cause of the slave. But still there are many that forget that freedom is as dear to the slave as to the master, whose fathers when oppressed armed in defence of liberty and with Washington at their head gained it. But to their shame they still hold slaves. But some countries have renounced slavery and I hope their example will be followed by [our] own." Hall, 27.

5. Hall, 42.

6. A. Hall, "My Connection with the Harvard Observatory and the Bonds—1857–1862," in *Memoirs of William Cranch Bond and of His Son George Phillips Bond*, by E. S. Holden (San Francisco: Holden-Day, 1897), 77–78.

7. Hall, *An Astronomer's Wife*, 72.

8. Hall, 73.

9. Asaph Hall to Angeline Hall, September 6, 1862, in Hall, 87.

10. Hall, 92–93.

11. Hall, 93.

12. For the history of the U.S. Naval Observatory, see S. J. Dick, *Sky and Ocean Joined: The US Naval Observatory 1830–2000* (Cambridge: Cambridge University Press, 2003).

13. Asaph Hall to Seth Carlo Chandler Jr., March 7, 1904; quoted in O. Gingerich, "The Satellites of Mars: Prediction and Discovery," *JHA* 1 (1970): 113.

14. Dick, *Sky and Ocean Joined*, 224. The mass of a planet, only with difficulty worked out from its mutual gravitational effects on the other planets, can be readily deduced using Newton's laws from the orbital period of a satellite. At this time Mars fell into the moonless category, and estimates of its mass were highly uncertain, ranging from 0.09 to 0.13 times that of Earth.

15. A. Hall, "Discovery of the Satellite of Mars," manuscript in U.S. Naval Observatory Archives, read before Washington Philosophical Society, February 16, 1878, quoted in Dick, 224. Even Alfred, Lord Tennyson, in the first version of "The Palace of Art" (1832) had referred to "the snows of moonless Mars," though the line was dropped from the revised version (1842).

16. A. Hall, "The Discovery of the Satellites of Mars," *MNRAS* 38 (1877–78): 205. Hall was keenly aware that others might be on the lookout for satellites. The most advantageously placed observers—given Mars's far southern declination of 12° south of the celestial equator—were those using the Great Melbourne reflector in Australia, which boasted a 120-centimeter speculum mirror. Though the report of the Melbourne Observatory for 1877 shows that it was used in a search for Martian satellites, the observers did not see them. Hall was also aware that a 69-centimeter refractor was on order by the Vienna Observatory, but it was not delivered until 1878.

17. Asaph Hall to E. C. Pickering, February 7, 1888, Director's Papers, Harvard College Observatory Archives.

18. Dick, *Sky and Ocean Joined*, 225.

19. Hall, *An Astronomer's Wife*, 112.

20. Dick, *Sky and Ocean Joined*, 227–228.

21. Asaph Hall to Arthur Searle, October 9, 1877, Harvard College Observatory Archives.

22. Asaph Hall to E. C. Pickering, October 30, 1877, Director's Papers, Harvard College Observatory Archives.

23. R. A. Proctor, "Note from Mr. Proctor," *Sidereal Messenger* 6 (1887): 260.

24. S. Newcomb, *Reminiscences of an Astronomer* (Boston: Houghton, Mifflin, 1903), 141–142.

25. Asaph Hall to Seth Carlo Chandler Jr., March 7, 1904, quoted in Gingerich, "The Satellites of Mars," 114.

26. B. E. Schafer, "Sherlock Holmes and Some Astronomical Connections," *JBAA* 103, no. 1 (1993): 30–34.

27. D. P. Todd, "Professor Todd's Own Story of the Mars Expedition," *Cosmopolitan Magazine*, March 11, 1908, 343.

28. Henry was the brother of Falconer Madan (1852–1935), the librarian of the Bodleian Library of the University of Oxford, whose granddaughter Venetia Burney would later suggest the name Pluto for the planet discovered by Clyde Tombaugh in 1930. In his *Observations and Orbits of the Satellites of Mars*, Hall credits Henry Madan for the suggested names and cites the Iliad translation by William Cullen Bryant, "He spake, and summoned Fear and Flight to yoke / His steed and put his glorious armor on." The same lines, translated by astronomer John Herschel into English hexameters, read: "To Fear and Flight his command he issued, his chariot / Quick to prepare, while himself in his radiant arms he invested." See A. Hall, *Observations and Orbits of the Satellites of Mars: With Data for Ephemerides in 1879* (Washington, D.C.: Government Printing Office, 1878), 6; J. F. W. Herschel, *The Iliad of Homer: Translated into English Accentuated Hexameters* (London: Macmillan, 1866), 318–319.

29. It was not only the smaller aperture of d'Arrest's telescope that was at fault here, since after he discovered them, Hall found the satellites visible in the U.S. Naval Observatory's 24-centimeter refractor, which was slightly smaller than the 28-centimeter refractor d'Arrest had used in 1862. Moreover, as Hall himself calculated, for 42 days, from September 4 to October 16, the satellites had been brighter in 1862 than in 1877, and Mars was also 15° higher in the northern hemisphere sky. He noted that they should have been easily within reach of the Harvard 38-centimeter refractor, but they went unnoticed at the time. Hall's conclusion about why they had been missed previously was that "astronomers did not search for them in the right place and in the right manner. Probably everyone who searched looked too far away from the planet, and the fact appears to be that in recent times very few searched at all." Dick, *Sky and Ocean Joined*, 230.

30. See J. Westfall and W. Sheehan, *Celestial Shadows: Eclipses, Transits, and Occultations* (New York: Springer-Verlag, 2015), 234–236.

31. Dick, *Sky and Ocean Joined*, 229.

32. Hall's discovery of the Martian satellites was recognized with many honors, including the Lalande Prize from the French Academy of Sciences (1877), the Gold Medal of the Royal Astronomical Society (1879), and the Arago Medal of the French Academy of Sciences (1893). He was also named a chevalier in the French Legion of Honor (1896).

33. Hall, *An Astronomer's Wife*, 51.

Chapter 7

1. For a sketch of Trouvelot's life and work, see R. Rosenfeld and W. Sheehan, "How an Artist Brought the Heavens to Earth," *Astronomy*, January 2011, 52–57.

2. The standard scheme for classifying Martian dust storms is that of L. J. Martin and R. W. Zurek, "An Analysis of the History of Dust Activity on Mars," *JGR* 98 (1993): 3221–3246.

3. The account here follows R. McKim, "Nathaniel Everett Green: Artist and Astronomer," *JBAA* 114 (2004): 13–23.

4. The technology for coating glass telescope mirrors with silver was introduced in 1857 by Karl August von Steinheil and Léon Foucault, and silver-on-glass mirrors quickly became state of the art. The main advantage over the old speculum metal mirrors was that such mirrors reflected 95 percent of the incident light, instead of only 65 percent as in the case of the speculum metal mirrors. They were also, however, lighter in weight, and less prone to tarnish.

5. A. Ferrari, "Between Two Halley's Comet Visits," *Mem. S. A. It.* 82 (2011): 232.

6. Ferrari, 232. It is significant that Schiaparelli mentions each of his parents in these early reminiscences. Though not particularly well off financially, they were always strongly committed to the education of their children, several of whom

were possessed of great talent and headed on to careers of distinction. In addition to Giovanni, a brother, Celestino, became an Arabic scholar, who collected books and manuscripts in Persian, Sanskrit, and Arabic, and would regularly exchange coins from his own wonderful collection with Umberto I, the second king of Italy, a keen numismatist. A sister of Giovanni and Celestino was in charge of all the convents in Italy; a cousin, Ernesto, was a famous Egyptologist, who helped make the discovery of Queen Nefertari's tomb in the Valley of the Queens.

7. G. V. Schiaparelli to Onorato Roux, April 29, 1907, Library of the Astronomical Observatory of Brera.

8. G. V. Schiaparelli, diary entry, December 29, 1855, Library of the Astronomical Observatory of Brera.

9. P. Tucci, "The Diary of Schiaparelli in Berlin (26 October 1857–10 May 1859): A Guide for His Future Scientific Activity," *Mem. S. A. It.* 82 (2011): 240–247.

10. A brother, Eugenio Schiaparelli, participated on the side of a combined French and Piedmontese force, which defeated the Austrians at the Battles of San Martino and Solferino in June 1859. The casualty rate at Solferino was appalling: the numbers involved were greater than those in the Crimean War, and sympathy for the great number of wounded and dying who littered the battlefields, and for whom no provisions seemed to have been made, inspired the philanthropist Henri Dunant to found the Red Cross.

11. G. V. Schiaparelli, *Intorno al corso ed all'origine probabile delle Stelle Meteoriche: Lettere di G. V. Schiaparelli al P. A. Secchi* (Rome: Scienze mathematiche e fisiche, 1866). Before this time, ideas about meteors were still vague, and little was known about their nature and origin. Neither was it clear what their connection might be with meteorites, which are the rocky or iron-rich remnants of meteors that have survived passage through Earth's atmosphere. Schiaparelli began his attack on the problem by making his own observations of the directions of arrival and calculated speeds of the Perseid meteors, which peak between August 9 and 14 each year. With the skill he had acquired in calculating comet orbits under Encke, he was able to calculate the orbits of these meteors, deducing that they followed elongated elliptical paths similar to those of comets. He surmised that the Perseids and other meteor showers occur when Earth in its orbit crosses the trail of particles and dust thrown off by a comet vaporized by solar radiation as it approaches perihelion, and he clinched the suggestion when he identified a specific comet—Swift-Tuttle—with the Perseids. He then went on to show the identity of another comet—Tempel-Tuttle—associated with the November Leonids.

12. Schiaparelli's niece, Elsa Schiaparelli, remembered as one of the most prominent figures in haute couture fashion design between the two world wars, had some

curious things to say about her astronomer uncle in her autobiographical *Shocking Life* (New York: E. P Dutton, 1954). For instance, on p. 25:

> My father's brother, Giovanni Schiaparelli, was an amazing person. He was the Director of the Brera Observatory in Milan, and, like my father, was so accustomed to being alone that in contact with new faces he bristled out like a hedgehog—but this did not prevent him from producing a very large family.
>
> He was appallingly absent-minded. After his marriage he took his young bride to Vienna. On the evening of their arrival he exclaimed:
>
> "Excuse me, I must go and see an astronomer who lives in this town. I won't be long."
>
> He rushed off with his mind already full of telescopes and stars while his poor little wife stayed sobbing in the hotel bedroom. She waited for dinner. He did not come back. The hour for supper passed. Midnight struck. And then the whole night went by without any sign of Giovanni. The next morning my uncle, quite unconcerned, returned to the hotel, asked for the key, and went up. There on the floor, a pathetic bundle, was his sweet wife still sobbing.
>
> The astronomer let out a cry of surprise.
>
> "Oh!" he exclaimed, running to her. "I completely forgot I was married!"

13. Ferrari, "Between Two Halley's Comet Visits," 234.

14. G. Andrissi, introduzione to G. V. Schiaparelli, *Corrispondenza su Marte*, vol. 1, *1877–1889* (Pisa: Domus Galilaeana, 1963), xx. It is rather remarkable that Schiaparelli, from the center of Milan, enjoyed such dark skies that he was able to see the Milky Way, zodiacal light, and gegenschein. All that was to change within only a few years. All English translations from this and other works by Schiaparelli by W. Sheehan, unless otherwise noted.

15. G. V. Schiaparelli, *Osservazioni Astronomiche e fisiche sull'asse di rotazione e sulla topograpfia del pianeta Marte. Memoria Reale Accademia dei Lincei, Anno CCLXXV (1877–78)* (Rome: Coi Tipi del Salviucci, 1878); translated into English by W. Sheehan, *Astronomical and Physical Observations of the Axis of Rotation and the Topography of the Planet Mars. First Memoir, 1877–1878* (San Francisco: Association of Lunar and Planetary Observers, 1996), 1.

16. Schiaparelli, *Astronomical and Physical Observations of the Axis of Rotation and the Topography of the Planet Mars*, 1.

17. Ferrari, "Between Two Halley's Comet Visits," 235.

18. Schiaparelli, *Astronomical and Physical Observations of the Axis of Rotation and the Topography of the Planet Mars*, 9.

19. Schiaparelli, 10.

20. G. V. Schiaparelli to N. E. Green, October 27, 1879, cited in a note by Green in "Mars and the Schiaparelli Canals," *Observatory* 3 (1879): 252.

21. Quoted in C. Flammarion, *Camille Flammarion's The Planet Mars*, trans. P. Moore, ed. W. Sheehan (Cham, Switzerland: Springer, 2015), 374.

22. Green was in the school of the great art critic John Ruskin, who, in *The Elements of Drawing* (1904; repr., New York: Dover, 1971), wrote (pp. 14–15): "No pupil in my class [is] ever allowed to draw an outline, in the ordinary sense. It is pointed out to him, from the first, that Nature relieves one mass, or one tint, against another; but outlines none. . . . Everything that you can see in the world around you, presents itself to your eyes only as an arrangement of patches of different colors variously shaded. . . . I believe . . . that [you] never *ought* to be able to draw a straight line. . . . A great draughtsman can, as far as I have observed, draw every line *but* a straight one."

23. T. W. Webb, "Planets of the Season: Mars," *Nature* 21 (January 1, 1880): 212–213.

24. W. Whitman, "Some Diary Notes at Random: Negro Slaves in New York—Canada Nights—Country Days and Nights—Central Park Notes—Plate Glass, St. Louis, Missouri, November, '79," in *Complete Prose Works* (Philadelphia: David McKay, 1892), 414.

25. G. V. Schiaparelli to N. E. Green, October 27, 1879, in N. E. Green, "Mars and the Schiaparelli Canals," 252.

26. G. V. Schiaparelli, *Osservazioni Astronomiche e fisiche sull'asse di Rotazione e sulla Topographia del Pianeta Marte, Memoria Seconda, Reale Accademia dei Lincei Anno CCLXXVIII (1880–81)* (Rome: Coi Tipi del Salviucci, 1881), 4.

27. Schiaparelli, *Osservazioni, 1880–81*, 75. Nix Olympica was the first of a number of whitish patches recorded on the Martian surface. Its name suggests that Schiaparelli recognized kinship between it and the polar cap. Also in 1879, he noted an even smaller white patch, between Syrtis Major and Thoth, which he called Nix Atlantica (Snow Plain of Mount Atlas); it was seen again in 1881 but then not until 1888. Other areas that frequently appear whitish included Elysium, Hellas, and Argyre. Nix Olympica corresponds to the towering shield volcano Olympus Mons, discovered by Mariner 9 in 1971–72 and often capped by orographic clouds. Though the season was northern hemisphere winter when Schiaparelli discovered Nix Olympica, which has not historically been the usual season for the Olympus Mons orographic clouds, probably he just saw an unusual off-season orographic cloud due to some unusual off-season weather.

28. Schiaparelli, *Osservazioni, 1880–81*, 66.

29. N. E. Green, "On Some Changes in the Markings of Mars, since the Opposition of 1877," *MNRAS* 40 (1880): 332.

30. N. E. Green to G. V. Schiaparelli, March 15, 1878, in Schiaparelli, *Corrispondenza su Marte*, 1:14.

31. N. E. Green to G. V. Schiaparelli, March 15, 1878.

32. G. V. Schiaparelli to F. Terby, November 24, 1879, in Schiaparelli, *Corrispondenza su Marte*, 1:28.

33. "Report of the Meeting of the Association Held Dec. 31, 1890," *JBAA* 1 (1890): 112.

34. G. V. Schiaparelli to O. von Struve, February 19, 1879, in Schiaparelli, *Corrispondenza su Marte*, 1:20–22.

35. G. V. Schiaparelli to F. Terby, May 3, 1880, in Schiaparelli, *Corrispondenza su Marte*, 1:148.

36. Flammarion, *Camille Flammarion's The Planet Mars*, 299.

37. Flammarion, 300. Schiaparelli dedicated his third memoir on Mars to Professor Tito Vignoli, director of the Museum of Natural History in Milan, in an elegant poem in which he showed off his knowledge of classical mythology and his fluency in producing Latin verses: "Dedica della Memoria Terza Sopra Marte al Prof. Tito Vignoli," in G. V. Schiaparelli, *Le Opere di G. V. Schiaparelli*, vol. 2 (Milan: Hoepli, 1930), 477–78. A few lines, as translated by W. Sheehan, read as follows:

> Not Cadmus, nor the prince that did destroy
> The windy walls and lofty tow'rs of Troy,
> Nor ev'n Aeneas, who left his burning home
> And wandered far to found a city, Rome,
> Shall kindle in this day the poet's fire . . .
> The truth about the stars Urania brings,
> And from our own Milan the glory springs,
> New features showing on the face of Mars.

38. See W. Sheehan, *Mercury* (London: Reaktion Books, 2018).

39. See R. J. McKim, W. P. Sheehan, and R. Rosenfeld, "Étienne Leopold Trouvelot and the Planet-Encircling Martian Dust Storm of 1877," *JBAA* 119 (2009): 349–350.

40. G. V. Schiaparelli to F. Terby, June 7, 1888, in Schiaparelli, *Corrispondenza*, 1:203. The Latin phrase means, "Much changed from that."

41. G. V. Schiaparelli, materials for a chronicle of the Brera Observatory, 1893, Library of the Astronomical Observatory of Brera.

42. G. V. Schiaparelli to F. Terby, January 27, 1895, in Schiaparelli, *Corrispondenza*, 2:167.

43. The following are Schiaparelli's published Mars memoirs, listed by opposition:

    - 1877/78—G. V. Schiaparelli, "Osservazioni astronomiche e fisiche sull'asse di rotazione e sulla topografia del pianeta Marte," in *Atti della R. Accademia del Lincei, Memoria della classe di scienze fisiche*, Memoria 1, ser. 3, vol. 2, 1877–78.

- 1879/80—Memoria 2, ser. 3, vol. 10, 1880–81.
- 1881/82—Memoria 3, ser. 4, vol. 3, 1885–86.
- 1883/84—Memoria 4, ser. 5, vol. 2, 1895–96.
- 1886—Memoria 5, ser. 5, vol. 2, 1895–96.
- 1888—Memoria 6, ser. 6, vol. 3, 1899.
- 1890—Memoria 7, ser. 5, vol. 8, 1910.

44. G. V. Schiaparelli, "Il Pianeta Marte," *Natura ed Arte*, no. 5 and 6 (February 1 and 15, 1893): 24.

45. Schiaparelli, 24.

46. Schiaparelli, 25.

47. Among Schiaparelli's later writings about Mars (exclusive of his memoirs), "La vita sul pianete Marte," *Natura ed Arte* 4, no. 11 (1895): 921–929, seems to have been written somewhat tongue in cheek, and contains designs for possible Martian irrigation systems to account for the geminations; and "Il pianete Marte," *Natura ed Arte* 19, no. 1 (1909): 39–45, in addition to reviewing his own series of observations, takes into account more recent publications by Flammarion, P. Lowell, and E. S. Morse.

48. Schiaparelli, "La vita sul pianeta Marte," 921–929. The hippogriff is first described by Virgil in Eclogue 8 of the *Eclogues*.

49. G. V. Schiaparelli, "On the Method of Intellectual Work," 1907, Library of the Astronomical Observatory of Brera.

50. Elsa Schiaparelli (in *Shocking Life*, 25) remembered spending time with her uncle there as a girl:

> His discovery of the canals of Mars was hailed as a great event in the world of astronomy. He . . . owned a Napoleonic villa near Milan where I used to spend many happy hours crouching in a corner of the hearth while the national dish of Lombardy, polenta, was being cooked.
>
> We ate our meals at a long family table, and then walked up and down an alley of cypresses which remains very vividly in my mind. A simple soul, with tremendous strength in a small body, always discovering new worlds in the sky, penetrating the mysterious relations between [falling] stars and comets, he would describe Mars to me as if he had just returned from a long visit.

Schiaparelli's country honored him by appointing him to the Senate of the Kingdom of Italy, but he attended only once, to take the oath showing that he was a patriot. Otherwise, his view was, "Senators make laws for men, but I only know a few of the laws of the skies." According to Elsa, "He was very shortsighted and at last could no longer see through his telescope. . . . On his death there was a great silence at the Observatory, a silence that invited meditation. His name was spoken of with the respect men show for Volta, Galileo, . . . The work of this small, great man continued to burn like a pure, bright light" (*Shocking Life*, 26).

Chapter 8

1.   C. Flammarion, *Camille Flammarion's The Planet Mars*, trans. P. Moore, ed. W. Sheehan (Cham, Switzerland: Springer, 2015), 436.

2.   K. M. D. Lane, *Geographies of Mars: Seeing and Knowing the Red Planet* (Chicago: University of Chicago Press, 2011), 39.

3.   R. McKim, "Nathaniel Everett Green: Artist and Astronomer," *JBAA* 114 (2004): 19.

4.   E. S. Holden, "Note on the Mount Hamilton Observations of Mars, June–August 1892," *Astronomy and Astro-Physics* 11 (1892): 667.

5.   Quoted in E. S. Holden, "The Lowell Observatory at Arizona," *PASP* 6 (1894): 165.

6.   Holden, 166.

7.   Quoted in B. Z. Jones and L. G. Boyd, *The Harvard College Observatory: The First Four Directorships* (Cambridge, Mass.: Belknap Press, 1971), 307.

8.   On William's later career after his return from Arequipa, see H. Plotkin, "William H. Pickering in Jamaica: The Founding of Woodlawn and Studies of Mars," *JHA* 24 (1993): 101–122.

9.   About Lowell, much has been written. There is the rather formal and "official" but still useful biography by his brother, A. L. Lowell, *Biography of Percival Lowell* (New York: Macmillan, 1935). Other works include F. Greenslet, *The Lowells and Their Seven Worlds* (Boston: Houghton Mifflin, 1946), which profiles Percival along with other members of the family and contains many interesting details based on Greenslet's access to the Lowell family correspondence; W. G. Hoyt, *Lowell and Mars* (Tucson: University of Arizona Press, 1976), which discusses in detail Lowell's obsession with Mars, based on access to material in the Lowell Observatory Archives; D. Strauss, *Percival Lowell: The Culture of a Boston Brahmin* (Cambridge, Mass.: Harvard University Press, 2001), the best biography overall, with special emphasis on Lowell's personal life and Far East period. As for other members of the Lowell family, the list is enormous, but the present writers found especially useful S. F. Damon, *Amy Lowell: A Chronicle* (Boston: Houghton Mifflin, 1935); C. D. Heymann, *American Aristocracy: The Lives and Times of James Russell Lowell, Amy Lowell, and Robert Lowell* (New York, Dodd, Mead, 1980); and K. R. Jamison, *Robert Lowell: Setting the River on Fire, a Study of Genius, Mania, and Character* (New York: Alfred A. Knopf, 2017).

10.   W. W. Campbell, review of *Mars*, by Percival Lowell, *PASP* 8 (1896): 208.

11.   Heymann, *American Aristocracy*, 34.

12.   Strauss, *Percival Lowell*, 14.

13. Abbott Lawrence, the future president of Harvard and Percival's future biographer, was born on December 13, 1856. Katherine, born 1858, married, first, banker James Alfred Roosevelt of Long Island (first cousin of Theodore, the future president of the United States), and then, after this first husband died in a railroad accident in 1891, T. James Bowlker, a Boston cotton mill owner. Elizabeth, born in 1862, whose twin, Roger, died before his second birthday, Elizabeth married the son of another Boston scion, William Lowell Putnam, a named partner in the law firm of Putnam, Putnam & Bell. May, born in 1870, lived one day. Amy, born in 1874, junior to Percival by almost 20 years and known in the family as the "Postscript," inherited Sevenels (the mansion on Heath Street) after Augustus's death in 1900, and became an avant-garde poet who dressed in men's clothes, wore a pince-nez, smoked cigars, and and openly cohabited with a former actress and divorcée, Ada Dwyer Russell, at a time when such behavior was regarded as highly scandalous.

14. Amy Lowell; quoted in Damon, *Amy Lowell*, 29–30.

15. Greenslet, *The Lowells and Their Seven Worlds*, 348.

16. P. Lowell to Barrett Wendell, January 16, 1877, Houghton Library, Harvard University. The expression of suicidal ideation, though veiled in a quip, may well have been genuine enough: it recalls the experience of his second cousin James Russell Lowell, who at age 20 had held a cocked pistol to his forehead and come close to pulling the trigger.

17. P. Lowell to Elizabeth Lowell, July 19, 1882, Houghton Library, Harvard University. Katherine was then engaged to marry Alfred Roosevelt (Theodore's cousin); she did so in December of that year.

18. C. W. Eliot to E. C. Pickering, November 22, 1894, Director's Correspondence, Harvard University Archives.

19. W. Shakespeare, *Coriolanus*, in *The Riverside Shakespeare*, ed. G. B. Evans et al. (Boston: Houghton Mifflin, 1974), act 3, scene 3, lines 134–135.

20. Early indications of Lowell's interest in Mars and contact with William Pickering going back to 1890 are noted in D. Strauss, "Percival Lowell, W. H. Pickering and the Founding of the Lowell Observatory," *Annals of Science* 51 (1994): 37–54. His contact with E. E. Barnard in San Francisco in December 1892 is inferred from Lowell's inscription, with a forwarding address care of Messrs. Walsh, Hale & Co, Yokohama, Japan, in Barnard's copy of the *American Ephemeris and Nautical Almanac* for 1892. This volume, which was the one he had beside him when he discovered the fifth satellite of Jupiter that August, was saved from among the items intended to be tossed by the U.S. Naval Observatory, by Richard Schmidt and was generously given to one of the authors (W. S.).

21. E. H. Gombrich, *Art and Illusion: A Study in the Psychology of Pictorial Representation*, 2nd rev. ed. (Princeton, N.J.: Princeton University Press, 1961), 82.

22. J. A. Eddy, "Pioneer of Dendochronology," *JHA* 17 (1986): 69.

23. P. Lowell, "The Lowell Observatory and Its Work," *Boston Commonwealth*, May 26, 1894, clipping in scrapbook in Mary Lea Shane Archives of the Lick Observatory.

24. P. Lowell to K. B. Lowell, May 28, 1894, quoted in Lowell, *Biography of Percival Lowell*, 72.

25. P. Lowell, observing logbook, June 1, 1894, Lowell Observatory Archives.

26. P. Lowell, *Mars* (Boston: Houghton Mifflin, 1895), 86–87.

27. This poem was unpublished until 2016, when author W. S. published a transcript of it: "Mars: A Poem by Percival Lowell," *Society for the History of Astronomy Bulletin*, no. 25 (Spring 2016): 24–29.

28. Hoyt, *Lowell and Mars*, 68.

29. P. Lowell to A. E. Douglass, March 25, 1895, Lowell Observatory Archives.

30. Lowell would describe the canals as a "mesh of lines and dots like a lady's veil" several years later, in his article "Our Solar System," *Popular Astronomy* 24 (1916): 424.

31. P. Lowell, *Annals of the Lowell Observatory*, vol. 1 (Boston: Houghton Mifflin, 1898), 83.

32. Lowell, 86.

33. E. E. Barnard, "Mars: His Moons and His Heavens," unpublished manuscript, Barnard Papers, Joseph Heard Library, Vanderbilt University, Nashville, Tenn.

34. E. E. Barnard, observing logbook, Lick Observatory.

35. E. E. Barnard to Simon Newcomb, September 11, 1894, Library of Congress.

36. E. E. Barnard, "Micrometric Measures of the Ball and Ring System of the Planet Saturn, and Measures of the Diameter of his Satellite Titan. . . . With Some Remarks on Large and Small Telescopes," *MNRAS* 56 (1896): 166.

37. R. McKim, "Telescopic Martian Dust Storms: A Narrative and Catalogue," *Mem. BAA* 44 (1999): 28–31.

38. Lane, *Geographies of Mars*, 45.

39. What constitutes "seeing," in the astronomical sense, is the way this never-ceasing movement of the air overhead gives rise to motion and distortion of images. The motion and distortion is produced by very small differences in the refractive index of the air due to small temperature fluctuations from point to point. Local variations in temperature of a few hundredths of a degree are produced throughout the atmosphere by wind, but larger inequalities occur near the ground, where air temperatures of a few tenths of a degree or even more exist near the surface. As a wave front from a star or planet passes through the turbulent layers of the atmosphere, the turbulence—which varies with the size of the turbulent elements (eddies)—causes the light ray to be bent and distorted. With a small telescope aperture, only a small piece of the wave front passes in front of the telescope; this small portion appears tilted but more or less planar, so the image remains sharp but bodily shifted in position as the wind moves the eddies past the aperture.

In a larger aperture, a larger sample of the wave front is included. The image tends not to move as much in bodily position, but the inclusion of larger and stronger eddies over the aperture produces a blurring of the image as if it were being viewed through a bad piece of window glass. See A. T. Young, "Seeing and Scintillation," *Sky & Telescope*, September 1971, 139–141 and 150.

40. "Revelation peeps" is from Lowell's observing logbook, June 9, 1894, Lowell Observatory Archives. For "tachistoscope flash," see W. Sheehan, *Planets and Perception: Telescopic Views and Interpretations* (Tucson: University of Arizona Press, 1988), 266.

41. G. C. Comstock, *A Text-Book of Astronomy* (New York: D. Appleton, 1901), 244.

42. Comstock, 244.

43. E. S. Morse, *Mars and Its Mystery* (Boston: Little, Brown, 1906), 79.

44. W. L. Leonard, *Percival Lowell: An Afterglow* (Boston: Richard G. Badger, 1921), 25.

45. Greenslet, *The Lowells and Their Seven Worlds*, 366.

46. Lowell, *Mars*, 201.

47. Lowell, 209–210.

48. H. G. Wells, *The War of the Worlds* (London: William Heinemann, 1898), 3.

49. Wells, 6.

50. R. Markley, *Dying Planet: Mars in Science and the Imagination* (Durham, N.C.: Duke University Press, 2005), 25.

51. W. J. Walter, *Space Age* (New York: Random House, 1992), 22. Because of his illnesses, Goddard had fallen two years behind in high school. He later studied at Worcester Polytechnic Institute, where he graduated with a degree on physics in 1908, and a year later he went to Clark University (also in Worcester), where he received his MA in Physics in 1910 and his PhD in 1911. In 1909, he heard Lowell give a vicennial lecture, "The Planet Venus," on the occasion of Lowell's receiving an honorary LLD. This lecture was published in *Popular Science Monthly* 35 (1909): 531–36.

52. Walter, *Space Age*, 22.

Chapter 9

1. The North American monsoon is caused by an area of high pressure that moves northward during the summer months and an area of low pressure (forming due to intense surface heating) over the Mexican Plateau and the American Southwest. This configuration draws moisture inland from the Gulf of California and eastern Pacific, leading to heavy rainfall. The monsoon begins in late May to early June in southern Mexico and quickly spreads along the western slopes of the Sierra Madre Occidental, reaching Arizona and New Mexico in early July. The departure dates are less consistent, but on average the monsoon season is over by mid-September.

2. As is partially described in chapter 3, chromatic aberration refers to the inability of a lens—even an achromat, in which two components of crown and flint glass are combined to try to mitigate the problem—to focus all the colors to a single point. Thus, a bright star or planet forms an overlapping series of concentric blue, yellow, and red images, with the specific color seen depending on how the eyepiece focuses this series. The star or planet image thus appears swathed in a haze of unfocused color, rather like the Moon on a misty night.

3. Quoted in T. E. R. Phillips, ed., *Splendour of the Heavens*, vol. 1 (London: Hutchinson, 1923), 295.

4. Douglass's pioneering investigations were published in A. E. Douglass, "Atmosphere, Telescope and Observer," *Pop. Ast.* 5 (1897): 64–84. Other accounts include P. Lowell, "Atmosphere: In Its Effect on Astronomical Research," lecture text, c. spring 1897, Lowell Observatory Archives; P. Lowell, "Means, Methods and Mistakes in the Study of Planetary Evolution," unpublished manuscript, April 13, 1905, Lowell Observatory Archives. This last was submitted to H. H. Turner, editor of *MNRAS*, but withdrawn for unknown reasons—possibly for its rather belligerent tone. Douglass's work has largely been confirmed by subsequent investigators. A. T. Young, "Seeing and Scintillation," *Sky & Telescope*, September 1971, 150, notes: "Evidently, there is an optimum aperture to use if maximum resolution is wanted. With too small a telescope, the seeing is good (apart from image motion), but the telescope size limits the resolution. With too big a telescope, larger and stronger turbulent areas are included, and seeing limits the resolution. The aperture must be matched to the seeing conditions."

5. C. H. Giffen and C. R. Chapman, "The Theory of Visual Lunar and Planetary Observation: Resolution, Contrast Perception, Color Perception, Estimating Intensities, and Methods for Measuring Atmospheric Transparency and Seeing Conditions," in *Observing the Moon, Planets, and Comets*, ed. C. R. Chapman and D. P. Cruikshank, unpublished manuscript.

6. Clyde Tombaugh to William Sheehan, pers. comm., December 5, 1986.

7. E. M. Antoniadi, *The Planet Mars*, trans. Patrick Moore (Devon, England: Keith Reid, 1975), 15.

8. H. Brown and K. J. Friston, "Free-Energy and Illusions: The Cornsweet Effect," *Frontiers in Psychology* 3 (February 2012): 1.

9. W. Sheehan, *Planets and Perception: Telescopic Views and Interpretations* (Tucson: University of Arizona Press, 1988), 85.

10. "Report of the Meeting of the Royal Astronomical Society," *Observatory* 19 (1896): 420.

11. E. M. Antoniadi, "The 'Canals' of Venus," *English Mechanic* 67 (July 8, 1898): 474.

12. D. Strauss, *Percival Lowell: The Culture of a Boston Brahmin* (Cambridge, Mass.: Harvard University Press, 2001), 173.

13. A. E. Douglass, "The Markings of Venus," *MNRAS* 58 (1898): 320.

14. A. E. Douglass to W. H. Pickering, March 8, 1901, Douglass Papers, University of Arizona Libraries.

15. A. E. Douglass to W. W. Campbell, August 23, 1901, Mary Lea Shane Archives of the Lick Observatory, University of California, Santa Cruz. In shock at his dismissal, Douglass began casting around in earnest for other positions. Unable to find anything in astronomy, he improvised for a time, traveling around northern Arizona while living off his savings, investing in local mining ventures, operating an assay office in Flagstaff for several years, and even receiving nomination as Republican candidate for a probate judge; he was elected, despite his lack of legal training and experience. What recommended him, according to the Republican newspaper *Coconino Sun* (Flagstaff), was "good judgment and sterling honesty," traits that, ironically, had not always won him favor as an astronomer (September 27, 1902). Eventually he found his way back into science. He would achieve special distinction as founder of the science of dendrochronology, the technique of determining dates from tree rings. Only a few months before Lowell's death in November 1916, he was appointed the first director of the University of Arizona's Steward Observatory in Tucson. During the rest of his career, he did nothing particular, however, with respect to Mars. See G. E. Webb, *Tree Rings and Telescopes: The Scientific Career of A. E. Douglass* (Tucson: University of Arizona Press, 1983).

16. P. Lowell to Elizabeth Lowell, April 12, 1901, Lowell Family Papers, Houghton Library, Harvard University.

17. The 1896–97 observations are discussed in great detail in C. Flammarion, *La Planète Mars et ses conditions d'habitabilité*, vol. 2 (Paris: Gauthier-Villars, 1909), 267–435.

18. E. M. Antoniadi, "Report of Mars Section, 1896," *Mem. BAA* 5 (1897): 82.

19. P. B. Molesworth, manuscript, Royal Astronomical Society Archives.

20. Antoniadi, "Report of Mars Section, 1896."

21. Quoted in E. M. Antoniadi, "Report of Mars Section, 1898–99," *Mem. BAA* 20 (1901): 68.

22. Quoted in E. M. Antoniadi, "Report of the Mars Section, 1905," *Mem. BAA* 17 (1911): 38.

23. E. W. Maunder, "The Tenuity of the Sun's Surroundings," *Knowledge*, March 1, 1894, 49.

24. Maunder, 49.

25. E. W. Maunder, "The Canals of Mars," *Knowledge*, November 1, 1894, 251.

26. E. W. Maunder, "A New Chart of Mars," *Observatory* 206 (1903): 353. He is commenting on Antoniadi's official map for the BAA for that year, from which all canals had been excluded.

27. Antoniadi, "Report of Mars Section, 1896," 105.

28. He worked out these ideas in two memoirs: V. Cerulli, *Marte nel 1896–97* (Collurania, Italy: Pubblicazioni dell'Osservatorio privato di Collurania [Teramo], 1898); V. Cerulli, *Nuove osservazione di Marte (1898–1899)* (Collurania, Italy: Pubblicazioni dell'Osservatorio privato di Collurania [Teramo], 1900).

29. P. Lowell, *Mars and Its Canals* (New York: Macmillan, 1906), 202.

30. What is known of this important Mars observer has been unearthed by Stéphane Lecomte of the Société Astronomique de France, and was summarized in an interesting talk for the International Workshop on One Century of Mars Observations (IWCMO) conference at Meudon, France, in 2009. Millochau was born in 1866, and on recommendation of Joseph Vinot, the editor of the *Journal du Ciel* (an early French astronomy newspaper), was hired by the pioneering solar astronomer Henri Deslandres (best remembered for inventing, independently of George Ellery Hale, the spectroheliograph) as an assistant in the spectroscopy service at the Paris Observatory. During this time Millochau was a regular visitor at Juvisy-sur-Orge, where he probably met Antoniadi. For many years he was the main collaborator of Deslandres, accompanying him on solar eclipse expeditions to Senegal in 1893, Japan in 1896, and Spain in 1900. Late in 1897, when Deslandres became director of the Meudon Observatory, Millochau followed him there, and thus he came to use the Henry Brothers telescope on Mars. From 1903, Millochau seems to have fallen out with Deslandres and left the spectroscopy team, and from 1904 he focused on the study of the Sun's temperature and the solar constant. For this research he made many observations from the Mont Blanc Observatory, including spending what must have been a grueling run of 13 consecutive days on the summit. He continued to work at the Paris Observatory until 1915, when, owing to some sort of shock related to the war, he disappeared from astronomical circles. Nothing is known of his fate, or even the year of his death (though it seems to have been sometime after 1919). It is pleasant to add that his pioneering work on Mars has been recognized by the International Astronomical Union (IAU), which has named a crater on Mars after him.

31. Quoted in E. M. Antoniadi, *La Planète Mars: Étude basée sur les résultats obtenus avec la Grande Lunette de l'observatoire de Meudon et exposé analytique de l'ensemble des travaux exécutés sur cet aster depuis 1659* (Paris: Librairie Scientique Hermann et Cie, 1930), 23–24.

32. E. S. Morse, *Mars and Its Mystery* (Boston: Little, Brown, 1906), 100.

33. P. Lowell to D. P. Todd, undated typed manuscript page, Lowell Observatory Archives. For Lowell's definitive statement regarding the way the canals made themselves known, see Lowell, *Mars and Its Canals*, 174–175.

34. P. Lowell, "New Photographs of Mars: Taken by the Astronomical Expedition to the Andes and Now First Published," *Century Magazine* 75 (1907): 304.

35. J. Wallace to G. E. Hale, January 30, 1904, Hale Papers, California Institute of Technology Archives.

36. Lowell, *Mars and Its Canals*, 275.

37. Lowell, 277.

38. E. E. Barnard to G. E. Hale, October 26, 1905, Hale Papers, California Institute of Technology Archives. Hale responded on November 9: "I would rather have your opinion of the planet than anyone else," and added that he regretted that "the facts are not generally known, since everyone will now be convinced of the reality of the canals on the supposition that photographs cannot lie." Albert Michelson, the Nobel laureate who had brought his interferometer to Flagstaff that year, also admitted to being unable to see canals on the photographs shown to him. As noted in G. E. Hale to W. W. Campbell, October 23, 1909, Campbell Papers, Mary Lea Shane Archives of the Lick Observatory.

39. See W. Sheehan and A. Misch, "The Great Mars Chase of 1907," *Sky & Telescope*, November 2007, 20–24.

40. Sheehan and Misch, 22.

41. Sheehan and Misch, 22.

42. K. M. D. Lane, *Geographies of Mars: Seeing and Knowing the Red Planet* (Chicago: University of Chicago Press, 2011), 56.

43. P. Lowell to D. P. Todd, September 6, 1907, Lowell Observatory Archives.

44. G. V. Schiaparelli to P. Lowell, September 2, 1909, Lowell Observatory Archives. For a full discussion, see R. McKim and W. Sheehan, "Schiaparelli's Final Words about Mars," *JBAA* 119, no. 5 (2009): 255–261.

45. *Wall Street Journal*, December 28, 1907.

46. See Simon Newcomb, "The Optical and Psychological Principles Involved in the Interpretation of the So-Called Canals of Mars," *ApJ* 26 (July 1907): 1–17; V. Cerulli, "Polemica Newcomb-Lowell-fotografie lunari," *Rivista di astronomia* 2 (1908): 13–23.

47. Strauss, *Percival Lowell*, 220. For Campbell on the "Lowell situation," see W. W. Campbell to G. E. Hale, May 11, 1908, Mount Wilson and Palomar Observatory Library.

48. Strauss, *Percival Lowell*, 220.

49. P. Lowell, "Temperature of Mars: A Determination of the Solar Heat Received," *Proceedings of the American Academy of Arts and Sciences* 42 (1906–1907): 651.

50. P. Lowell, "A General Method for Evaluating the Surface-Temperatures of the Planets; With Special Reference to the Temperature of Mars," *Philosophical Magazine*, ser. 6, vol. 14 (July 1907): 171.

51. A. R. Wallace, *Is Mars Habitable? A Critical Examination of Professor Lowell's Book "Mars and Its Canals," with an Alternative Explanation* (London: Macmillan, 1907), 46.

52. P. Lowell, *Mars as the Abode of Life* (New York: Macmillan, 1908), 238–241. Lowell began by considering what the total reflectivity of the Martian surface would be were there no atmosphere at all. Since almost five-eighths of the planet is desert (albedo 0.16), and three-eighths consists of the dark blue-green areas (albedo 0.07), this came out to around 0.13. However, the actual (measured) albedo of Mars was 0.27. Allowing for the fact that some sunlight would be absorbed by the atmosphere before it reached the ground, Lowell came up with 0.10 for the albedo of the bare surface, leaving 0.17 for the albedo of the atmosphere, which he compared with an estimated albedo of Earth's atmosphere of 0.75. Further, since albedo is a measure of the scattering of light, Lowell estimated that the mass of the atmosphere per unit area of Mars would be $0.17/0.75 = 0.23$ of the mass per unit area of the terrestrial atmosphere. Given that the force of gravity on Mars is only 0.38 times that of Earth, the atmospheric pressure on Mars should be about $0.23 \times 0.38 = 0.087$ of Earth's, or 87 millibars. Though this value is some 10 times greater than the actual value, it still made the thinness of the Martian atmosphere equivalent to that at some 17 kilometers above the surface of Earth, an altitude not reached by humans until the stratospheric balloon ascensions of the 1930s!

53. In 1926, D. H. Menzel, refining Lowell's calculation, came up with a value of some 68 millibars. In 1934, the Russian astronomers N. Barabashov and B. Semejkin, using a different method based on scattering of light from the bright areas of Mars in different wavelengths, came up with 50 millibars, though later Barabashov and other Russian astronomers preferred a higher value of 100 millibars. As late as the 1950s, the French astronomer Gérard de Vaucouleurs, using a method that depended on how patches of brightness varied with distance from the central meridian of the planet (since a feature near the limb would be seen through a greater thickness of atmosphere), came back to a figure right around 85 millibars. See G. de Vaucouleurs, *Physics of the Planet Mars*, trans. P. Moore (London: Faber and Faber, 1954). The fatal flaw in all these albedo-based methods is that they drastically underestimated the amount of dust in suspension in the Martian atmosphere, which is very efficient at scattering light.

54. On W. W. Campbell's 1909 observations, see Donald E. Osterbrock, "To Climb the Highest Mountain: W. W. Campbell's 1909 Mars Expedition to Mount Whitney," *JHA* 20 (1989): 77–97.

55. Wallace, *Is Mars Habitable?*, 14.

56. This letter was called to author W. S.'s attention by Luigi Prestinenza, and first published in English in W. Sheehan, *The Planet Mars: A History of Observation and Discovery* (Tucson: University of Arizona Press, 1996), 130. It is hard to believe that anyone capable of such penetrating analysis as Schiaparelli demon-

strates here should have subsequently retreated again into the realm of error and illusion, but apparently Schiaparelli did just that after studying some of Lowell's 1907 Mars photographs, and so apparently did Cerulli himself. Schiaparelli wrote to E. M. Antoniadi in 1909 (quoted in Sheehan, 131): "The polygonations and geminations for which you show so much horror (and, with you, so many others) are a proved fact, against which it is heedless to dissent. Dr. Cerulli was convinced some weeks ago."

57. The most complete account is R. McKim, "The Life and Times of E. M. Antoniadi, 1870–1944," 2 parts, *JBAA* 103 (1993): 164–170 and 219–227.

58. Abdul Hamid Khan II succeeded his uncle, Abdülaziz II, who became unpopular because of the crop failure of 1873 and lavish expenditures on his navy and new palaces. The French astronomer Henri Camichel knew Antoniadi somewhat, and recalled to one of the authors (W. S.) a story that Antoniadi had told him about the occasion of Abdülaziz's death. Antoniadi, who was only six at the time, felt quite jubilant, but his father warned him: "The next one will be worse." And so it proved to be.

59. Seldom has a more fateful collision of a book and a man occurred. Flammarion's book is still in some ways unsurpassed as a source of information about the history of Mars observations up to 1892, and is now available in the English translation of Patrick Moore as: *Camille Flammarion's The Planet Mars*, ed. W. Sheehan (Cham, Switzerland: Springer, 2015).

60. A. S. D. Maunder to P. B. Molesworth, August 31, 1900, Molesworth Papers, Royal Astronomical Society.

61. A. S. D. Maunder to P. B. Molesworth, December 13, 1900, Molesworth Papers, Royal Astronomical Society.

62. E. M. Antoniadi, "Second Interim Report of the Mars Section on the Observations of 1909, Dealing with the Meteorology of the Planet," *JBAA* 20 (October 1909): 79.

63. The situation of Lowell Observatory is rather similar: it is located on the northeast (leeward) corner of a plateau, with Mars Hill, the site of the 61-centimeter Clark refractor, positioned at its very edge, looking down a 100-meter drop to the valley below (which includes the city of Flagstaff). Despite some misgivings of Barnard and Douglass about the way the San Francisco Peaks might create turbulence affecting the seeing, and though air rising over the peaks does tend to ripple the air to the east, Mars Hill is well to the south, and doesn't seem much affected. In fact, the observatory seems to enjoy a high percentage of nights with good to excellent seeing.

Lowell and Antoniadi both used scales to attempt to estimate the quality of seeing. Lowell (and other observers in Flagstaff) used a scale of 1–10, which continues to be used by many American observers of the planets:

1. Very poor images, impossible to see details

2–3. Almost continuous distortion with occasional, brief good moments

4–6. More continuous distortions with short intervals of good seeing

7–8. Intervals of perfect seeing with fine-scale distortions in between

9–10. Perfect seeing with steady images at high magnification

Antoniadi published the following scale, the well-known Antoniadi scale, which has come into almost exclusive use by European (including British) observers:

I. Perfect seeing without a quiver, even with the highest magnifications

II. Slight undulations, with moments of calm lasting several seconds

III. Moderate seeing, with larger tremors permitting only medium powers to be used

IV. Poor seeing, with constant tremulous undulations

V. Very bad seeing, scarcely allowing the making of a rough sketch; blurred images even at low power

64. R. McKim, "E.-M. Antoniadi, 1870–1944," talk at IWCMO conference, Paris, September 18, 2009.

65. E. M. Antoniadi, "Fourth Interim Report for the Apparition of 1909, Dealing with the Appearance of the Planet Mars Between September 20 and October 23 in the Great Refractor of the Meudon Observatory," *JBAA* 20 (November 1909): 78–79.

66. Antoniadi, 79.

67. J. Ruskin, *Modern Painters*, vol. 3 (London: George Allen, 1906), part 4, chapter 16.

68. Many an observer, over a long career, has had the experience of nearly perfect planetary images, only to think with a sigh, as Alan Pennell Lenham (1930–66) did in 1958 when he and Dale P. Cruikshank were observing Mars with the 1.02-meter Yerkes refractor, "I wish I could draw like Antoniadi." There was, alas, only one Antoniadi. As a draftsman of planetary detail, he has never been surpassed. D. P. Cruikshank to W. Sheehan, pers. comm.

69. R. J. McKim to W. Sheehan, pers. comm., April 1, 2017.

70. E. B. Frost to W. W. Campbell, November 16, 1908, Mary Lea Shane Archives, Lick Observatory; E. B. Frost to W. W. Campbell, September 16, 1908, Mary Lea Shane Archives, Lick Observatory.

71. E. M. Antoniadi, "Fifth Interim Report of the Mars Section, Dealing with the Fact Revealed by Observation That Prof. Schiaparelli's 'Canal' Network Is the Optical Product of the Irregular Minor Details Diversifying the Martian Surface," *JBAA* 20 (December 1909): 140.

72. E. M. Antoniadi to W. H. Wesley, September 25, 1909, Royal Astronomical Society Letters, 1909.

73. E. M. Antoniadi, "Report of Mars Section 1909," *Mem. BAA* 20 (1916): 32.

74. Antoniadi, 32.

75. Antoniadi, "Fifth Interim Report," 137.

76. Antoniadi, "Fourth Interim Report," 79.

77. Antoniadi, 79.

78. Antoniadi, "Fifth Interim Report," 138.

79. E. E. Barnard to E. M. Antoniadi, November 27, 1909, quoted in Antoniadi, *La Planète Mars*, 24.

80. G. E. Hale to E. M. Antoniadi, January 3, 1910, quoted in E. M. Antoniadi, "Sixth Interim Report for 1909," *JBAA* 20 (1910): 191–192.

81. E. M. Antoniadi, "Le Retour de la Planète Mars," *Bulletin Société Astronomique de France* 40 (1926): 350.

82. E. M. Antoniadi to P. Lowell, September 9, 1909, Lowell Observatory Archives. He certainly drew well compared to Lowell. The proportions with which he represented the dark areas on Mars are almost perfect, whereas in Lowell's drawings of all the planets (not just Mars) they are systematically represented as too small relative to the size of the planetary disc. Moreover, Lowell's systematic shrinking of dark features may be a partial explanation for the narrowness of his canals. C. R. Chapman, "The Techniques of Observation and Methods of Reduction," in Chapman and Cruikshank, *Observing the Moon, Planets, and Comets*.

83. P. Lowell to E. M. Antoniadi, September 26, 1909, Lowell Observatory Archives.

84. E. M. Antoniadi to P. Lowell, October 9, 1909, Lowell Observatory Archives.

85. W. W. Campbell to F. R. Moulton, October 25, 1909, Campbell Papers, Mary Lea Shane Archives, Lick Observatory.

86. P. Lowell to E. M. Antoniadi, November 2, 1909, Lowell Observatory Archives.

87. P. Lowell to W. Kaempffert, January 31, 1916, Lowell Observatory Archives.

88. E. M. Antoniadi to P. Lowell, October 9, 1909, Lowell Observatory Archives.

89. Antoniadi, "Fifth Interim Report," 137.

90. The original sketch appears in E. M. Antoniadi to P. Lowell, November 15, 1909, Lowell Observatory Archives. It was first published in Sheehan, *Planets and Perception*, 243. The likeness to a specific area of Mars shown in Mariner 9 images was noted by optical designer Thomas Back (T. Back to W. Sheehan, pers. comm., January 12, 1992). The area in question is located near Orcus (longitude 187° and latitude 16° N) and appears in the MC-15 topographic map, Elysium quadrangle, in R. M. Batson, P. M. Bridges, and J. L. Inge, *Atlas of Mars: The 1:5,000,000 Map Series* (Washington, D.C.: NASA, 1979), 52.

91. E. E. Barnard, Mount Wilson observing book, Yerkes Observatory Archives, University of Chicago.

92. P. Lowell to C. O. Lampland, April 10, 1910, Lowell Observatory Archives.

93. Though Lowell continued to devote most of his time and effort to Martian studies around the times of opposition, these years also marked the peak of his effort

to calculate the position of a trans-Neptunian planet, "Planet X," a quest that—if successful—would, as he hoped, have increased his prestige in the astronomical community. The project, on which he engaged several human computers, was so rigorous and demanding that for a period in 1912–13 he collapsed from sheer exhaustion. In January 1915, when he presented his *Memoir on a Trans-Neptunian Planet* to the American Academy of Arts and Sciences in Boston, it was effectively an admission of defeat, and it ended his own involvement in the search for the remainder of his lifetime. In his last two years, he was a high-strung, bitter, and discouraged man, appearing much older than his stated age and feeling keenly the pain of failing to win the recognition of his astronomical peers that he so craved. His old Harvard friend Frederic J. Stimson wrote after Lowell's death, "Mars went back on him and was a disappointment" (F. Stimson to B. Wendell, n.d., Houghton Library, Harvard University), while his brother, Abbott Lawrence Lowell, called the search for Planet X the greatest disappointment of his life. On the search for Planet X, see D. P. Cruikshank and W. Sheehan, *Discovering Pluto: Exploration at the Edge of the Solar System* (Tucson: University of Arizona Press, 2018).

94. The suggestion that Lowell should view through the Meudon or Lick refractors was set forth by canal skeptic R. G. Aitken of the Lick Observatory in "A Review of the Recent Observations of Mars," *PASP* 22 (1910): 78–87; and by M. E. J. Gheury, "Prof. Lowell's Address on Mars," *JBAA* 20 (1910): 385–386. Lowell, speaking to the BAA at its meeting on March 30, 1910, had attempted to argue that Antoniadi's broken detail was spurious and produced when atmospheric disturbances caused a diffraction pattern to break into a mosaic. Gheury had suggested that if Antoniadi and Lowell did simultaneous work at each other's observatories, then this might "advance the question more than if each spent years of the most persevering labour, engaged in deepening his own particular furrow, while the world looked on, unable to judge." Gheury, "Prof. Lowell's Address on Mars," 385–386.

95. J. Keats, letter to G. and T. Keats, c. December 21–27, 1817, in *Letters of John Keats*, ed. Robert Gittings (Oxford: Oxford University Press, 1970), 42.

96. Hattie Lowell once said of her mother, Lowell's niece Katherine Lawrence Putnam Bundy, "Mother's sense of righteousness was very deep. . . . How well I remember our fights over the dining room. . . . For her, things were black and white. It's an outlook that descends directly from the Puritans and we all have it." Quoted in K. Bird, *Color of Truth* (New York: Touchstone, 1998), 36.

97. These were qualities that his sister Amy aspired to in her imagist poems. See A. Lowell, *Tendencies of Modern American Poetry* (New York: Macmillan, 1917), 239.

98. P. Lowell, "Our Solar System," *Pop. Ast.* 237 (1916): 419–427.

99. On Lowell's activities in 1916, see W. Sheehan, "Percival Lowell's Last Year," *Journal of the Royal Astronomical Society of Canada* 110, no. 6 (December 2016): 224–235.

100. W. G. Hoyt, *Lowell and Mars* (Tucson: University of Arizona Press, 1976), xiv.

Chapter 10

1. R. Markley, *Dying Planet: Mars in Science and the Imagination* (Durham, N.C.: Duke University Press, 2005), 113.

2. H. N. Russell, "Percival Lowell and His Work," *Outlook*, December 6, 1916, 781–782.

3. See, for instance, W. Sheehan, *Planets and Perception: Telescopic Views and Interpretations* (Tucson: University of Arizona Press, 1988).

4. E. M. Antoniadi, in "Second Interim Report of the Mars Section on the Observations of 1909, Dealing with the Meteorology of the Planet," *JBAA* 20 (October 1909): 28, presents contrasting views of the canals in the Grand Lunette and in his 22-centimeter Calver reflector, which he had moved from Juvisy-sur-Orge to a spot on the lawn in front of the dome at Meudon: "The fugitive impressions of straight lines were too manifestly illusive, since they had the same breadth and intensity in the [83-centimeter] O.G. as in the [22-centimeter] mirror! In fact, the first glance thrown on Mars by the Director in the Meudon refractor on September 20 last showed him that the visibility of the average straight 'canal' did not increase *pari passu* with that of the real markings in the great telescope; and that while the latter showed wonderful details which could not be dreamt of in the small reflector—details held most steadily—the appearances of straight lines were as faint, fugitive, and cowardly in the great instrument as in the small one; and, of course, such is the deportment of illusions only. Besides, if the 'canal' network were an objective reality at the limit of visibility, we ought to see it by glimpses *as a network*, and not glimpse *severally* its various components."

5. P. Lowell, *Mars and Its Canals* (New York: Macmillan, 1906), 174–175.

6. S. Massey, T. A. Dobbins, and E. Douglass, *Video Astronomy*, 2nd ed. (Cambridge, Mass.: Sky, 2004), 10.

7. To give a historical, nonplanetary example, the rhythm and successive modifications of a galloping horse's action long remained elusive for painters such as the once-popular pre-impressionist Ernest Meissonier (1815–91), and were finally captured only by the stop-action photographs of Eadweard Muybridge (1830–1904). See R. Solnit, *River of Shadows: Eadweard Muybridge and the Technological Wild West* (New York: Penguin, 2003); R. King, *The Judgment of Paris: The Revolutionary Decade That Gave the World Impressionism* (New York: Walker, 2006). For astronomers, struggling to make out detail that was elusive and wavering, Mars was a kind of a celestial galloping horse.

8. M. C. Potter, B. Wyble, C. E. Hagmann, and M. S. McCourt, "Detecting Meaning in RSVP at 13 ms per Picture," *Attention, Perception and Pyschophysics* 76 (February 2014): 270–279.

9. G. Berlucchi, "Some Historical Crossroads Between Astronomy and Visual Neuroscience," *Memorie della Società Astronomica Italiana* 82 (2011): 229. Much of the information about perception described in outline here was learned from experiments with tachistoscopes, apparatuses used experimentally to expose images for very short periods of time in order to study the way perceptions are constructed. The first such apparatus, designed by the German physiologist Alfred Volkmann in 1850, used a series of falling doors to interrupt the exposure, though within a few years this contraption was generally replaced by a device in which the short exposure was produced by means of illumination by an electric spark produced from the secondary spiral of a large induction coil connected with the terminals of a Leyden jar. Hermann von Helmholtz used such a device to establish the finding described by Berlucchi: in the cases of certain figures prone to illusion, the illusory effect was heightened by eye movements, and "there is a prevailing tendency in moving our eyes to follow the direction of the more conspicuous lines in the field of view." See H. von Helmholtz, *Treatise on Physiological Optics*, ed. J. P. C. Southall, vol. 3 (Rochester, New York: Optical Society of America, 1924), 197.

10. E. Clerke, "The Planet Mars," *Month* 76 (1892): 188.

11. M. J. Sharps, *Processing Under Pressure: Stress, Memory and Decision-Making in Law Enforcement* (Flushing, New York: Looseleaf Law, 2010). See also E. F. Loftus, *Eyewitness Testimony* (Cambridge, Mass: Harvard University Press, 1979), which includes a discussion of the doubtful testimony used to convict Sacco and Vanzetti of the Braintree, Massachusetts, murders in 1921. Loftus is writing about the testimony of witnesses to the murder, but she could equally be writing about observers' testimony of canals on Mars. She asks (pp. 3–4), "Why . . . did so many witnesses, once so uncertain, make positive identifications of Sacco and Vanzetti at the trial? Were these witnesses improperly influenced by the police, and if so, how was this influence achieved? Why did the jurors believe Sacco's and Vanzetti's eyewitness accusers, even in the face of plausible alibis? Was the jury correct to give so much credence to these eyewitness accounts?" See W. Sheehan, "Eyewitness Testimony and the Canals of Mars," *Mercury* (July–August 2003): 35. Ironically, Percival's brother, Abbott Lawrence Lowell, was appointed to an advisory committee to consider the fairness of the trial in which the immigrants Sacco and Vanzetti were convicted and sentenced to death, and he largely wrote the report that, while criticizing the judge, concluded that the trial had been fair. The three-person committee was criticized because, despite members' high status in the community, they were not really qualified to perform the difficult

task assigned to them, and they—and especially Lowell—imagined that the committee could use its fresh and more powerful analytic abilities to outperform the efforts of those who had worked on the case for years. Lowell would be dogged by the affair for the rest of his life, and the case elicited widespread international comment and criticism, including even from H. G. Wells in the *New York Times,* October 16, 1927.

12. M. Shermer, *The Believing Brain: From Ghosts and Gods to Politics and Conspiracies—How We Construct Beliefs and Reinforce Them as Truths* (New York: Henry Holt, 2011), 59.

13. Lowell had a penchant for Type I errors. Another example was his attempt to use small residuals in the motion of the planet Uranus to deduce the existence and calculate the orbit of a trans-Neptunian planet, which led him to instigate the search for "Planet X," a search that eventually led (serendipitously) to the discovery of Pluto.

14. Antoniadi, "Second Interim Report," 28.

15. See G. Basalla, *Civilized Life in the Universe: Scientists on Intelligent Extraterrestrials* (Oxford: Oxford University Press, 2007), 67ff.

16. W. Stegner, *Beyond the Hundredth Meridian: John Wesley Powell and the Second Opening of the West* (Boston: Houghton Mifflin, 1954), 7.

17. M. Amundsen. "Seeing Arizona, Imagining Mars: Deserts, Canals, Global Climate Change, and the American West," *Journal of Arizona History* 58, no. 4 (Winter 2017): 331–350.

18. Markley, *Dying Planet,* 6.

19. Typical was a 1909 trip with his closest Flagstaff friend, Judge Edward M. Doe, by train to Adamana, a town that no longer exists but was then the port of entry to the North Sigillaria Forest, just discovered by naturalist John Muir. "When we left Winslow for the banks of the Little Colorado," Lowell wrote in a never-published manuscript, invoking perhaps unconsciously some of the same Martian color contrasts seen in the eyepiece, "the cottonwoods along its banks stood clad in ineffable yellow mixed here and there with an orange tint. Against the blue the effect was singularly beautiful, perhaps the more so for the contrast of the ocherish desert all about." P. Lowell, "The Newly Discovered Petrified Forest of Arizona," unpublished manuscript, Lowell Observatory Archives.

20. Lowell, *Mars and Its Canals,* 149–150.

21. E. R. Burroughs, *A Princess of Mars* (New York: Ballantine Books. 1963), 11.

22. P. Lowell, *Mars as the Abode of Life* (New York: Macmillan, 1908), 134; Markley, *Dying Planet,* 69.

23. R. Miller and F. C. Durant III, with M. Schuetz, *The Art of Chesley Bonestell* (London: Paper Tiger, 2001), 167.

24. R. L. Waterfield, *A Hundred Years of Astronomy* (London: Duckworth, 1938), 50–51.

25. See J. N. Tatarewicz, *Space Technology and Planetary Astronomy* (Bloomington: Indiana University Press, 1990), 2–6. Also see R. E. Doel, *Solar System Astronomy in America: Communities, Patronage, and Interdisciplinary Science, 1920–1960* (Cambridge: Cambridge University Press, 1996), xi.

26. W. H. Pickering, "Monthly Report on Mars—no. 6," *Pop. Ast.* 22 (1914): 421.

27. H. Plotkin, "William H. Pickering in Jamaica: The Founding of Woodlawn and Studies of Mars," *JHA* 24 (1993): 101–122.

28. W. H. Haas to W. Sheehan, pers. comm., May 31, 1994. In 1935, Haas had been offered funding to attend college or to spend time observing with Pickering in Jamaica. He chose to do the latter.

29. Antoniadi was financially independent during these years, and described himself simply as the "astronome volontaire à l'Observatoire de Meudon." The latter part of his career has been described by R. McKim in "The Life and Times of E. M. Antoniadi, 1870–1944, Part 2: The Meudon Years," *JBAA* 103, no. 5 (1993): 219–227. According to McKim, "Two Chessmen of Mars: Edgar Rice Burroughs and Eugène-Michel Antoniadi," *AA*, no. 12 (June 2018): 38n34, "Antoniadi was considered a rather unapproachable character by the young French astronomers in the early 1940s. But together with them he did attend reunions of the [Societe Astronomique de France], where its members stood at Flammarion's tomb at Juvisy to mark each anniversary of the great astronomer's death [Flammarion had died in 1925]. The reunion in the summer of 1943 (whose participants were, as was usual then, listed in the SAF *Bullétin*) was attended by Audouin Dollfus, Jean Dragesco, and Gerard de Vaucouleurs, all then in their twenties and destined to become as famous as Antoniadi. Years later [McKim] reminded them of this event, and each told [him] either verbally or by correspondence that no-one had dared to approach Antoniadi! Only de Vaucouleurs had ever corresponded with him, when Antoniadi had declined to attend a meeting being organized to test observational skills in a laboratory environment."

30. E. M. Antoniadi, *La Planète Mars: Étude basée sur les résultats obtenus avec la Grande Lunette de l'observatoire de Meudon et exposé analytique de l'ensemble des travaux exécutés sur cet aster depuis 1659* (Paris: Librairie Scientique Hermann et Cie, 1930), 16. Though in general he adopted the standard pitiless desert view of Mars, it is interesting to recall that on his evening of unsurpassed seeing, September 20, 1909, he had described the land between Syrtis Major and Hellas almost in terms of an Alpine scene. This area had appeared "like a green meadow, sprinkled with tiny white spots of various sizes, and diversified with darker or lighter shades of green." E. M. Antoniadi, "On the Advantages of Large over Small Telescopes in Revealing Delicate Planetary Detail," *JBAA* 21 (1910): 104.

31. R. McKim, "Telescopic Martian Dust Storms: A Narrative and Catalogue," *Mem. BAA* 44 (1999): 46. The idea that the yellow clouds might be dust clouds was first

proposed by A. E. Douglass in 1899, and his suggestion was adopted by Lowell—perhaps grudgingly, since the latter continued to believe that these dust clouds were exceedingly rare on Mars, which continued to be the official line taken by the observatory. Thus, McKim (p. 46) notes that in the margins of the Lowell Observatory library copy of Antoniadi's paper on the December 1924 cloud, E. C. Slipher penciled these remarks: "Greatly exaggerated" and "Bosh—see ECS photos at the same time." This annotation seems rather cryptic, however, as no visual or photographic work was done from Flagstaff at the time of the storm. Slipher was at this time probably still prejudiced by the Lowellian orthodoxies about the planet.

32. Antoniadi, *La Planète Mars*, 67.
33. W. L. Putnam III, *The Explorers of Mars Hill* (Kennebunkport, Maine: Phoenix, 1994), 101.
34. E. M. Antoniadi to Gabrielle Flammarion (Camille's second wife, following Sylvie's death in 1919), August 23, 1928, Fonds Camille Flammarion de l'observatorie de Juvisy-sur-Orge, France.
35. E. C. Slipher to John A. Miller, December 10, 1915, V. M. Slipher Papers, Lowell Observatory Archives.
36. V. M. Slipher, "Spectrum Observations of Mars," *PASP* 6 (1924): 261–262.
37. W. G. Hoyt, "Vesto Melvin Slipher, 1875–1969," *Biographical Memoirs of the National Academy of Sciences* 52 (1980): 431.
38. V. M. Slipher to F. O. Grover, January 23, 1923, Lowell Observatory Archives.
39. Herschel published a series of papers on these heat rays, of which the first was "Investigations of the Powers of the Prismatic Colours to Heat and Illuminate Objects; with Remarks, That Prove the Different Refrangibility of Radiant Heat. To Which Is Added, an Inquiry into the Method of Viewing the Sun Advantageously, with Telescopes of Large Apertures and High Magnifying Powers," *Philosophical Transactions of the Royal Society of London* 90 (1800): 255–283.
40. See W. G. Scaife, *From Galaxies to Turbines: Science, Technology and the Parsons Family* (London: Institute of Physics, 2000), 108–114.
41. For details, see W. M. Sinton, "Taking the Temperature of the Moon and Planets," *Astronomical Society of the Pacific Leaflets* 7, no. 345 (1958): 361–368.
42. The principle of the thermocouple is based on an effect discovered by the German physicist Thomas Johann Seebeck in 1821: when different metals are joined at the ends and there is a temperature difference between the joints, a thermoelectric current is produced, which can be measured with a galvanometer. See J. B. Hearnshaw, *The Measurement of Starlight: Two Centuries of Astronomical Photometry* (Cambridge: Cambridge University Press, 1996), 68. On the work at Lowell Observatory, see W. W. Coblentz and C. O. Lampland, "Measurements of Planetary Radiation," *Lowell Observatory Bulletins* 3, no. 10 (1923): 91–134.

43. Like the Slipher brothers, Lampland was also a convinced Lowellian. In a manuscript produced for a Mars book the three were collaborating on in the 1920s (it was never finished), he wrote: "The reality of the 'canals' has been a much-discussed question. Over a long period of years the matter was indeed a highly controversial subject. But the advent of the photographic plate . . . fully confirmed the results obtained with the eye and quite removes the basis of the arguments of some who attributed these strange markings to the imagination of the observer or optical illusions. Having established the reality of these markings it seems without profit to attempt to explain and evaluate the personal equation of different visual observers in their delineation of these markings. The differences in the draftsmanship and rendering of say landscape views by different artists is recognized and all the more might one expect marked differences on so difficult a subject as planetary observations." C. O. Lampland, "Planetary Radiometry: New Temperature Determinations of the Planet Mars," unfinished manuscript, Lowell Observatory Archives.

44. W. W. Coblentz and C. O. Lampland, "Radiometric Measurements of Mars," *PASP* 36 (1924): 272–274. The results were consistent with those obtained at Mount Wilson by E. Pettit and S. B. Nicholson, as described in "Measurements of the Radiation from the Planet Mars," *Pop. Ast.* 32 (1924): 601–608.

45. W. W. Coblentz, "Climactic Conditions on Mars," *Pop. Ast.* 33 (May 1925): 310–316, 363–367. For a general discussion of the history of Mars observations related to the existence of life there, see S. J. Dick, *The Biological Universe: The Twentieth-Century Extraterrestrial Life Debate and the Limits of Science* (Cambridge: Cambridge University Press, 1996), esp. 105–135.

46. E. Pettit and S. B. Nicholson, "Radiation Measures on the Planet Mars," *Publications of the Astronomical Society of the Pacific* 36 (1924): 269–272.

47. C. O. Lampland, "Mars," unpublished draft, c. 1927, Lowell Observatory Archives.

48. In addition to the work at Flagstaff, Slipher had led the effort to photograph Mars on the Lowell Observatory Expedition to the Andes in 1907, as discussed earlier, he would also do so in 1939, 1954, and 1956 on trips to the Lamont-Hussey Observatory at Bloemfontein, South Africa. His photographic work on Mars is summarized in: E. C. Slipher, *Mars: The Photographic Story* (Cambridge, Mass.: Sky, 1962).

49. G. de Vaucouleurs, "The Slow Progress of Martian Studies," *Graduate Journal* 7 (Winter 1965): 181–193.

50. Lowell's calculation is given in "The Mean Temperature of Mars," in Lowell, *Mars as the Abode of Life*, 240ff. De Vaucouleurs gives 80–90 millibars in *Physics of the Planet Mars*, trans. Patrick Moore (London: Faber and Faber, 1954).

51. W. S. Adams and T. Dunham, "The B Band of Oxygen in the Spectrum of Mars," *ApJ* 79 (1934): 308–316; W. S. Adams and T. Dunham, "Water-Vapor Lines in the Spectrum of Mars," *PASP* 49 (1937): 209–211.

52. G. P. Kuiper, "Carbon Dioxide on Mars," *Harvard College Observatory Announcement Card*, no. 851 (1947).

53. De Vaucouleurs, *The Physics of the Planet Mars*, 127.

54. Lowell, *Mars and Its Canals*, 165.

55. C. E. Burton, "On the Aspect of Mars at the Oppositions of 1871 and 1873," *Transactions of the Royal Irish Academy* 26, part 11 (1878): 427–430.

56. L. J. Martin and R. W. Zurek, "An Analysis of the History of Dust Activity on Mars," *JGR* 98 (1993): 3221–3246; R. W. Zurek and L. J. Martin, "Interannual Variability of Planet-Encircling Dust Storms on Mars," *JGR* 98 (1993): 3247–3259. The first of these papers is a compilation of past lists of dust storms, with new data. The second addresses the incidence of great storms, based on data including that from Viking. The authors concluded that overall the chance of a planet-encircling or global dust storm occurring in an arbitrary Mars year is about one in three, a statistic that seems to have been borne out by the record since 1993.

57. R. J. McKim to W. Sheehan, pers. comm., January 1, 2019.

58. McKim, "Telescopic Martian Dust Storms," 76.

59. W. K. Hartmann to W. Sheehan, pers. comm., February 26, 2020.

60. H. Aspaturian, California Institute of Technology Oral History Project Interview with Robert B. Leighton, 1995, California Institute of Technology Archives.

61. R. J. McKim to W. Sheehan, pers. comm., February 12, 2021.

62. Although, for a while in the modern era, Mars atmospheric science expert and noted baseball fan Robert Haberle from NASA's Ames Research Center noted a striking correlation between the years in which a planet-encircling dust storm would occur and the years in which a U.S. Major League Baseball pitcher threw a "perfect game" (there have only been 23 such games out of more than 200,000 played), Bob's correlation has sadly not held into the 21st century. R. Haberle to J. Bell, pers. comm., 1998.

63. Slipher, *Mars*, 22.

64. T. A. Dobbins, D. C. Parker, and C. F. Capen, *Observing and Photographing the Solar System: A Practical Guide for the Amateur Astronomer* (Richmond, Va.: Willmann-Bell, 1988), 71.

65. In September 1909, Count Aymar de la Baume Pluvinel, a wealthy scientist who had his own research laboratory in a castle at Marcoussis, together with amateur astronomer Fernand Baldet, climbed 2,800-meter Pic du Midi in the Pyrenees to attempt to photograph Mars with a 50-centimeter Newtonian reflector and a 23-centimeter refractor set up on the site by the Baillaud brothers. This was two years after the Lowell Expedition to the Andes, but in contrast to E. C. Slipher, who had obtained 14,000 photographic images of Mars showing the planet at all longitudes by continuously recording long sequences of images all night long,

Pluvinel and Baldet chose to watch the seeing at the eyepiece of the telescope and to make exposures only at the best moments. Twenty images were exposed on each plate, the plate was processed in the darkroom and scrutinized, and then the whole operation was repeated an hour later. This first expedition to study Mars from Pic du Midi demonstrated the favorable conditions for planetary studies at Pic. A permanent program of planetary research began in 1941, when Bernard Lyot replaced the 50-centimeter reflector with a 38-centimeter reflector, replaced a few years later with a 60-centimeter folded refractor. The historical planetary work at Pic was brilliant, and many important contributions were made by such legendary observers as Lyot, M. Gentili, Henri Camichel, Jean Focas, and Audouin Dollfus. Excellent planetary work has continued into modern times. For details, refer to H. Camichel, M. Gentili, and B. Lyot, "Observations planetaires au Pic du Midi," *Bull. Soc. Astron. France* 57 (1943): 49–72; B. Lyot, "L'aspect des planètes au Pic du Midi dans une lunette de 60 d'ouverture," *Bull. Soc. Astron. France* 67 (1953): 3–21; A. Dollfus, "Visual and Photographic Studies of Planets at Pic du Midi," in *The Solar System III—Planets and Satellites*, ed. G. P. Kuiper and B. M. Middlehurst (Chicago: University of Chicago Press, 1961), 534–571. An interesting personal account by a leading observer at Pic du Midi is: A. Dollfus, *50 Ans d'Astronomie* (Les Ulis, cedex A, France: EDP Sciences, 1998).

66. T. E. Thorpe, "Viking Orbiter Observations of the Mars Opposition Effect," *Icarus* 36 (1978): 204–215.

67. Lowell, *Mars and Its Canals*, 39–40.

68. G. de Vaucouleurs, *Physics of the Planet Mars*, 26.

69. G. Fournier, quoted in de Vaucouleurs, 63.

70. E. M. Antoniadi, *The Planet Mars*, trans. Patrick Moore (Devon, England: Keith Reid, 1975), 27.

71. Antoniadi, 29.

72. E. E. Barnard, observing notebook, Yerkes Observatory Archives, University of Chicago.

73. G. P. Kuiper, "Visual Observations of Mars, 1956," *ApJ* 125 (1957): 307–317; A. P. Lenham, unpublished manuscript, 1958–59, in W. Sheehan's possession.

74. For example, in his analysis of 1956 and 1958 Mars telescopic observations, Sinton reported three prominent infrared absorption bands at 271, 273, and 289.8 nanometers, which appeared strongest in spectra of the Martian dark regions Syrtis Major, Pandorae Fretum, Mare Sirenum, and Mare Cimmerium. These bands were interpreted as originating from the C-H stretching fundamental in vegetation or carbonate minerals on the Martian surface. Observations in 1963 by the noted Russian planetary astronomer Vasily Moroz provided some corroboration for these results by identifying IR bands at similar wavelengths in whole-disc spectra. Subsequent critiques of these interpretations, however—

accepted eventually even by Sinton himself—revealed that several of the bands were caused by previously unrecognized Earth atmospheric gas isotope bands; still, some of the "Sinton bands" were never fully explained. But the debate was effectively quashed by the 1965 Mariner 4 Mars flyby, which returned images of a barren, lunarlike region of the Martian southern highlands, as well as data confirming the extremely low temperatures and atmospheric pressure at the surface. The possibilities of life on Mars, even vegetation, or of aqueous mineralogy, seemed too remote and speculative to warrant additional consideration. One wonders whether this debate would not have unfolded differently if Mariner 4 had instead passed over some of the spectacular canyons or fluvial valley landforms discovered by the later Mariner and Viking missions. For details, see W. M. Sinton, "Spectroscopic Evidence for Vegetation on Mars," *ApJ* 126 (1957): 231–239; W. M. Sinton, "Further Evidence of Vegetation on Mars," *Science* 130 (1959): 1234–1237; W. M. Sinton, "Identification of Aldehyde in Mars Vegetation Regions," *Science* 134 (1961): 529; V. I. Moroz, "The Infrared Spectrum of Mars (λ 1.1–4.1 μ)," *Sov. Astron. AJ* 8 (1964): 273–281; and the review (and different interpretation of the features) by J. F. Bell, E. A. Cloutis, D. R. Klassen, and R. N. Clark, "Spectroscopic Evidence For Diaspore (α-AlOOH) on Mars" (31st Lunar and Planetary Science Conference, Houston, Tex., 2000), abstract #1227.

75. See D. B. McLaughlin, "Volcanism and Aeolian Deposition on Mars," *Geological Society of American Bulletin* 65 (1954): 715–717; D. B. McLaughlin, "Interpretation of Some Martian Features," *PASP* 66 (1954): 161–170; D. B. McLaughlin, "Wind Patterns and Volcanoes on Mars," *Observatory* 74 (1954): 166–168; D. B. McLaughlin, "Further Notes on Martian Features," *PASP* 66 (1954): 221–229; D. B. McLaughlin, "Additional Evidence of Volcanism of Mars," *Bulletin of the American Geological Society* 66 (1955): 769–772; D. B. McLaughlin, "Changes on Mars, as Evidence of Wind Deposition and Volcanism," *AJ* 60 (1955): 261–270; D. B. McLaughlin, "The Volcanic-Aeolian Hypothesis of Martian Features," *PASP* 68 (1956): 211–218; D. B. McLaughlin, "A New Theory of Mars," *Michigan Alumnus Quarterly Review* 62 (1956): 301–307.

76. E. N. Wells, J. Veverka, and P. Thomas, "Mars: Experimental Study of Albedo Changes Caused by Dust Fallout," *Icarus* 58 (1984): 331–338.

77. Tsuneo Saheki, "Martian Phenomena Suggesting Volcanic Activity," *Sky & Telescope*, February 1955, 144–146.

78. G. P. Kuiper, "Note on Dr. McLaughlin's Paper," *PASP* 68 (1956): 304–305.

79. Kuiper, "Visual Observations of Mars, 1956."

80. Quoted in S. Glasstone, *The Book of Mars* (Washington, D.C.: NASA, 1968), 119.

81. C. Sagan et al., "Variable Features on Mars, 2, Mariner 9 Global Results," *JGR* 78 (1973): 4163–4196.

82. Tombaugh also believed that "the Martian canals were global fracture fault lines, perhaps produced by internal heating and expansion, bursting a thick rigid crust at the round spots known as oases, or triggered by asteroid impacts at the round spots known as oases." C. W. Tombaugh to W. Sheehan, pers. comm., December 6, 1986.

83. A. Dollfus, "Mèsure de la quantité de vapeur d'eau contenue dans l'atmosphere de la planète Mars," *Comptes Rendu Académie Sciences* 256 (1963): 3009–3011. Dollfus had been in quest of this result for a long time, and had gone to great lengths—or rather heights—to achieve it. In 1958, he carried his instruments aloft in a stratospheric gondola, lifted by 104 large sounding balloons clustered along a vertical cable, to a height of 14 kilometers, hoping to detect water vapor and oxygen in the atmosphere of Mars. The results from the Jungfraujoch Scientific Station in 1963 were made at a height of only 3.5 kilometers, but were better than the balloon results. A description of the innovative telescope used is found in: A. Dollfus, "Observation of Water Vapor on Mars and Venus," in *The Origin and Evolution of Atmospheres and Oceans*, ed. P. J. Brancazio and A. G. W. Cameron (New York: John Wiley, 1964), 257–266.

84. H. Spinrad, G. Münch, and L. D. Kaplan, "The Detection of Water Vapor on Mars," *ApJ* 137 (1963): 1319–1321. The reason for the greater accuracy of these results is that while Dollfus's estimate required calibration on the assumption of a specific barometric pressure at the surface of Mars—he chose 85 millibars, which was a typical value for the time—the Mount Wilson team, which identified the bands of water vapor using the Doppler effect, made a direct measurement.

Chapter 11

1. D. R. Williams, "Chronology of Mars Exploration," NSSDCA, last updated December 15, 2020, https://nssdc.gsfc.nasa.gov/planetary/chronology_mars.html.

2. Quoted in J. N. Wilford, *We Reach the Moon* (New York: Bantam, 1969), 37.

3. Quoted in A. A. Siddiqi, *Sputnik and the Soviet Space Challenge* (Gainesville: University Press of Florida, 2003), 5. The phrase had been commonly used by Tsander going back to the 1920s.

4. "Believes Rocket Can Reach Moon. Smithsonian Institution Tells of Prof. Goddard's Invention to Explore Upper Air," *New York Times*, January 12, 1920.

5. Siddiqi, *Sputnik and the Soviet Space Challenge*, 5.

6. W. A. Anders to W. Sheehan, pers. comm., July 2018.

7. This design bureau had begun as a plant for producing anti-tank and air-defense guns for use in World War II and was later rededicated to the production of rockets, launch vehicles, and spacecraft. It was located in what was then known as Kaliningrad (not to be confused with the Baltic port), a suburb of Moscow. Since 1997, it has been known as Korolev, in the chief designer's honor.

8. W. von Braun, *The Mars Project* (1953; repr., Urbana: University of Illinois Press, 1991), xv. The German edition, *Das Marsprojekt*, was published in 1952 by Bechtle Verlag, Esslingen, Germany.

9. S. Khruschev, foreword to V. Hardesty and G. Eisman, *Epic Rivalry: The Inside Story of the Soviet and American Space Race* (Washington, D.C.: National Geographic, 2007), xi.

10. Siddiqi, *Sputnik and the Soviet Space Challenge*, 299.

11. Siddiqi, 333.

12. Khruschev, foreword to *Epic Rivalry*, xvi.

13. Siddiqi, *Sputnik and the Space Challenge*, 386–387.

14. Siddiqi, 478.

15. Siddiqi, 385.

16. In the West, very little information was available about what the Soviets were up to, and for a long time it was rumored that one of the Mars missions had resulted in the catastrophic explosion (not officially acknowledged until 1989) that occurred at Baikonur on October 24, 1960, costing 78 lives, including that of the chief marshal of the artillery, Mitrofan Ivanovich Nedelin. In fact, it was one of Mikhail Yangel's hypergolic fuel ICBMs, the R-16, that blew up, rather than one of Korolev's rockets. Siddiqi observes (*Sputnik and the Soviet Space Challenge*, 290), "Although it did not have any direct connection to the piloted space effort, there was clearly a repercussive delay on the Vostok program," which sought to put a human into orbit. Remarkably, it did not delay the launch of the first Venus probe, the following February.

17. W. T. Huntress and M. Ya. Marov, *Soviet Robots in the Solar System: Mission Technologies and Discoveries* (New York: Springer-Praxis Books, 2011), 90.

18. Earth and Mars line up in their orbits around the Sun roughly every 26 months. During this time, there is a window of opportunity, informally called a "launch window" and typically lasting three to four weeks, to send a spacecraft from Earth to Mars (or vice versa) for the least amount of energy, much less than at any other times in their orbits.

19. S. Khruschev, in R. das Saswato, "The Moon Landing Through Soviet Eyes," *Scientific American*, July 16, 2009.

20. Quoted in K. Schindler and W. Sheehan, *Northern Arizona Space Training* (Charleston, S.C.: Arcadia, 2017), 8.

21. D. E. Wilhlems, *To a Rocky Moon: A Geologist's History of Lunar Exploration* (Tucson: University of Arizona Press, 1993), 50.

22. See, for example, S. S. Limaye et al., "Venus' Spectral Signatures and the Potential for Life in the Clouds," *Astrobiology* 18 (2018): 1181–1198; J. S. Greaves et al., "Phosphine Gas in the Cloud Decks of Venus," *Nature Astronomy*, 2020, https://doi.org/10.1038/s41550-020-1174-4.

23. *To Mars: The Odyssey of Mariner IV*, JPL Technical Memorandum no. 33–229 (Pasadena: JPL-Caltech and NASA, 1965), 16–17, https://ntrs.nasa.gov/citations /19650018349.

24. If the orbits of Earth and Mars were exactly circular, the Hohmann transfer ellipse would be a path in which the spacecraft left Earth at an angle tangential to its orbit and arrived at an angle tangential to the orbit of Mars. This orbit would have its perihelion at the launch point and its aphelion at the orbit of Mars; the spacecraft's period of revolution around the Sun would be 520 days, and in getting from Earth to Mars it would travel halfway around this ellipse, so that the transit time from Earth to Mars would be 260 days. In that time Mars would have moved a distance of $(260/687) \times 360° = 136°$ around the Sun. It follows that in order for the spacecraft to reach Mars, the relative positions at launch must be such that the Earth-Sun-Mars angle is $180° − 136° = 44°$, or some 50 days before each opposition. Of course, the actual orbits of Earth and Mars are not exactly circular, nor do the two orbits lie in exactly the same plane; thus the actual conditions vary from launch window to launch window. In particular, less energy is required to reach Mars during launch windows that occur during perihelic opposition years (the late 1964 launch window was before an aphelic opposition). Since, in fact, the minimum energy requirement for the spacecraft to reach Mars is actually rather modest, it is possible for the actual trajectory to depart considerably from the ideal case of the Hohmann transfer ellipse.

25. Leighton's main research interests, in fact, were in other areas, such as cosmic ray research and, later, solar physics and infrared astronomy. He also played a role in organizing the notes of Richard Feynman's (1918–88) lectures in physics, which led to publication in 1963 as *Feynman Lectures in Physics*.

26. Quoted in "Special Supplement: Mariner 4 Photographs of Mars," *Sky & Telescope*, September 1965, 158.

27. R. B. Leighton et al., "Mariner IV Photography of Mars: Initial Results," *Science* 149 (1965): 629–630.

28. K. Davidson, *Carl Sagan: A Life* (New York: John Wiley, 1999), 179.

29. Quoted in W. Sheehan and S. J. O'Meara, *Mars: The Lure of the Red Planet* (Amherst, N.Y.: Prometheus, 2001), 277–278.

30. Leighton et al., "Mariner 4 Photography of Mars," 630.

31. "Special Supplement," 158. TIROS was the first satellite capable of remote sensing of Earth (that is, of viewing Earth from space). The program ran from TIROS-1, launched in 1960, through TIROS-10, launched in 1965, with the last TIROS spacecraft being deactivated by NASA in 1968.

32. It should be pointed out that the shock of Mariner 4's discovery was not universal. A number of astronomers had predicted the existence of craters on Mars,

including Clyde Tombaugh, Ernst Julius Öpik, Ralph Belknap Baldwin, Fred L. Whipple, and others.

33. Leighton et al., "Mariner 4 photography of Mars," 627.

34. "Special Supplement," 158–159.

35. D. Portree, "The First Voyager (1967)," *No Shortage of Dreams* (blog), February 4, 2019, http://spaceflighthistory.blogspot.com/2019/02/the-first-voyager-1967.html.

36. Huntress and Marov, *Soviet Robots in the Solar System*, 206–215.

37. NSSDCA, "Mariner 6," accessed February 10, 2012, https://nssdc.gsfc.nasa.gov/nmc/spacecraftDisplay.do?id=1969-014A.

38. Wikipedia, "NASA Exceptional Bravery Medal," last edited November 23, 2020, https://en.wikipedia.org/wiki/NASA_Exceptional_Bravery_Medal.

39. R. B. Leighton et al., "Mariner 6 and 7 Television Pictures: Preliminary Analysis," *Science* 166 (1969): 49–67.

40. Leighton et al., 62.

41. Leighton et al., 63.

42. R. B. Leighton and B. C. Murray, "Behavior of Carbon Dioxide and Other Volatiles on Mars," *Science* 153 (1966): 136–144.

43. Author J. B. had it assigned in class by Bruce Murray himself, in 1987. It's still an elegant and powerful demonstration of simple physics and thermodynamics applied to make an important, testable scientific prediction. Not until Mariner 9 would it become clear that, while the seasonal caps are mostly frozen CO_2 and there is a residual but degrading CO_2 cover over the south polar cap in summer, the residual north polar cap surface in summer is composed of "dirty" water ice. And it wouldn't be until years into the Mars Reconnaissance Orbiter mission, more than 40 years after the flyby Mariners, that it would be discovered that the permanent south polar cap actually stores enormous amounts of deeply buried CO_2 ice, enough to double the surface pressure if it sublimed back into the atmosphere.

44. See, for example, G. C. Pimentel, P. B. Forney, and K. C. Herr, "Evidence About Hydrate and Solid Water in the Martian Surface from the 1969 Mariner Infrared Spectrometer," *JGR* 79 (1974): 1623–1634.

45. Leighton et al., "Mariner 6 and 7 Television Pictures," 65.

Chapter 12

1. A. A. Siddiqi, *The Soviet Space Race with Apollo* (Gainsville: University Press of Florida, 2003), 677.

2. For a summary of what was known at the beginning of the space probe era, see P. Moore, *The Planet Venus* (London: Faber and Faber, 1961). Within a little over a year it was completely out of date.

3. C. Sagan, *The Radiation Balance of Venus*, NASA technical report no. 32–34 (Pasadena: JPL-Caltech, 1960).

4. R. Reeves, *The Superpower Space Race: An Explosive Rivalry Through the Solar System* (New York: Plenum, 1994), 170.

5. A wonderfully Saganesque view of the world's response to the threat of global warming—from 1990 but still poignant today—can be found in C. Sagan, "Croesus and Cassandra: Policy Response to Global Warming," *American Journal of Physics* 58 (1990): 721.

6. NSSDCA, "Mariner-H," accessed February 10, 2021, https://nssdc.gsfc.nasa.gov/nmc/spacecraftDisplay.do?id=MARINH.

7. NSSDCA, "Cosmos 419," accessed February 10, 2021, https://nssdc.gsfc.nasa.gov/nmc/spacecraftDisplay.do?id=1971-042A.

8. C. R. Capen, "Martian Yellow Clouds—Past and Future," *Sky & Telescope*, February 1971, 120.

9. For example, see C. Capen and L. J. Martin, "Mars' Great Storm of 1971," *Sky & Telescope*, May 1972, 276.

10. For a detailed review, see R. J. McKim, "Telescopic Martian Dust Storms: A Narrative and Catalogue," *Mem. BAA* 44 (June 1999): 85–91.

11. Quoted in R. Bradbury et al., *Mars and the Mind of Man* (New York: Harper and Row, 1973), 23.

12. Bradbury et al., 35.

13. Bradbury et al., 24–25.

14. NSSDCA, "Mars 2," accessed February 10, 2021, https://nssdc.gsfc.nasa.gov/nmc/spacecraftDisplay.do?id=1971-045A; see also W. T. Huntress and M. Ya. Marov, *Soviet Robots in the Solar System: Mission Technologies and Discoveries* (New York: Springer-Praxis Books, 2011), 258.

15. Huntress and Marov, *Soviet Robots in the Solar System*, 248–259.

16. See the story by NASA MRO HiRISE principal investigator Alfred McEwen, "Could This Be the Soviet Mars 3 Lander?," *LPL: HiRISE*, April 11, 2013, https://www.uahirise.org/ESP_031036_1345; as well as additional details compiled and translated from the original Russian by The Planetary Society's Emily Lakdawalla, "Russia's Mars 3 Lander Maybe Found by Russian Amateurs," *The Planetary Society*, April 12, 2013, https://www.planetary.org/articles/0412-how-we-searched-for-mars-3.

17. Huntress and Marov, *Soviet Robots in the Solar System*, 260–262.

18. Details about the mission and some of its results can be found (in Russian) through the Lavochkin Institute's website: "Avtomaticheskie mezhplanetnye stantsii 'Mars-2,3,'" *Nauchno-Proizvodstvennoe Obedinenie im. S. A. Lavochkina*, accessed February 10, 2021, http://www.laspace.ru/projects/planets/mars-2_3/.

19. W. K. Hartmann to W. Sheehan, pers. comm., 2019.

20. Pollack was author J. B.'s postdoctoral research advisor at NASA's Ames Research Center, just north of San Jose, Calif.

21. See, for example, R. P. Turco et al., "Nuclear Winter: Global Consequences of Multiple Nuclear Explosions," *Science* 222 (1983): 1283–1292.

22. For an excellent history of the discovery of global warming, see S. Weart, *The Discovery of Global Warming* (Cambridge, Mass.: Harvard University Press, 2003).

23. W. K. Hartmann to W. Sheehan, pers. comm., January 17, 2020.

24. W. K. Hartmann and O. Raper, *The New Mars: The Discoveries of Mariner 9* (Washington, D.C.: NASA, 1974), v.

25. R. Bradbury, "Dark They Were, and Golden-Eyed," originally published as "The Naming of Names," *Thrilling Wonder Stories*, August 1949.

26. W. K. Hartmann to W. Sheehan, pers. comm., 2020.

27. "16. Commission pour l'etude physique des planetes et des satellites," *Transactions IAU* 13, no. 2 (B) (1967): 99.

28. "Commission 16: Physical Study of Planets and Satellites," *Transactions IAU* 14, no. 2 (B) (1970): 129.

29. "Commission 16: Physical Study of Planets and Satellites," *Transactions IAU* 15, no. 2 (B) (1973): 105–107; "Lunar and Martian Nomenclature," *Transactions IAU* 15, no. 2 (B) (1973): 217–221.

30. See, for instance, T. Mutch et al., *The Geology of Mars* (Princeton, N.J.: Princeton University Press, 1976); M. H. Carr, *The Surface of Mars*, rev. and updated ed. (Cambridge: Cambridge University Press, 2006); V. R. Baker, *The Channels of Mars* (Austin: University of Texas Press, 1982); H. H. Kieffer et al., eds., *Mars* (Tucson: University of Arizona Press, 1982); J. Bell, ed., *The Martian Surface: Composition, Mineralogy, and Physical Properties* (Cambridge: Cambridge University Press, 2008); and N. Barlow, *Mars: An Introduction to Its Interior, Surface, and Atmosphere* (Cambridge: Cambridge University Press, 2008). For a more popular-level account, see W. K. Hartmann, *Mars: The Mysterious Landscapes of the Red Planet* (New York: Workman, 2003).

31. R. McKim, "Telescopic Martian Dust Storms: A Narrative and Catalogue," *Mem. BAA* 44 (1999): 140.

32. W. K. Hartmann to W. Sheehan, pers. comm., March 1, 2021.

33. McKim, "Telescopic Martian Dust Storms," 133–134.

34. R. W. Zurek, "Martian Great Dust Storms: An Update," *Icarus* 50 (1982): 288–310.

35. McKim, "Telescopic Martian Dust Storms," 134.

36. Bradbury et al., *Mars and the Mind of Man*, 19.

37. An oft-cited paper based on Mariner 9 findings is C. Sagan and P. Fox, "The Canals of Mars: An Assessment After Mariner 9," *Icarus* 25 (1975): 602–612, in which the authors concluded (p. 609) "that while a small subset of the classical Lowellian canals corresponds to topographic or albedo features on Mars, the

bulk of the canals do not. Indeed there are many canals where there are no real surface features, and many real surface features where there are no canals. Although we have not pursued the relevant statistical study, we have the impression that there exists an anticorrelation between the cartographic accuracy of a map and the number of canals it displays. The vast majority of the canals appear to be largely self-generated by the visual observers of the canal school, and stand as monuments to the imprecision of the human eye-brain-hand system under difficult observing conditions." The canals that do have at least some basis in reality are generally associated with albedo rather than topographic features.

38. C. Sagan et al., "Variable Features on Mars: Preliminary Mariner 9 Television Results," *Icarus* 17 (1973): 346–372.

39. See V. R. Baker and D. J. Milton, "Erosion by Catastrophic Floods on Mars and Earth," *Icarus* 23 (1974): 27–41. See also Baker, *The Channels of Mars*.

40. The valleys have received the names that Mars had in different cultures; thus, Ma'adim is Hebrew for Mars; the dry river valleys associated with Margaritifer chaos, Ares, Tiu, and Simud, are the Greek, German, and Sumerian names for Mars; and so on. See also appendix E.

41. The floor of Hellas is the lowest point on Mars, located some five kilometers below the Martian datum. There, the atmospheric pressure can be as much as 14 millibars. By comparison, the average atmospheric pressure is 6 or 7 millibars, depending on the season, while at the summit of Olympus Mons, the pressure is only 0.7 millibar. A reminder: the standard pressure on Earth is 1,013 millibars, and even at the top of Mount Everest is still some 250 millibars. The atmosphere of Mars is very thin!

42. C. Sagan, O. B. Toon, and P. J. Gierasch, "Climatic Change on Mars," *Science* 181 (1973): 1045–1049.

43. A. C. Clarke, *The Sands of Mars* (London: Sidgwick and Jackson, 1951).

44. C. Sagan, *The Cosmic Connection* (New York: Doubleday, 1973), 69.

45. K. S. Robinson, *Red Mars* (New York: Bantam Spectra, 1993), 83.

Chapter 13

1. C. Sagan, "The Solar System," *Scientific American*, January 1975.

2. H. Shapley, *The View from a Distant Star: Man's Future in the Universe* (New York: Basic Books, 1963), 33.

3. Quoted in P. Raeburn and M. Golombek, *Mars: Uncovering the Secrets of the Red Planet* (Washington, D.C.: National Geographic Society, 1998), 76.

4. A. C. Clarke, *Interplanetary Flight: An Introduction to Astronautics* (New York: Harper and Row, 1951), 135.

5. C. Sagan, *The Demon-Haunted World: Science as a Candle in the Dark* (New York: Ballantine Books, 1996), xiv.

6. For more details about Miller's experiments and their application to the chemistry of planets, see D. P. Cruikshank and W. Sheehan, *Discovering Pluto: Exploration at the Edge of the Solar System* (Tucson: University of Arizona Press, 2018), 312ff.

7. On Kuiper, see D. W. G. Sears, *Gerard P. Kuiper and the Rise of Modern Planetary Science* (Tucson: University of Arizona Press, 2019).

8. Quoted in C. Flammarion, *Camille Flammarion's The Planet Mars*, trans. P. Moore, ed. W. Sheehan (Cham, Switzerland: Springer, 2015), 436.

9. R. S. Richardson, *Exploring Mars* (New York: McGraw-Hill, 1954), 127.

10. K. Davidson, *Carl Sagan: A Life* (New York: John Wiley, 1999), 69.

11. C. Sagan, "The Planet Venus," *Science* 133 (1961): 849.

12. W. Sheehan and S. J. O'Meara, *Mars: The Lure of the Red Planet* (Amherst, N.Y.: Prometheus, 2001), 276.

13. G. P. Kuiper, *Condon Report: Scientific Study of Unidentified Flying Objects* (Denver: University of Colorado, 1969), 1312.

14. C. Sagan, "Mars: A New World to Explore," *National Geographic*, December 1967, 835.

15. D. McNab and J. Younger, *The Planets* (New Haven, Conn.: Yale University Press, 1999), 91.

16. C. Sagan, "The Solar System," in *The Solar System* (San Francisco: W. H. Freeman, 1975), 1–11.

17. H. Masursky, "Mars," in *The New Solar System*, ed. J. K. Beatty, C. C. Petersen, and Andrew Chaikin (Cambridge, Mass.: Sky, 1981), 83.

18. Davidson, *Carl Sagan*, 177.

19. Sagan, "The Solar System," 7.

20. Apollo 18 would have landed in Schroeter's Valley or perhaps Gassendi crater, but that booster was instead repurposed to launch the Skylab space station in 1973. Apollo 19 and 20 had been intended for possible landing sites like Hyginus rille, Tycho crater, Marius Hills, or Copernicus crater but were never flown; pieces were split up and are currently on display at the Johnson Space Center in Texas, the Kennedy Space Center in Florida, and the Smithsonian National Air and Space Museum in Washington, D.C.

21. D. J. Shayler, *Apollo: The Lost and Forgotten Missions* (New York: Springer-Praxis, 2002).

22. J. Bell, *The Interstellar Age: Inside the Forty-Year Voyager Mission* (New York: Dutton, 2016).

23. See E. C. Ezell, *On Mars: Exploration of the Red Planet, 1958–1978*, NASA special publication 4212 (Washington, D.C.: Government Printing Office, 1984), chapter 4, https://history.nasa.gov/SP-4212/contents.html.

24. Ezell, chapter 4.

25. Ezell, chapter 5.

26. Ezell, chapter 6.

27. For a summary, see, for example, C. A. Scharf, "The Great Martian Dust Storm of '71," *Life, Unbounded* (blog), *Scientific American*, October 21, 2013, https://blogs.scientificamerican.com/life-unbounded/the-great-Martian-storm-of-e2809971.

28. See M. H. Carr et al., *Viking Orbiter Views of Mars*, NASA special publication SP-441 (Washington, D.C.: Government Printing Office, 1980).

29. Nadine Barlow of Northern Arizona University and H. Jay Melosh of Purdue University were two noteworthy pioneers in the modern study of impact cratering on Mars and elsewhere in the solar system. Barlow built an impressive career around counting, classifying, and assessing the geological implications of planetary impacts. Her book *Mars: An Introduction to Its Interior, Surface and Atmosphere* (Cambridge: Cambridge University Press, 2008) has become a standard reference not only on Martian impact cratering but also on other planetary surface geological and surface-atmosphere processes. Melosh, a gifted and internationally recognized theorist, wrote the book that outlined the theoretical basis for our understanding of the impact cratering process (*Impact Cratering: A Geologic Process* [Oxford: Oxford University Press, 1996]), expanding that work more recently into a deeper observational and theoretical treatise on the ways that planetary surfaces are modified in general (*Planetary Surface Processes* [Cambridge: Cambridge University Press, 2011]). Sadly, both colleagues passed away in 2020, both while still making active and important contributions to the study of impact cratering and other planetary surface processes.

30. See, for example, Wikipedia, "Milankovitch cycles," last edited February 8, 2021, https://en.wikipedia.org/wiki/Milankovitch_cycles.

31. Ezell, *On Mars*, chapter 9. See also the outstanding summary and illustrations in P. J. Stooke, *The International Atlas of Mars Exploration: The First Five Decades* (Cambridge: Cambridge University Press, 2012), 52–74.

32. Ezell, *On Mars*, chapter 6.

33. Ezell, chapter 10.

34. Celestial navigation experts and spacecraft engineers had identified a number of so-called landing ellipses toward which to target the Viking Landers. Each ellipse represented a statistically estimated surface area over 700 kilometers long and 300 kilometers wide within which the team felt confident it could deploy each lander. For graphical representations of the potential landing ellipses, see, for example, D. F. S. Portree, "The Earliest Candidate Viking Landing Sites (1970)," *Wired* (blog), November 22, 2012, https://www.wired.com/2012/11/early-candidate-viking-landing-sites.

35. R. Arvidson to J. Bell, pers. comm., 2020.

36. Ezell, *On Mars*, chapter 10.

37. G. A. Soffen and C. W. Snyder, "The First Viking Mission to Mars," *Science* 193 (1976): 759–766.

38. Quoted in Raeburn and Golombek, *Mars*, 86.

39. N. H. Horowitz, *To Utopia and Back: The Search for Life in the Solar System* (New York: W. H. Freeman, 1986), 16–17.

40. "Indigenous" here means complex organic compounds that had formed through prebiotic or biotic chemical reactions, as within life-forms on Earth. These are contrasted to "exogenous" organic compounds like those created abiotically and identified in giant molecular clouds or even in certain classes of meteorites. Low levels of exogenous organics are expected to be found on Mars and other planets because of the constant rain of cosmic dust, asteroids, and comets that fall to planetary surfaces. While they were not identified in the Viking Lander mass spectrometer data, organic molecules were later discovered by a more advanced spectrometer on the Curiosity rover, providing a hypothesis for how meteoritic organics are converted to other carbon-bearing molecules in the presence of strong oxidizers like perchlorate.

41. See, for example, the Viking Lander Imaging Team, *The Martian Landscape*, NASA special publication 425 (Washington, D.C.: NASA Scientific Information and Technical Office, 1978); R. E. Arvidson, J. L. Gooding, and H. J. Moore, "The Martian Surface as Imaged, Sampled, and Analyzed by the Viking Landers," *Reviews of Geophysics* 27 (1989): 39–60.

42. For many decades, the Soil Science Society of America, in its official definition of "soil," required that life be present for soil to exist. More recently, the society has adopted a more general definition of soil that appears to enable a more expansive, not-so-Earthcentric view, defining it anew as "the layer(s) of generally loose mineral and/or organic material that are affected by physical, chemical, and/or biological processes at or near the planetary surface, and usually hold liquids, gases and biota and support plants." So, Mars (and lunar) "soil" it is! See H. van Es, "A New Definition of Soil," *CSA News* 62 (October 5, 2017): 20–21, https://acsess.onlinelibrary.wiley.com/doi/full/10.2134/csa2017.62.1016.

43. F. O. Huck et al., "Spectrophotometric and Color Estimates of the Viking Lander Sites," *JGR* 82 (1977): 4401–4411; see also M. D. Gunn and C. R. Cousins, "Mars Surface Context Cameras Past, Present, and Future," *Earth & Space Science* 3 (2016): 144–162.

44. Imaging scientists work hard to get good signal levels through all color filters in a camera system. In the case of Mars, because the surface and atmosphere are quite reddish in color, images taken through a blue filter have to use a much longer exposure time—up to five times or more the exposure time needed through a red filter—to get to the same signal levels (this is because reddish colors involve less blue light being reflected back). Merging separate raw color

filter images of red, green, and blue into an RGB color composite will yield a color view with way too much blue (and marginally too much green) because the exposure times for the blue and green filters were longer than for the red filter. When the differences in exposure time are properly accounted for as part of the image calibration, and the red, green, and blue data is all compared at the same exposure time (like what happens in the human eye, naturally), the colors come out properly balanced.

45. Ezell, *On Mars*, chapter 10.

46. R. Arvidson to J. Bell, pers. comm., 2020.

47. R. Arvidson to J. Bell, pers. comm., 2020.

48. For a great starting point to understand more about this unique and interesting kind of geological terrain, see Wikipedia, "Patterned Ground," last edited February 1, 2021, https://en.wikipedia.org/wiki/Patterned_ground.

49. Viking Lander 2 only dug trenches down to about 15 centimeters deep. A more recent analysis of the likely depth of buried ground ice at that latitude reveals that if the team could have trenched only about another 10 centimeters down, it might have discovered buried ice 30 years before the NASA Phoenix lander did! See S. Byrne et al., "Distribution of Mid-Latitude Ground Ice on Mars from New Impact Craters," *Science* 325 (2009): 1674–1676.

50. S. Wall, "Analysis of Condensates Formed at the Viking 2 Lander Site: The First Winter," *Icarus* 47 (1981): 173–183.

51. Quoted on Wikipedia, "Gerald Soffen," last edited October 20, 2020, https://en.wikipedia.org/wiki/Gerald_Soffen.

52. For a fun and fanciful take on that idea, see author J. B.'s book *The Ultimate Interplanetary Travel Guide* (New York: Sterling, 2018).

Chapter 14

1. For details on the mineralogical characteristics of SNCs, see O. R. Norton, *The Cambridge Encyclopedia of Meteorites* (Cambridge: Cambridge University Press, 2002), 202.

2. See, for example, M. Kramer, "Mars Meteorite May Be Missing Link to Red Planet's Past," *Space.com*, January 3, 2013, https://www.space.com/19117-mars-meteorite-martian-missing-link.html.

3. For a comprehensive account, see K. Sawyer, *The Rock from Mars: A Detective Story on Two Planets* (New York, Random House, 2002).

4. W. Sheehan and S. J. O'Meara, *Mars: The Lure of the Red Planet* (Amherst, N.Y.: Prometheus, 2001), 294.

5. D. S. McKay et al., "Search for Past Life on Mars: Possible Relic Biogenic Activity in Martian Meteorite ALH84001," *Science* 273 (1996): 924–930.

6. Sheehan and O'Meara, *Mars*, 295.

7. C. Sagan, *Broca's Brain: Reflections on the Romance of Science* (New York: Random House, 1979), 62. Sagan was of course phenomenally well-read. While he may have gotten the idea for his "standard" from Occam's razor, he might also have been drawing from Théodore Flournoy's *From India to the Planet Mars: A Study of a Case of Somnambulism*, trans. Daniel R. Vermilye (New York: Harper and Brothers, 1900), 369–370. Flournoy was a Swiss physician and psychologist of psychic phenomena who transcribed the utterances of the medium Hélène Smith and published communications purporting to be from past lives that she spent in India and on Mars. These extraordinary claims obviously admitted more mundane explanations, such as that Smith suffered from hysteria, a diagnosis that was just then becoming subject to serious study. Flournoy himself, citing Pierre-Simon Laplace's writings on probability, wrote: "The weight of evidence for an extraordinary claim must be proportioned to its strangeness."

8. I. Halevy, W. W. Fischer, and J. M. Eiler, "Carbonates in the Martian Meteorite Allan Hills 84001 Formed at 18°±4°C in a Near-Surface Aqueous Environment," *Publications of the National Academy of Sciences* 108 (2001): 16895–16899.

9. NSSDCA, "Phobos Project Information," last updated July 22, 2019, https://nssdc.gsfc.nasa.gov/planetary/phobos.html.

10. For a review of ISM and other contemporary pre-MGS mineralogic inferences about Mars, see W. M. Calvin and J. F. Bell III, "Historical Context: The Pre-MGS View of Mars' Surface Composition," in *The Martian Surface: Composition, Mineralogy, and Physical Properties*, ed. J. F. Bell III (Cambridge: Cambridge University Press, 2008), 20–34.

11. R. Reagan, first inaugural address, January 20, 1981, *Avalon Project*, Yale Law School, Lillian Goldman Law Library, https://avalon.law.yale.edu/20th_century/reagan1.asp; G. Norquist, interview, National Public Radio, *Morning Edition*, May 25, 2001.

12. For a comprehensive historical overview of Reagan-era space policies and achievements, see J. Logsdon, *Ronald Reagan and the Space Frontier* (London: Palgrave Macmillan, 2019).

13. For more details on the origin and early role of The Planetary Society in helping save NASA funding, see J. Bell, *The Interstellar Age: Inside the Forty-Year Voyager Mission* (New York: Dutton, 2016), 21–26; and visit the society's website for more details about its current mission: http://www.planetary.org.

14. See E. M. Conway, *Exploration and Engineering: The Jet Propulsion Laboratory and the Quest for Mars* (Baltimore, Md.: Johns Hopkins University Press, 2015). See also J. Eberhart, "New Starts and Tough Choices at NASA," *Science News*, October 19, 1985.

15. D. Vaughn, *The Challenger Launch Decision: Risky Technology, Culture, and Deviance at NASA* (Chicago: University of Chicago Press, 1996), 20.

16. J. A. Van Allen, "Space Science, Space Technology, and the Space Station," *Scientific American*, January 1986, 22.

17. For more history, details, and mission highlights, see J. F. Bell, *Hubble Legacy: 30 Years of Images and Discoveries* (New York, Sterling, 2020).

18. These goals included: (1) determine the global elemental and mineralogical character of the surface material; (2) define globally the topography and gravitational field; (3) establish the nature of the Martian magnetic field; (4) determine the temporal and spatial distribution, abundance, sources, and sinks of volatiles and dust over a seasonal cycle; and (5) explore the structure and circulation of the atmosphere. See, for example, NSSDCA, "Mars Observer," accessed February 10, 2021, https://nssdc.gsfc.nasa.gov/nmc/spacecraft/display.action?id=1992-063A.

19. *The Mars Observer Mission*, NASA contractor report 197707 (Washington, D.C.: Government Printing Office, 1993), https://ntrs.nasa.gov/search.jsp?R=19950019916.

20. S. Hubbard, *Exploring Mars: Chronicles from a Decade of Discovery* (Tucson: University of Arizona Press, 2011), 9.

21. M. C. Malin et al., "An Overview of the 1985–2006 Mars Orbiter Camera Science Investigation," *Mars* 5 (2010): 1–60.

22. In the absence of any smoking-gun telemetry or other information from the spacecraft, this is the most likely scenario for the mission's failure according to an independent investigative report delivered to the NASA administrator in January 1994. For a summary of this and other potential failure scenarios, and links to more information on the investigation, see Malin Space Science Systems, "The Loss of Mars Observer," accessed February 10, 2021, http://www.msss.com/mars/observer/project/mo_loss/moloss.html.

23. Hubbard, *Exploring Mars*, 9.

24. Hubbard, 9.

25. K. Edgett to J. Bell, pers. comm., February 2020.

26. Author J. B. was trained as one of those ground-based telescopic and HST observers, using such facilities first for his thesis work trying to constrain the iron-bearing mineralogy of Mars in the late 1980s and early 1990s, and then later to help provide some basic information on surface properties that could be used to optimize the design and operation of future orbital and landed instruments eventually sent to Mars.

27. For a summary of the kinds of work being done telescopically during this time, see the proceedings of the two Mars Telescopic Observations Workshops, convened in 1995 and 1997: J. F. Bell III and J. E. Moersch, *Workshop on Mars Telescopic Observations*, LPI technical report no. 95–04 (Houston, Tex.: Lunar and Planetary Institute, 1995), http://hdl.handle.net/2060/19960027472; A. L. Sprague and J. F. Bell III, *Mars Telescopic Observations Workshop II*, LPI technical report

no. 97–03 (Houston, Tex.: Lunar and Planetary Institute, 1997), http://hdl.handle
.net/2060/19980151104.

28. K. Edgett to J. Bell, pers. comm., February 2020.

29. See A. L. Albee et al., "Overview of the Mars Global Surveyor mission," *JGR* 106 (2001): 23291–23316.

30. M. C. Malin et al., "Design and Development of the Mars Observer Camera," *International Journal of Imaging Systems and Technology* 3 (1991): 76–91.

31. The classic reference is the following peer-reviewed research journal article, which is so long (142 pages!) and detailed that it is actually more like a short book: M. C. Malin and K. S. Edgett, "Mars Global Surveyor Mars Orbiter Camera: Interplanetary Cruise Through Primary Mission," *JGR* 106 (2001): 23429–23570.

32. K. Edgett to J. Bell, pers. comm., February 2020.

33. Malin and Edgett, "Mars Global Surveyor Mars Orbiter Camera," 23566.

34. Another classic peer-reviewed journal article on this topic is M. C. Malin and K. S. Edgett, "Sedimentary Rocks of Early Mars," *Science* 290 (2000): 1927–1937.

35. Malin and Edgett, "Mars Global Surveyor Mars Orbiter Camera," 23566.

36. M. C. Malin and K. S. Edgett, "Evidence for Persistent Flow and Aqueous Sedimentation on Early Mars," *Science* 302 (2003): 1931–1934.

37. Regarding the terminology, as approved by the International Astronomical Union (IAU): with the exception of Valles Marineris (the Valley of Mariner), large valleys on Mars receive the names of Mars in various languages. Smaller ones are named for rivers on Earth. The Coogoon is a river in Queensland, Australia. See appendix E.

38. M. C. Malin and K. S. Edgett, "Evidence for Recent Groundwater Seepage and Surface Runoff on Mars," *Science* 288 (2000): 2330–2335.

39. W. K. Hartmann, T. Thorsteinsson, and F. Sigurdsson, "Martian Hillside Gullies and Icelandic Analogs," *Icarus* 162 (2003): 259–277.

40. M. C. Malin et al., "Present-Day Impact Cratering Rate and Contemporary Gully Activity on Mars," *Science* 314 (2006): 1573–1577.

41. Malin et al.

42. See, for example, C. Dundas et al., "The Formation of Gullies on Mars Today," *Geological Society, London: Special Publications* 467 (2017): 67–94.

43. Malin et al., "Present-Day Impact Cratering Rate," 1577.

44. "Mars Orbiter Laser Altimeter (MOLA) Elevation Map," JPL press release, May 27, 1999, https://mars.jpl.nasa.gov/mgs/sci/mola/mola-may99.html.

45. See, for example, D. E. Smith and M. T. Zuber, "The Shape of Mars and the Topographic Signature of the Hemispheric Dichotomy," *Science* 271 (1996): 184–188; and D. E. Smith et al., "Mars Orbiter Laser Altimeter: Experiment Summary After the First Year of Global Mapping of Mars," *JGR* 106 (2001): 23689–23722.

46. Malin et al., "Present-Day Impact Cratering Rate," 1577.

47. See review by M. H. Acuña, G. Kletetschka, and J. E. P. Connerney, "Mars' Crustal Magnetization: A Window into the Past," in Bell, *The Martian Surface*, 242–262.

48. See E. Lakdawalla, "Why Is Only Half of Mars Magnetized?," *The Planetary Society* (blog), October 24, 2008, http://www.planetary.org/blogs/emily-lakdawalla/2008/1710.html.

49. Acuña, Kletetschka, and Connerney, "Mars' Crustal Magnetization," 242.

50. P. R. Christensen et al., "Global Mineralogy Mapped from the Mars Global Surveyor Thermal Emission Spectrometer," in Bell, *The Martian Surface*, 195–220.

51. See review by V. E. Hamilton et al., "Thermal Infrared Spectral Analyses of Mars from Orbit Using the Thermal Emission Spectrometer and Thermal Emission Imaging System," in *Remote Compositional Analysis: Techniques for Understanding Spectroscopy, Mineralogy, and Geochemistry of Planetary Surfaces*, ed. J. L. Bishop, J. Moersch, and J. F. Bell III (Cambridge: Cambridge University Press, 2020), 484–498.

52. Hamilton et al.

53. Christensen et al., "Global Mineralogy Mapped," 214.

54. M. T. Mellon, R. L. Fergason, and N. E. Putzig, "The Thermal Inertia of the Surface of Mars," in Bell, *The Martian Surface*, 399–427.

55. P. E. Geissler et al., "Orbital Monitoring of Martian Surface Changes," *Icarus* 278 (2016): 279–300.

56. H. Wang et al., "Cyclones, Tides and the Origin of a Cross-Equatorial Dust Storm," *GRL* 30 (2003): 1488.

57. C. Pellier to W. Sheehan, pers. comm., November 25, 2006.

58. Malin and Edgett, "Mars Global Surveyor Mars Orbiter Camera," 23566.

Chapter 15

1. The quote is from A. G. W. Cameron, chair of the National Academy of Sciences Space Science Board, in a letter of November 23, 1976, to Fletcher, as cited in D. S. F. Portree, "Mars 1984 Rover-Orbiter-Penetrator Mission (1977)," *Wired.com*, July 30, 2012, https://www.wired.com/2012/07/mars-1984-rover-mission-1977.

2. Mars Science Working Group, *A Strategy for the Scientific Exploration of Mars*, NASA/JPL document no. D-8211 (1991).

3. G. S. Hubbard et al., *Mars Environmental Survey (MESUR) Science Objectives and Mission Description* (Moffett Field, Calif.: NASA Ames Research Center, 1991).

4. For more details on the makeup of the Mars Pathfinder team, see D. B. Shirley with D. Morton, *Managing Martians* (New York: Broadway Books, 1998); B. K. Muirhead and W. L. Simon, *High Velocity Leadership: The Mars Pathfinder Approach to Better, Faster, Cheaper* (New York: HarperCollins, 1999).

5. A. Mishkin, *Sojourner* (Berkeley, Calif.: Berkeley, 2003).

6. Author J. B. was fortunate to be among those scientists selected to participate, and what a thrill ride it was!

7. M. P. Golombek et al., "Selection of the Mars Pathfinder Landing Site," *JGR* 102 (1997): 3967–3988.

8. M. P. Golombek, ed., *Mars Pathfinder Landing Site Workshop*, LPI technical report 94–04 (Houston, Tex.: Lunar and Planetary Institute, 1994).

9. Golombek et al., "Selection of the Mars Pathfinder Landing Site."

10. M. P. Golombek et al., "Overview of the Mars Pathfinder Mission and Assessment of Landing Site Predictions," *Science* 278 (1997): 1743–1748.

11. See V. R. Baker, *The Channels of Mars* (Austin: University of Texas Press, 1982).

12. "Mars Pathfinder Entry, Descent, and Landing," *NASA* (website), accessed February 10, 2021, https://mars.nasa.gov/MPF/mpf/edl/edl1.html.

13. "Mars Pathfinder Entry, Descent, and Landing."

14. G. E. Wood, S. W. Asmar, and T. A. Rebold, *Mars Pathfinder Entry, Descent, and Landing Communications*, TDA progress report 42–131, November 15, 1997, https://tmo.jpl.nasa.gov/progress_report/42-131/131I.pdf.

15. Golombek et al., "Overview of the Mars Pathfinder Mission."

16. See "Mars Pathfinder Mission—LIVE Coverage—1997—Part 1," uploaded May 19, 2012, by MoosePower740, YouTube video, 59:11, https://www.youtube.com/watch?v=zUaalbRC7KA, starting at about 8:00.

17. "Mars Pathfinder Update," NASA/JPL press conference, uploaded July 5, 1997, C-SPAN video, 57:00, https://www.c-span.org/video/?87512-1/mars-pathfinder-update.

18. A. C. Clarke, *The City and the Stars; and The Sands of Mars* (New York: Warner Books, 2001), 381. In the introduction, Clarke notes that Donna Shirley sent her autobiography *Managing Martians* to him with the dedication: "To Arthur C. Clarke, who inspired my summer vacation on Mars." She had found his work inspiring when she was a 12-year-old girl (how often these things happen in late preadolescence): "searching for my own place in the world, I'd read *The Sands of Mars*, a book that pointed me towards the sky." Shirley and Morton, *Managing Martians*, 16.

19. Shirley and Morton, *Managing Martians*, 15.

20. Shirley and Morton, 199.

21. On humans' social responses to technology, see, for instance, B. Reeves and C. Nass *The Media Equation: How People Treat Computers, Television, and New Media Like Real People and Places* (Stanford, Calif.: Center for the Study of Language and Information, 1992).

22. H. J. Eisen et al., "Mechanical Design of the Mars Pathfinder Mission," in *7th European Space Mechanisms and Tribology Symposium, Proceedings of the Con-*

ference Held 1–3 October 1997 at ESTEC, Noordwijk, the Netherlands, ed. B. H. Kaldeich-Schürmann (Paris: European Space Agency, 1997), 293–301; see also additional details in Mishkin, *Sojourner.*

23. This and lots of other time-lapse movies of the rover, taken with the IMP camera and most of them assembled by IMP team member and JPL imaging expert Justin Maki, can be found at "Mars Pathfinder," *NASA* (website), accessed February 10, 2021, https://mars.nasa.gov/MPF/ops/rvrmovie.html.

24. I. Asimov, *Science, Numbers and I* (New York: Doubleday, 1968).

25. Golombek et al., "Overview of the Mars Pathfinder Mission."

26. J. F. Bell III et al., "Mineralogic and Compositional Properties of Martian Soil and Dust: Results from Mars Pathfinder," *JGR* 105 (2000): 1721–1755.

27. M. B. Madsen et al., "The Magnetic Properties Experiments on Mars Pathfinder," *JGR* 104 (1999): 8761–8780.

28. See, for example, R. Rieder, "The Chemical Composition of Martian Soil and Rocks Returned by the Mobile Alpha Proton X-ray Spectrometer: Preliminary Results from the X-ray Mode," *Science* 278 (1997): 1771; H. Y. McSween Jr. et al., "Chemical, Multispectral, and Textural Constraints on the Composition and Origin of Rocks at the Mars Pathfinder Landing Site," *JGR* 104 (1999): 8679–8716.

Chapter 16

1. B. DeVoto, introduction to *The Journals of Lewis and Clark*, ed. B. DeVoto (Boston: Houghton Mifflin, 1953), xv.

2. As William K. Hartmann, one of the most prolific Mariner 9 and Viking scientists, recalls, "Some of my colleagues and I have lamented that a lot of the earlier work, some of it questioned at the time, is accepted now—or even independently 'rediscovered' by people who didn't know the early literature—but the initial papers (original source of the idea) are often forgotten. Interestingly this problem is unusually common with Mars because of the 20-year gap between Viking and later close-up Mars missions. The result was that when Mars work 'geared up again' in the 90s it was a new generation. But worse yet, the first generation work in the 60s and 70s was published in paper journals. In the 90s those journals were all properly stored in libraries, but the new work was beginning to be transmitted online—and the early journals had not yet been digitized! The result (as I sensed happening at the University of Arizona) was that a lot of graduate students never 'went across campus' to the musty archives of the library to chase down the early literature. This is what led to the awkward rediscovery of facts and effects that had already been discussed in the 60s and 70s." W. K. Hartmann to W. Sheehan, pers. comm., February 18, 2020.

3. S. Squyres, *Roving Mars: Spirit, Opportunity, and the Exploration of the Red Planet* (New York: Hyperion, 2006), 2–3.

4. M. C. Malin et al., "The Mars Color Imager (MARCI) on the Mars Climate Orbiter," *JGR* 106 (2001): 17651–17672.

5. See, for example, NSSDCA, "Pressure Modulated Infrared Radiometer (PMIRR)," accessed February 10, 2021, https://nssdc.gsfc.nasa.gov/nmc/experiment/display .action?id=1998-073A-02.

6. NASA Public Lessons Learned System, "Mars Climate Orbiter Mishap Investigation Board—Phase I Report," lesson 641, November 30, 1999, https://llis.nasa .gov/lesson/641.

7. NASA Public Lessons Learned System.

8. See, for example, L. Grossman, "Metric Math Mistake Muffed Mars Meteorology Mission," *Wired Magazine*, November 10, 1999, https://www.wired.com/2010/11 /1110mars-climate-observer-report.

9. NASA Public Lessons Learned System, "Mars Climate Orbiter Mishap Investigation Board," 7.

10. NASA, "1998 Mars Missions: Press Kit," December 1998, 33, https://www2.jpl .nasa.gov/files/misc/mars98launch.pdf.

11. JPL Special Review Board, *Report on the Loss of the Mars Polar Lander and Deep Space 2 Missions*, JPL D-18709 (Pasadena, Calif.: JPL, 2000), https://spaceflight .nasa.gov/spacenews/releases/2000/mpl/mpl_report_1.pdf.

12. JPL Special Review Board, xii.

13. NSSDCA, "Nozomi," accessed February 10, 2021, https://nssdc.gsfc.nasa.gov/nmc /spacecraft/display.action?id=1998-041A.

14. Author J. B. was a young member of that frustrated team, whose experiences are recorded in Squyres, *Roving Mars*. Undaunted, this team would soon also rise from the ashes and have the privilege of exploring Mars with modified versions of their previously selected payload instruments on the Spirit and Opportunity rovers.

15. See review by W. V. Boynton et al., "Elemental Abundances Determined via the Mars Odyssey GRS," in *The Martian Surface: Composition, Mineralogy, and Physical Properties*, ed. J. F. Bell III (Cambridge: Cambridge University Press, 2008), 105–124.

16. See review by W. C. Feldman et al., "Volatiles on Mars: Scientific Results from the Mars Odyssey Neutron Spectrometer," in Bell, *The Martian Surface*, 125–152.

17. Boynton et al., "Elemental Abundances Determined via the Mars Odyssey GRS."

18. Boynton et al., "Elemental Abundances Determined via the Mars Odyssey GRS"; Feldman et al., "Volatiles on Mars."

19. See G. J. Taylor and L. M. V. Martel, "Exploring the Mantle of Mars," *Planetary Science Research Discoveries*, October 26, 2012, http://www.psrd.hawaii.edu /Oct12/Mantle-of-Mars.html.

20. Boynton et al., "Elemental Abundances Determined via the Mars Odyssey GRS."

21. J. F. Bell III et al., "Visible to Near-IR Multispectral Observations of Mars," in Bell, *The Martian Surface*, 169–194.

22. P. R. Christensen et al., "The Compositional Diversity and Physical Properties Mapped from the Mars Odyssey Thermal Emission Imaging System," in Bell, *The Martian Surface*, 221–241.

23. M. T. Mellon, R. L. Fergason, and N. E. Putzig, "The Thermal Inertia of the Surface of Mars," in Bell, *The Martian Surface*, 399–427.

24. See also the review by B. L. Ehlmann and C. S. Edwards, "Mineralogy of the Martian Surface," *Annual Review of Earth and Planetary Sciences* 42 (2014): 291–315.

25. See M. Williams, "How Bad Is the Radiation on Mars?," *Universe Today*, November 19, 2016, https://www.universetoday.com/14979/mars-radiationl.

26. See Wikipedia, "European Space Agency," last edited February 16, 2021, https://en.wikipedia.org/wiki/European_Space_Agency.

27. J.-P. Bibring and Y. Langevin, "Mineralogy of the Martian Surface from Mars Express OMEGA Observations," in Bell, *The Martian Surface*, 153–168.

28. V. Formisano et al., "Detection of Methane in the Atmosphere of Mars," *Science* 306 (2004): 1758–1761.

29. See, for example, Y. L. Yung et al., "Methane on Mars and Habitability: Challenges and Responses," *Astrobiology* 18 (2018): 1–22.

30. For an outline of the mission, see R. W. Zurek and S. E. Smrekar, "An Overview of the Mars Reconnaissance Orbiter (MRO) Science Mission," *JGR* 112 (2007), https://doi.org/10.1029/2006JE00270.

31. On MARCI, see J. F. Bell III et al., "Mars Reconnaissance Orbiter Mars Color Imager (MARCI): Instrument Description, Calibration, and Performance," *JGR* 114 (2009): E08S92, https://doi.org/10.1029/2008JE003315.

32. For mission photos and details, see NASA's website for the Mars Reconnaissance Orbiter, accessed February 10, 2021, https://mars.jpl.nasa.gov/mro.

33. A. S. McEwen et al., "Mars Reconnaissance Orbiter's High Resolution Imaging Science Experiment (HiRISE)," *JGR* 112 (2007), https://doi.org/10.1029/2005JE002605.

34. See, for example, A. S. McEwen, C. Hansen-Koharcheck, and A. Espinoza, *Mars: The Pristine Beauty of the Red Planet* (Tucson: University of Arizona Press, 2017).

35. See "Science Nuggets," *LPL HiRISE*, accessed February 10, 2021, https://www.ua hirise.org/epo/nuggets.

36. See, for example, I. J. Daubar et al., "The Current Martian Cratering Rate," *Icarus* 225 (2013): 506–516.

37. The topic of aeolian activity on Mars has recently reviewed by M. A. Bishop, "Dark Dunes of Mars: An Orbit-to-Ground Multidisciplinary Perspective of Aeolian Science," in *Dynamic Mars: Recent and Current Landscape Evolution of the Red Planet*, ed. R. Soare, S. J. Conway, and S. M. Clifford (Amsterdam: Elsevier, 2018), 317–362.

38. See, for example, C. Pilorget and F. Forget, "Formation of Gullies on Mars by Debris Flows Triggered by CO_2 Sublimation," *Nature Geosci.* 9 (2016): 65–69; C. Dundas, "The Formation of Gullies on Mars Today," *Geological Society, London: Special Publications* 467 (2017): 67–94.

39. See, for example, L. Ojha et al., "Spectral Evidence for Hydrated Salts in Recurring Slope Lineae on Mars," *Nature Geosci.* 8 (2015): 829–833.

40. See, for example, C. Dundas et al., "Granular Flows at Recurring Slope Lineae on Mars Indicate a Limited Role for Liquid Water," *Nature Geosci.* 10 (2017): 903–907.

41. See C. Q. Choi, "Red Planet Heats Up: Ice Age Ending on Mars," *Space.com*, May 26, 2016, https://www.space.com/33001-mars-ice-age-ending-now.html.

42. See NASA Science Mars Exploration Program, "Mars Global Coverage by Context Camera on MRO," March 29, 2017, https://mars.nasa.gov/resources/8334/mars-global-coverage-by-context-camera-on-mro.

43. See, for example, J. H. Shirley et al., "Orbit-Spin Coupling and the Triggering of the Martian Planet-Encircling Dust Storm of 2018," *JGR* 125 (2020), https://doi.org/10.1029/2019JE006077. According to those authors' analysis, over the next decade a planet-encircling dust event is seen as most probable in 2025–26.

44. For example, J. J. Plaut et al., "Subsurface Radar Sounding of the South Polar Layered Deposits of Mars," *Science* 316 (2007): 92–95.

45. R. J. Phillips et al., "Massive CO_2 Ice Deposits Sequestered in the South Polar Layered Deposits of Mars," *Science* 332 (2011): 838–841.

46. P. B. Buhler et al., "Coevolution of Mars's Atmosphere and Massive South Polar CO_2 Ice Deposit," *Nature Astronomy* 4 (2019): 364–371.

47. A detailed description of the CRISM instrument and investigation was published in S. Murchie et al., "Compact Reconnaissance Imaging Spectrometer for Mars (CRISM) on Mars Reconnaissance Orbiter (MRO)," *JGR* 112 (2007), https://doi.org/10.1029/2006JE002682.

48. See, for example, CRISM, "CRISM's Investigations and New Discoveries (2006–Present)," accessed February 10, 2021, http://crism.jhuapl.edu/science/themes.

49. A comprehensive review of recent CRISM results like these and others described here can be found in S. L. Murchie et al., "Visible to Short-Wave Infrared Spectral Analyses of Mars from Orbit Using CRISM and OMEGA," in *Remote Compositional Analysis: Techniques for Understanding Spectroscopy, Mineralogy, and Geochemistry of Planetary Surfaces*, ed. J. L. Bishop, J. Moersch, and J. F. Bell III (Cambridge: Cambridge University Press, 2020), 453–483.

50. The presence of opal on Mars brings to mind Percival Lowell's reference to "the opaline tints of the planet" in the telescope. P. Lowell, *Mars as the Abode of Life* (New York: Macmillan, 1908), 134.

51. See, for example, B. L. Ehlmann et al., "Clay Minerals in Delta Deposits and Organic Preservation Potential on Mars," *Nature Geosci.* 1 (2008): 355–358.

Chapter 17

1. S. Hubbard, *Exploring Mars: Chronicles from a Decade of Discovery* (Tucson: University of Arizona Press, 2011).

2. Hubbard, 68–81.

3. S. Squyres, *Roving Mars: Spirit, Opportunity, and the Exploration of the Red Planet* (New York: Hyperion, 2006), 75.

4. According to JPL's Rob Manning, Squyres didn't yet know that his science payload was being proposed for a shotgun wedding with a scaled-up Pathfinder rover. Still, Adler and Manning shared Squyres's vision of a highly capable, highly mobile, roving robotic geologist exploring Mars. R. Manning to J. Bell, pers. comm., 2020.

5. R. Manning to J. Bell, pers. comm., 2020.

6. This was the mission option that eventually became the Mars Reconnaissance Orbiter (MRO), what Manning refers to as a large "spy satellite" around Mars that can obtain high-resolution images for science and future landing site safety assessment at the scale of a lander—around 30 centimeters per pixel, as needed by Manning and other future lander and rover mission planners.

7. R. Manning to J. Bell, pers. comm., 2020.

8. Hubbard, *Exploring Mars*, 83.

9. When Manning and JPL rover development leaders Pete Theisinger, Richard Cook, and others heard Goldin's question, the first response they wanted to give was, "Because we aren't idiots!" Ultimately, though, they said yes to the idea, even though "we really didn't know if we could build *one*, let alone *two*, in that time frame." R. Manning to J. Bell, pers. comm., 2020.

10. Rob Manning points out that ultimately having two rovers was *essential* for meeting the 2003 launch date. For example, at one point one rover was used to test the launch/cruise/EDL phase of the mission, and the other rover was used to test the surface phase of the mission, in parallel, using a second team. R. Manning to J. Bell, pers. comm., 2020.

11. Squyres, *Roving Mars*, 29–33.

12. See "Pres. Clinton's Remarks on the Possible Discovery of Life on Mars (1996)," uploaded July 2, 2015, by clintonlibrary42, YouTube video, 9:57, https://www.youtube.com/watch?v=pHhZQWAtWyQ.

13. J. Bell, *Postcards from Mars* (New York: Dutton, 2006), 16–25; and see details of payloads in S. W. Squyres et al., "The Spirit Rover's Athena Science Investigation at Gusev Crater, Mars," *Science* 305 (2004): 794–799.

14. M. P. Golombek et al., "Selection of the Mars Exploration Rover Landing Sites," *JGR* 108 (2003): 8072.

15. R. Manning to J. Bell, pers. comm., 2020.

16. Bell, *Postcards from Mars*, 37–44.

17. See, for example, "Six Minutes of Terror," uploaded June 23, 2007, by John Beck-Hofmann, YouTube video, 5:12, https://www.youtube.com/watch?v=tZRXwRybb1I.

18. See appendixes C and F for details on Mars rotation and orbital parameters and timekeeping.

19. See, for example, R. E. Arvidson et al., "Overview of the Spirit Mars Exploration Rover Mission to Gusev Crater: Landing Site to Backstay Rock in the Columbia Hills," *JGR* 111 (2006), https://doi.org/10.1029/2005JE002499.

20. Rob Manning has pointed out that 600 meters driving distance was the "Level-1 requirement," while 1 kilometer was a "Level-1 goal." Prelaunch lifetime testing of the rover mechanism design showed that it could drive three times the 600-meter goal requirement, or 1.8 kilometers. So the decision to try to drive 3 kilometers to Husband Hill was indeed based on a strategy of "hope." Rob also pointed out that "of course not to try would have been wrong too but it was certainly a gamble. And indeed, the first mobility mechanism failure (Spirit's right front wheel motor) occurred just after 3 times its expected 'Level-1' number of revolutions. People often think we sandbag these designs. We don't." R. Manning to J. Bell, pers. comm., 2020.

21. See the image and announcement at JPL, "You Are Here: Earth as Seen from Mars," March 11, 2004, https://www.jpl.nasa.gov/spaceimages/details.php?id=PIA05547.

22. See J. Bell, "Backyard Astronomy from Mars," *Sky & Telescope*, August 2006, 41–44.

23. For details, see R. E. Arvidson et al., "Spirit Mars Rover Mission to the Columbia Hills, Gusev Crater: Mission Overview and Selected Results from the Cumberland Ridge to Home Plate," *JGR* 113 (2008), https://doi.org/10.1029/2008JE003183.

24. See details in the final Spirit rover high-level mission summary research paper: R. E. Arvidson et al., "Spirit Mars Rover Mission: Overview and Selected Results from the Northern Home Plate Winter Haven to the Side of Scamander Crater," *JGR* 115 (2010), https://doi.org/10.1029/2010JE003633.

25. See S. W. Ruff and J. D. Farmer, "Silica Deposits on Mars with Features Resembling Hot Spring Biosignatures at El Tatio in Chile," *Nature Communications* 7 (2016): article 13554. Note that although certainly not definitive, the evidence for a hot spring environment with potential biosignatures in Gusev was sufficiently compelling to get Columbia Hills on the shortlist of candidate landing sites for NASA's Mars 2020 rover Perseverance. Ultimately, though, a more compelling case was judged to have been presented for sending that rover to Jezero crater to search for evidence of ancient life there.

26. What is a "habitable environment"? See, for example, J. Bell, "The Search for Habitable Worlds: Planetary Exploration in the 21st Century," *Daedalus* 141, no. 3 (2012): 8–22.

27. Rob Manning likes to say that working on "Mars time" allows the team to sleep in an extra 39 minutes, every day. "Ironically, I loved the schedule," he said.

"While annoying to my family, Mars ops was *far* less stressful for me than the years before landing. I got far less sleep working MER before surface ops!" R. Manning to J. Bell, pers. comm., 2020.

28. See, for example, J. Vertesi, *Seeing Like a Rover* (Chicago: University of Chicago Press, 2015).

29. S. W. Squyres et al., "The Opportunity Rover's Athena Science Investigation at Meridiani Planum, Mars," *Science* 306 (2004): 1698–1703.

30. S. W. Squyres et al., "Overview of the Opportunity Mars Exploration Rover Mission to Meridiani Planum: Eagle Crater to Purgatory Ripple," *JGR* 111 (2006), https://doi.org/10.1029/2006JE002771.

31. S. W. Squyres et al., "In Situ Evidence for an Ancient Aqueous Environment at Meridiani Planum, Mars," *Science* 306 (2004): 1709–1714.

32. J. P. Grotzinger et al., "Stratigraphy and Sedimentology of a Dry to Wet Eolian Depositional System, Burns Formation, Meridiani Planum, Mars," *Earth and Planetary Science Letters* 240 (2005): 11–72.

33. S. W. Squyres et al., "Exploration of Victoria Crater by the Rover Opportunity," *Science* 324 (2009): 1058–1061.

34. See, for example, S. W. Squyres et al., "Ancient Impact and Aqueous Processes at Endeavour Crater, Mars," *Science* 336 (2012): 570–576; R. E. Arvidson et al., "Ancient Aqueous Environments at Endeavour Crater, Mars," *Science* 343 (2014), https://doi.org/10.1126/science.1248097.

35. Squyres et al., "Ancient Impact and Aqueous Processes."

36. Rob Manning points out that "it wasn't a modified version . . . it was the very same lander that had been mothballed in 2000. The Mars Surveyor '01 lander had lived the last 3–4 years under a mylar blanket in a Denver high bay clean room before being brought back to life as Phoenix. It needed some TLC and some upgrades for a new set of science instruments on its payload top deck but it was essentially the same Mars Surveyor '01 lander hardware—which itself was only a bit modified from the (lost) Mars Polar Lander design." R. Manning to J. Bell, pers. comm., 2020.

37. R. Shotwell, "Phoenix—The First Mars Scout Mission," *Acta Astronautica* 57 (2005): 121–134.

38. P. H. Smith et al., "H_2O at the Phoenix Landing Site," *Science* 325 (2009): 58–61.

39. See, for example, J. R. Minkel, "Phoenix Gas Analyzer Confirms Water on Mars," *Scientific American*, August 1, 2008, https://www.scientificamerican.com/article/phoenix-confirms-water-mars.

40. See review in J. F. Bell III, "Iron, Sulfate, Carbonate, and Hydrated Minerals on Mars," in *Mineral Spectroscopy: A Tribute to Roger G. Burns*, ed. M. D. Dyar, C. McCammon, and M. W. Schaefer (Washington, D.C.: Geochemical Society, 1996), 359–380.

41. See B. Sutter et al., "The Detection of Carbonate in the Martian soil at the Phoenix Landing Site: A Laboratory Investigation and Comparison with the Thermal and Evolved Gas Analyzer (TEGA) Data," *Icarus* 218 (2012): 290–296, and references therein.

42. S. P. Kounaves et al., "Wet Chemistry Experiments on the 2007 Phoenix Mars Scout Lander Mission: Data Analysis and Results," *JGR* 115 (2010), https://doi.org/10.1029/2009JE003424.

43. E. Howell, "Salts on Mars May Turn Ice into Liquid Water," *Space.com*, July 2, 2014, https://www.space.com/26424-mars-salt-turns-ice-water-video.html.

Chapter 18

1. National Research Council, *New Frontiers in the Solar System: An Integrated Exploration Strategy* (Washington, D.C.: National Academies Press, 2003), https://www.nap.edu/catalog/10432/new-frontiers-in-the-solar-system-an-integrated-exploration-strategy.

2. National Research Council, tables ES.2 and 8.1.

3. See JPL, "Mars Science Laboratory Mission: Project Science Integration Group (PSIG), Final Report," June 6, 2003, https://mepag.jpl.nasa.gov/reports/PSIG_Final_Full_Report4.pdf.

4. For an engaging personal account of the travails of the MSL program, leading up to the triumph of the Mars rover Curiosity, see R. Wiens, *Red Rover: Inside the Story of Robotic Space Exploration, from Genesis to the Mars Rover Curiosity* (New York: Basic Boooks, 2013).

5. National Research Council, *New Frontiers in the Solar System*, 7.

6. E. Lakdawalla, *The Design and Engineering of Curiosity: How the Mars Rover Performs Its Job* (Cham: Springer-Praxis Books, 2018).

7. For technical details see, for example, Wikipedia, "Radioisotope Thermoelectric Generator," last edited February 4, 2021, https://en.wikipedia.org/wiki/Radioisotope_thermoelectric_generator.

8. Although, ironically, and of course not yet knowable to MSL mission designers in the early 2000s, Spirit and Opportunity were able to survive and thrive for more than 6 and 14 years, respectively, using solar power, rechargeable batteries, and occasionally lucky gusts of dust-clearing wind. However, it was indeed the starvation of electricity from the solar panels, because of the inexorable effects of atmospheric and airfall dust, that ultimately killed both missions.

9. See JPL, "NASA Selects Investigations for the Mars Science Laboratory," December 14, 2004, https://www.jpl.nasa.gov/news/news.php?feature=699.

10. M. C. Malin et al., "The Mars Science Laboratory (MSL) Mast Cameras and Descent Imager: I. Investigation and Instrument Descriptions," *Earth & Space Science* 4 (2017), https://doi.org/10.1002/2016EA000252.

11. R. C. Wiens et al., "The ChemCam Instrument Suite on the Mars Science Laboratory (MSL) Rover: Body Unit and Combined System Performance," *Space Sci. Rev.* 170 (2012): 167–227.

12. K. S. Edgett et al., "Curiosity's Mars Hand Lens Imager (MAHLI) Investigation," *Space Sci. Rev.* 170 (2012): 259–317.

13. Malin et al., "The Mars Science Laboratory (MSL) Mast Cameras and Descent Imager."

14. Wiens et al., "The ChemCam Instrument Suite on the Mars Science Laboratory (MSL) Rover."

15. R. Gellert et al., "In Situ Compositional Measurements of Rocks and Soils with the Alpha Particle X-ray Spectrometer on NASA's Mars Rovers," *Elements* 11 (2015): 39–44.

16. P. R. Mahaffey et al., "The Sample Analysis at Mars Investigation and Instrument Suite," *Space Sci. Rev.* 170 (2012): 401–478.

17. D. Blake et al., "Characterization and Calibration of the CheMin Mineralogical Instrument on Mars Science Laboratory," *Space Sci. Rev.* 170 (2012): 341–399.

18. I. Mitrofanov et al., "Dynamic Albedo of Neutrons (DAN) Experiment Onboard NASA's Mars Science Laboratory," *Space Sci. Rev.* 170 (2012): 559–582.

19. J. Gómez-Elvira et al., "REMS: The Environmental Sensor Suite for the Mars Science Laboratory Rover," *Space Sci. Rev.* 170 (2012): 583–640.

20. D. Hassler et al., "The Radiation Assessment Detector (RAD) Investigation," *Space Sci. Rev.* 170 (2012): 503–558.

21. R. Manning and W. L. Simon, *Mars Rover Curiosity: An Inside Account from Curiosity's Chief Engineer* (Washington, D.C.: Smithsonian Books, 2017).

22. See, for example, A. Brown, "Mars Science Laboratory: The Technical Reasons Behind Its Delay," *Space Review*, March 2, 2009, https://www.thespacereview.com/article/1319/1.

23. NASA HQ, "NASA Memo to Space Science Community: Mars Science Laboratory Project Changes Respond to Cost Increases, Keep Mars Program on Track," *SpaceRef*, September 16, 2007, http://www.spaceref.com/news/viewsr.html?pid=25415.

24. NASA HQ.

25. For a more detailed account of the history and impact of the MSL descope process for MSL and for NASA overall, see Manning and Simon, *Mars Rover Curiosity*, 129–138; Lakdawalla, *The Design and Engineering of Curiosity*, 28–31; and Wiens, *Red Rover*, chapter 14.

26. Lakdawalla, *The Design and Engineering of Curiosity*, 24–38.

27. NASA, "NASA Selects Student's Entry as New Mars Rover Name," press release, May 27, 2009, https://www.nasa.gov/mission_pages/msl/msl-20090527.html.

28. M. P. Golombek et al., "Selection of the Mars Science Laboratory Landing Site," *Space Sci. Rev.* 170 (2012): 641–737.

29. Science and engineering presentations and details from this and the four sub-sequent public workshops that were held to narrow down the selections for the MSL landing site are archived online at NASA, *Marsoweb* (website), accessed February 10, 2021, https://marsoweb.nas.nasa.gov/landingsites/landingsites.html.

30. See "Latest Quad Charts of Landing Sites," PowerPoint file, posted June 9, 2011, to the Marsoweb site.

31. NASA, "NASA's Next Mars Rover to Land at Gale Crater," press release, July 22, 2011, https://www.nasa.gov/mission_pages/msl/news/msl20110722.html.

32. Manning and Simon, *Mars Rover Curiosity*, 55–63.

33. Lakdawalla, *The Design and Engineering of Curiosity*, 78–101.

34. In fact, the Curiosity Engineering Camera team, analyzing hazard avoidance camera (hazcam) images taken immediately after landing, noticed a short-lived plume on the horizon in the direction of the crashed descent stage. See J. Maki et al., "Mars Science Laboratory Navcam/Hazcam Operations and Results" (44th Lunar and Planetary Science Conference, Woodlands, Tex., 2013), abstract #1236.

35. R. M. E. Williams et al., "Martian Fluvial Conglomerates at Gale Crater," *Science* 340 (2013): 1068–1072.

36. See, for example, JPL, "NASA Rover Finds Old Streambed on Martian Surface," press release, September 27, 2012, https://www.jpl.nasa.gov/news/news.php?release =2012-305.

37. Many of the most exciting initial results from Yellowknife Bay are gathered in a series of peer-reviewed research papers; see J. P. Grotzinger, ed., "Exploring Martian Habitability," special issue, *Science* 343, no. 6169 (2014), https://science .sciencemag.org/content/343/6169.

38. K. A. Farley et al., "In Situ Radiometric and Exposure Age Dating of the Martian Surface," *Science* 343, no. 6169 (2014), https://doi.org/10.1126/science.1247166.

39. J. P. Grotzinger et al., "A Habitable Fluvio-Lacustrine Environment at Yellow-knife Bay, Gale Crater, Mars," *Science* 343 (2013), https://doi.org/10.1126/science .1242777.

Chapter 19

1. National Research Council, *New Frontiers in the Solar System: An Integrated Exploration Strategy* (Washington, D.C.: National Academies Press, 2003), https:// www.nap.edu/catalog/10432/new-frontiers-in-the-solar-system-an-integrated -exploration-strategy.

2. National Research Council, *Vision and Voyages for Planetary Science in the Decade 2013–2022* (Washington, D.C.: National Academies Press, 2011), https://www.nap .edu/catalog/13117/vision-and-voyages-for-planetary-science-in-the-decade-2013 -2022.

3. *Commerce, Justice, Science, and Related Agencies Appropriations for 2008: Hearings Before a Subcommittee of the Committee on Appropriations*, U.S. House of Representatives, 110th Congress, First Session (Washington, D.C.: Government Printing Office, 2008), 141; see also E. Lakdawalla, "NASA Announces Delay of Mars Scout Launch Until 2013," *The Planetary Society* (blog), December 21, 2007, https://www.planetary.org/blogs/emily-lakdawalla/2007/1268.html.

4. See University of Colorado Boulder Laboratory for Atmospheric and Space Physics, "Science Orbit," *MAVEN* (website), accessed February 10, 2021, http://lasp.colorado.edu/home/maven/science/science-orbit.

5. B. M. Jakosky et al., "Mars' Atmospheric History Derived from Upper-Atmosphere Measurements of $^{38}Ar/^{36}Ar$," *Science* 355 (2017): 1408–1410.

6. B. M. Jakosky et al., "MAVEN Explores the Martian Upper Atmosphere," *Science* 350 (2015): 643.

7. K. Chang, "Mars' Atmosphere Stripped by Solar Winds, NASA Says," *New York Times*, November 5, 2015, https://www.nytimes.com/2015/11/06/science/space/mars-atmosphere-stripped-away-by-solar-storms-nasa-says.html.

8. G. Orwell, "In Front of Your Nose," *Tribune*, March 22, 1946, https://www.orwellfoundation.com/the-orwell-foundation/orwell/essays-and-other-works/in-front-of-your-nose/.

9. See University of Colorado Boulder Laboratory for Atmospheric and Space Physics, "Science Orbit."

10. More highlights about MAVEN's scientific discoveries at Mars are summarized at NASA, "1,000 Days in Orbit: MAVEN's Top 10 Discoveries at Mars," June 16, 2017, https://www.nasa.gov/feature/goddard/2017/maven-1000-days.

11. S. M. Ahmed, "MENCA Brings Divine Wealth from Mars: First Science Results from the Mars Orbiter Mission," *The Planetary Society* (blog), March 2, 2016, https://www.planetary.org/blogs/guest-blogs/2016/0225-menca-brings-divine-wealth.html.

12. *Mars Orbiter Mission (MOM) Mars Atlas* (Ahmedabad, India: Space Applications Centre, Indian Space Research Organization, 2015), https://planetary.s3.amazonaws.com/assets/resources/ISRO/Mars-atlas-MOM.pdf.

13. U. Tejonmayam, "After Mars, Venus on ISRO's Planetary Travel List," *Times of India*, May 18, 2019, https://timesofindia.indiatimes.com/india/after-mars-venus-on-isros-planetary-travel-list/articleshow/69381185.cms.

14. See, for example, T. Owen, "The Composition and Early History of the Atmosphere of Mars," in *Mars*, ed. H. Kieffer et al. (Tucson: University of Arizona Press, 1992), 818–834.

15. V. Formisano et al., "Detection of Methane in the Atmosphere of Mars," *Science* 306 (2004): 1758–1761.

16. M. J. Mumma et al., "Strong Release of Methane on Mars in Northern Summer 2003," *Science* 323 (2009): 1041–1045.

17. Y. L. Yung et al., "Methane on Mars and Habitability: Challenges and Responses," *Astrobiology* 18 (2018): 1–22.

18. P. R. Mahaffey et al., "The Sample Analysis at Mars Investigation and Instrument Suite," *Space Sci. Rev.* 170 (2012): 401–478.

19. C. R. Webster et al., "Low Upper Limit to Methane Abundance on Mars," *Science* 342 (2013): 355–357.

20. C. R. Webster et al., "Mars Methane Detection and Variability at Gale Crater," *Science* 347 (2015): 415–417.

21. C. R. Webster et al., "Background Levels of Methane in Mars' Atmosphere Show Strong Seasonal Variations," *Science* 360 (2018): 1093–1096.

22. See, for example, K. Zahnle, R. S. Freedman, and D. C. Catling, "Is There Methane on Mars?," *Icarus* 212 (2011): 493–503; F. Lefèvre and F. Forget, "Observed Variations of Methane on Mars Unexplained by Known Atmospheric Chemistry and Physics," *Nature* 460 (2009): 720–723.

23. K. Zahnle and D. C. Catling, "The Paradox of Mars Methane" (Ninth International Conference on Mars, Houston, Tex., 2019), abstract #6132.

24. See, for example, R. Zurek and A. Chicarro, "Report to MEPAG on the ESA-NASA Joint Instrument Definition Team (JIDT) for the Proposed 2016 Orbiter-Carrier," NASA, July 29, 2009, https://web.archive.org/web/20090730203811/http://mepag .jpl.nasa.gov/meeting/jul-09/JIDT_for_MEPAG.pdf.

25. See, for example, K. Kremer, "Experts React to Obama Slash to NASA's Mars and Planetary Science Exploration," *Universe Today*, February 17, 2012, https:// www.universetoday.com/93512/experts-react-to-obama-slash-to-nasas-mars-and -planetary-science-exploration/.

26. O. Korablev et al., "No Detection of Methane on Mars from Early ExoMars Trace Gas Orbiter Observations," *Nature* 568 (2019): 517–520.

27. Korablev et al.

28. Substantial details about the history, design, and mission of Schiaparelli, as well as links to many additional resources, can be found at Wikipedia, "Schiaparelli EDM," last edited January 14, 2021, https://en.wikipedia.org/wiki/Schiaparelli _EDM.

29. S. Hubbard, *Exploring Mars: Chronicles from a Decade of Discovery* (Tucson: University of Arizona Press, 2011), 7.

Chapter 20

1. "Notes," *Observatory* 1 (1877): 181.

2. E. E. Barnard, "Mars: His Moons and His Heavens," unpublished manuscript, Barnard Archives, Joseph Heard Library, Vanderbilt University; see also W. Sheehan, "E.E. Barnard: The Early Years," *JBAA* 103 (1993): 34–36.

3. "The Moons of Mars," *Cornhill Magazine* 36 (1877): 425.

4. D. Brower and G. M. Clemence, "Orbits and Masses of Planets and Satellites," in *The Solar System*, vol. 3, ed. G. P. Kuiper and B. M. Middlehurst (Chicago: University of Chicago Press, 1961), 31–94.

5. "Notes," *Observatory* (1877): 182.

6. G. P. Kuiper, "Limits of Completeness," in Kuiper and Middlehurst, *The Solar System*, 3:575–591.

7. B. A. Smith, "Phobos: Preliminary Results from Mariner 7," *Science* 168 (1970): 828–830.

8. Smith, 830.

9. T. C. Duxbury et al., "Spacecraft Exploration of Phobos and Deimos," *Planet. Space Sci.* 102 (2014): 9–17.

10. J. Veverka and J. A. Burns, "The Moons of Mars," *Annual Review of Earth and Planetary Sciences* 8 (1980): 527–558.

11. T. C. Duxbury, G. H. Born, and N. Jerath, "Viewing Phobos and Deimos for Navigating Mariner 9," *Journal of Spacecraft and Rockets* 11 (1974): 215–222.

12. See, for example, Press Trust of India Washington, "MAVEN Spacecraft Avoids Collision with Mars' Phobos," *Hindu Business Line*, March 3, 2017, https://www.thehindubusinessline.com/news/science/maven-spacecraft-avoids-collision-with-mars-phobos/article9569878.ece.

13. For a description of IAU's naming schemes for features across the solar system, see appendix E; IAU, "Categories (Themes) for Naming Features on Planets and Satellites," accessed February 10, 2021, https://planetarynames.wr.usgs.gov/Page/Categories.

14. J. Blunck, *Mars and Its Satellites: A Detailed Commentary on the Nomenclature* (Smithtown, N.Y.: Exposition Press, 1982), 158. Beyond all doubt, both Swift and Voltaire derived the inspiration for their tales from Kepler: in 1610, when Galileo announced to him the discovery of the four moons of Jupiter, Kepler supposed, on numerological grounds, that Mars must have two moons.

15. See, for example, Veverka and Burns, "The Moons of Mars," 539; A. T. Basilevsky et al., "The Surface Geology and Geomorphology of Phobos," *Planet. Space Sci.* 102 (2014): 95–118; N. Hirata, "Spatial Distribution of Impact Craters on Deimos," *Icarus* 288 (2017): 69–77.

16. Veverka and Burns, "The Moons of Mars," 535–537.

17. Veverka and Burns, 537.

18. M. W. Busch et al., "Arecibo Radar Observations of Phobos and Deimos," *Icarus* 186 (2007): 581–584.

19. See, for example, E. Asphaug and H. J. Melosh, "The Stickney Impact of Phobos—A Dynamical Model," *Icarus* 101 (1993): 144–164.

20. See review and recent analysis by K. R. Ramsley and J. W. Head, "Origin of Phobos Grooves: Testing the Stickney Crater Ejecta Model," *Planet. Space Sci.* 165 (2019): 137–147.

21. J. B. Murray and D. C. Heggie, "Character and Origin of Phobos' Grooves," *Planet. Space Sci.* 102 (2014): 119–143.

22. T. A. Hurford et al., "Tidal Disruption on Phobos as the Cause of Surface Fractures," *JGR* 121 (2016): 1054–1065.

23. P. Thomas and J. Veverka, "Down-Slope Movement of Material on Deimos," *Icarus* 42 (1980): 234–250.

24. Veverka and Burns, "The Moons of Mars," 545–546.

25. P. C. Thomas, "Ejecta Emplacement on the Martian Satellites," *Icarus* 131 (1998): 78.

26. Thomas, 105.

27. G. A. Avanesov et al., "Television Observations of Phobos," *Nature* 341 (1989): 585–587.

28. See, for example, the gallery at MSSS, "Mars Global Surveyor Mars Orbiter Camera," accessed February 10, 2021, https://mars.jpl.nasa.gov/mgs/msss/camera/images/9_11_98_phobos_rel.

29. B. G. Bills et al., "Improved Estimate of Tidal Dissipation Within Mars from MOLA Observations of the Shadow of Phobos," *JGR* 110 (2005), https://doi.org/10.1029/2004JE002376.

30. O. Witasse et al., "Mars Express Investigations of Phobos and Deimos," *Planet. Space Sci.* 102 (2014): 18–34.

31. See NASA, "MAVEN Observes Mars Moon Phobos in the Mid- and Far-Ultraviolet," February 29, 2016, https://www.nasa.gov/feature/goddard/2016/maven-observes-phobos-in-ultraviolet.

32. For example images, see A. McEwen, "Deimos, Moon of Mars," *LPL: HiRISE*, March 9, 2009, https://www.uahirise.org/deimos.php; N. Bridges, "Phobos Imaged by HiRISE," *LPL: HiRISE*, April 9, 2008, https://www.uahirise.org/phobos.php.

33. An excellent recent review that touches on many of these themes can be found in S. L. Murchie et al., "Phobos and Deimos," in *Asteroids IV*, ed. P. Michel, F. E. DeMeo, and W. F. Bottke (Tucson: University of Arizona Press, 2015), 451–467.

34. Veverka and Burns, "The Moons of Mars," 531–532.

35. Shklovsky was a sometime collaborator with Carl Sagan on intelligent life in the universe, and author of a memoir with the incomparable title, *Five Billion Vodka Bottles to the Moon: Tales of a Soviet Scientist* (New York: W. W. Norton, 1991). The book with Sagan is: I. S. Shklovsky and C. Sagan, *Intelligent Life in the Universe* (San Francisco: Holden-Day, 1966). See also B. P. Sharpless, "Secular Accelerations in the Longitudes of the Satellites of Mars," *Astronomical Journal* 51 (1945): 185–186.

36. G. A. Wilkins, "Motion of Phobos," *Nature* 224 (1969): 789.

37. Veverka and Burns, "The Moons of Mars," 533–535.

38. See, for example, A. S. Rivkin et al., "Near-Infrared Spectrophotometry of Phobos and Deimos," *Icarus* 156 (2002): 64–75.

39. See review by S. Murchie and S. Erard, "Spectral Properties and Heterogeneity of Phobos from Measurements by Phobos 2," *Icarus* 123 (1996): 63–86.

40. See, for example, B. E. Clark, F. P. Fanale, and J. W. Salisbury, "Meteorite–Asteroid Spectral Comparison: The Effects of Comminution, Melting, and Recrystallization," *Icarus* 97 (1992): 288–297.

41. Basilevsky et al., "The Surface Geology and Geomorphology of Phobos," 107–108.

42. M. J. Gaffey, T. H. Burbine, and R. P. Binzel, "Asteroid Spectroscopy: Progress and Perspectives," *Meteoritics* 28 (1993): 161–187.

43. Murchie and Erard, "Spectral Properties and Heterogeneity of Phobos," 81–82.

44. See A. A. Fraeman et al., "Analysis of Disk-Resolved OMEGA and CRISM Spectral Observations of Phobos and Deimos," *JGR* 117 (2012), https://doi.org/10.1029/2012JE004137.

45. Fraeman et al.

46. A.A. Fraeman et al., "Spectral Absorptions on Phobos and Deimos in the Visible/Near Infrared Wavelengths and Their Compositional Constraints," *Icarus* 229 (2014): 196–205.

47. See, for example, J. I. Lunine, G. Neugebauer, and B. M. Jakowsky, "Infrared Observations of Phobos and Deimos from Viking," *JGR* 87 (1982): 10297–10305.

48. T. D. Glotch et al., "MGS-TES Spectra Suggest a Basaltic Component in the Regolith of Phobos," *JGR* 123 (2018): 2467–2484.

49. J. L. Bandfield et al., "Mars Odyssey THEMIS Observations of Phobos: New Spectral and Thermophysical measurements" (49th Lunar and Planetary Science Conference, Houston, Tex., 2018), abstract #2643.

50. For an early review, see J. A. Burns, "Contradictory Clues as to the Origin of the Martian Moons," in *Mars*, ed. H. Kieffer et al. (Tucson: University of Arizona Press, 1992), 1283–1301. For a more recent review, see the introduction and references cited in Fraeman et al., "Analysis of Disk-Resolved OMEGA and CRISM Spectral Observations."

51. See, for example, Fraeman et al., "Spectral Absorptions on Phobos and Deimos," and references therein.

52. See, for example, Glotch et al., "MGS-TES Spectra Suggest a Basaltic Component," 2479–2481.

53. Fraeman et al., "Analysis of Disk-Resolved OMEGA and CRISM Spectral Observations," 1.

54. Glotch et al., "MGS-TES Spectra Suggest a Basaltic Component," 2467.

55. A. J. Hesselbrock and D. A. Minton, "An Ongoing Satellite-Ring Cycle of Mars and the Origins of Phobos and Deimos," *Nature Geosci.* 10 (2017): 266–269.

56. J. A. Burns, "The Evolution of Satellite Orbits," in *Satellites*, ed. J. A. Burns and M. S. Matthews (Tucson: University of Arizona Press, 1986), 117–158; See also Veverka and Burns, "The Moons of Mars," 529–530.

57. Duxbury, Born, and Jerath, "Viewing Phobos and Deimos."

58. J. F. Bell III et al., "Solar Eclipses of Phobos and Deimos Observed from the Surface of Mars," *Nature* 436 (2005): 55–57.

59. Witasse et al., "Mars Express Investigations of Phobos and Deimos," 21.

60. Veverka and Burns, "The Moons of Mars," 530, 551–555.

61. Veverka and Burns, 530, 551–555.

Chapter 21

1. J. Bell, "Digging Deep into Mars," *Astronomy*, October 2019, 18–27.

2. M. Golombek et al., "Selection of the InSight Landing Site," *Space Sci. Rev.* 211 (2017): 5–95.

3. M. Golombek et al., "Geology of the InSight Landing Site on Mars," *Nature Geosci.* 11 (2020).

4. Y. Nakamura, "Rebirth of Extraterrestrial Seismology," *Nature Geosci.* 11 (2020): article 1014.

5. See W. B. Banerdt et al., "Initial Results from the InSight Mission on Mars," *Nature Geosci.* 13 (2020): 183–189, and other InSight mission results cited therein.

6. C. L. Johnson et al., "Crustal and Time-Varying Magnetic Fields at the InSight Landing Site on Mars," *Nature Geosci.* 13, no. 3 (2020), 199–204.

7. B. Langlais et al., "A New Model of the Crustal Magnetic Field of Mars Using MGS and MAVEN," *JGR* 124 (2019): 1542–1569.

8. See, for example, JPL, "Mars Cube One (MarCO)," accessed February 10, 2021, https://www.jpl.nasa.gov/cubesat/missions/marco.php.

9. See NASA, "NASA Space Launch System's First Flight to Send Small Sci-Tech Satellites into Space," press release, February 2, 2016, https://www.nasa.gov/press-release/nasa-space-launch-system-s-first-flight-to-send-small-sci-tech-satellites-into-space.

10. See, for example, B. Berger, "UAE Unveils Science Goals for 'Hope' Mars Probe," *Space News*, May 6, 2015, https://spacenews.com/uae-positions-2020-mars-probe-as-catalyst-for-a-new-generation-of-arab-scientists-and-engineers.

11. See W. X. Wan et al., "China's First Mission to Mars," *Nature Astronomy* 4 (2020), https://doi.org/10.1038/s41550-020-1148-6; L. Xin, "Chinese Spacecraft Poised for First Mars Mission," *Scientific American*, July 15, 2020, https://www.scientificamerican.com/article/chinese-spacecraft-poised-for-first-mars-mission.

12. See NASA's mission website for Perseverance, accessed February 10, 2021, https://mars.nasa.gov/mars2020; and JPL, *Mars 2020* (website), accessed February 10, 2021, https://www.jpl.nasa.gov/missions/mars-2020-perseverance-rover.

13. Kennda Lynch's research on Pilot Valley Basin is found in "Geolobiological Investigation of the Hypersaline Sediments of Pilot Valley, Utah: A Terrestrial Analog to Ancient Lake Basins on Mars" (PhD diss., Colorado School of Mines,

2015). Sarah Stewart Johnson has published a charming memoir describing her lifelong romance with the Red Planet: *The Sirens of Mars, Searching for Life on Another World* (New York: Crown, 2020).

14. European Space Agency, "The Exomars Programme: 2016–2022," last updated March 12, 2020, https://exploration.esa.int/web/mars/-/46048-programme -overview.

15. See Japan Aerospace Exploration Agency, *MMX: Martian Moons Exploration* (website), accessed February 10, 2021, http://mmx.isas.jaxa.jp/en/index.html.

16. See, for example, B. Pallava, "India Eyes a Return to Mars and a First Run at Venus," *Science*, February 17, 2017, https://www.sciencemag.org/news/2017/02 /india-eyes-return-mars-and-first-run-venus; P. Gupta, "Everything You Need to Know About Mangalyaan 2, India's Second Mars Mission," *Swarajya*, December 4, 2016, https://swarajyamag.com/technology/everything-you-need-to-know -about-mangalyaan-2-indias-second-mars-mission.

17. M. Wall, "NASA's Mars Sample-Return Plans Get a Boost in 2021 Budget Request," *Space.com*, February 11, 2020, https://www.space.com/nasa-mars-sample-return -2021-budget.html.

18. M. Wall, "Bringing Pieces of Mars to Earth in 2031: How NASA and Europe Plan to Do It," *Space.com*, July 29, 2019, https://www.space.com/mars-sample-return -plan-nasa-esa.html.

Chapter 22

1. The phrase "nod me nay" in the epigraph recalls a passage in Shakespeare's *Two Gentlemen of Verona*, act 1, scene 1. Speed has just delivered a love letter for Proteus:

> Proteus: So what did she say?
> Speed (nodding): Ay.
> Proteus: Nod-ay? Well, that's "naughty."
> Speed: You misunderstood, sir. I said she nodded, and you asked me if she nodded, and I said, "Ay."

Lowell suggests, then, that he is not likely to be as fortunate with Mars. For this scene, see W. Shakespeare, *Two Gentlemen of Verona*, in *The Riverside Shakespeare*, ed. G. B. Evans et al. (Boston: Houghton Mifflin, 1974), act 1, scene 1, lines 112–114.

2. C. Sagan, *The Cosmic Connection* (New York: Doubleday, 1973), 69.

3. W. P. Webb, *The Great Frontier* (Austin: University of Texas Press, 1951), 280.

4. H. G. Wells, *The War of the Worlds* (London: Penguin Classics, 2005), 7.

5. P. Lowell, *Mars* (Boston: Houghton Mifflin, 1895), 212.

6. W. von Braun, *The Mars Project* (1952; repr., Urbana: University of Illinois Press, 1991).

7. N. Mailer, *Of a Fire on the Moon* (Boston: Little, Brown, 1969), 79.

8. B. Aldrin, "Fly Me To L1," *New York Times*, December 5, 2003.

9. Obama was speaking to an audience (that included author J. B.) at the Kennedy Space Center on April 15, 2010. While his remarks have often been taken out of context to indicate a lack of interest in NASA and space exploration, they were instead actually part of a much more forward-looking strategy. Almost immediately after his blunt remark, he went on to say, "There's a lot more of space to explore, and a lot more to learn when we do. So I believe it's more important to ramp up our capabilities to reach—and operate at—a series of increasingly demanding targets, while advancing our technological capabilities with each step forward." Transcript available online at NASA, "President Barack Obama on Space Exploration in the 21st Century," April 15, 2010, https://www.nasa.gov/news/media/trans/obama_ksc_trans.html.

10. T. O. Paine, in von Braun, *The Mars Project*, xlii.

11. See F. T. Hoban, W. M. Lawbaugh, and E. J. Hoffman, *Where Do You Go After You've Been to the Moon?* (Malabar, Fla.: Krieger, 1997).

12. A much more complete and comprehensive summary of possible crewed missions to Mars studied seriously by NASA in the second half of the 20th century appears in D. S. F. Portree, *Humans to Mars: Fifty Years of Mission Planning, 1950–2000*, NASA SP-2001–4521 (Washington, D.C.: Government Printing Office, 2001), https://history.nasa.gov/monograph21/humans_to_Mars.htm.

13. See "Aldrin Mars Cycler," *Buzz Aldrin* (website), accessed February 10, 2021, https://buzzaldrin.com/space-vision/rocket_science/aldrin-mars-cycler.

14. R. Zubrin, *The Case for Mars* (New York: Simon & Schuster, 1996).

15. See S. Hubbard et al., *Humans Orbiting Mars: A Critical Step Toward the Red Planet* (Pasadena, Calif.: The Planetary Society, 2015), https://hom.planetary.org/.

16. See The Planetary Society, "Principles for Human Spaceflight," accessed March 12, 2021, https://www.planetary.org/advocacy/principles-for-human-spaceflight.

17. T. Cichan et al., "Mars Base Camp: An Architecture for Sending Humans to Mars by 2028," *New Space* 5 (2017), https://doi.org/10.1089/space.2017.0037; Lockheed Martin, *Mars Base Camp* (website), accessed February 10, 2021, https://www.lockheedmartin.com/en-us/products/mars-base-camp.html.

18. See SpaceX, *Mars & Beyond* (website), accessed February 10, 2021, https://www.spacex.com/mars.

19. C. A. Scharf, "Death on Mars," *Life, Unbounded* (blog), *Scientific American*, January 20, 2020, https://blogs.scientificamerican.com/life-unbounded/death-on-mars1/.

20. Portree, *Humans to Mars*.

21. NASA, *The Vision for Space Exploration* (Washington, D.C.: Government Printing Office, 2004), http://www.nasa.gov/pdf/55583main_vision_space_exploration2.pdf.

22. D. Hassler et al., "Mars' Surface Radiation Environment Measured with the Mars Science Laboratory's Curiosity Rover," *Science* 343 (2013), https://doi.org/10.1126/science.1244797.

23. D. B. Shirley with D. Morton, *Managing Martians* (New York: Broadway Books, 1998), 17.

24. And even though we have been living in a remarkably mild and stable period of Earth's climate for the past 5,000 years—at least until the sudden warming of the past few decades—even the tempestuous and violent episodes of the past on Earth have been more habitable. See P. Brannen, *The Ends of the World: Volcanic Apocalypses, Lethal Oceans, and Our Quest to Understand Earth's Past Mass Extinctions* (New York: HarperCollins, 2017).

25. Shirley and Morton, *Managing Martians*, 2.

26. B. Jakosky and C. Edwards, "Inventory of CO_2 Available for Terraforming Mars," *Nature Astronomy* 2 (2018): 634–639.

27. T. S. Eliot, "Little Gidding," in *Four Quartets* (1943; New York: Houghton Mifflin Harcourt, 1971), 59.

28. Quoted here are Tennyson's c. 1886 poem "Locksley Hall Sixty Years After," and Milton's *Paradise Lost*, bk. 2, lines 1052–1053.

Appendix A

1. This chronology is modified from D. R. Williams, "Chronology of Mars Exploration," NSSDCA, last updated December 15, 2020, https://nssdc.gsfc.nasa.gov/planetary/chronology_mars.html.

Appendix F

1. Scaliger chose the date because it was when three cycles of calendrical events—the 28-year cycle of days of the week repeating on calendar days, the 19-year cycle of phases of the Moon landing on the same calendar days, and the 15-year imperial Roman tax cycles—converged. See Wolfram Alpha, "Julian Date," accessed August 30, 2020, https://scienceworld.wolfram.com/astronomy/JulianDate.html.

2. See equation 27 in M. Allison and M. McEwen, "A Post-Pathfinder Evaluation of Areocentric Solar Coordinates with Improved Timing Recipes for Mars Seasonal/Diurnal Climate Studies," *Planet. Space Sci.* 48, no. 2–3 (2000): 215–235, https://doi.org/10.1016/S0032-0633(99)00092-6.

3. T. Reichardt, "The Man Who Named the Martian Day," *Air & Space*, November 20, 2015, http://www.airspacemag.com/daily-planet/man-who-named-martian-day-180957350/. The community is not consistent about the pronunciation of "sol"; some rhyme it with "soul," others with "call."

4. Allison and McEwen, "A Post-Pathfinder Evaluation of Areocentric Solar Coordinates."

5. R.T. Clancy et al., "An Intercomparison of Ground-Based Millimeter, MGS TES, and Viking Atmospheric Temperature Measurements: Seasonal and Interannual Variability of Temperatures and Dust Loading in the Global Mars Atmosphere," *Journal of Geophysical Research* 105, no. E4 (2000): 9553–9571, https://doi.org/10.1029/1999JE001089.

6. S. Piqueux et al., "Enumeration of Mars Years and Seasons Since the Beginning of Telescopic Exploration," *Icarus* 251 (2015): 332–338, https://doi.org/10.1016/j.icarus.2014.12.014.

7. See, e.g., M. A. Kahre et al., "The Mars Dust Cycle," in *The Atmosphere and Climate of Mars*, ed. R. M. Haberle et al. (Cambridge: Cambridge University Press, 2017), 295–337, https://doi.org/10.1017/9781139060172.010.

Appendix G

1. NASA, "Mars24 Sunclock—Time on Mars," accessed February 10, 2021, https://www.giss.nasa.gov/tools/mars24.

2. AM2000 does, however, define a Mars Sol Date (MSD) analogous to JD: the sequential count of Mars solar days elapsed since December 29, 1873, at 12:00 (JD 2405522.0). This date was before almost all detailed observations of Mars. It corresponds to a Mars L_s of 277°.

3. T. C. Duxbury et al., "Mars Geodesy/Cartography Working Group Recommendations on Mars Cartographic Constants and Coordinate Systems" (Symposium on Geospatial Theory, Processing and Applications, Symposium sur la théorie, les traitements et les applications des données Géospatiales, Ottawa 2002).

4. For more detail on the specifics of how LMST was defined for various missions, visit M. Allison and R. Schmunk, "Technical Notes on Mars Solar Time as Adopted by the Mars24 Sunclock," NASA, updated March 8, 2020, https://www.giss.nasa.gov/tools/mars24/help/notes.html.

Appendix H

1. NASA, "NASA FY19 Inflation Tables—to Be Utilized in FY20," spreadsheet, accessed February 10, 2021, https://www.nasa.gov/sites/default/files/atoms/files/2019_nasa_new_start_inflation_index_for_fy20_final2.xlsx.

Index

Page numbers in *italics* indicate figures; page numbers in **bold** indicate tables.

About the Authors

William Sheehan is a retired psychiatrist, astronomical historian, and amateur astronomer. His books on Mars include *Planets and Perception* (University of Arizona Press, 1988), a study of the Martian canal phenomenon in the light of perceptual psychology, which was an Astronomical Society of the Pacific Book of the Year in 1988, and *The Planet Mars* (University of Arizona Press, 1996), a Book-of-the-Month Club alternate selection and Astronomy Book Club main selection. In addition, he has published many articles on Mars in journals, most notably in *Sky & Telescope*, where he is a contributing editor. He was awarded the Goldwork on Mars in 2004. He is also a 2001 Fellow of the John Simon Guggenheim Memorial Foundation for his project "Structure and Evolution of the Galaxy." Besides his work on Mars, he has written a number of other books on astronomy.

Jim Bell is a professor in the School of Earth and Space Exploration at Arizona State University, an adjunct professor in the Department of Astronomy at Cornell University, and a Distinguished Visiting Scientist at NASA's Jet Propulsion Laboratory. He has been and continues to be heavily involved in many NASA solar system exploration missions, including the Mars rovers Spirit, Opportunity, Curiosity, and Perseverance. In 2011 he received the Carl Sagan Medal for Excellence in Public Communication from the American

Astronomical Society. He is an avid writer for space-related magazines and blogs, and often appears in media interviews on space-related topics. His popular science and space photography books include *Postcards from Mars* (Dutton, 2006), *Mars 3-D* (Sterling, 2008), *The Space Book* (Sterling, 2013), *The Interstellar Age* (Dutton, 2015), *The Earth Book* (Sterling, 2019), and most recently *Hubble Legacy: 30 Years of Images and Discoveries* (Sterling 2020). He served as president of The Planetary Society from 2008 to 2020. His website is http://jimbell.sese.asu.edu.